THE COLLOIDAL DOMAIN

ADVANCES IN INTERFACIAL ENGINEERING SERIES

Microstructures constitute the building blocks of the interfacial systems upon which many vital industries depend. These systems share a fundamental knowledge base—the molecular interactions that occur at the boundary between two materials.

Where microstructures dominate, the manufacturing process becomes the product. At the Center for Interfacial Engineering, a National Science Foundation Research Center, researchers are working together to develop the control over molecular behavior needed to manufacture reproducible interfacial products.

The books in this series represent an intellectual collaboration rooted in the disciplines of modern engineering, chemistry, and physics that incorporates the expertise of industrial managers as well as engineers and scientists. They are designed to make the most recent information available to the students and professionals in the field who will be responsible for future optimization of interfacial processing technologies.

Other Titles in the Series

THE COLLOIDAL DOMAIN

Where Physics, Chemistry, Biology, and Technology Meet

Second Edition

D. Fennell Evans
and
Håkan Wennerström

⊛ WILEY-VCH

NEW YORK · CHICHESTER · WEINHEIM · BRISBANE · SINGAPORE · TORONTO

D. Fennell Evans
Center for Interfacial Engineering
University of Minnesota
179 Union Street S.E.
Minneapolis, MN 55455

Håkan Wennerström
Division of Physical Chemistry 1
Center for Chemistry and
  Chemical Engineering
Lund University
SE-22100 Lund, Sweden

Library of Congress Cataloging-in-Publication Data:

Evans, D. Fennell.
   The colloidal domain : where physics, chemistry, biology, and
technology meet / by D. Fennell Evans and Håkan Wennerström -- 2nd
ed.
      p.   cm. -- (Advances in interfacial engineering series)
   Includes index.
   ISBN 0-471-24247-0 (cloth : alk. paper)
   1. Colloids. 2. Surface chemistry. I. Wennerström, Håkan.
II. Title. III. Series.
QD549.E93   1999
541.3'45--dc21   98-23227

Printed in the United States of America.

10 9 8 7 6 5 4 3

DEDICATION

We dedicate this book to our wives and children (Joyce and Christopher Evans and Maria, Hanna, Hjalmar, Maja, and Ebba Wennerström), who graciously encouraged us over the period in which this book developed.

CONTENTS

2 / Surface Chemistry and Monolayers   45

3 / Electrostatic Interactions in Colloidal Systems 99

4 / Structure and Properties of Micelles   153

5 / Forces In Colloidal Systems   *217*

8 / Colloidal Stability 401

9 / Colloidal Sols 443

11 / Micro- and Macroemulsions 539

12 / Epilogue 601

Index *613*

PREFACE TO THE FIRST EDITION

Colloidal interactions are omnipresent. They influence almost all technological processes, yet they are frequently invisible and often ignored. When scientists trained in the classical study of atoms and molecules, or engineers armed with correlations that focus on bulk properties confront a problem, they may not recognize its colloidal source. Yet the colloidal domain forms a critical interface between micro and macroscopic regimes. Because colloids have a major impact on our ability to understand biological processes and to control technology, we must deal with colloidal phenomena to maintain a competitive technological base. This book offers a gateway to the colloidal domain.

For several decades the study of colloids was an intellectual backwater, but significant advances in experiment and theory have combined to create a renaissance in colloidal science.

D. Fennell Evans
Håkan Wennerström
January 1994

PREFACE TO THE SECOND EDITION

The second edition of *The Colloidal Domain* incorporates new topics that have emerged over the past five years and includes modifications that increase the clarity of the presentation and new additions that promote integration and coherence of the topics. Major changes include the addition and/or expansion of several chapters

Epilogue (Chapter 12)

This new chapter seeks to provide an integrative prospective on *The Colloidal Domain.* By its very nature, the colloidal domain encompasses entities that contain many molecules. Thus, it occupies an intermediate position between the molecular and the macroscopic worlds. A major goal in this book has been to show how molecular interactions are manifested in the colloidal domain as well as how colloidal phenomena are manifested in the macroscopic world that we perceive with our senses. In establishing these links, we were guided by the need to address four questions:

- How molecular, mesoscale, and macroscopic structures merge in the colloidal domain

- How molecular interaction forces determine colloidal behavior

- How the interplay between entropy and energy organizes colloidal entities

- How dynamic, kinetic, and transport properties influence colloidal processes

By using the second edition in manuscript form, we have encouraged our students to read this chapter both at the beginning of the course—knowing the answer before you start is a real help—and at the end of the course, where a perspective rooted in detailed analysis helps.

Solutes and Solvents, Self-Assembly of Amphiphiles (Chapter 1)

The sections on regular solutions, chemical potential, and transport properties, which constitute the fundamental building blocks of the colloidal domain, have been expanded in order to provide a deeper understanding of the topic.

Electrostatic Interactions in Colloidal Systems (Chapter 3)

In response to feedback and evaluations from students, we have completely rewritten and reorganized this chapter. Chapter 3 provides the foundations for understanding long-range electrostatic interactions, which dominate in colloidal systems.

Forces in Colloidal Systems (Chapter 5)

The second half of this chapter has been extensively revised and now includes sections on electrostatic correlation forces, forces due to density variations, forces due to entropic effects, and hydrophobic forces.

Bilayer Systems (Chapter 6)

The focus in this chapter is on biological membranes with an emphasis on relationships between bilayer structure and function and between energetics and transmembrane transport processes.

Polymer Molecules in Colloidal Systems (Chapter 7)

The first part of this chapter has been reorganized and a new section that analyzes polymer solubility in terms of the entropic contribution associated with polymer chain conformations has been added.

Micro- and Macroemulsions (Chapter 12)

A section on foams has been added and the presentation on emulsion stability has been reanalyzed in terms of new insights and data.

D. FENNELL EVANS
HÅKAN WENNERSTRÖM

ACKNOWLEDGMENTS

In preparing the second edition of *The Colloidal Domain*, we again acknowledge the helpful comments, suggestions, and evaluations provided by students and colleagues. We thank Jane Schmillen for typing the manuscript and managing the chaos that inevitably arises with a revised text. We also thank Sheila Hoover, who patiently drew and redrew the graphics, and Hertha Schulze, who provided value added by increasing the clarity and readability of the text. H. W. acknowledges the support from the Goran Gustavsson Foundation.

SYMBOLS

Symbol	Name	Value	Units	Defining Equation
A	Helmholtz free energy		J	
E	energy		J	
G	Gibbs free energy		J	1.4.1
H	enthalpy		J	
S	entropy		J/K	
R	molar gas constant	8.314×10^3	J/mol K	
U	internal energy		J	
c	number of components		10.1.3	
f	degrees of freedom		10.1.3	
k	Boltzmann's constant	1.380×10^{-23}	J/K	
p	number of phases			10.1.3
A	area		m^2	
a_0	surfactant head group area		m^2	1.3.1
B_2	second virial coefficient		m^3	1.5.11
C	concentration		M (m^{-3})	2.3.13
C	capacitance		F	3.5.1
C^*	polymer overlap concentration		M (m^{-3})	7.2.2
C_p	heat capacity		J/K	10.4.1
c	electrolyte concentration		M (m^{-3})	
c^*	electrolyte concentration		m^{-3}	3.8.1
c_m	solute concentration		M (m^{-3})	4.2.11
CCC	critical coagulation concentration		(m^{-3})	8.2.11
CMC	critical micelle concentration		M (m^{-3})	4.1.5
D	diffusion coefficient		$m^2 s^{-1}$	1.6.9
$D(r)$	diffusion coefficient		$m^2 s^{-1}$	5.7.5
E	electromotive force		V	4.2.3
E^o	standard electromotive force		V	4.2.3
e	charge of an electron	1.602×10^{-19}	C	3.2.6
F	force		N	
F	Faraday's constant	96,500	C	4.2.3
G_0	static shear modulus		N/m^2	7.2.16
$G(r)$				4.5.5
g	acceleration due to gravity	9.81	ms^{-2}	2.2.3
$g(r)$	radial distribution function			7.1.28
H	mean curvature		m^{-1}	1.3.8
H	Hamaker constant		J	5.2.5

Symbol	Name	Value	Units	Defining Equation
H_o	spontaneous curvature		M^{-1}	11.1.2
h	separation distance		m	5.1.1
I	current		A	8.4.14
I_0	intensity of incident radiation			4.2.4
$I(Q)$	intensity of scattering			7.1.27
J_A	molecular flux		$m^{-2}s^{-1}$	1.6.12
K_a	area expansion elastic modulus		J/m^2	11.1.1
K_n	equilibrium constant			4.1.2
K_0	optical constant		m^{-2}	4.2.5
k_b	bulk conductivity			8.4.20
k_n^+	association rate constant		$M^{-1}s^{-1}$	4.2.24
k_n^-	dissociation rate constant		s^{-1}	4.2.24
k_q	fluoresence decay constant		s^{-1}	4.2.22
k_r	diffusion-controlled rate constant		$m^3s^{-1}; M^{-1}s^{-1}$	8.3.11
k_s	barrier-controlled rate constant		$m^3s^{-1}; M^{-1}s^{-1}$	8.3.27
k_s	surface conductivity			8.4.21
$\langle L_r^2 \rangle$	mean square length of polymer		m^2	7.1.11
l_p	Kuhn length		m	7.1.13
l_r	Random walk step length		m	1.6.4
l	surfactant hydrocarbon chain length		m	1.3.3
M	molecular weight			4.2.11
$\langle M \rangle_n$	polymer molecular weight (number averaged)			7.1.5
$\langle M_n \rangle_{wa}$	polymer molecular weight (weight averaged)			7.1.4
m_i	dipole moment		D (Cm)	3.2.7
N	micelle aggregation number			4.1.6
N_{Av}	Avogadro's number	6.022×10^{23}		
$\langle N_p \rangle_{na}$	number averaged degree of polymerization			7.1.1
$\langle N_p \rangle_{wa}$	weight averaged degree of polymerization			7.1.2
N_s	surfactant parameter			1.3.1
n	refractive index			4.2.5
n_i	number of moles			2.3.8
P_S	permeability		$m\ s^{-1}$	6.3.1
$P(\theta)$	scattering form factor			7.1.27
p	pressure		N/m^2 (P)	2.2.3
Q	scattering vector		m^{-1}	7.1.25
q	charge		C	3.2.6
$\bar{q}$	quenchers/micelle			4.2.16
R	electrical resistance		ohm (V/A)	4.2.2

Symbol	Name	Value	Units	Defining Equation
R	radius		m	2.1.17
R_g	radius of gyration		m	7.1.6
R_1, R_2	radii of curvature		m	1.3.8
R_{1N}	polymer end-to-end distance		m	
R_θ	Rayleigh ratio		m^{-1}	4.2.8
$\vec{r}$	distance		m	3.2.2
S	spreading coefficient			2.1.16
S_{CD}	NMR order parameter			6.2.1
S_n	surfactant aggregate			4.1.1
$S(Q)$	scattering structural factor			7.1.27
S_T	total surfactant			4.1.7
s	dimensionless charge parameter			3.9.14
T	temperature		K	
u	electrophoretic mobility		$m^2 s^{-1} V^{-1}$	8.4.2
V_A	molar volume		m^3	
$V(r)$	potential energy		J	3.1.1
v	surfactant hydrocarbon chain volume		m^3	1.3.2
w	interaction parameter		J/mol	1.4.16
W	colloidal stability ratio			8.3.28
W_{AA}, W_{BB}	microscopic interaction energy		J	1.4.12
$\boldsymbol{w}$	work		J	2.1.6
X_A	mole fraction			1.4.8
z	distance		m	3.8.5
Z	ion valency			3.2.6
z_b	coordination number of nearest neighbors in bulk			1.4.12
z_s	coordination number of nearest neighbors at surface			2.1.2
Φ	electrostatic potential		V	3.2.1
Φ	volume fraction			7.2.3
Φ_0	surface potential		V	3.8.14
Λ	equivalent conductance			4.2.3
Θ	fractional surface coverage			2.5.8
Π	osmotic pressure		N/m^2	1.5.10
Π_s	surface pressure		N/m	2.5.8
Γ_i	surface concentration		m^{-2}	2.4.4
Γ_0	Gouy–Chapman coefficient			3.8.14
τ	time of coagulation for colloids		s	8.3.16
τ_0	lifetime		s	4.2.21
Ω	solid angle			3.3.2
Ω	number of states accessible to a system			1.4.3

Symbol	Name	Value	Units	Defining Equation
χ	chi-parameter for polymer solutions			7.1.10
α	polarizability		Fm^2	3.4.2
ε_o	permittivity in vacuum	8.854×10^{-12}	F/m, C/V·m	3.2.1
ε_r	relative dielectric permittivity			3.5.12
γ	surface tension		J/m^2, N/m	2.1.6
γ	shear strain			7.2.15
$\dot{\gamma}$	shear rate		s^{-1}	7.2.17
δ	distance of closest approach		m	5.3.2
λ	wavelength		m	4.2.8
λ_i	decay length		m	5.4.2
κ	inverse Debye length		m^{-1}	3.8.13
$\bar{\kappa}$	specific conductance			4.2.1
κ	bending rigidity		J	6.2.3
$\bar{\kappa}$	saddle splay constant		J	11.1.4
ϕ	angle			
η	viscosity		$Nm^{-2}s$	5.4.2
η_s	solvent viscosity		$Nm^{-2}s$	7.2.34
$[\eta]$	intrinsic viscosity			7.2.34
θ	angle			
$\rho(r)$	charge density		$C\,m^{-3}$	3.2.1
σ	surface charge density		$C\,m^{-2}$	3.5.1
σ	standard deviation			
σ	shear stress			
μ	friction coefficient		$J\,s\,m^{-2}$	5.7.1
μ_i	chemical potential of species i		J	1.5.5
μ_i^{θ}	standard state chemical potential of i		J	1.5.6
ζ	zeta-potential		V	8.4.5

REFERENCES

Adamson, W. A., and Gasy, A. P. 1997. *Physical Chemistry of Surfaces*, 6th ed. New York: John Wiley & Sons.

Davis, H. T. 1996. *Statistical Mechanics of Phase, Interfaces, and Thin Films*, New York: Wiley-VCH.

Hiemenz, P. C., and Rajagopalan, R. 1997. 3rd ed., New York: Marcel Dekker, Inc.

Hunter, R. J. 1987. *Foundations of Colloid Science*. New York: Oxford University Press.

Hunter, R. J. 1993. *Modern Colloid Science*, New York: Oxford University Press.

Israelachvili, J. 1991. *Intermolecular and Surface Forces*, 2nd ed. San Diego: Academic Press, Inc.

Kruyt, H. R. 1952. *Colloid Science*, Vol. 1. New York: Elsevier Publishing Company.

Meyers, D. 1991. *Surfaces, Interfaces and Colloids*. New York: Wiley-VCH.

Shaw, D. J. 1980. *Introduction to Colloid and Surface Chemistry*, 3rd ed. London: Butterworths.

Shinoda, K. 1978. *Principles of Solution and Solubility*. New York: Marcel Dekker, Inc.

Tanford, C. 1980. *The Hydrophobic Effect. Formation of Micelles and Biological Membranes*, 2nd ed. New York: John Wiley & Sons, Inc.

Vold, M. J., and Vold, R. D. 1983. *Colloid and Interface Chemistry*. Reading, MA: Addison-Wesley, Inc.

AUTHOR BIOGRAPHIES

D. Fennell Evans is Director of the Center for Interfacial Engineering and professor of chemical engineering and materials science at the University of Minnesota. He is the author of more than 180 publications on self-assembly processes in water and nonaqueous solvents, microemulsions, diffusion in liquids, and characterization of surfaces using scanning tunneling and atomic force microscopies.

Håkan Wennerström is professor of physical chemistry at the Chemical Center, University of Lund. He is the author of more than 170 publications, mainly in the area of phase behavior and surface forces in colloidal systems, surfactant self-assembly, electrostatic interactions, and nuclear magnetic resonance spectroscopy.

WHY COLLOIDAL SYSTEMS ARE IMPORTANT

The Colloidal Domain Encompasses Many Biological and Technological Systems

Evidence of man's use of colloids dates back to the earliest records of civilization. Stone Age paintings in the Lascaux caves of France and written records of Egyptian pharaohs were produced with stabilized colloidal pigments. Many of our earliest technological processes, such as papermaking, pottery making, and the fabrication of soaps and cosmetics, involved manipulation of colloidal systems. During the Middle Ages artisans codified their knowledge and formed guilds that served as the prototypes of present-day professional organizations.

We can trace the establishment of colloid science as a scientific discipline to the mid-nineteenth century. In 1845 Francesco Selmi described the first examples of colloidal particles. By defining their common properties, he assigned these clear or slightly turbid "pseudosolutions" of silver chloride, sulfur, and Prussian blue in water to a class that also included alumina and starch.

In the 1850s Michael Faraday made extensive studies of colloidal gold sols, which involve solid particles in water. He found that these sols are thermodynamically unstable. Once they have coagulated, the process cannot be reversed. For this reason, such insoluble dispersions are called *lyophobic* ("liquid-hating") colloids. Faraday's observations imply that such dispersions must be stabilized kinetically; if properly prepared, however, they can exist for many years. In fact, some of the colloidal systems Faraday prepared are still on display in the British Museum in London.

In 1861 Thomas Graham coined the term *colloid* (which means "glue" in Greek) to describe Selmi's "pseudosolutions." The term emphasizes their low rate of diffusion and lack of crystallinity. Graham deduced that the low diffusion rates of colloidal particles implied that the particles were fairly large— at least 1 nm in diameter in modern terms. On the other hand, the failure of the particles to sediment under the influence of gravity implied that they had an upper size limit of approximately 1 μm. Graham's definition of the range of particle sizes that characterize the colloidal domain is still widely used today.

Colloidal phenomena played an important role in the genesis of physical chemistry by establishing a connection between descriptive chemistry and theoretical physics. For example, the discovery of Brownian motion resulted from the observation of colloidal-sized pollen particles by light microscopy. Einstein developed the relationship between Brownian motion and diffusion coefficients. In 1909 Jean Baptiste Perrin used this relationship to determine a definite value of Avogadro's number. Subsequently, Marian Smoluchowski derived an expression that related the kinetics of rapid coagulation of colloidal particles—in which each Brownian encounter between two particles resulted in permanent contact—to the formation of a larger dimer particle. His expression was extended to explain the role of diffusion in bimolecular reactions in general.

These fundamental developments paralleled increasingly sophisticated industrial use of colloidal systems. Formulations of paints and other coatings were improved, more sophisticated ceramics were developed, use of emulsions increased, and novel uses were found for newly discovered polymer molecules. However, the complexity of these practical systems far exceeded the ability of theorists to explain them, and to a large extent colloids remained an empirically descriptive field.

Significant advances in explaining the stability of colloidal sols were made in 1945. Publication of the Derjaguin–Landau and Verwey–Overbeek (DLVO) theory provided a quantitative relationship between the attractive van der Waals forces, which lead to coagulation, and the repulsive electrostatic forces, which stabilize colloidal dispersions. The DLVO theory was studied extensively and verified; it now constitutes a cornerstone of colloid science.

Colloidal domains encompass much more than just the classical sols on which research was focused for almost a century. Polymers that have at least one molecular dimension greater than 1 nm exhibit many features of colloidal solutions. Unlike sols, polymer solutions form spontaneously and are thermodynamically stable. Such macromolecules (which include biopolymers like DNA and proteins) are called *lyophilic* ("liquid-loving") colloids. The great technical importance of polymers has led to the development of polymer science as a separate discipline .

Another major area of colloidal science involves association colloids formed by amphiphilic (both "oil and water-loving") molecules. It developed more slowly than the study of colloidal dispersions involving sols—partly because it depended on the synthesis and isolation of amphiphilic molecules in reasonably pure form. A key development in the study of association colloids was the observation by James McBain that the osmotic pressure of alkali metal fatty acid salts displays a pronounced break in concentration curve beyond which the osmotic coefficient remains almost constant. McBain attributed this fact to the self-association of these molecules into structures he called micelles. This theory did not meet with immediate success, however. When McBain first presented his observation at a symposium, the chairman is said to have remarked, "Nonsense, McBain."

Despite this unenthusiastic reception, McBain's observation stimulated considerable discussion concerning the structure of micelles. G. S. Hartley proposed that the structures are spherical, with a radius equal to that of the fully extended hydrocarbon chain. William Harkins considered the possibility of cylindrical micelles. McBain himself believed that micelles might be flat laminar sheets. In fact, when surfactants aggregate, they form structures of all these types. But in the case of the micelles McBain studied, Hartley was correct. Debate about the structure of amphiphilic assemblies continues to this day. In many ways, amphiphilic self-organizing systems are more complex than colloidal dispersions. One reason for this is that amphiphiles are associated physically, not chemically. Consequently, their microstructural size and shape can change in response to subtle variations in concentration or temperature. This facile response to changing environmental conditions contrasts strongly with the relatively immutable behavior of colloidal sols.

As the properties of association colloids were characterized more thoroughly, it became clear that the DLVO theory could not satisfactorily describe the interaction between such aggregates. In fact, our understanding of the complex interactions displayed by these systems has expanded significantly during the past two decades and in the process has brought new richness and sophistication to colloid science.

Large surface areas associated with the characteristic size of colloidal particles form an intrinsic property of all colloidal systems. For example, a typical micellar solution containing 0.1 M amphiphile has $\sim 4 \times 10^4$ m^2 of micellar–water interfacial area per liter of solution. It is often convenient to view such systems as being comprised of two phases: a dispersed phase (the colloidal particles) and the dispersing medium (the solvent) separated by a well-defined interface.

These interfaces are generally regions of high free energy, and many surface phenomena, such as adsorption and the formation of electrical double layers, occur in them. Because surface properties dominate the behavior of colloidal systems, surface phenomena play an integral role in colloidal science. The organized study of surface phenomena can be traced back to Benjamin Franklin's observation of the effect of pouring oil on turbulent waters. In 1757, while watching a military fleet sail out of New York Harbor, Franklin observed that the wakes behind two of the ships were calmed after the ships' cooks dumped greasy refuse. Franklin later reproduced this phenomenon himself by pouring oil on Clapham Pond in England. (We now know that he used three times as much oil as he needed.) Franklin's reports stimulated considerable scientific interest, which led to a new field of inquiry. Maritime patents for the process were issued, but the attempt to exploit the effect commercially proved impractical.

One of the more enigmatic characters in nineteenth-century scientific history is Agnes Pockels. A German woman with no formal scientific training, Pockels devised a series of experiments that remain a hallmark of surface science. Having developed a rudimentary surface balance in her kitchen sink, Pockels used this apparatus to determine surface pressure versus area curves. She sent her observations to Lord Rayleigh, who submitted them to Nature magazine. Publication of Pockels's work in 1891 set the stage for Langmuir's quantitative work on mono layers and laid the foundation for our ability to characterize them.

Understanding of Colloidal Phenomena Is Advancing Rapidly

Why should we be interested in colloidal systems? Everyday experience provides a variety of reasons.

The first reason is purely epistemological. As scientists we are interested both in knowledge and in how we structure what we know. If we consider the three states of matter—gas, liquid, and solid—we can observe colloidal systems in all possible combinations. As Table I.1 illustrates, manifestations

TABLE I.1 / Types of Colloidal System

Medium	Particle	Name	In Nature	Biological	Technical
			Examples		
Liquid	Solid	Colloidal sol	River water, glacial runoff, muddy water		Paint, ink Sol–Gel Processing
Liquid	Liquid	Emulsion (nonequilibrium) solution (equilibrium)		Fat digestion, biological membranes	Drug delivery, emulsion polymerization
Liquid	Gas	Foam	Polluted rivers	Vacuoles, insect excretions	Fire extinguishers, production of porous plastics
Gas	Solid	Aerosol	Volcanic smoke	Pollen	Inhalation of solid pharmaceuticals
Gas	Liquid	Aerosol	Clouds	Result of coughing	Hair spray, smog
Solid	Solid	Solid suspension	Wood	Bone	Composites
Solid	Liquid	Porous material	Oil reservoir rock, opals	Pearl	High impact plastics
Solid	Gas	Solid foam	Pumice	Loofah	Styrofoam, zeolites

of the colloidal domain surround us every day. Our ability to understand, use, and control colloids depends on our mastery of their properties. In the past few decades, what we know about matter at the molecular level has increased dramatically. Recent progress in our understanding of the forces that mediate colloidal interactions provides a new basis for further scientific advance.

The second reason is anthropological. Living systems are made up largely of proteins, polyelectrolytes, and amphiphilic molecules contained in an aqueous medium. By their very size, such molecules possess colloidal properties. More importantly, they self-assemble into a great variety of organized structures that perform extremely sophisticated chemical transformations. Life processes involve the control and transformation of colloidal assemblies, and many diseases are associated with malfunctions at the colloidal level. The new level of knowledge made available by colloidal science has considerably enhanced our understanding of interaction forces that control biological processes.

The third reason is technological. Nearly all industrial processes involve colloidal systems. The partial list given in Table I.1 makes their importance in a highly industrialized society seem almost self-evident. However, the challenge facing contemporary scientific research goes beyond these traditional applications. In many rapidly advancing areas of technology, such as fabrication of new high strength composite materials, development of high performance ceramics for use at elevated temperatures, and control of targeted drug delivery and medical diagnostic systems, progress depends on our ability to control colloidal interactions.

Association Colloids Display Key Concepts That Guided the Structure of This Book

Writing a textbook is an active process. An author must make difficult choices among subjects that vie for inclusion like characters clamoring to take center stage during a play. Just as the play's action determines how long each character is allowed to remain in the spotlight or what staging is required, the author's creative philosophy determines the ideas in the text and the approach that is chosen. And just as the playwright becomes invisible once the actors have begun to perform, publication usually draws a curtain over the decision-making process that shaped the content balance of a text. Because we believe our readers deserve to see backstage, so to speak, we want to explain the thought process that lies behind our work.

Emerging information about fundamental forces and interactions in the colloidal domain stimulated the writing of this book. We decided to focus on association colloids rather than on the more traditional lyophobic systems for two reasons. First, association colloids more readily illustrate the new concepts which we believe are the hallmark of modern colloid science. Second, concentrating on association colloids establishes a more immediate link between fundamental concepts, which are the organizing feature of this book, and their manifestations in the colloidal domain.

Most textbooks on colloids begin with lengthy discussions of the physical techniques, such as diffusion, light scattering, and viscosity, used to characterize colloidal systems. We find teaching from such books frustrating because they generally avoid addressing the real subject in a straightforward way. For this reason, we have chosen to include a succinct discussion of relevant experimental techniques in the appropriate chapters, but we focus our attention on describing fundamental physical phenomena, defining appropriate equations, and illustrating their applications through examples in the text and through ex ercises at the end of each chapter. In addition, we cite relevant references at the end of each chapter, indicating the portions of the text to which each applies; additional literature citations ap pear as credits to figures and tables.

We have made a particular effort to articulate key concepts that form the basis of our understanding of colloidal phenomena. A major goal is to show how molecular interactions lead to colloidal behavior. We find that this connection between the molecular world and the colloidal domain emerges most clearly with association colloids. In some instances we employ theoretical derivations that depart from those usually given. For example, instead of simply deriving expressions for free energy, we often obtain separate equations for the energy and entropy so that we can clarify the underlying fundamentals as we combine them. We also go to some lengths to point out where and how averaging our physical properties transforms energy expressions into expressions for free energy.

Our presentation presupposes mastery at an upper under graduate level of physics, mathematics, and physical chemistry. While we have made every effort to provide linkages in the derivations, there is no substitute for writing them out as you read along. Learning is not a passive process. Forging enduring neurological connections in the brain requires a lot of groaning and sweating; it's just plain hard work.

Chapter 1 discusses the general characteristics of amphi philic self-organizing systems. We describe their structural characteristics as well as the solvophobic interactions that drive ag gregation. We also provide a simple geometric relationship between molecular and aggregate structures. This chapter in cludes a section on regular solutions which summarizes many useful thermodynamic equations that we employ repeatedly throughout this book.

Chapter 2 treats fundamental surface science, focusing on the aspects of the subject—particularly surface tension—that provide the background necessary to understand colloidal inter actions. Amphiphilic monolayers, which hold considerable in terest in their own right, are used in this chapter to illustrate many of the relevant principles of surface science.

Chapters 3 through 11 employ an alternating pattern in which a chapter that develops fundamental concepts is followed by one or more chapters focusing on colloidal systems that illustrate these concepts.

For example, in Chapter 3, we derive the Gouy–Chapman theory and related equations that describe the properties of charged electrical double layers. Because the words *charged electrical double layer* describe both a charged surface and its adjacent solutions, we obtain relationships between the distribution of ions in a solution adjacent to a charged surface, then examine how the distribution varies with ionic valency and concentration and the surface potential or charge density. These relationships and the concepts they encompass form one of the bases of colloid science.

In Chapter 4 we discuss micellar systems in general but emphasize ionic micelles, because they illustrate electrical double-layer properties with particular clarity. In many ways, micelles are the simplest and best understood of all amphiphilic self-organizing systems. Equations describing micellar aggregation processes and micellar properties, dynamics, and kinetics provide a framework for dealing with more complex self-organizing structures, such as bilayers, vesicles, and emulsions, which we consider in subsequent chapters.

Chapter 5 develops equations that describe the repulsive and attractive forces that operate in colloidal systems. The picture that emerges reflects the myriad interaction forces that come into play as colloidal particles approach one another. This is a difficult chapter, not only because of the inherent intricacy of the phenomena, but because here we merge older classical concepts with newly emerging ones.

In Chapter 6 we discuss bilayer systems as an example of the interaction forces described in Chapter 5. The discussion also deals with association processes leading to the formation of liquid crystals, vesicles, and gels. Chapter 7 examines self-organizing properties of polymers, proteins, and polyelectrolytes.

Integrating the electrostatic and London dispersion forces gives an expression for the potential energy that we use in Chapter 8 to obtain the DLVO theory. We also consider the critical concentration of salt required for rapid coagulation and derive an expression for the kinetics of coagulation. This chapter develops the concept of the zeta potential and describes how to measure it.

Continuing our system of alternating theory with applications, Chapter 9 focuses on colloidal sols that exhibit typical properties and exemplify many of the concepts developed in previous chapters. Chapter 10 considers fundamental rules for understanding phase diagrams, and Chapter 11 concludes with a discussion of micro- and macroemulsions that illustrate these ideas.

As an aid to the reader, we have written all section headings as complete sentences. When gathered together in the Contents, these sentences provide a succinct summary of the contents of each chapter. Key concept words appear in boldface the first time they are used. Symbols, units, and the equations that define them are listed following Chapter 11. In addition, all chapters begin with concept maps, which should be used to guide the detailed reading. We have found these maps of great help to students approaching the material for the first time. Spending 20 to 30 minutes internalizing the content of a concept map can significantly increase the information gained from reading the chapter.

Two final issues pertaining to our philosophy of authorship must be addressed. The first concerns the balance we have created between concepts and the systems that illustrate them. Our coverage of colloidal systems is clearly selective; the concepts under discussion dictated our choice of examples. However, references at the end of each chapter provide citations of alternative surveys of colloidal systems.

The second issue concerns our treatment of concepts that are still the subject of lively debate. We have tried to indicate which ideas are controversial, but we have not pursued the controversies in detail. All too often, a scholar who engages in such polemics is reminiscent of a player who "struts and frets his hour upon the stage . . . and then is heard no more." Rather than attempting to present all the intellectual options currently available, we identified the strategies we believe are most critical to the forward movement of the play and allowed them to assume the leading roles in our text.

1

SOLUTES AND SOLVENTS, SELF-ASSEMBLY OF AMPHIPHILES

CONCEPT MAP

Self-Assembly Processes

- Amphiphilic molecules possess polar head groups and nonpolar chains (Table 1.1); to minimize unfavorable solvophobic (solvent-hating) interactions, they spontaneously aggregate to form a variety of microstructures. The sodium dodecyl sulfate (SDS) spherical micelle shown in Figure 1.2 is a specific example.

- Three general characteristics of the self-organizing capabilities of amphiphilic molecules—as illustrated by SDS spherical micelles—are:

  1. Aggregates form spontaneously—SDS micelles form at a well-defined critical micelle concentration (CMC).

  2. Aggregation is a start–stop process—adding more SDS results in the formation of more micelles of the same size.

  3. Aggregates have well-defined properties—the maximum radius of a spherical micelle is set by the length of the hydrocarbon chain.

- Amphiphilic molecules are associated physically, not chemically, and aggregate size and shape range in response to variations in solution conditions (temperature, pH, salt, etc.).

Krafft Temperature

- At the Krafft temperature T_{Kr} a surfactant's solubility equals the CMC as shown for SDS in Figure 1.3.

- T_{Kr} is a characteristic temperature for each surfactant; below T_{Kr} surfactant precipitates out of solution as hydrated crystals.

- NMR relaxation measurements above T_{Kr} show that the surfactant hydrocarbon chains in aggregates are in a melted state and display local chain motions similar to those of liquid hydrocarbons.

The Surfactant Number

- The surfactant number $N_S = v/la_0$ (eq. 1.3.1) relates amphiphilic molecular structure (Table 1.1) to aggregate architecture (Figure 1.6), where v and l represent the volume and length of the hydrocarbon chains and a_0 is the area per head group.

- N_S relates properties of the amphiphile to preferred curvature of the aggregate as given in eqs. 1.3.4–1.3.7.

Entropy of Mixing and Regular Solutions

Although entropy of mixing and regular solutions are traditionally part of physical chemistry, they are reviewed here because they play an important role in understanding colloidal behavior.

- From elementary statistical mecanics, we obtain $\Delta S_{mix} = -k\Sigma N_i \ln X_i$ (eq. 1.4.6) for the ideal entropy of mixing.

- For regular binary solutions, $\Delta E_{mix} = n_{AB}w$ (eq. 1.4.15), where the interaction parameter w depends on the microscopic interaction energies W_{AA}, W_{BB}, and W_{AB} (eq. 1.4.16).

- From $\Delta G_{mix} = \Delta H_{mix} - T\Delta S_{mix}$ (eq. 1.4.1b) we can predict the temperatures and compositions under which a binary solution will be homogeneous or form two phases.

Chemical Potentials

- The chemical potential μ_i (eq. 1.5.5a) gives the free energy change in adding a molecule i to the system. It is the thermodynamic property which determines a component's ability to redistribute and is used throughout this text to analyze colloidal systems.

- For a solution the chemical potential of the solvent is often expressed in terms of the osmotic pressure Π_{osm} (eq. 1.5.11). In analogy with the nonideal gas the second virial coefficient B_2 (eq. 1.5.13) accounts for the interactions between the solute molecules.

- Chemical potentials (eq. 1.5.22) can be generalized to show how an external gravitational field (eq. 1.5.26) or an electrical field (eq. 1.5.30) affect the distribution of components in a chemical system.

Brownian Motion, Intermolecular Interactions and Diffusion

- In the Boltzmann distribution (eq. 1.5.24) the interplay between Brownian motion as expressed by kT and an organizing force arising from intermolecular interactions is a key element in understanding interactions in the colloidal domain.

- Collisions between molecules can be described by a random walk model which leads to the equation, $\langle r^2 \rangle = 6Dt$ (eq. 1.6.6) which relates the average distance traveled to the time and the diffusion coefficient.

- The flux of molecules in the presence of a concentration gradient is described by Fick's first law (eq. 1.6.9) for the molecular flow and by Fick's second law (eq. 1.6.11) for the concentration changes.

Solvophobicity

- Microphasic separation accompanying amphiphilic aggregation is driven by solvophobic interactions and requires (1) an amphiphile with well-defined polar and nonpolar groups and (2) either a very polar or nonpolar solvent.

- The Gordon parameter ($\gamma/V^{1/3}$ N/m^2, where γ and V are the solvent's surface tension and molar volume) provides a useful measure of solvent cohesive energy, as illustrated in Figure 1.11.

- Micelle formation is observed in polar (water, hydrazine, and ethylene glycol, $1.3 < \gamma/V^{1/3}$) and in nonpolar (oil and fluorocarbons, $\gamma/V^{1/3} < 0.3$) solvents, but not in intermediate solvents (alcohols, esters, and amides).

1

Most colloidal phenomena involve a liquid phase as at least one of the actors. In this liquid we can typically identify a *solvent* and one or more *solutes.* Molecular interactions involving colloidal entities, solutes, and solvents determine the properties of the system.

Throughout this book we try to maintain this molecular perspective on the colloidal phenomena. In this chapter we apply this strategy to a description of the general characteristics of amphiphilic assemblies. As we shall see, these colloidal self-assembled structures arise from a delicate interplay between solute–solvent and solute–solute interactions. The term **amphiphilic** indicates that one part of the molecule likes the solvent, while the other does not.

1.1 Amphiphilic Self-Assembly Processes Are Spontaneous, Are Characterized by Start–Stop Features, and Produce Aggregates with Well-Defined Properties

Amphiphilic molecules spontaneously self-organize into a variety of structures. The simplest and best understood of these is the micelle. To characterize the amphiphilic aggregation process, we can begin by considering how adding surfactant to water leads to the formation of this typical structure.

One example of a dual-character molecule possessing the well-defined polar head and nonpolar tail needed to produce amphiphilic behavior is sodium dodecyl sulfate (SDS), whose structure is given in Table 1.1. In aqueous solutions up to 8×10^{-3} M, most of the properties SDS displays are similar to those we can observe for a typical electrolyte such as NaCl. Figure 1.1 illustrates an exception: decreasing surface tension, denoted as γ.

TABLE 1.1 Typical Amphiphilic Structures

Types and Class	Structure	Amphiphil[a]	Symbol	CMC M $(T°G)$[b]
Ionic				
(sodium alkylsulfate)	$C_8H_{17}OSO_3Na$	sodium octyldulfate	C_8SO_4Na	1.33×10^{-1}
	$C_{10}H_{21}OSO_3Na$	sodium decylsulfate	$C_{10}SO_4Na$	3.3×10^{-2}
	$C_{12}H_{25}OSO_3Na$	sodium dodecylsulfate	$C_{12}SO_4Na$ (SDS)	8.1×10^{-3}
(sodium alkylcarboxylate)	$C_{12}H_{25}CO_2Na$	sodium dodecanoate	$C_{11}CO_2Na$	2.5×10^{-3}
sodium 1,4-dialkyl-3-sulfonate-succinate	$(CH_3CH_2)_2-CH-(CH_2)_3-O-C$ $O=C$ $CHSO_3$ CH_2 $(CH_3CH_2)_2-CH(CH_2)_3-O-C$ $O=C$	Aerosol OT	AOT	
alkyltrimethyl-ammonium bromide	$C_{12}H_{25}N(Me)_3Br$	dodecyl trimethyl-ammonium bromide	$C_{12}N(Me)_3Br$	1.5×10^{-2}
	$C_{16}H_{32}N(Me)_3Br$	hexadecyl trimethyl-ammonium bromide	$C_{16}N(Me)_3Br$ CTAB	9.2×10^{-4}
dialkyldimethyl-ammonium bromide	$(C_{12}H_{25})_2N(Me)_2Br$	didecyldimethyl-ammonium bromide	$(C_{12})_2N(Me)_2Br$ (DDAB)	
dialkyldimethyl-ammonium acetate	$(C_{16}H_{33})_2N(Me)_2CH_3CO_2$	dihexadecyldimethyl-ammonium acetate	$(C_{16})_2N(Me)_2OAc$ (DHDAA)	

Zwitterionic

Name	Structure	Common name	Symbol	CMC
dialkylphosphatidylcholine	$CH_2-O-C(=O)-C_nH_{2n+1}$ $CH_2-O-C(=O)-C_nH_{2n+1}$ $CH_2-O-P(=O)(O^-)-O-(CH_2)_2-N^+(Me)_3$	dimyristoyl-lecithin ($n=14$) DMPC dipalmitoyl-lecithin ($n=16$) DPPC distearoyl-lecithin ($n=18$) DSPC		5×10^{-10} (41°C)
Alkyldimethylpropane sultaine	$C_{12}H_{25}N(CH_3)_2(CH_2)_3SO_3$	dodecanedimethyl-propanesultaine	$C_{12}N(Me)_2(CH_2)_3SO_3$	3.6×10^{-3} (30°C)
Nonionic				
alkylethylene glycol	$C_{10}H_{21}(OCH_2CH_2)_8OH$ $C_{12}H_{25}(OCH_2CH_2)_6OH$		$C_{10}E_8$ $C_{12}E_6$	1.0×10^{-3} 8.7×10^{-5}
alkyldimethylamine oxides	$C_{12}H_{23}(CH_2)NO$	dodecyldimethyl-amine oxide	$C_{12}(Me)_2NO$	1.9×10^{-2}
alkylglucoside	$C_{10}H_{21}C_6O_6H_{11}$	β-d-decylglucoside		2.2×10^{-3}

(a) The naming of organic compounds is often a source of considerable frustration to the uninitiated. To add to the confusion, two or three names are used interchangeably for many common alkyl chains. For example, the former IUPAC (International Union of Pure and Applied Chemistry) name for the carboxylic acid, $C_5H_{11}COOH$, is hexanoic acid, while the common name is caproic acid because its rancid odor suggested the smell of goats to early chemists. For the $C_{16}H_{32}$ alkyl chain, the IUPAC name is hexadecyl. However, surface chemists often use "cetyl," while lipid chemists use "palmitoyl." Adding to the confusion are the conventions for specifying the presence of unsaturation leading to cis and trans configurations or [the presence] of substituents along the chain. Throughout this book, we will use the symbols given in column 4. Table 6.1 and Figure 6.2 give additional information on the names and structures of lipids.
(b) 25°C unless otherwise noted.

Figure 1.1
Solution surface tension versus sodium dodecyl sulfate (SDS) concentration. As SDS concentration increases, the surface tension γ of the solution initially decreases and then becomes almost constant. (Note the logarithmic scale.) The critical micelle concentration (CMC) is determined as the intersection of two linearly extrapolated lines, A and B. The lowering of γ results from the concentration of the surfactant at the air–water surface.

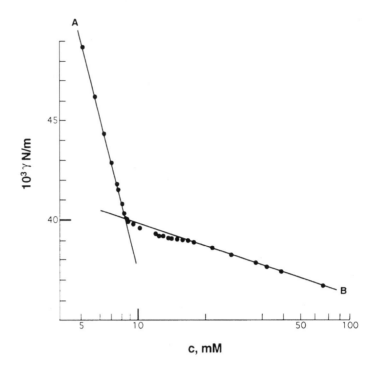

The insolubility of SDS's hydrocarbon chain in water causes the molecule to concentrate at the air–water interface, with its hydrocarbon chain oriented toward the vapor phase. Since the surface tension of water is higher than that of hydrocarbons, adding surfactant results in a decrease in γ.

We can write the distribution of surfactant between the air–water interface and the solution as an equilibrium, but above $8 \times 10^{-3}\,M$ the surface tension becomes almost constant. As we add more surfactant to the solution, the surface tension, and thus the surface, remain virtually unchanged. So what is happening?

One possibility is analogous to the phase separation that occurs when water is saturated with nonpolar or apolar compounds such as benzene or dodecanol. The same force works to diminish unfavorable hydrocarbon–water interactions, but for SDS, formation of two bulk phases also would produce an energetically unfavorable result because it would remove the polar ionic head groups from the water.

What actually happens, therefore, is a compromise between the extremes of a complete phase separation and molecular disperse solution. The surfactant molecules self-assemble to create a microphase in which the hydrocarbon chains sequester themselves inside the aggregate and the polar head groups orient themselves toward the aqueous phase, as illustrated in Figure 1.2. In the case of SDS, groups of approximately 60 dodecyl sulfate molecules self-assemble in this way to form spherical micelles. We denote the point at which we can

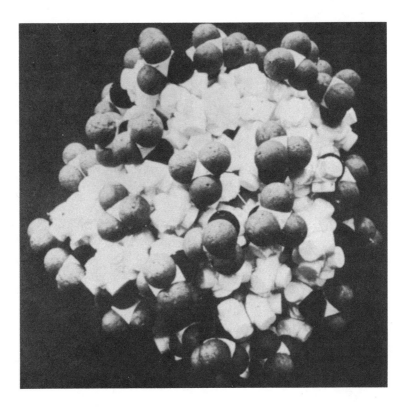

Figure 1.2
Space-filling model of an ionic spherical micelle. (B. Lindman and H. Wennerström, *Topics in Current Chemistry*, Vol. 87, Springer-Verlag, Berlin, 1980.)

discern this micellization process as the **critical micelle concentration** (CMC).

The occurrence of a CMC is the result of two competing factors. Transfer of the hydrocarbon chain out of water and into the oillike interior of the micelle drives micellization. Repulsion between head groups as they come close together opposes it. Under these conditions the growth process has specific limits, and aggregates cease to grow when they have reached a certain size. Above the CMC, adding more surfactant simply produces more micelles over a considerable concentration range rather than further growth of existing micelles.

The way SDS behaves when added to water provides a specific example of the self-organizing capabilities of amphiphilic molecules, but we can generalize about the characteristics of these systems as follows:

1. Aggregates form spontaneously.

2. Aggregation is a start–stop process.

3. Aggregates have well-defined properties.

Self-assembly must be understood as a physicochemical process. Amphiphilic molecules are associated physically, not chemically, and for this reason, they can change the size or shape of their microstructure in response to small changes in concentration, salt content, temperature, pH, and pressure. However, they still retain their start–stop characteristics, and the size of the molecule limits the aggregate's growth in at least one dimension.

1.2 Amphiphilic Molecules Are Liquidlike in Self-Assembled Aggregates

As we have seen, amphiphilic behavior requires schizophrenic molecules that possess discrete polar and nonpolar regions. They should not be readily soluble individual molecules in solvents that are either strongly polar or strongly apolar. To achieve this property, the nonpolar tail of the typical molecule is a hydrocarbon or a fluorocarbon with at least eight carbons. The polar head group can be ionic, zwitterionic, or dipolar. Table 1.1 lists typical amphiphiles along with the abbreviations we will use to designate them in the text.

Functional groups that are not "schizophrenic" enough to serve as effective amphiphilic head groups undergo phase separation instead of forming micelles or other association structures. These groups include alcohols, ketones, aldehydes, and simple ethers.

Temperature plays an important role in the behavior of amphiphilic molecules. For example, the result is quite different if we measure the surface tension of SDS both at a temperature below 25°C and at a higher temperature. At lower temperatures the surfactant precipitates out of solution as a hydrated crystal instead of forming micelles. In other words, at 18°C its solubility limit lies below the CMC, as illustrated in Figure 1.3.

The point at which solubility equals the CMC is called the **Krafft temperature**, or T_{Kr}. Just above this point, solubility increases rapidly, and a solution of almost any composition becomes a single homogeneous phase (or a mixture of two phases with similar composition). Surfactant solutions cooled below their T_{Kr} can remain in a metastable homogeneous state for days.

The Krafft phenomenon reflects an equilibrium between surfactants in solution and in hydrated crystals. A familiar complication is the "ring around the tub" caused by precipitation of Ca^{2+} and Mg^{2+} soaps in hard water. The Krafft temperature is much higher for divalent than for monovalent counterions, and it is also sensitive to small changes in the chemical structure of the surfactant because of the rather stringent packing requirements of the molecules in the solid (crystalline) phase.

Saturated crystalline hydrocarbons display an all-trans configuration shown in Figure 1.4. Table 1.2 summarizes the thermodynamics for the melting of crystalline hydrocarbons. These values illustrate packing differences between molecules with odd and even numbers of carbons in the solid state. Disturbance of the packing caused by a *cis* (see Figure 1.4), and to a lesser extent by a *trans* double bond, weakens the stability of the ordered phase, thus decreasing both the chain melting

Figure 1.3
Krafft temperature T_{Kr} is the point at which surfactant solubility equals the critical micelle concentration. Above T_{Kr}, surfactants form a dispersed phase; below T_{Kr}, they crystallize out of solution as hydrated crystals. (B. Lindman and H. Wennerström, *Topics in Current Chemistry*, Vol. 87, Springer-Verlag, Berlin, 1980.)

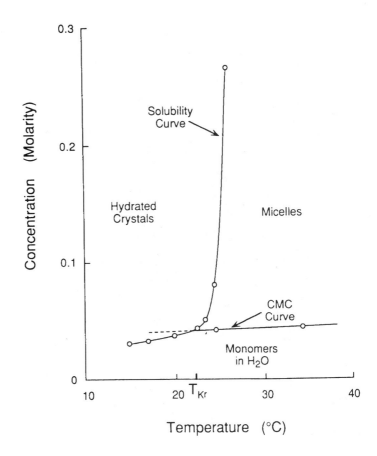

TABLE 1.2 Thermodynamics of Hydrocarbon Chain Melting[a]

Structure	Melting Point (°C)	ΔH° (kJ/mol)	ΔS° (J/K·mol)	ΔG° at 25°C (kJ/mol)
$C_{13}H_{28}$	−5.4	28.5	106	−3.2
$C_{14}H_{30}$	5.9	45	162	−3.1
$C_{15}H_{32}$	10.0	35	122	−1.8
$C_{16}H_{34}$	18.2	54	183	−1.3
$C_{17}H_{36}$	22.0	41	137	−0.4
$C_{18}H_{38}$	28.2	62	206	+0.7
$C_{19}H_{40}$	32.0	50	151	+1.0
$C_{20}H_{42}$	36.7	70	227	+2.6

[a]Data are for saturated *n*-paraffins based on compilations by the American Petroleum Institute (1953). Values of ΔH° and ΔS° refer to the melting point. Values of ΔG° at 25°C are calculated assuming constant ΔH° and ΔS° between 25°C and the melting point. Crystals formed by unsaturated hydrocarbons melt at much lower temperatures.

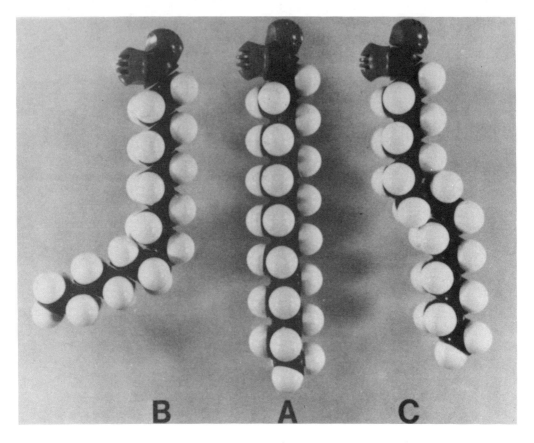

Figure 1.4
Configurations of
hydrocarbon chains.
Space-filling models of
hydrocarbon chains
displaying (A) an all-trans
configuration as it would
exist in the crystalline
state, (B) one gauche
orientation in an otherwise
all-trans chain, and (C)
two gauche orientations
resulting in a kink in the
chain. (C. Tanford, *The
Hydrophobic Effect,
Formation of Micelles and
Biological Membranes*, 2nd
ed., Wiley, New York,
1980, p. 50.)

temperature and T_{Kr}. Strong head group interactions, on the other hand, contribute substantially to a crystal's stability. Thus, divalent counterions stabilize the crystals of carboxylates, resulting in high T_{Kr} and precipitation at ambient temperatures. For sulfate and sulfonate head groups with their more delocalized charge, this difference between monovalent and divalent counterions is less pronounced.

As hydrocarbons melt, they may take many conformations. Rotation around the individual C–C bonds results in either retention of the *trans* configuration or a change into one of the two possible gauche states that are 3 to 4×10^{-21} J higher in energy than the *trans* state. This energy is very close to $kT = 4.1 \times 10^{-21}$ J at room temperature, and a balance between energetic and entropic factors determines the overall conformation of the chain. This raises questions about the state of the alkyl chains in micelles. Are they solidlike or liquidlike? Are they coiled or straight?

One way to determine the state of hydrocarbon chains in the pure phase and in the interior of surfactant aggregates is to track the rate of motion of the hydrocarbon molecules themselves or of dissolved molecules. In a medium of low viscosity, diffusional rotation of small molecules occurs with a characteristic time on the order of 10^{-11} s. This time can define the relevant time scale for a spectroscopic measurement.

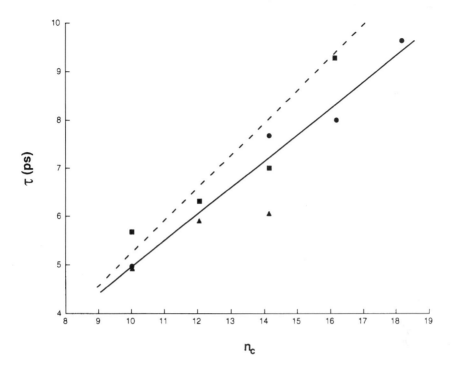

Figure 1.5
Micelle interiors show liquid phase properties. Nuclear magnetic resonance data showing characteristic time τ (ps) for rotation as a function of hydrocarbon chain length n_c for the micellar surfactants: ●, $C_nH_{2n+1}N(CH_3)_3Cl$; ■, $C_nH_{2n+1}N(CH_3)Br$; ▲, $C_nH_{2n+1}SO_4Na$; and hydrocarbons, --- $C_nH_{2n+1}C_6H_5$. The micellar times (solid line) are on the same scale as those in anhydrous alkylbenzene solution (dashed line), demonstrating the liquidlike properties of the micelle's hydrocarbon interior. If the micellar hydrocarbons were in the solid phase, the characteristic time would be many orders of magnitude longer. (P. Stilbs, H. Walderhaug, and B. Lindman, *J. Phys. Chem.* **87**, 4762 (1983).)

Several spectroscopic techniques are available. Some, like fluorescence depolarization and electron spin resonance (ESR), require the addition of a suitable probe molecule. Others like nuclear magnetic resonance (NMR), can measure hydrocarbon chain motions directly.

As Figure 1.5 illustrates, relaxation measurements using ^{13}C and ^2H NMR (with specifically deuterated chains) show that, to a good approximation, local chain motions in a micelle are as rapid as in a corresponding hydrocarbon. These direct measurements definitively confirm an inference already made by G. S. Hartley in the 1930s that micelles have a liquidlike interior—that is, the molecular motion in the interior is as rapid as in a corresponding liquid hydrocarbon.

1.3 Surfactant Numbers Provide Useful Guides for Predicting Aggregate Structures

So far we have limited our discussion of amphiphile self-assembly to spherical micelles. However, a variety of aggregates can actually form. Possible structures are restricted by an amphiphile's need to keep its nonpolar and polar parts surrounded in a favorable way. To make a priori predictions about the size and shape of amphiphilic aggregates, we need a conceptual frame-work that relates molecular parameters (such as hydrophobic volume, chain length, and head group area) and intensive variables (such as temperature and ionic strength) to microstructure. Such theories attempt to determine the **optimal aggregate or aggregates** for a given set of molecular parameters and intensive variables—sometimes called *field variables*—through free energy minimization arguments.

As in the example of SDS micellization discussed in Section 1.1, we can assume that the free energy of surfactant self-assembly in dilute solution is made up of three terms:

- A favorable **hydrophobic contribution**, due to the hydrocarbon chains sequestering themselves within the interior of the aggregates.

- A **surface term** that reflects the opposing tendencies of the surfactant head groups to crowd close together to minimize hydrocarbon–water contacts and to spread apart, as a result of electrostatic repulsion, hydration, and steric hindrance.

- A **packing term** which, at its simplest level, requires that the hydrophobic interior of the aggregate exclude water and head groups, thus limiting the geometrically accessible forms available to the aggregate.

Surface and packing terms take on different functional forms for each specific aggregated geometry. The form giving the minimum free energy for a given set of conditions determines the optimal aggregate.

For dilute solutions in which interactions between aggregates are not important, we can subsume these ideas conveniently into the **surfactant parameter**

$$N_s = \frac{v}{la_0} \tag{1.3.1}$$

where v stands for the volume of the hydrophobic portion of the surfactant molecule, l is the length of the hydrocarbon chains, and a_0 is the effective area per head group.

The volume of the hydrocarbon core of a saturated hydrocarbon chain, in cubic nanometers, is

$$v = 0.027(n_c + n_{Me}) \tag{1.3.2}$$

where n_c is the total number of carbon atoms per chain and n_{Me} is the number of methyl groups which are twice the size of a CH_2 group. For amphiphilic structures with closed hydrocarbon cores, such as micelles and vesicles, the total hydrocarbon volume is given by Nv (for single-chain surfactants) and $2Nv$ (for double-chain surfactants), where N is the number of amphiphiles contained in the aggregate.

We can estimate the maximum length (nm) of a fully extended hydrocarbon chain from

$$l = 0.15 + 0.127n_c \tag{1.3.3}$$

The 0.15 nm in this equation comes from the van der Waals radius of the terminal methyl group (0.21 nm) minus half the bond length of the first atom not contained in the hydrocarbon core (0.06 nm). The 0.127 nm is the carbon–carbon bond length (0.154 nm) projected onto the direction of the chain in the all-*trans* configuration.

Micelles can reach $r_{micelle} \approx l$ without a major free energy cost because only a few of their chains need to extend the length of the radius. For bilayers, the average length of a chain is half the bilayer thickness, which seldom exceeds $1.8\,l$ for chains in the liquid state.

The most problematic quantity in the definition of the surfactant number, N_s, is the area per polar group, a_0. For ionic surfactants, a_0 depends on both electrolyte and surfactant concentration, and the surfactant number has only limited usefulness for quantitative descriptions. For nonionic or zwitterionic surfactants, a_0 is more insensitive to external conditions, but it is still difficult to make an a priori calculation of its value as we did for v and l. However, in all cases, we can predict the trends in a_0 for a specific change in the system, and this ability can be extremely useful for qualitative interpretation of experimental data.

Surfactant numbers relate the properties of the molecule to the preferred curvature properties of the aggregates. Small values of N_s imply highly curved aggregates, but when N_s is close to unity, planar bilayers usually form. For spheres, cylinders, and bilayers, a simple relation exists between volume, area, and radius (thickness), and a comparison of the surfactant numbers shows that optimal stability of the different aggregates occurs as follows:

$$\text{Spherical micelles } N_s = 0.33 \quad (1.3.4)$$

$$\text{Infinite cylinders } N_s = 0.5 \quad (1.3.5)$$

$$\text{Planar bilayers } N_s = 1 \quad (1.3.6)$$

$$\text{Inverted cylinders and micelles } N_s > 1 \quad (1.3.7)$$

Various amphiphilic structures are illustrated in Figure 1.6.

Several things can happen when the value of N_s lies between 0.33 and 0.5 or between 0.5 and 1. The molecules may assemble into a highly symmetrical aggregate that is slightly off optimal condition or into aggregates of lower symmetry, or they may undergo phase separation. All these variations occur, and one of the challenges in research on surfactant systems is to determine when and why one of these things happens rather than another.

Obviously, surfactant numbers can represent only an approximate model of the nature of surfactant self-assembly. In spite of this, the theory provides valuable insight into the way that changes in solution conditions and molecular structures affect an aggregate's size and shape.

For example, surfactant numbers enable us to predict the effect of adding a second hydrophobic chain to a polar head

Figure 1.6

Amphiphilic aggregate structures. (A) Spherical micelles ($N_s = 0.33$) in which the micelle's radius $R_{micelle} \approx l_{max}$. For SDS micelles $R_{micelle} = 20$Å, the micelle aggregation number N is 60, and the head group area a is 60 Å^2. The actual cross-sectional area of the sulfate head group is 27 Å^2. Thus, 55% of the micelle's surface involves direct contact between hydrocarbon and water. (B) Cylindrical micelles ($N_s = 0.5$) in which $R_{micelle} = l_{max}$. The ends of the cylinder are capped by hemispheres covered by surfactant head groups. Cylindrical micelles are usually polydisperse because the rod can grow to varying lengths by simply incorporating more surfactants into the cylindrical portion of the micelle. (C) Planar bilayers ($N_s = 1$) in which the bilayer thickness is $\approx 1.6 l_{max}$ in the liquid state. Such structures can grow in the bilayer plane without limit and usually curve back on themselves to minimize exposure of hydrocarbons at the edges of the bilayer to water.

A

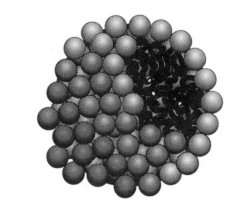

B

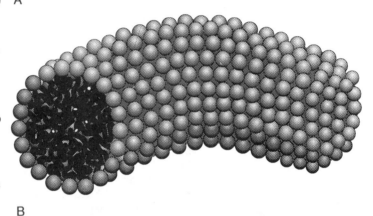

C

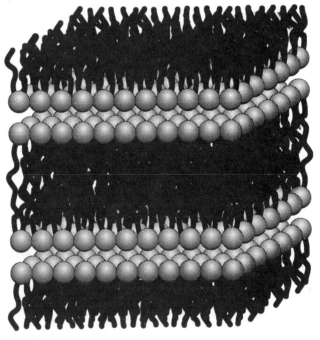

(D) Inverted micelles ($N_s > 1$) in which the head groups point into an aqueous environment and the hydrocarbon tails point out into the continuous oil medium. Such inverted structures can be spheres, cylinders, or vesicles. (E) Bicontinuous structures ($N_s \geqslant 1$) in which the two radii of curvature are equal but of opposite sign, leading to a small mean curvature. (F) Vesicles ($N_s \approx 1$) formed by small regions of bilayers closing back on themselves to form a hollow spherical structure in which the interaqueous compartment is isolated from the surroundings.

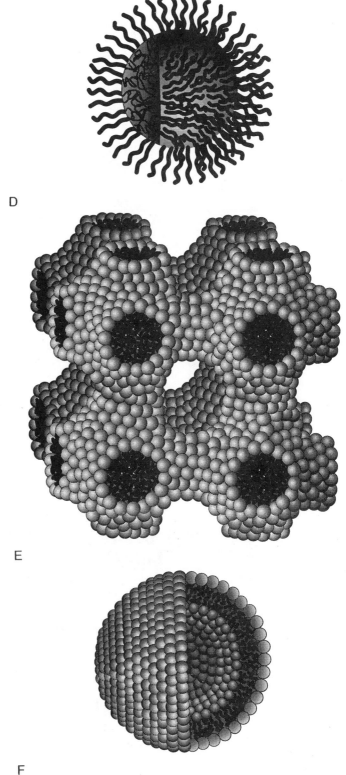

D

E

F

group. Because amphiphiles with single hydrophobic chains have surfactant parameters of less than 0.5, they are constrained — at least in dilute solution — to form micellar aggregates. Adding a second hydrocarbon chain doubles v, while a_0 and l ramain essentially the same. In effect, the addition doubles v/a_0l into the range 0.5–1. Hence, double-chain surfactants tend to form bilayer structures such as vesicles and lamellar liquid crystals, whose curvatures are inherently lower than those formed by their single-chain counterparts.

A comparable, although quantitatively different, approach to analyzing aggregate structures used the curvature concept explicitly. In this approach, the crucial parameter is not the surfactant number N_s but the preferred mean curvature of a surfactant film. We can define the **mean curvature** H at a point on a surface as

$$H = \tfrac{1}{2}\left(\frac{1}{R_1} + \frac{1}{R_2}\right) \tag{1.3.8}$$

where R_1 and R_2 are the radii of curvature in two perpendicular directions, as illustrated in Figure 1.7. For a sphere, $R_1 = R_2 = R$ and $H = 1/R$, for a cylinder, $R_1 = R$, $R_2 = \infty$ and $H = 1/2R$, while for a planar bilayer, $H = 0$. A value of $H = 0$ also can occur on a saddle-shaped surface in which $R_1 = -R_2$. To assign a sign to the radii of curvature, we must define a normal direction, n, for the surface. By convention, $\bar{n}$ is usually positive, pointing toward the polar region, and therefore the curvature of inverted aggregates is negative.

We will consider the specific properties of the various amphiphilic aggregates shown in Figure 1.6 in later chapters. The rest of this chapter deals with the general features of solution and solvents.

1.4 Understanding the Origin of Entropy and Enthalpy of Mixing Provides Useful Molecular Insight into Many Colloidal Phenomena

Throughout this book we will be concerned with the thermodynamics of the self-assembly of surfactant molecules into aggregates, the properties of polymer solutions, and the behavior of colloidal entities such as emulsions. To obtain a basis for these more complex systems we start by considering the mixing of two pure liquids. Figure 1.8 shows three possible outcomes of mixing two components:

1. Formation of a homogeneous solution resulting from complete miscibility at all compositions. An example is t-BuOH and water at 25°C.

2. Formation of two phases containing different concentrations of the two components. An example is n-BuOH and water at 25°C.

Figure 1.7
Radius of curvature. For a surface in three dimensions, two mutually perpendicular radii of curvature, R_1 and R_2, can be specified at each point. On a saddle-shaped surface, the two radii of curvature have opposite sign. Here R_1 and R_2 are shown at two different points on the surface.

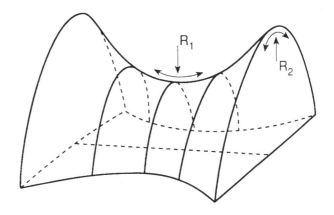

3. Retention of two unmixed phases, illustrated by water and mercury.

In a thermodynamic description, the outcome of these mixing processes differs because the free energies of mixing, ΔG_{mix}, are different. At constant T and P, the Gibbs free energy is

$$\Delta G = \Delta H - T\Delta S \tag{1.4.1a}$$

and for a mixing process

$$\Delta G_{mix} = \Delta H_{mix} - T\Delta S_{mix} \tag{1.4.1b}$$

where $\Delta G_{mix} = G_{final\ solution} - G_{initial\ solution}$. We can define ΔH_{mix}

Figure 1.8
Three possible outcomes of mixing equal amounts of two liquids A and B: (a) Formation of a homogeneous mixture; (b) Formation of two phases, one solution of B in A and one solution of A in B; (c) retention of the two unmixed phases with the second component present in small amounts.

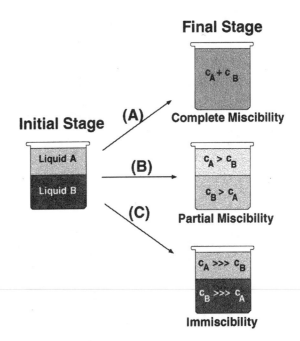

and ΔS_{mix} in an analogous manner. Our goal in this section is to develop expressions for ΔH_{mix} and ΔS_{mix} which reveal the reasons for the behavior illustrated in Figure 1.8. This process leads us naturally to introduce a basic thermodynamic concept that we use repeatedly throughout the book, the chemical potential.

1.4.1 The Ideal Mixing Model Provides a Basis for Understanding the Formation of a Miscible Phase

We focus first on the $T\Delta S_{mix}$ term in eq. 1.4.1b. For the moment we assume that ΔH_{mix} is sufficiently small that it can be ignored so that

$$\Delta G_{mix} = -T\Delta S_{mix} \qquad (1.4.2)$$

The relationship is called ideal mixing and it applies only when we mix two molecular species that are virtually identical, such as CH_3OH and CD_3OH. However, in the same way that the ideal gas model is very useful for the description of real gases, the ideal mixing model provides a basis for the understanding of real liquid mixtures.

When two miscible liquids such as H_2O and t-BuOH come into contact, the concentration gradient across the initial sharp interface causes t-BuOH molecules to diffuse into water and vice versa. We can gain insight into this molecular process by examining the Brownian motion and the resulting diffusion. In 1827, Robert Brown observed that pollen grains displayed a continuous random motion when viewed under a light microscope. All species in solution, whether colloidal or molecular, experience such Brownian motion. In an ideal solution, specific molecular interactions are negligible and the Brownian motion is completely random. Given enough time, each molecule then moves throughout the entire liquid volume in a diffusion process. The mixing process results in an increased disorder which means an increased entropy. If we assume that enthalpy changes could be neglected, ΔG_{mix} is negative for the mixing process, as it should be for a spontaneous process.

Even if it is qualitatively clear that the ideal mixing process leads to an increase in entropy, later applications require a quantitative expression for the ideal entropy of mixing. To derive this expression, we start from one of the fundamental equations of statistical mechanics

$$\mathbf{S} = \boldsymbol{k}\ln\Omega \qquad (1.4.3)$$

which the Austrian scientist, Boltzmann, first formulated in 1896. In this equation, $\boldsymbol{k}$ is the Boltzmann constant and Ω stands for the number of states accessible to the system under the given constraints.

The simplest approach to the task of calculating the entropy of mixing N_A molecules A with N_B molecules B in the liquid state is to adopt a so-called lattice model. In that case, we divide the total volume V into $N = N_A + N_B$ cells. Figure 1.9

Figure 1.9
Lattice model for calculating the entropy of mixing. Nine crosses are mixed with 16 rings to form a mixture of 64% rings and 36% crosses. The figure of the mixed state shows one of the 25!/(16!9!) = 2,042,975 possible configurations.

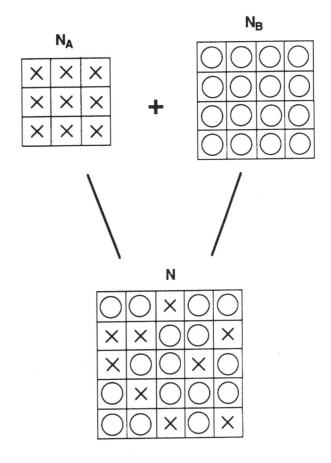

shows, using a two-dimensional analog, how the A and B molecules initially are compartmentalized in the unmixed state and are arranged in a unique way. When the barrier is removed, the molecules interchange positions by diffusion. Since we assume that the interaction between all molecules is identical ($\Delta H_{mix} = 0$), Brownian motion leads to a random distribution of molecules which depends only on the fraction $X_A(X_B)$ of $A(B)$ molecules.

Given these rules, calculating Ω becomes an exercise in statistics. We start by distributing the $N_A A$ molecules. There are

$$N!/(N_A)!(N - N_A)! \tag{1.4.4}$$

ways to distribute N_A identical objects on N identical positions.

All that remains is to place the $N_B B$ molecules. However, since these molecules are equivalent and there are now N_B positions left, we have no choice except to put the Bs on the empty positions. We can do this in only one way. The total number of ways of distributing the molecules is then

$$\Omega = N!/(N_A!)(N - N_A)! = N!/(N_A)!(N_B)! \tag{1.4.5}$$

and using eq. 1.4.3 the entropy is

$$\mathbf{S} = \mathbf{k}(\ln N! - \ln N_A! - \ln N_B!) \qquad (1.4.6)$$

For the specific mixing process shown in Figure 1.9, with $N = 25$, $N_A = 9$ and $N_B = 16$, $\Omega_{final} = 2{,}042{,}975$ and since $\Omega_{initial} = 1$ then $\Delta S_{mix} = k \ln \Omega_{final}/\Omega_{initial} = 15k$. With systems where N is very large, we can simplify further by using the Stirling approximation

$$\ln N! = N(\ln N - 1) \qquad (1.4.7)$$

Thus, eq. 1.4.6 becomes

$$S_{mix} = -k\left\{N_A \ln \frac{N_A}{N} + N_B \ln \frac{N_B}{N}\right\} = -k(N_A \ln X_A + N_B \ln X_B) \qquad (1.4.8)$$

where X_A and X_B are the respective mole fractions. Equation 1.4.8 represents the desired expression for the ideal entropy of mixing of species A and B. Note that ΔS_{mix} has a maximum at $X_B = 0.5$ because the largest number of ways of arranging the molecules in the solution occurs when there are equal numbers of the two species.

In deriving eq. 1.4.8, we made a number of assumptions that do not seem to apply, even approximately, to a real solution. However, the situation is not as bad as one might suspect. In fact, eq. 1.4.8 has much more general validity than the restricted lattice model shown in Figure 1.9 would suggest. For example, it correctly describes the mixing of two ideal gases, where the notion of a lattice is totally irrelevant.

The minimal assumption needed to obtain eq. 1.4.8 as an exact result is that when we distribute the molecules in space, the same distribution rule must apply to both the A and B species. In our derivation, we used the specific constraint of placing one and only one molecule into each cell, with no correlations. This restriction applied to both A and B. For an ideal gas mixture, we use the rule that we can place any molecule anywhere within the volume V irrespective of the location of the other molecules. Again, this rule applies equally to the A and B species.

We also can obtain eq. 1.4.8 as a limiting case at high dilution for inherently nonideal systems. Nonideality formally enters such systems in two ways. First, selective interactions between the molecules develop correlations in the solution so that A molecules prefer to be surrounded by A molecules and B molecules by other B molecules. However, when $N_B \ll N_A$, dilution ensures that solvent A molecules surround solute B molecules anyway.

When the two species have different sizes, the geometric factor creates a second source of nonideality. In the derivation of eq. 1.4.8, we assumed that A and B occupied equal volumes. However, at high dilution, unequal volumes $V_A \leqslant V_B$ change eq. 1.4.8 by a linear term $N_B \ln(V_B/V_A)$, which is concentration independent and can be included in the standard free energy.

1.4.2 The Regular Solution Model Provides a Simple Description of Nonideal Mixing which Ultimately Leads to the Formation of a Liquid Two-Phase System

If we repeat the mixing experiment using water and n-BuOH instead of t-BuOH, the following sequence of events occurs:

1. Mutual diffusion results in the smearing of the initially sharp interface, just as with t-BuOH.

2. Instead of forming a homogeneous solution, a sharp interface separates the water-rich and n-BuOH-rich phases. This constitutes the equilibrium state.

In this case, the driving force to randomize that arises from the entropy of mixing is balanced by a second driving force that organizes the system. Direct molecular interactions contribute to making $\Delta \mathbf{H}_{mix}$ positive which favors segregation.

The regular solution theory is the simplest model that describes liquid mixtures, including interactions that lead to a nonzero $\Delta \mathbf{H}_{mix}$, and a deviation from ideal behavior. In its simplest form, this theory deals with an incompressible mixture of two molecular species of equal size. These molecules interact only with their nearest neighbors whose number is fixed to z_b (the bulk coordination number). As a result, the total interaction energy, which is assumed to be pairwise additive, depends exclusively on the numbers n_{AA}, n_{BB} and n_{AB} of the three different types of near-neighbor pairs. These, in turn, can be related to the numbers N_A and N_B of A and B molecules in the system.

Due to the (near) incompressibility of liquids the energy and enthalpy only differ by constant term pV. Therefore, if we compare two states at the same pressure, the enthalpy change equals the change in internal energy

$$\Delta \mathbf{H} = \Delta \mathbf{U} \qquad (1.4.9)$$

Because we usually consider incompressible systems, eq. 1.4.9 applies in most of this book. It also follows that changes in Helmholtz, $\mathbf{A}$, and Gibbs, $\mathbf{G}$, free energies are equal under these circumstances. Furthermore, for a molecular system the thermodynamic internal energy, $\mathbf{U}$, equals the total intermolecular interaction energy, $\mathbf{E}$ plus a kinetic energy term that is unaffected on mixing. For pairwise additive interactions, such as in the regular solution theory, $\mathbf{E}$ is the sum over all pair interactions W_{ij}

$$\mathbf{E} = \sum_{\substack{i > j \\ \text{all molecules}}} W_{ij} \qquad (1.4.10)$$

Figure 1.9 illustrates the mixing of two pure liquids into a randomized mixture. The energy of mixing is

$$\Delta \mathbf{E}_{mix} = \mathbf{E}_{final} - \mathbf{E}_{initial} = \mathbf{E}_{mix} - \mathbf{E}_0 \qquad (1.4.11)$$

where $\mathbf{E}_0$ is the energy of the unmixed liquids

$$\mathbf{E}_0 = \frac{1}{2} z_b N_A W_{AA} + \frac{1}{2} z_b N_B W_{BB} \qquad (1.4.12)$$

since $(\frac{1}{2}) z_b N_A ((\frac{1}{2}) z_b N_b)$ gives the number of $AA(BB)$ pairs in the liquids and W_{AA} and W_{BB} are the microscopic pair interaction energies. We can estimate these energies from the enthalpy of vaporization, $\Delta\mathbf{H}_{vap}$, since in this process molecules are separated from their nearest neighbors in the liquid state to isolated molecules in the vapor phase.

Problem: Given that the enthalpy of vaporization, $\Delta\mathbf{H}_{vap}$, for carbon tetrachloride is 29.7 kJ/mol, calculate W_{AA}, assuming that each molecule has six nearest neighbors, $z_b = 6$.

Solution: $-z_b W_{AA}/2 = \Delta\mathbf{E}_{vap} = \Delta\mathbf{H}_{vap} - RT = 27.2$ kJ/mol where the factor of 2 avoids double counting the interactions. Counted per molecule $z_b W_{AA} = -2 \times 27.2 \times 10^3/6 \times 10^{23} = -9.1 \times 10^{-20}$ J, which corresponds to $-22 \, kT$ at room temperature. With $z_b = 6$, W_{AA} is $-3.7 \, kT$ per pair interaction.

We obtain the value of $\mathbf{E}_{mix}$ by summing over all interactions on a pairwise basis

$$\mathbf{E}_{mix} = n_{AA} W_{AA} + n_{BB} W_{BB} + n_{AB} W_{AB} \qquad (1.4.13)$$

The values of n_{AA}, n_{BB}, and n_{AB} are related to the total number of molecules by

$$2n_{AA} + n_{AB} = z_b N_A \qquad (1.4.14a)$$

$$2n_{BB} + n_{AB} = z_b N_B \qquad (1.4.14b)$$

The first equality follows, for example, by realizing that an A molecule is engaged in z_b pairs, either of AA or AB molecules. When we sum over all $N_A A$ molecules, we double count the AA pairs but not the AB ones. Solving eqs. 1.4.14 for n_{AA} and n_{BB} and substituting back into eq. 1.4.13 yields after some algebraic manipulation

$$\Delta\mathbf{E}_{mix} = \mathbf{E}_{mix} - \mathbf{E}_0 = n_{AB}\left(W_{AB} - \frac{1}{2} W_{AA} - \frac{1}{2} W_{BB}\right) \equiv n_{AB} w/(z_b \mathbf{N}_{Av}) \qquad (1.4.15)$$

where a single effective interaction parameter

$$w = z_b \mathbf{N}_{Av}\left(W_{AB} - \frac{1}{2} W_{AA} - \frac{1}{2} W_{BB}\right) \qquad (1.4.16)$$

emerges naturally. Here N_{Av} represents Avogadro's number so that w is the interaction per mole. Even though all the interaction energies are negative, w can easily be positive. In fact, this is normally the case. This case requires only that the heterogeneous interaction, W_{AB}, be weaker than the average of the interactions in the respective pure phases.

The thermodynamic properties of the solution are determined by the free energy rather than the energy. Thus, we need an estimate of the entropy of mixing. In a real system, this mixing involves changes in internal conformations, in the density, and in orientational correlations as well as in the mixing of the two species. The regular solution theory makes the simplifying assumption that mixing affects only the positional order at constant density. Furthermore, it assumes that the molecules mix in a fully random way.

These assumptions produce two consequences. First, the entropy of mixing is that of an ideal mixture, as in eq. 1.4.8. Second, we can calculate the number of n_{AB} of A–B pairs in eq. 1.4.15 using the assumed random distribution. Of the z_b nearest neighbors to an A molecule, $z_b X_B$ are B molecules. Since there are $N_A = X_A(N_A + N_B)$ of A molecules, the number of AB pairs is

$$n_{AB} = z_b X_A X_B (N_A + N_B) \tag{1.4.17}$$

Combining eqs. 1.4.8, 1.4.15, and 1.4.17 permits us to calculate the Gibbs free energy of mixing as $\Delta G_{mix} = \Delta A_{mix} = \Delta E_{mix} - T S_{mix}$, so that

$$\Delta G_{mix} = (n_A + n_B)[X_A X_B w + RT(X_A \ln X_A + X_B \ln X_B)] \tag{1.4.18}$$

for n_A and n_B moles of A and B. This free energy expression forms the basis of the regular solution theory. It contains only one unknown parameter, w. Figure 1.10 shows plots of ΔG_{mix}, ΔH_{mix}, and $T\Delta S_{mix}$ versus composition obtained within the regular solution model that demonstrate the three types of behavior shown in Figure 1.8. A totally miscible system implies that not only is ΔG_{mix} negative for all mixing ratios, it is also minimal for the equimolar mixture. This qualitative behavior is observed for $w < 2RT$. With a partially miscible system (Figure 1.10b) when w is slightly larger than $2RT$, a local maximum appears in ΔG_{mix} at the equimolar point. For mixing ratios between $X_B = 0.2$ and $X_B = 0.8$, the total free energy for the system is minimized when two separate phases with a composition at the two minima in the free energy form. Here complete mixing does not occur in spite of the fact that ΔG_{mix} is negative at all compositions. With a completely immiscible system (Figure 1.8c), at least one component has strong internal interactions, W_{AA} and/or W_{BB}, while the W_{AB} is relatively weak. As a result, w is large and positive and ΔG_{mix} positive except for dilute solutions not discernible in the diagram.

1.5 The Chemical Potential is a Central Thermodynamic Concept in the Description of Multicomponent Systems

At constant temperature and pressure, the equilibrium state is the one that minimizes the Gibbs free energy. We used this fact

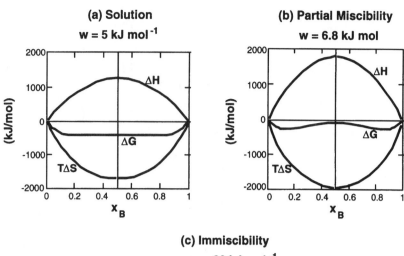

(a) Solution

w = 5 kJ mol^{-1}

(b) Partial Miscibility

w = 6.8 kJ mol

(c) Immiscibility

w = 30 kJ mol^{-1}

Figure 1.10
Predictions of the regular solution theory for ΔG_{mix}, ΔH_{mix} and ΔS_{mix} as a function of the composition for three values of the interaction constant w and at $T = 298$ K. The three w values correspond to the three mixing cases illustrated in Figure 1.8 and are: (a) $w = 4.9$ kJ/mol leading to miscibility at all compositions; (b) $w = 6.8$ kJ/mol where a two-phase region occurs in the interval $0.10 < X_B < 0.90$. Note that ΔG_{mix} is negative at all compositions, but it displays a maximum at $X_B = 0.5$ and the free energy can be made lower by forming two separate phases of $X_B = 0.10$ and $X_B = 0.90$ corresponding to the minima in ΔG_{mix}; (c) $w = 30$ kJ/mol leading to immiscibility.

(along with the free energy functions given in Figure 1.10) to find that n-BuOH and H$_2$O phases separate at room temperature. In systems with two or more components, the individual constituents always have the option to rearrange into a separate phase or just spatially respond to an external potential. To arrive at the chemical potential, the thermodynamic property that determines a component's ability to redistribute, let us analyze the n-BuOH–H$_2$O system in detail.

At equilibrium, molecules continue to move across the interface, but since the system's composition does not change with time, the net fluxes of each component must be zero. Now let us determine the change in free energy associated with transferring one BuOH molecule from the upper to the lower phase at constant T and P. The free energy before transfer is

$$\mathbf{G}_i = \mathbf{G}(N_{H_2O}^l, N_{BuOH}^l) + \mathbf{G}(N_{H_2O}^u, N_{BuOH}^u) \qquad (1.5.1)$$

while after transfer it becomes

$$\mathbf{G}_f = \mathbf{G}(N_{H_2O}^l, N_{BuOH}^l + 1) + \mathbf{G}(N_{H_2O}^u, N_{BuOH}^u - 1) \quad (1.5.2)$$

where the superscripts l and u signify the upper and lower phases. The change in free energy is $\Delta \mathbf{G} = \mathbf{G}_f - \mathbf{G}_i$

$$\begin{aligned}
\Delta \mathbf{G} &= \mathbf{G}(N_{\text{H}_2\text{O}}^l, N_{\text{BuOH}}^l + 1) + \mathbf{G}(N_{\text{H}_2\text{O}}^u, N_{\text{BuOH}}^u - 1) \\
&\quad - (\mathbf{G}(N_{\text{H}_2\text{O}}^l, N_{\text{BuOH}}^l) + \mathbf{G}(N_{\text{H}_2\text{O}}^u, N_{\text{BuOH}}^u)) \\
&= \mathbf{G}(N_{\text{H}_2\text{O}}^l, N_{\text{BuOH}}^l + 1) - \mathbf{G}(N_{\text{H}_2\text{O}}^l, N_{\text{BuOH}}^l) \\
&\quad + (\mathbf{G}(N_{\text{H}_2\text{O}}^u, N_{\text{BuOH}}^u - 1) - \mathbf{G}(N_{\text{H}_2\text{O}}^u, N_{\text{BuOH}}^u)) \quad (1.5.3)
\end{aligned}$$

Now N_{BuOH} is a large number in the thermodynamic limit. Adding or subtracting one molecule represents an infinitesimal change in the concentration of the respective phases. This enables us to interpret the two differences given in eq. 1.5.3 as two derivatives that define the chemical potential μ_{BuOH} so that per molecule

$$\Delta \mathbf{G} = \left(\frac{\partial \mathbf{G}}{\partial N_{\text{BuOH}}^l} \right)_{T,p,N_{\text{H}_2\text{O}}} - \left(\frac{\partial \mathbf{G}}{\partial N_{\text{BuOH}}^u} \right)_{T,p,N_{\text{H}_2\text{O}}} = \mu_{\text{BuOH}}^l - \mu_{\text{BuOH}}^u \qquad (1.5.4)$$

where μ_i^α is the chemical potential of component i in phase α with the general definition

$$\mu_i^\alpha = \left(\frac{\partial \mathbf{G}^\alpha}{\partial N_i} \right)_{T,p,N_{j \neq i}} \qquad (1.5.5a)$$

per molecule or

$$\mu_i^\alpha = \left(\frac{\partial \mathbf{G}^\alpha}{\partial n_i} \right)_{T,p,n_{j \neq i}} \qquad (1.5.5b)$$

per mole. We will use the definitions 1.5.5a or 1.5.5b, which differ by a factor of $\mathbf{N}_{\text{Av}}$, alternatively throughout this text.

Equation 1.5.4 shows that the free energy change in transferring a butanol molecule from the upper to the lower phase is simply the difference in chemical potential of the butanol in the two phases. At equilibrium, $\Delta \mathbf{G}$ should be zero, and then the chemical potentials are equal in the two phases.

1.5.1 Having an Expression for the Free Energy We Can Determine the Chemical Potential by Differentiation

It is usually convenient to write the chemical potential as a sum of one term that is independent of concentration plus one or more dependent on concentration. For example, for an ideal solution

$$\mu_i = \mu_i^\theta + \mathbf{R}T \ln X_i \qquad (1.5.6)$$

The concentration-independent standard chemical potential μ_i^θ is an essential quantity that enters, for example, into the expression for chemical equilibrium constants and partition coefficients. The concentration-dependent term $\mathbf{R}T \ln X_i$ arises from the changes in the free energy on mixing.

Problem: Derive eq. 1.5.6 for a binary system $A + B$ with ideal mixing.

Solution: The free energy of the mixture is

$$\mathbf{G} = n_A\mathbf{G}_A + n_B\mathbf{G}_B + n_A\mathbf{R}T \ln X_A + n_B\mathbf{R}T \ln X_B$$

where $\mathbf{G}_A$ and $\mathbf{G}_B$ is the free energy per mole of pure A and B, respectively. Taking the derivative with respect to n_A yields

$$\mu_A = \left(\frac{\partial \mathbf{G}}{\partial n_A}\right)_{n_B} = \mathbf{G}_A + \mathbf{R}T \ln X_A + n_A\mathbf{R}T\frac{1}{X_A}\left(\frac{\partial X_A}{\partial n_A}\right)_{n_B} + n_B\mathbf{R}T\frac{1}{X_B}\left(\frac{\partial X_B}{\partial n_A}\right)_{n_B}$$

$$= \mathbf{G}_A + \mathbf{R}T \ln X_A + (n_A + n_B)\mathbf{R}T\left\{\left(\frac{\partial X_A}{\partial n_A}\right)_{n_B} + \left(\frac{\partial(1 - X_A)}{\partial n_A}\right)_{n_B}\right\} = \mathbf{G}_A + \mathbf{R}T \ln X_A$$

which reduces to eq. 1.5.6 with $\mathbf{G}_A$ identified with μ_A^θ.

In the regular solution model, an additional interaction term, wX_AX_B, occurs in the expression for the free energy of mixing in eq. 1.4.18. This expression introduces an extra term into the chemical potential and

$$\mu_A = \mu_A^\theta + \mathbf{R}T \ln X_A + wX_B^2 \tag{1.5.7}$$

We obtain the corresponding expression for μ_B by exchanging the subscripts A and B.

Problem: For the regular solution model evaluate the contribution to μ_A from the term $w(n_A + n_B)X_AX_B$.

Solution:

$$w\left(\frac{\partial(n_A + n_B)X_AX_B}{\partial n_A}\right)_{n_B} = wn_B\left(\frac{\partial n_A/(n_A + n_B)}{\partial n_A}\right)_{n_B}$$

$$= wn_B[1/(n_A + n_B) - n_A/(n_A + n_B)^2]$$

$$= wn_B[n_B/(n_A + n_B)^2] = wX_B^2$$

1.5.2 Mixtures and Solutions Differ Through the Choice of Standard States

Intermolecular interactions fundamentally determine the value of the concentration independent term μ^θ of the chemical potential. However, μ^θ is also affected by two choices in the description. One needs to choose a standard state, which for liquids is normally the pure liquid of the component or the component totally surrounded by solvent molecules; this is termed the infinite dilution standard state. The choice of the concentration variable also affects the value of the standard chemical potential, mole fraction units, molar or possibly molal concentrations. This point is a classical source of confusion for students!

We can use the regular solution model to illustrate the factors that determine the value of the standard chemical potential. In this model, the natural standard state is the neat liquid, and we use mole fraction units to characterize the concentration. In this case the standard chemical potential is simply the free energy of the liquid divided by the number of molecules, so for example

$$\mu_A^\theta = \mathbf{G}_A(\text{neat liquid})/N \qquad (1.5.8)$$

In eq. 1.4.12, we obtained an expression for the energy in the neat liquid. The strict lattice model exhibits no entropy for a pure system but it is clear that a real liquid has substantial disorder, and we assign an entropy $\mathbf{S}_{\text{liq}}^A$ to this effect. The regular solution model implicitly assumes that this contribution is constant under mixing. Combining the energy and entropy contributions.

$$\mu_A^\theta = \frac{z_b}{2} \mathbf{N}_{\text{Av}} W_{AA} - T\mathbf{S}_{\text{liq}}^A + pV_A + U_A(\text{int}) \qquad (1.5.9)$$

where V_A is the molar volume of A and p is the pressure and $U_A(\text{int})$ the intramolecular energy of A.

In the regular solution model, the standard chemical potential of B is as in eq. 1.5.9 with sub- and superscript A substituted with B. However, in some cases we prefer not to treat A and B on an equal footing but to consider A as the solvent and B as the solute. Then the infinite dilute becomes the standard state for B and

$$\mu_B^\theta(X_B \to 0) = z_b \mathbf{N}_{\text{Av}} \left(W_{AB} - \frac{1}{2} W_{AA} \right) - T\mathbf{S}_{\text{liq}}^B + pV_B + U_B(\text{int}) \qquad (1.5.10)$$

still measuring concentrations in mole fractions. If we measure the concentrations in moles per liter, an additional term $\mathbf{R}T \ln V_A$ enters $\mu_B^\theta(c_B \to 0)$.

1.5.3 The Chemical Potential of the Solvent is Often Expressed in Terms of the Osmotic Pressure

The chemical potential of the solvent often is expressed in terms of the osmotic pressure, Π_{osm}. For an incompressible liquid the relation between Π_{osm} and μ is

$$V_A \Pi_{\text{osm}} = (\mu_A^\theta - \mu_A) \qquad (1.5.11)$$

where V_A represents the molar volume of the solvent, A. In Chapter 5, we will see that the osmotic pressure plays a seminal role in the description of forces between particles in a liquid.

In the same way as we can write the pressure, p, of a nonideal gas in terms of a so-called virial expansion in the density $\rho = N/V$

$$p = \mathbf{R}T\rho(1 + B_2\rho + \cdots) \qquad (1.5.12)$$

we can expand the osmotic pressure in terms of the solute concentration $c_B = n_B/V$

$$\Pi_{\text{osm}} = \mathbf{R}T c_B(1 + B_2 c_B + \cdots) \qquad (1.5.13)$$

Here B_2 is called the second virial coefficient. The general statistical mechanical expression for B_2 is

$$B_2 = -\frac{1}{2} \int_0^\infty [\exp(-V(\bar{r})/kT) - 1]d\bar{r}^3 \qquad (1.5.14)$$

where $V(\bar{r})$ is the pair interaction potential. An identification between eqs. 1.5.13, 1.5.7 and 1.5.11 yields

$$B_2 = \left(\frac{1}{2} - w/RT\right)V_A \qquad (1.5.15)$$

for B_2 in the regular solution theory. For $w/RT > 1/2$, B_2 is negative. In Chapter 7 we will see that the situation $w/RT = 1/2$ has a particular significance for polymer systems.

The second virial coefficient; which often is determined in light scattering measurements of colloidal systems, reveals information about the pair interaction between two colloidal particles. As seen from eq. 1.5.14, B_2 goes negative for attractive potentials $(V(\bar{r}) < 0)$ but is positive when $V(\bar{r}) > 0$.

1.5.4 The Chemical Potential Enters into Many Thermodynamic Equalities

For further reference we now state how the chemical potential enters into some central thermodynamic relations. Textbooks such as those on physical chemistry contain more complete discussions.

T, p, and n_i are the natural independent variables for the Gibbs free energy, and by formal differentiation

$$d\mathbf{G} = (d\mathbf{G}/dp)_{T,n_i}\,dp + (d\mathbf{G}/dT)_{p,n_i}\,dT + \sum_i (d\mathbf{G}/dn_i)_{T,p,n_j}\,dn_i \qquad (1.5.16)$$

where the thermodynamic relations provide the expression for the partial derivatives and

$$d\mathbf{G} = V\,dp - \mathbf{S}\,dT + \sum_i \mu_i\,dn_i \qquad (1.5.17)$$

Since the chemical potentials are intensive, that is independent of size, quantities defined in eq. 1.5.17 can be integrated at constant T and p to

$$\mathbf{G} = \sum n_i \mu_i \qquad (1.5.18)$$

Problem: Show explicitly that the chemical potentials of eq. 1.5.7 for the regular solution model obey eq. 1.5.18.
Solution: $\mu_i = \mu_i^0 + RT\ln X_i + (1 - X_i)^2 w$ so that

$$n_A\mu_A + n_B\mu_B = n_A\mu_A^0 + n_B\mu_B^0 + n_A RT \ln X_A + n_B RT \ln X_B + (n_A X_B^2 + n_B X_A^2)w$$

$$= \mathbf{G}^0 + RT(n_A \ln X_A + n_B \ln X_B) + (n_A + n_B)X_A X_B w$$

$$= \mathbf{G}^0 + \Delta\mathbf{G}_{\text{mix}}(\text{regular solution})$$

since

$$n_A X_B^2 + n_B X_A^2 = (1/(n_A + n_B)^2(n_A n_B^2 + n_B n_A^2)$$

$$= (n_A + n_B)(n_A n_B)/(n_A + n_B)^2 = (n_A + n_B)X_A X_B$$

Using the defining relation $\mathbf{G} = \mathbf{U} + pV - T\mathbf{S}$ between internal and Gibbs free energy, we find

$$\mathbf{U} = T\mathbf{S} - pV + \sum_i n_i \mu_i \qquad (1.5.19)$$

We can also formally differentiate eq. 1.5.18 and arrive at the Gibbs-Duhem equation at constant T and p

$$\sum n_i d\mu_i = 0 \qquad (1.5.20)$$

This equation shows, for example, that the chemical potentials for a two-component system cannot be varied independently, but a prescribed change $d\mu_1$ of the chemical potential of component 1 necessarily changes the chemical potential of component 2 by $-(n_2/n_1)d\mu_2$.

1.5.5 Chemical Potentials Can Be Generalized to Include the Effect of External Fields

So far we have considered the chemical potential in terms of a system's intermolecular interactions expressed, for example, in terms of w and the entropy of mixing. Clearly, external fields also can exert forces on molecules that affect their distribution in a system. Two external forces that are important in colloids are gravitational and electrical fields.

Suppose we prepare an aqueous suspension containing Fe_2O_3 particles and examine it first on a space shuttle and again when it returns to earth. In space the distribution of particles is uniform and random, but on earth gravitational forces generate a vertical concentration gradient of particles; that is, they create a more ordered system. We can use the chemical potential concept to derive the concentration profile of the suspension. An external potential acts to give an additional energy, $E_{\text{ext}}^i(\bar{r})$, to a molecule i at position r. Consider a volume element containing N_i' molecules of component i. If it is small enough that the variation in the external potential over the volume element is negligible, then its free energy, $\mathbf{G}$, is

$$\mathbf{G} = \mathbf{G}_n'(\{X_i'\}) + \sum_i N_i E_{\text{ext}}^i(\bar{r}) \qquad (1.5.21)$$

$\mathbf{G}_n'$ is the free energy of the corresponding system in the absence of the external field and at the concentrations $X_i' = N_i'/\Sigma_i N_i'$. We obtain the chemical potential of component j by differentiating with respect to N_j'

$$\mu_j(\bar{r}) = \mu_j^b(\{X_i'\}) + E_{\text{ext}}^j(\bar{r}) \qquad (1.5.22)$$

as a sum of one term derived from the bulk liquid and one due to the external field.

Through Brownian motion, molecules seek out the concentration profile that minimizes the free energy. This occurs when the chemical potential is uniform throughout the system and $\mu_j^b(\{X_i(\bar{r})\}) + E_{ext}^j(\bar{r}) = constant.$ In the case of an ideal bulk solution

$$\mu_j^b = \mu_j^o + kT \ln X_j \qquad (1.5.23)$$

and the condition of constant chemical potential results in a concentration

$$X_j(\bar{r}) = const \cdot \exp(-E_{ext}^j(\bar{r})/kT) \qquad (1.5.24)$$

that varies exponentially in the external potential. This form of the Boltzmann equation describes how the interplay between entropy of mixing (as expressed by kT) and the external field generates a concentration variation in the system at equilibrium.

In a vertical constant gravitational field

$$E_{ext}^i = m^i g z \qquad (1.5.25)$$

where m^i is the mass of molecule i and g is the standard acceleration of gravity. For a gas where X_i is proportional to the partial pressure p_i

$$p_i(z) = p_i(z = 0) \exp(-m^i g z/kT) \qquad (1.5.26)$$

This equation is often called the barometric formula. When we apply the same argument to a liquid, a complication arises. Due to the (near) incompressibility of a liquid, we cannot move a liquid molecule from one position to another without moving another molecule in the opposite direction. In the simple case of perfect incompressibility with the total volume given as a sum of concentration independent molecular volumes, we can take care of the interdependence by writing

$$E_{ext}^i = m_{eff}^i g z \qquad (1.5.27)$$

where m_{eff} is the effective mass corrected for the bouyancy effect of the medium.

Problem: Calculate the effective mass for both components in an equimolar mixture of CH_3OH and CD_3OH.
Solution: In effect, this is an ideal solution in which both components have the same molecular volumes. The masses differ by 3.0 Dalton of unit masses, $u = 1.66 \times 10^{-27}$ kg. CH_3OH will float so that relative to the average composition it has $m_{eff} = -0.5 \times 3.0 \times 1.66 \times 10^{-27}$ kg $= -2.5 \times 10^{-27}$ kg while for CD_3OH $m_{eff} = 2.5 \times 10^{-27}$ kg.

The ultracentrifuge was probably the first advanced instrument specifically designed to study colloidal systems. In this case, the gravitational potential is

$$E_{ext}^i = m_{eff} \omega^2 r^2 \qquad (1.5.28)$$

where ω is the rotational velocity and r is the distance from the rotational axis. To conduct accurate studies, one must include effects due to concentration dependent partial molal volumes in calculating the effective mass.

Electrical fields play an important role in colloidal systems. They influence all charged species, and for an ion j of charge $z_j e$ at position r the external interaction energy is

$$E_{ext}^j(r) = -z_j e \Phi(\vec{r}) \tag{1.5.29}$$

where e is the fundamental charge 1.62×10^{-20} C, z_j the ion valency ($z_j = +1$ for Na$^+$ and $z_j = -1$ for Cl$^-$), and Φ is the electrostatic potential at r.

Assuming ideal mixing in the bulk, the chemical potential is

$$\mu_j = \mu_j^0 + kT \ln c_j(r) + z_j e \Phi(\vec{r}) \tag{1.5.30}$$

In many textbooks, μ_j as given in eq. 1.5.30 is called the electrochemical potential. Such a distinction between the electrochemical and the chemical potential presupposes that electrostatic interactions provide a very special contribution to the free energy. This point of view is not compatible with the spirit of this book, and we will avoid using the concept of an electrochemical potential.

Sections 1.5.1 through 1.5.5 have introduced and developed the abstract thermodynamic concept of the chemical potential. We will use it extensively in this book and hopefully demonstrate that this concept is the gateway for understanding the thermodynamics of multicomponent systems. Using the chemical potential as a key concept we can unify the description of a range of phenomena of the colloidal domain.

1.6 Understanding Brownian Motion Provides an Important Enabling Concept in Analyzing Colloidal Systems

1.6.1 The Diffusional Motion of Individual Molecules Can be Analyzed in Terms of a Random Walk

In the Boltzmann distribution the interplay between Brownian motion as expressed by kT and an organizing force arising from intermolecular interactions as utilized in the regular solution or external gravitational or electrical fields is a key element in understanding molecular interactions in the colloidal domain. Because of the ubiquitous role that Brownian motion plays, we analyze it in more detail below.

Understanding Brownian motion involves three key concepts which are most readily grasped by first considering the transport properties of molecules in the gas phase.

1. Molecules possess kinetic energy that according to the equilibrium theorem is

$$1/2m\langle v^2 \rangle = 3/2kT \qquad (1.6.1)$$

At room temperature the root-mean-square speed for typical molecules ranges from 1.921 m/s for hydrogen to 192 m/s for mercury.

2. In the vapor phase, molecules move long distances, compared to their diameter σ, before they collide with other molecules. The mean free path λ decreases with pressure according to

$$\lambda = 0.2kT/\sigma^2 p \qquad (1.6.2)$$

For water vapor at 298 K, λ varies from 1 m at $p = 10^{-7}$ atm. to 10^{-7} m at $p = 1$ atm.

3. In the vapor phase the collision frequency Z per molecule increases with the pressure and the molecular diameter σ

$$Z = 7\sigma^2 p(mkT)^{-1/2} \qquad (1.6.3)$$

Over the pressure range 10^{-7} to 1 atm. Z changes from 6×10^2 to 6×10^9 for water.

Thus, because of their kinetic energy, molecules move with an average velocity $(\langle v^2 \rangle)^{1/2}$ and average distance λ before they collide with another molecule, a process which occurs an average Z times per second. As a result of the collisions the molecules change direction. If we follow the trajectory of one particular molecule, we find that multiple collisions randomize the direction of its motion with respect to the initial direction. This physical picture suggests a modeling of a molecule's motion by a sequence of equally sized steps of length in which the direction of each successive step is randomly chosen; a random walk.

By analyzing a particle's Brownian motion in terms of a random walk process we can get further insight into the effective translational motion of a molecule subject to many collisions. It is an inherent property of a sequence of random events that the net result is proportional to the square root of the number of such events. For a random walk this implies that the average distance traveled

$$\langle x^2 + y^2 + z^2 \rangle^{1/2} = \langle r^2 \rangle^{1/2} \sim (N_r)^{1/2} l_r \qquad (1.6.4)$$

is proportional to the individual step length l_r times the square root of the number of steps N_r.

During a time t a molecule collides $N_r = Zt$ times and the average step length $l_r \simeq \lambda$ so by combining eqs. 1.6.2–1.6.4

$$\langle r^2 \rangle^{1/2} \sim (kT)^{3/4}/(\sigma m^{1/4} p^{1/2}). \qquad (1.6.5)$$

This regime where the molecules experience many collisions is called diffusional motion and we can introduce the diffusion coefficient D through the relation

$$\langle r^2 \rangle = 6Dt \qquad (1.6.6)$$

For a molecule in the gas phase we find by comparing eqs. 1.6.5 and 1.6.6 that

$$D = 0.2(kT)^{3/2}/(\sigma^2 m^{1/2} p) \qquad (1.6.7)$$

where we have also specified the numerical coefficient. For water vapor D varies from $3.4 \times 10^2\,\mathrm{m^2/s}$ at $p = 10^{-7}\,\mathrm{atm}$. to $3.4 \times 10^{-5}\,\mathrm{m^2/s}$ at $p = 1\,\mathrm{atm}$.

On a short timescale, a molecule in a liquid is surrounded by nearest neighbors with which it continuously collides. The adjacent molecules form a "cage" which confines the central molecule. However, the positions of all the molecules fluctuate and this results in a translational motion in fundamentally the same way as in a gas. The collisions are so frequent that a molecule collides many times with neighboring ones in the cage before it has moved to a new position.

In liquids, molecules have the same velocity as molecules in the gas phase for a given temperature. However, the average separation between molecules is much less than a molecular diameter and molecules collide with their neighbors approximately 10^{13} times per second. For a small molecule the thermal velocity is in the range 100 to 1000 m/s. The average distance traveled between collisions is then less than $1000 \times 10^{-13} = 10^{-10}$ m which is smaller than the molecular size and much shorter than in the gas phase. Since the average velocity is the same the length of a trajectory, $L_t = Zt \cdot \lambda$, during a time t, is also the same in the liquid and in the gas. The distance traveled in space rather than along the trajectory differs since

$$\langle r^2 \rangle^{1/2} \sim (Zt)^{1/2}\lambda = L^{1/2}\lambda^{1/2} \qquad (1.6.8)$$

from eq. 1.6.4. Now combining the relation $\lambda(liquid) \ll \lambda(gas)$ and an identification between eqs. 1.6.6 and 1.6.8 shows that the diffusion coefficient is much lower in a liquid than in a gas. Typical diffusion coefficients in neat liquids are $D_{water} = 2.8 \times 10^{-9}$, $D_{benzene} = 2.2 \times 10^{-9}$ and $D_{tetradecane} = 6 \times 10^{-10}\,\mathrm{m^2/s}$. The essential point is that the effective translational motion of a molecule in a liquid is slower, not because the individual molecular velocities are lower, but because the road from here to there is more tortuous.

In a crystalline solid where molecules are confined to distinct positions in a close packed periodic array, the mechanism leading to diffusion is somewhat different. In a perfect crystal, diffusion could involve an exchange of positions between two nearest neighbors or a more complex concerted positional exchange. This is normally a rare event and diffusion by this mechanism is slow indeed. In practice the diffusion usually takes place at defects such as unoccupied lattice sites or stacking faults. Successive motion of molecules into neighbor empty lattice sites leads to a diffusional transport of molecules. Typical values at 25°C are $D_{Cu} = 6 \times 10^{-40}$ and $D_{Na} = 6 \times 10^{-13}\,m^2/s$. For solids the diffusion coefficient is much more system dependent than for liquids and gases. For example, the large difference between copper and sodium reflects mainly the fact that sodium has a lower melting temperature.

1.6.2 Diffusional Motion Leads to a Net Transport of Molecules in a Concentration Gradient

We have seen that the thermal excitations lead to constant transitional displacements of molecules in a gas, in a liquid and to a lesser extent in solids. For a system in which there are nonequilibrium concentration gradients, diffusional motion results in a net flow J_i of molecules i and according to Fick's first law

$$J_i(z) = -D_i(dc_i(z)/dz) \qquad (1.6.9)$$

assuming that the concentration gradient is in the z direction. Thus, if we can measure the flow J_i in a known concentration gradient we can determine the diffusion coefficient D_i.

Fick's first law in the form of eq. 1.6.9 implicitly assumes ideal solution behavior. However, the fundamental driving force for the flux is a gradient in the chemical potential so that a more general form of Fick's law is

$$J_i = -(D_i/kT)c_i(d\mu_i/dz) \qquad (1.6.10)$$

At equilibrium in an external field, as discussed above, there are concentration gradients while the chemical potential is homogeneous.

Problem: In a gravitational field

$$c_i = c_{io} \exp(-m^i_{eff}gz/kT)$$

and $\mu_i = \mu_o^i + kT \ln c_i + m^i_{eff}gz$ show by explicit calculation that J_i in eq. 1.6.10 is zero.
Solution:

$$\frac{d\mu_i}{dz} = \frac{d\mu_{io}}{dz} + kT\frac{d}{dz}(\ln c_i) + \frac{d}{dz}(m^i_{eff}gz)$$

$$= 0 + kT\left(\frac{d\ln c_{io}}{dz} - \frac{d}{dz}(m^i_{eff}gz/kT)\right) + m^i_{eff}g$$

$$= -kT(m^i_{eff}g/kT) + m^i_{eff}g = 0$$

The flow J_i of molecules described by either eq. 1.6.9 or eq. 1.6.10 results generally in a change in the local concentration $c_i(z)$ of molecular species i. By combining eq. 1.6.9 with the equation of continuity, one arrives at Fick's second law

$$\frac{dc_i(t,z)}{dt} = D_i \frac{d^2 c_i(t,z)}{dz^2} \qquad (1.6.11)$$

When the concentration changes in both space and time we have to use Fick's second law.

In a liquid the diffusion of one species i in a concentration gradient necessarily involves the diffusion of another species in the opposite direction in order to keep the total density virtually constant. Thus, the D's entering eqs. 1.6.9 to 1.6.11 are referred to as mutual diffusion coefficients.

In an equilibrium system, the molecular motion is as rapid as under typical nonequilibrium conditions. To monitor the resulting molecular transport we can tag some molecules either by radioactive labeling or by some spectroscopic techniques. The diffusional motion of the molecules is still described by Fick's laws. In this case we consider the motion of individual molecules so there is no net compensating macroscopic flow. The measured D is called the self-diffusion coefficient and it can be different from the mutual diffusion coefficient, but they are the same in the limit of high dilution.

1.7 Solvophobicity Drives Amphiphilic Aggregation

Amphipile aggregation results from the low solubility of either the amphiphile's hydrocarbon chain or its polar head group. This means that to promote the self-organization of amphiphilic molecules, the solvent must be either very nonpolar (like oil) or very polar (like water). Because water is widely used in industrial applications and biological processes, we will consider its solvent properties first.

Low solubility of the amphiphilic hydrocarbon chains drives aggregation in polar solvents like water. It arises because we try to introduce a weakly interacting species into a liquid of high cohesive energy. The Gordon parameter, which is defined as $\gamma/V_m^{1/3}$ (J/m^3), where γ is the surface tension and V_m the molar volume, is very useful for characterizing the cohesive energy of solvents. It even allows us to also consider fused salts such as ethylammonium nitrate (EAN). Figure 1.11 summarizes the free energy of transfer of various gases from the vapor phase to selected solvents. A roughly linear relationship between ΔG and $\gamma/V^{1/3}$ appears in all of them except water, whose solute order shows a different sequence than in other solvents. We will return to this point in Chapter 5.

Table 1.3 summarizes the physical properties of water, hydrazine, ethylammonium nitrate, formamide, and ethylene glycol, the polar solvents in which micelle formation has been documented. Gordon parameters decrease from 2.7 to 1.3 J/m^3. As a solvent's cohesive energy density decreases, the driving

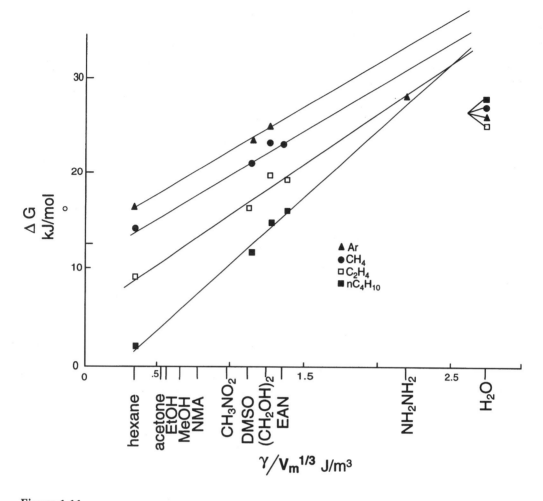

Figure 1.11
Free energy $\Delta G°$ of transferring nonpolar gases into solvents increases almost linearly with the Gordon parameter $\gamma/V^{1/3}$ (which is a measure of solvent cohesive energy density). Substantially lower values in water reflect its unusual response to nonpolar groups. (D.F. Evans, *Langmuir* **4**, 6 (1988).)

force for aggregation also decreases. This causes the CMC to increase and become less well-defined. No evidence of amphiphilic aggregation exists for solvents with a Gordon parameter below $1.3\,J/m^3$.

We will consider oil solvents briefly. Here insolubility drives the polar head groups inward, and the nonpolar chains point out into the oil. For the most part, these inverted structures form gradually, and we do not observe sharply defined CMCs. In the case of ionic surfactants, the concentration of a third polar component, such as water, governs the formation and size of inverted micelles. Nevertheless, well characterized structures result.

The behavior of the partially fluorinated hydrocarbon $F(CF_2)_8(CH_2)_{12}H$ in the solvent perfluorotributylamine $[CF_3\text{-}(CF_2)]_3N$ clearly illustrates the role of solvophobicity in driving amphiphilic aggregation. We can detect a Krafft temperature at 24°C, and well-defined changes in NMR chemical shifts, light scattering, and fluorescence at 6 wt% $F(CF_2)_8(CH_2)_{12}H$ define a CMC consistent with the formation of small spherical micelles.

TABLE 1.3 Properties of Polar Solvents

Polar Solvents	ε_r^a	γ $(10^3 N/m)$	m (D)	$\gamma/V^{1/3}$ (J/m^3)
Water	78.4	72.5	1.85	2.75
Hydrazine	51.7	66.4	1.86	2.10
Ethylammonium nitrate	∞	46.0	—	1.40
Formamide	109.0	57.2	3.73	1.50
Etylene glycol	37.7	47.0	2.28	1.20
3-Methylsydnone	144.0	57.0	7.30	1.40

From A. H. Beesley, D. F. Evans, and R. G. Laughlin, *J. Phys. Chem.* **92**, 791 (1988).
[a]Where ε_r is the relative dielectric permittivity, γ is the surface tension, m is the molecular dipole moment in Debye units, and $\gamma/V^{1/3}$ is the Gordon parameter for the cohesive energy density.

Fluorocarbons and hydrocarbons are mutually insoluble. In a fluorinated solvent, a partially fluorinated hydrocarbon behaves as a dual-character molecule and aggregates to form a micelle with a hydrocarbon core containing fluorocarbon head groups.

These observations clearly establish that insolubility of one part of the amphiphilic molecule drives the physicochemical process of amphiphilic aggregation. Such an aggregation can appear for many combinations of solutes and solvents. However, the most common case is that of apolar molecules or groups in water. In this case, the general term solvophobic is specialized to hydrophobic, which is a central concept for understanding many physicochemical properties of aqueous systems. Unfortunately, as used in the scientific literature, hydrophobicity has meant different things to different people at different times. Some distinctions between various aspects of this loose concept may be useful.

Hydrophobic hydration refers to the water structure around a single apolar solute in water. Many experimental and theoretical studies manifest the existence of such a hydration structure. *Hydrophobic interactions* refer to the attractive forces between two or more apolar solutes in water. They simply manifest the insolubility of oil in water, whatever the molecular mechanism. The third term, *hydrophobic effect*, has more than one meaning. Sometimes it refers to the combined phenomena observed for aqueous solutions of apolar substances and sometimes it describes specific anomalous thermodynamic properties, particularly partial molar enthalpies, heat capacities, and volumes of apolar solutes in water.

Hydrophobic interaction occurs as a general phenomenon when apolar substances are dissolved in water. However, hydrocarbon chains have particular importance in applications, especially in any discussion of surfactants. For this reason, this chapter concludes by summarizing some quantitative data on the thermodynamics of the dissolution of hydrocarbon compounds in water.

Table 1.4 summarizes data for the transfer of alkanes, aromatics, and alkanols from water to the neat liquid. The

TABLE 1.4 Thermodynamic Data for the Transfer of Small Apolar Compounds from a Dilute Aqueous Solution (*aq*) to the Neat Liquid (*l*) at 25°C

Compound	$\mu^{\theta}(l) - \mu^{\theta}(aq)$ (kJ/mol)	$H^{\theta}(l) - H^{\theta}(aq)$ (kJ/mol)	$S^{\theta}(l) - S^{\theta}(aq)$ (J/mol·K)	$C_p^{\theta}(aq)$ (J/mol·K)	$C_p^{\theta}(l)$ (J/mol·K)
C_4H_{10}	−25.0	3.3	96	414	142
C_5H_{12}	−28.7	2.1	104	573	171
C_6H_{14}	−32.4	0.0	108	635	196
C_6H_6	−19.2	−2.1	59	359	134
$C_6H_5CH_3$	−22.6	−1.7	71	418	155
$C_6H_5C_2H_5$	−25.9	−2.0	79	501	184
$C_6H_5C_3H_8$	−28.8	−2.3	88	602	209
C_3H_7OH	−6.6	10.1	56	384	148
C_4H_9OH	−10.0	9.4	65	478	178
$C_5H_{11}OH$	−13.5	7.8	71	558	209

From C. Tanford, *The Hydrophobic Effect: Formation of Micelles and Biological Membranes*, Wiley, New York, 1980, Tables 4-1 and 4-2.

table's first column gives the difference in standard chemical potential, which we can view as a manifestation of the hydrophobic interaction. Note, for example, the substantial difference in the magnitude for benzene and hexane. The remaining four columns show manifestations of the hydrophobic effect. The change in enthalpy is small. In fact, at 25°C it is so small that the chemical variations between the hydrocarbons result in a change of the sign. Having independently measured $\mu^{\theta}(l) - \mu^{\theta}(aq)$ and $H^{\theta}(l) - H^{\theta}(aq)$, the values of the third column follow from the relation

$$\mathbf{S}^{\theta}(l) - \mathbf{S}^{\theta}(aq) = \{H^0(l) - H^0(aq) - \mu^0(l) + \mu^0(aq)\}/T$$

A comparison between the two last columns shows that the partial molal heat capacity is anomalously high in the aqueous solution. It can be more than a factor of three larger than in the neat liquid! At this stage, we can accept the data given in Table 1.4 as established experimental thermodynamic facts and postpone our analysis of the underlying molecular cause until Chapter 5.

An important observation in Table 1.4 is that for homologous series of compounds, like the alkanes, aromatics, and alkanols, the thermodynamic values vary linearly with the number of carbon atoms. This relationship indicates that a simple additivity rule applies that we can extrapolate to molecules with longer chains, where direct measurements are hard to come by. Most importantly, the difference in standard chemical potential can be written as

$$\mu^{\theta}(neat) - \mu^{\theta}(H_2O) = A - Bn_c \qquad (1.7.1)$$

where n_c represents the number of carbon atoms. Table 1.5 summarizes values of A and B for a series of apolar compounds. The constant B depends only weakly on the nature of the end

TABLE 1.5 The Standard Free Energy of Transfer $\mu^\theta(\text{neat}) - \mu^\theta(\text{H}_2\text{O})$ of Compounds $\text{R(CH}_2)_{n-1}\text{CH}_3$ from a Dilute Aqueous Solution to the Neat Liquid[a]

R	A(kJ/mol)	B(kJ/mol)
$-\text{CH}_3$	-10.2	3.70
$\text{H}_2\text{C}=\text{CH}-$	-6.3	3.70
$\text{H}(-\text{CH}=\text{CH}-)_2$	-3.1	3.60
$-\text{OH}$	3.5	3.43
COOH	17.8	3.45

From C. Tanford, *The Hydrophobic Effect: Formation of Micelles and Biological Membranes*, Wiley, New York, 1980, Tables 4-1 and 4-2.
[a]Values of the coefficients A and B in eq. 1.7.7; concentrations in mol fraction units, n_c, is the total number of carbon atoms.)

group, R, showing that a substitution at the end of the chain does not significantly influence the chain-length dependence. In contrast, substitution strongly influences the constant term, A, and this term contains a large contribution from the free energy of solvation of the group, R.

Exercises

1.1. Calculate the aggregation numbers of micelles formed by single hydrocarbon chain surfactants with 10, 12, and 16 carbon atoms by assuming that the hydrocarbon core forms a sphere with a radius equal to the extended length of the hydrocarbon chain.

1.2. Calculate the average area per polar group for micelles in Exercise 1.1 by assuming that the centers of the polar groups are located 0.25 nm outside the hydrocarbon core.

1.3. The table shows the chemical potential of a fluorocarbon, C_4F_{10}, in a mixture with the corresponding hydrocarbon, C_4H_{10}. The neat liquid is used as the standard state and the chemical potential μ is expressed in terms of the activity, $a = \exp\{(\mu - \mu^\theta)/RT\}$

Mole Fraction	Activity	Mole Fraction	Activity
0.90	0.91	0.30	0.68
0.80	0.86	0.20	0.62
0.70	0.81	0.10	0.54
0.60	0.77	0.075	0.47
0.50	0.75	0.050	0.40
0.40	0.72	0.025	0.25

Obtain a fit of the data to the regular solution theory. What is the value of the interaction parameter? Estimate the change in standard chemical potential if the reference state is chosen as infinite dilution in C_4H_{10}.

1.4. Calculate the second virial coefficient for a spherical solute with radius R so that the interaction potential is $V(r) = \infty$ for $r < 2R$ and $V(r) = 0$ otherwise.

1.5. Water is transported from the root to the leaves in the canopy of a tree. If the canopy is 25 meters above the root, what is the difference in osmotic pressure required to make water flow spontaneously? What is the concentration difference of a molecular solute that produces such a difference in the osmotic pressure? Assume ideal solution.

1.6. The table shows the osmotic pressure of a series of aqueous glucose solutions prepared from a crystalline hydrate, $(C_6H_{12}O_6)(H_2O)_n$. What is the number n? (T = 298 K).

Crystal % (w/w)	Osmotic Pressure $(\times 10^{-5}\,Pa)$
1.1	1.46
2.2	2.97
3.3	4.53
4.4	6.14
6.6	9.02
8.8	12.9
11.0	16.9

1.7. Calculate the aqueous solubilities of decane, 1-decene, decanol, and decanoic acid (at low pH) using Table 1.5 (T = 298 K).

1.8. Determine at which temperature $\mu^\theta(l) - \mu^\theta(aq)$ has its smallest value (i.e., where the hydrophobic interaction is strongest) for benzene and hexane. Neglect the temperature variation of the heat capacity.

Literature

Reference[a]	Sections 1.1—1.3: Amphiphilic Self-Assembly	Sections 1.4—1.5: Solution Thermo-dynamics	Section 1.6 Browian Motion Diffusion	Sections 1.7: Solvophobicity
Davis		CHAPTER 6.4 Regular solutions		
Hiemenz & Rajagopalan	CHAPTER 8 Association Colloins	CHAPTER 3 Osmotic pressure		
Hunter (II)			CHAPTER 2.1 Brownian Motion Diffusion	
Israelachvili	CHAPTER 16.1–16.4 Surfactant numbers			CHAPTERS 8.5–8.7 Hydrophobic effects
Myers	CHAPTER 3 Surfactants			
Shinoda		CHAPTERS 1–6 Solution thermo-dymanics and regular solutions		CHAPTER 7 Aqueous solutions
Tanford	CHAPTER 3 Solubility			CHAPTER 2 Solubility of hydrocarbon in water
	CHAPTER 4 Amphiphiles in water			CHAPTERS 4–5 Water effects
Vold and Vold		CHAPTER 7.4 Osmotic forces		

[a]For complete reference citations, see alphabetic listing, page xxix.

2

SURFACE CHEMISTRY AND MONOLAYERS

CONCEPT MAP

Surface Chemistry and Monolayers

- By analogy with the thermodynamic equation for bulk material, we can write $d\mathbf{G} = \gamma\,dA + \Sigma\mu_i\,dn_i$ for surfaces and interfaces at constant temperature and pressure. A system can minimize its free energy by:

  1. Decreasing its surface area—this branch of surface science leads to capillarity.

  2. Changing its composition—this phenomenon leads to the Gibbs adsorption isotherm.

- We pursue each topic sequentially; in multicomponent systems, both area and composition can change at an interface to minimize $\Delta\mathbf{G}$.

Surface Tension

Surface tension reflects the unfavorable change in interaction energy, $\mathbf{E}_S \propto W_{AA}(z_s - z_b)$ (eq. 2.1.3) that occurs upon moving a molecule from the bulk to the surface, where the number of nearest neighbors decreases from z_b to z_s.

Surface tension can be written either as the change in free energy per unit surface area, $\gamma = (\partial\mathbf{G}/\partial A))_{TP}$ (eq. 2.1.10) or as the force per unit length, $\gamma = F/2l$ (eq. 2.1.7), as defined in Figure 2.1. These two points of view provide useful perspectives in interpreting γ.

Interfacial tension γ_{AB} between two immisible phases A and B can be expressed (eq. 2.1.5) in terms of the interaction parameter w (eq. 1.5.4), and the number of A–B interactions.

Work of Cohesion and Adhesion

The work of cohesion, $w_{AA} = 2\gamma_A$, measures the free energy associated with creating unit areas of surface by cleaving a homogeneous phase, as illustrated in Figure 2.2a.

The work of adhesion, $w_{AB} = \gamma_A + \gamma_B - \gamma_{AB}$, measures the free energy required to separate the unit area of the interface into two liquid–air interfaces, as illustrated in Figure 2.2b.

The spreading coefficient, $S = w_{AB} - w_{AA}$ (eq. 2.1.16), permits us to predict whether a drop of one liquid will spontaneously spread on the surface of a second liquid or solid.

Young–Laplace Equation

The Young–Laplace equation, $\Delta p = \gamma(1/R_1 + 1/R_2)$ (eq. 2.1.24), relates pressure differences across a curved interface to radii of curvature and surface tension and explains why liquids rise in capillary tubes. Useful applications of this equation include explaining:

- How surface tension is measured using capillaries, sessile and pendant drops, and the Wilhelmy plate method.

- Why and under what conditions vapors will condense into capillaries, into surface pores and cracks, or onto rough surfaces at vapor pressures p below the equilibrium value p_0.

- Why small particles in heterogeneous colloidal dispersions decrease in size and disappear, while larger ones grow at their expense, a process called Ostwald ripening.

- Why, in terms of the Kelvin equation (eq. 2.3.12), the nucleation of a liquid phase from a pure vapor phase requires vapor pressures considerably greater than the equilibrium value p_0.

Gibbs Adsorption Isotherm

- It is difficult to assign a volume to an interfacial region because physical properties change continuously across the interface, as illustrated in Figure 2.11.

- In the Gibbs model, we treat the interface as a mathematical plane of area A, containing n_i^σ molecules, and we define a surface concentration by $\Gamma_i = n_i^\sigma/A$ (eq. 2.4.4).

- We can relate the change in surface tension with $\ln c_i$ of an added component to a surface excess concentration, Γ_2 (eq. 2.4.14).

- In a binary solution, if we choose $\Gamma_1 = 0$, we focus on the behavior of the solute.

- Γ_i is an algebraic quantity reflecting concentration or depletion at the interface.

Langmuir Isotherm

The Langmuir isotherm is the simplest model for adsorption of molecules onto solid surfaces.

Assuming that solvent and solute molecules occupy the same adsorption area, that the surface solution is ideal, and that the solution is dilute, we can relate bulk solute concentration to fractional surface coverage by eq. 2.4.18.

Surface Monolayers

Monolayers are two-dimensional self-organizing systems.

- When surface-active molecules are in equilibrium with those in the adjacent solution, we can use surface tension measurements and the Gibbs adsorption isotherm to calculate surface concentrations and surface areas per molecule.

- With insoluble amphiphiles, we can use the Langmuir balance to measure the surface pressure, $\Pi_s = \gamma^0 - \gamma$, where γ stands for the surface tension with amphiphile and γ_0 with clean surface. Plots of Π_s versus surface area per molecule, a_0 yield two-dimensional isotherms.

- Insoluble monolayers are two-dimensional analogs of bulk systems that display gaslike, liquid and solid phases.

2

A typical colloidal system shows a large surface-to-volume ratio. These surfaces may consist of proper interfaces between different phases, as in an emulsion, or they may constitute boundaries between regions that possess distinctly different molecular properties, as in a micellar solution. Liquid–liquid, liquid–solid, liquid–gas, and solid–gas interfaces all appear in different applications of colloid science.

Although surface and colloid science are often treated together in a book, this text concentrates specifically on colloid science. However, surfaces and interfaces play such an important role in colloidal science that we present some of the more fundamental aspects of surface chemistry with an emphasis on those of particular importance for understanding colloids.

2.1 We Can Comprehend Surface Tension in Terms of Surface Free Energy

2.1.1 Molecular Origins of Surface Tension Can Be Understood in Terms of the Difference in Interaction Between Molecules in the Bulk and at the Interface

Differences between the energies of molecules located at the surface and in the bulk phase of a material manifest themselves as surface tension. Historically, this concept has been a key factor in understanding molecular behavior in liquids.

In Chapter 1, we examined the interaction of molecules in the bulk phase in terms of a pair interaction energy W_{AA}. If we now take as an example a pure liquid or solid in equilibrium with its vapor and assume its interaction energies to be pairwise additive, the energy $E_{A,\text{bulk}}$ per molecule in the bulk phase is (cf. eq. 1.5.1)

$$E_{A,\text{bulk}} = \frac{z_b}{2} W_{AA} \qquad (2.1.1)$$

where z_b stands for the number of nearest neighbors in the bulk phase.

The corresponding expression for molecules at a surface is

$$E_{A,s} = \tfrac{1}{2} z_s W_{AA} \qquad (2.1.2)$$

where z_s is the number of near neighbors for a surface molecule.

Because W_{AA} is negative and $z_b > z_s$, moving a molecule from the bulk to the surface increases the internal energy. In other words, work must be done to create a new surface. The energy $\mathbf{E}_s$ required to create an area A is then

$$\frac{\mathbf{E}_s}{A} = \left(\frac{E_{A,s} - E_{A,\text{bulk}}}{a_0} \right) = \tfrac{1}{2} \frac{W_{AA}(z_s - z_b)}{a_0} \qquad (2.1.3)$$

where a_0 equals the area per molecule at the surface.

Problem: Calculate the surface tension for carbon tetrachloride using the value of $z_b W_{AA} = -9.1 \times 10^{-20}\,\text{J}$ obtained from $\mathbf{\Delta H}_{\text{vap}}$. Assume that CCl_4 can be treated as a cube with $z_b = 6$.
Solution: In going from the bulk to the surface, one face of the cube is exposed so $z_b = 6$ and $z_s = 5$. Using the density of CCl_4, $\rho = 1.6 \times 10^3\,\text{kg/m}^3$, we can calculate the area for each side of the cube to be $3 \times 10^{-19}\,\text{m}^2$. Thus

$$\mathbf{E}_s/A = 4.5 \times 10^{-20}\,\text{J}/6 \times 3 \times 10^{-19}\,\text{m}^2 = 25\,\text{mN/m}$$

a value which is fortuitously close to the experimental value of $26\,\text{mN/m}$ for the surface tension of CCl_4. This calculation illustrates the major source of surface tension in a liquid.

If we consider the interface between two phases containing molecules A and B, we can see that when an A molecule moves to the interface, it loses interactions with A but gains about an equal number of interactions with B. Similarly, B molecules lose interactions with B and gain interactions with A. The net energy change for N_s surface molecules of each kind is

$$\mathbf{E}_s = N_s(z_s - z_b)\left\{ \tfrac{1}{2} W_{AA} + \tfrac{1}{2} W_{BB} - W_{AB} \right\} = \frac{N_s}{N_{Av}} \frac{z_b - z_s}{z_b} w \qquad (2.1.4)$$

with w defined as in eq. 1.4.16. We have assumed that the

coordinate number z_b is the same in the two bulk phases and at the interface, a reasonable assumption for molecules of similar size. The energy per unit area is then

$$\frac{\mathbf{E}_s}{A} = \frac{1}{a_0}\left(\frac{z_b - z_s}{z_b}\right)\frac{w}{N_{Av}} \qquad (2.1.5)$$

Problem: For two strongly immiscible liquids with $w = 6\mathbf{R}T$ estimate the interfacial tension and compare it to the value of the surface tension calculated for CCl_4. Assume that $a_0 = 3 \times 10^{-19}\,\mathrm{m}^2$ and room temperature.
Solution: Similar to the problem with carbon tetrachloride

$$\mathbf{E}_s/A = 6 \times 8.31 \times 300\,\mathrm{J}/(6 \times 3 \times 10^{-19}\,\mathrm{m}^2)(N_{Av}) = 14\,\mathrm{mN/m}$$

This example demonstrates the general rule that interfacial energies are smaller than surface energies.

In deriving eq. 2.1.4, we assumed that A and B are pure liquids. Then $\mathbf{E}_s$ is the energy of the instantaneous formation of the interface. At equilibrium, some B molecules will mix into the A liquid and vice versa. This exchange decreases the surface energy, and for miscible liquids the interface disappears altogether.

We have estimated the energy contribution to the formation of a surface or an interface. To calculate the surface free energy or surface tension, we must also consider an entropy contribution that arises, for example, from the orientation of surface molecules and from capillary waves at the surface. However, the energy contribution is usually the largest.

2.1.2 Two Complementary Concepts Define Surface Tension: Line of Force and Energy Required to Create New Surface Area

As the analysis of eq. 2.1.3 indicates, creating a new surface requires work. This work, $\boldsymbol{w}$, is proportional to the number of molecules transported to the surface and thus to the area of the new surface, which we can express in a basic linear defining equation as

$$\boldsymbol{w} = \gamma\Delta A \qquad (2.1.6)$$

where the proportionality constant γ is called the surface tension.

Therefore, we can view γ two ways: Either as the free energy per unit area (energy/(length)^2) or as a surface tension (force/length). Historically, the existence of these different perspectives has posed a dilemma, but they represent two ways of looking at the same thing, and both have their uses. Understanding them allows us to determine when one point of view works better and when the other provides a more effective cognitive interpretation of the data.

Figure 2.1 shows a simple device consisting of a wire loop with a movable slide that can clarify these two ways of viewing surface tension. We can assume that the device behaves like an idealized frictionless apparatus.

Dipping the wire loop into a liquid forms a liquid film. Surface tension causes the slide wire of the apparatus to move in the direction of decreasing film area unless we apply an opposing force. This force, F, operates along the entire edge of the film, varies linearly with the length l of the slide wire, and has a characteristic value for each liquid. If γ represents an intrinsic property of the liquid surface, the apparatus measures the surface tension as

$$\gamma = \frac{F}{2l} \tag{2.1.7}$$

because the film has two sides.

The loop-and-slide wire device is a two-dimensional analog of the cylinder–piston apparatus used to evaluate pressure·volume products in gas cycles in thermodynamics. This analogy suggests that γ represents a two-dimensional pressure, a point we will develop in Section 2.6. However, one major difference separates the two situations: if we remove the restraining force in the three-dimensional system, the piston

Figure 2.1
Wire loop with a slide wire on which a liquid film can be formed and stretched by an applied force F.

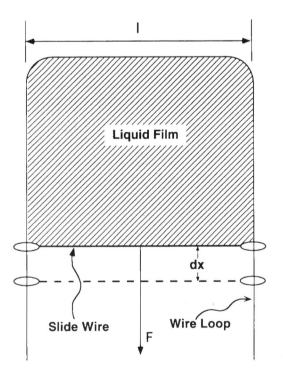

expands, yet in the two-dimensional system, the slide wire contracts.

We can write the work associated with expanding the interfacial area as

$$dw = F\,dx = \gamma 2l\,dx = \gamma\,dA \qquad (2.1.8)$$

At constant temperature and pressure the work of increasing the area contributes to the differential Gibbs free energy. Now it is possible to relate eq. 2.1.8 to thermodynamic quantities and write

$$dG = \gamma\,dA \qquad (2.1.9)$$

where

$$\gamma = \left(\frac{\partial G}{\partial A}\right)_{T,p} \qquad (2.1.10)$$

This expression identifies surface tension as the increase in Gibbs free energy per unit increment in area.

2.1.3 The Work of Adhesion and Cohesion Is Related to Surface Tension and Can Determine the Spontaneous Spreading of One Liquid on Another

To understand how liquids behave when they come into contact, we need to grasp the concepts of adhesion and cohesion. In a single liquid, the work of cohesion corresponds to the work required to pull apart a volume of unit cross-sectional area, as shown in Figure 2.2 and the equation

$$\Delta G = w_{AA} = 2\gamma_\alpha \qquad (2.1.11)$$

Interpreting γ_α as half the work of cohesion indicates that surface tension measures the free energy change involved when molecules from the bulk of a sample are moved to its surface.

The work of adhesion between two immiscible liquids equals the work required to separate the unit area of the interface into two liquid–air interfaces, as shown in Figure 2.2 and expressed by the Dupré equation

$$w_{AB} = \gamma_{\alpha v} + \gamma_{\beta v} - \gamma_{\alpha\beta} \qquad (2.1.12)$$

When we place a drop of an insoluble liquid, such as oil, on a clean liquid, such as water, or a solid surface, it may behave in one of three ways:

1. Remain as a nonspreading lens, as in Figure 2.3; in this case the liquid does not wet the surface.

2. Spread uniformly over the surface as duplex film thick enough for each liquid and air interface to remain independent and possess characteristic surface tensions (i.e., the liquid wets the surface).

Figure 2.2
(a) The work of cohesion, $w_{AA} = 2\gamma_{\alpha v}$, in a liquid and (b) the work of adhesion $w_{AB} = \gamma_{\alpha v} + \gamma_{\beta v} - \gamma_{\alpha\beta}$, between two different liquids. Breaking the column creates two new interfaces in the first case; in the second case it also destroys one. Sometimes we denote a surface tension γ_α as $\gamma_{\alpha v}$ to denote that the second phase is a vapor. γ_α is the low pressure limit of $\gamma_{\alpha v}$.

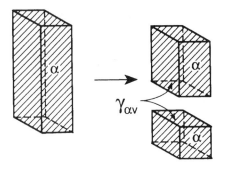

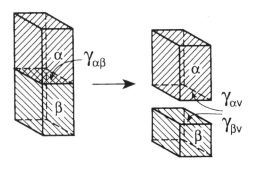

3. Spread as a thin film, leaving excess liquid as lenses at equilibrium, as shown in Figure 2.4.

Figure 2.3
Shape of an oil lens on a water surface. The shape may change with time if the oil is able to spread or if the two liquids are mutually soluble to some extent. The same analysis applies to any pair of immiscible liquids α and β.

The shape of a droplet of liquid α on a solid surface β in a vapor v can be determined by a force balance at the three-phase contact line in the direction parallel to the surface (see Figure 2.5) so that

$$\gamma_{sv} = \gamma_{sl} + \gamma_{lv} \cos \theta \qquad (2.1.13)$$

This is the Young equation. Thus, we can experimentally

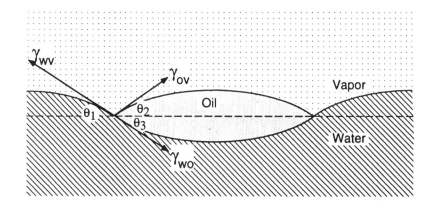

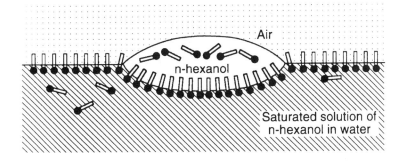

Figure 2.4
When a drop of hexanol is placed on water, the drop initially spreads; that is, the spreading coefficient (eq. 2.1.15) is positive. However, when equilibrium is achieved, the water surface contains a monolayer of hexanol, and the spreading coefficient becomes negative as a consequence of the lowering of γ_{wv}. Then the hexanol forms a lens as shown. (D.J. Shaw, *Introduction to Colloid and Surface Chemistry*, 4th ed., Butterworths, London, 1992, p. 90.)

determine the interfacial tension $\gamma_{\alpha\beta}$ by measuring the contact angle θ and obtain the surface tensions, $\gamma_{\beta v}$ and $\gamma_{\alpha v}$ separately. Clearly in eq. 2.1.13 the third phase need not be a vapor but could be a second immiscible liquid, provided we change the interpretation of $\gamma_{\beta v}$ and $\gamma_{\alpha v}$.

If the substrate phase is another liquid, the lateral force balance involves three angles, as shown in Figure 2.3 for oil on water. The resulting relation between the surface tensions is

$$\gamma_{wv} \cos\theta_1 = \gamma_{wo} \cos\theta_3 + \gamma_{ov} \cos\theta_2 \qquad (2.1.14)$$

What determines whether the lens shown in Figure 2.5 will spread over the surface to form thick wetting film? In the case of a wetting film, we have formed two macroscopic interfaces/surfaces, but if the lens remains, only the original $\alpha-v$ surface extends macroscopically. To have two interfaces/surfaces is clearly advantageous if the sum of the two surface free energies, $\gamma_{sl} + \gamma_{lv}$, is smaller than the initial surface free energy γ_{sv}. Thus, it is natural to introduce a spreading coefficient

$$S = \gamma_{sv} - (\gamma_{sl} + \gamma_{lv}) \qquad (2.1.15)$$

If $S \geqslant 0$, spreading will occur, while for negative S the liquid forms a finite lens. We see from eq. 2.1.13 that a solution for the contact angle θ exists only when $S < 0$, showing that our interpretation of S as a spreading coefficient is consistent.

Using eqs. 2.1.11 and 2.1.12, we can relate the spreading coefficient to the work of adhesion and cohesion so that for A

Figure 2.5
Contact angle on a solid surface.

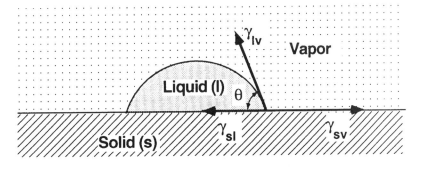

on B

$$S = w_{AB} - 2\gamma_A = w_{AB} - w_{AA} \qquad (2.1.16)$$

Applying eq. 2.1.15 to describe the wetting or nonwetting of a liquid on a surface creates a practical complication, which is particularly apparent if the substrate phase is a liquid. Even if two liquids, A and B, form two phases in equilibrium, some mutual solubility always exists. Initially, we typically place a drop of pure A on a pure B liquid. At this moment, the relevant surface tension, γ_{AB}, is that of a pure A–pure B interface. Because a balance of mechanical forces determines the shape of the lens, it adopts a mechanical equilibrium on a relatively short time scale. As time goes on, an exchange of A and B molecules across the A–B interface leads to a change in all three surface energy terms entering into S in eq. 2.1.15. This change can lead to the striking phenomenon that a drop first spreads on the surface and then slowly re-forms as full equilibrium is achieved.

Problem: Given that $\gamma_{H_2O/vapor} = 72.8$, $\gamma_{water/benzene} = 35.0$ and $\gamma_{benzene/vapor} = 28.9$ calculate the initial spreading coefficient. After equilibrium is reached, $\gamma_{H_2O/vapor}$ decreases to a value of 62.4 because the small amount of benzene that dissolves in water is surface active. Will the benzene remain spread after equilibrium is achieved?

Solution: $S(\text{init}) = 72.8 - (28.9 + 35.0) = 8.9\,\text{mJ/m}^2$ which corresponds to initial positive spreading coefficient. At equilibium

$$S(\text{eq}) = 62.4 - (28.9 + 35.0) = -1.5\,\text{mJ/m}^2$$

which means that the initial continuous benzene film will roll back to form a flat lens.

The effect is pronounced if we consider the spreading of hexanol on water. Initially, the spreading coefficient is strongly positive

$$S(\text{init}) = 72.8 - (24.8 + 6.8) = 41.2\,\text{mJ/m}^2$$

but at equilibrium it becomes negative

$$S(\text{eq}) = 28.5 - (24.8 + 6.8) = -3.1\,\text{mJ/m}^2$$

as a result of the pronounced surface activity of hexanol in the aqueous phase. In the final states both benzene and hexanol form very flat lenses, which go over into a near monolayer at the aqueous–vapor interface.

Trace amounts of impurity can cause dramatic changes in spreading behavior. For oil on water, impurities in the oil phase can reduce γ_{ow} enough to make S positive. Impurities in the aqueous phase generally also reduce S, since the impurity lowers γ_{wv} more than γ_{ow}, especially if γ_{ow} is low already. Thus, n-octane will spread on a clean water surface but not on a contaminated one.

2.1.4 The Young–Laplace Equation Relates Pressure Differences Across a Surface to Its Curvature

Surface tension operates in a liquid film, and to obtain mechanical equilibrium this tension must be balanced by some equal and opposite force. For example, to blow a bubble in a liquid medium, an excess pressure is applied. In an isolated particle, such as a droplet, the balancing force comes from stresses within the particle. How is the excess pressure related to surface tension and surface shape?

Figure 2.6 illustrates a circular cross section of a bubble with radius R. When the bubble is expanded infinitesimally to a radius $R + dR$, the concomitant area change is

$$dA = 4\pi\{(R + dR)^2 - R^2\} = 8\pi R\, dR \qquad (2.1.17)$$

The corresponding change in free energy is

$$dG = \gamma\, dA = \gamma 8\pi R\, dR \qquad (2.1.18)$$

At equilibrium this free energy is balanced by a pressure–volume work, $dw = \Delta p\, dV$, because of the pressure difference across the film. The volume change is

$$dV = \frac{4\pi}{3}\{(R + dR)^3 - R^3\} = 4\pi R^2\, dR \qquad (2.1.19)$$

Figure 2.6
Cross section of a gas bubble of radius R suspended in a liquid used to calculate the increase in free energy as the bubble expands from R to $R + dR$.

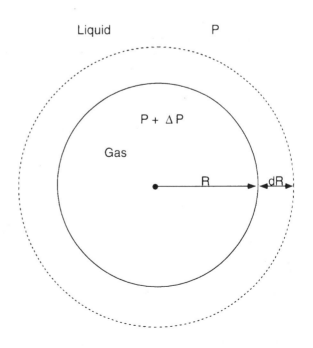

Liquid P

P + ΔP

Gas

R dR

Balancing the work of increasing the surface with the pressure–volume work leads to a pressure difference

$$\Delta p = \gamma \frac{2}{R} \qquad (2.1.20)$$

per interface. The pressure difference in a soap bubble is twice this value, due to the two surfaces.

Problem: What is the excess pressure in a soap bubble of radius 1 cm blown from an SDS solution above the CMC?
Solution:

$$\gamma \approx 38\,\text{mJ/m}^2 \text{ (Fig. 1.1)}$$

$$\Delta p = 2 \times \gamma \times 2/R = 38 \times 10^{-3} \times 4/10^{-2}$$

$$\simeq 15\,\text{Pa} \simeq 1.5 \times 10^{-4}\,\text{atm}.$$

If we consider Figure 2.6 as the cross section of a uniform cylinder of length L, the area change is

$$dA = 2\pi L(R + dR - R) = 2\pi L\,dR \qquad (2.1.21)$$

and the volume change is

$$dV = \pi L\{(R + dR)^2 - R^2\} = 2\pi L R\,dR \qquad (2.1.22)$$

A balance between the surface and the pressure–volume work now results in

$$\Delta p = \frac{\gamma}{R} \qquad (2.1.23)$$

per interface for the cylindrical geometry.

We can generalize the equations to interfaces of arbitrary shape and the result is

$$\Delta p = \gamma \left(\frac{1}{R_1} + \frac{1}{R_2}\right) = \gamma 2H \qquad (2.1.24)$$

where R_1 and R_2 are the radii of curvature of the interface and H represents the mean curvature as defined in eq. 1.3.8. Equation 2.1.24 is called the **Young–Laplace** equation. For a sphere, $R_1 = R_2 = R$, while for a cylinder $R_1 = R$ and $R_2 = \infty$, and eq. 2.1.24 reduces to the eqs. 2.1.20 and 2.1.23, for spheres and cylinders, respectively.

If we form a tubelike soap film between two circular wires, the pressure difference across the film is zero, since the ends of the tube are open. According to eq. 2.1.24, the mean curvature of the film then becomes zero everywhere. In fact, when the centers of the circles are on a common axis we obtain the catenoid surface of which a patch is shown in Figurer 1.7.

Many interesting applications of the equations contained in Section 2.1 involve systems with small radii whose structures are comparable in size to molecules. To apply these equations to such systems, we employ macroscopic concepts, such as density, surface tension, and radius of curvature. How far can we trust their validity?

Recent experiments using the surface forces apparatus (see Chapter 6) show that for simple liquids, such as cyclohexane, the macroscopic approach seems to be valid at least down to a radius of curvature that corresponds to about seven times the molecular diameter. Other experimental tests confirm this finding for simple liquids. An important exception is water, where deviations from macroscopic behavior have been reported for larger radii of curvature. This exception may reflect the fact that surface tension in water involves longer range interactions and consequently more water molecules.

2.2 Several Techniques Measure Surface Tension

2.2.1 Surface Tension Governs the Rise of a Liquid in a Capillary Tube

We can use the phenomenon of capillary rise illustrated in Figure 2.7 to measure the surface tension of a liquid. A simple, albeit somewhat approximate, relation exists between surface tension γ, capillary rise h, capillary radius r, and contact angle θ.

In the capillary tube the liquid surface curves as a spherical cap. As Figure 2.7 shows, the radius of curvature R is related to the radius of the tube r, and the contact angle:

$$R \cdot \cos \theta = r \qquad (2.2.1)$$

Across the interface is a Laplace pressure

$$\Delta p = \frac{-2\gamma}{R} = \frac{-2\gamma \cos \theta}{r} \qquad (2.2.2)$$

which is balanced by a hydrostatic pressure

$$\Delta p = \Delta \rho g h \qquad (2.2.3)$$

where $\Delta \rho$ represents the density difference, $\rho_{liq} - \rho_{vap} \simeq \rho_{liq}$, and g is the acceleration due to gravity. Combining the two equations we find

$$\gamma = \frac{\Delta \rho g h r}{2 \cos \theta} \qquad (2.2.4)$$

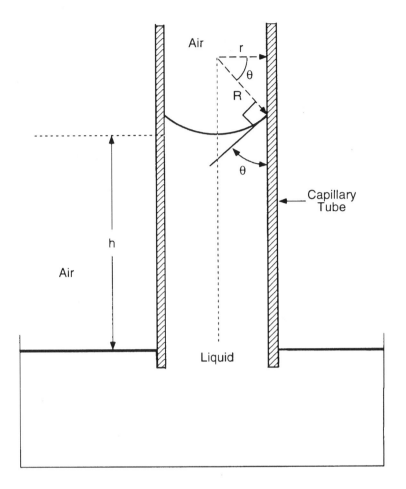

Air

r

θ

R

θ

Capillary
Tube

h

Air

Liquid

Figure 2.7
Schematic illustration of
capillary rise in a
cylindrical tube seen in the
cross section of radius r.

This equation is particularly simple to apply when the liquid wets the solid surface so that $\cos \theta = 1$. In this case it suffices to know $\Delta \rho$ and r and measure h to be able to calculate the liquid–vapor surface tension, γ.

In using the Young–Laplace equation to obtain the pressure forcing the liquid up the capillary tube, we have sidestepped the question of the molecular cause of the capillary rise. Even though the surface tension of the liquid is finally measured, the phenomenon is basically the result of the interfacial properties of both the solid and the liquid.

Consider an infinitesimal rise dh of the liquid in the tube at a given contact angle. The liquid–vapor contact area remains constant, while the solid–liquid interface increases as

$$dA = 2\pi r \, dh \qquad (2.2.5)$$

and the solid–vapor interface displays a corresponding decrease. The change in surface free energy is

$$d\mathbf{G}_s = (\gamma_{sl} - \gamma_{sv})2\pi r \, dh \qquad (2.2.6)$$

where γ_{sl} and γ_{sv} are the respective interfacial tensions. A change in the gravitational energy

$$dG_{grav} = \Delta\rho gh\, dV = \pi r^2\, \Delta\rho gh\, dh \qquad (2.2.7)$$

also occurs. At equilibium these two free energy contributions add to zero so that

$$\gamma_{sv} - \gamma_{sl} = \frac{\Delta\rho ghr}{2} \qquad (2.2.8)$$

To arrive at eq. 2.2.4, we recall that according to eq. 2.1.13

$$\gamma_{sv} - \gamma_{sl} = \gamma_{lv}\cos\theta$$

This derivation explains eq. 2.2.4 in terms of surface free energies for nonwetting conditions so that $\theta > 0$. When the liquid wets the surface, the argument is slightly different. In this case a wetting film of thickness δ will rise higher than h in the tube. The exact height to which the wetting film rises is determined by the difference $\gamma_{sv} - \gamma_{sl}$, but for the measurement of the liquid–vapor surface tension, it suffices to know that the rise is higher than for the liquid in the center of the capillary.

As bulk liquid rises in the tube a distance dh, the change in interfacial free energy is

$$dG_s = \gamma\, dA = \gamma 2\pi(r - \delta)\, dh \qquad (2.2.9)$$

because the solid surface is already covered. At equilibrium this energy balances with the gravitational energy and

$$\gamma = \frac{\Delta\rho gh(r - \delta)}{2} \qquad (2.2.10)$$

Provided the film thickness δ is negligible relative to radius r, we regain eq. 2.2.4 for the case $\cos\theta = 1$.

By deriving eq. 2.2.4 in two different ways, we have illustrated the complementarity between seeing γ mechanically as a surface tension and seeing it as a surface free energy. The former point of view is usually advantageous when determining the mechnical equilibrium, while the latter reveals more about the molecular mehanisms behind the observed phenomena.

Table 2.1 summarizes methods of measuring surface tension.

2.2.2 The Wilhelmy Plate Method Measures the Change in Weight of a Plate Brought into Contact with a Liquid

When a microscope glass slide or other thin plate attached to a balance is brought into contact with a fluid, surface tension manifests itself in the meniscus that forms around the perimeter of the plate (Figure 2.8). If we balance the apparatus before bringing the plate into contact with the liquid, the increase in weight is a result of the entrained meniscus.

TABLE 2.1 Methods of Measuring Surface Tension

Method	Principle	Performance
Capillary height	Capillary rise	Best when contact angle θ is 0
Wilhelmy plate	Capillary force on a plate	Convenient, but susceptible to atmospheric contamination. Best when $\theta = 0$.
Drop profile	Analysis of the geometric drop shape	Works also when a long time is needed to reach equilibrium.
DuNouy ring	Capillary force on a ring	Convenient. Best when $\theta = 0$. Often used for surfactant systems.
Spinning drop tensiometer	Analysis of the geometric shape of a centrifugally distorted drop	Liquid–liquid systems. Appropriate only for very low surface tensions (not measurable with the other methods).

Figure 2.8
The Wilhelmy plate used for measuring γ. A thin plate, seen from the side, is dipped into the liquid.

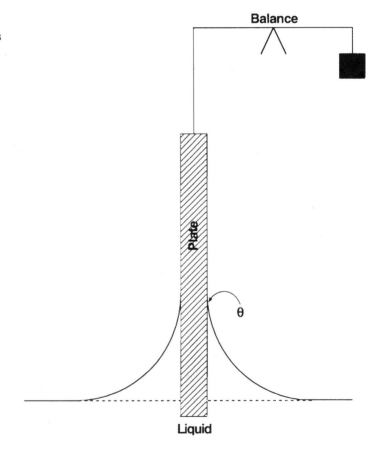

A vertical force balance shows that the weight of the meniscus must equal the upward force provided by the surface. This force in turn equals the vertical component of the surface tension ($\gamma \cos \theta$, where θ equals the contact angle) times the perimeter P of the plate. Since the surface tension is a force per unit length of the surface edge, we have

$$w = \gamma P \cos \theta \qquad (2.2.11)$$

The Wilhelmy plate technique is particularly simple to use if the contact angle is zero.

2.2.3 The Shapes of Sessile and Pendant Drops Can Be Used to Determine Surface or Interfacial Tension

Surface tension and gravity determine the shape of a drop. Surface forces depend on area, which varies as the square of the linear dimension, and gravitational forces depend on volume, proportional to the cube of the linear dimension. As a result, very small drops, bubbles, or a drop of liquid in a second liquid with the same density as the first will take on spherical shapes. However, when the two forces become equally important, sessile (sitting) and pendant (hanging) drops take on shapes determined by their size, the differene in density between the two liquids, and the surface tension. Figure 2.9 shows some examples. We can calculate surface tension from a drop's dimensions by using known solutions for the equilibrium shapes.

A variation of this method is useful for determining very low interfacial tension. In this experiment, a capillary filled with oil containing a small droplet of water is positioned horizontally and spun about its axis. We can obtain the surface tension from the droplet's length and radius at different angular velocities.

Figure 2.9
Shapes of sessile and hanging drops and of bubbles: (a) pendent drop, (b) sitting drop, (c) hanging bubble, and (d) sessile bubble. Detailed analysis of the shapes of these drops makes it possible to determine surface tension.

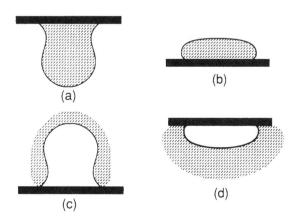

2.2.4 Contact Angles Yield Informaton on Solid Surfaces

From eq. 2.1.13 we see that by measuring the contact angle θ we can obtain the difference $\gamma_{sv} - \gamma_{sl}$. This contact angle measurement provides a useful method for surface characterization. It is performed by observing the shape of liquid drop on a surface through microscope. By connecting the drop to a pipette, the drop can be made smaller or larger. A typical practical complication is that the contact angle changes as the contact line advances or recedes on the surface. This phenomenon is called contact angle hysteresis, and in general, it suggests the importance of measuring both the advancing and the receding contact angle.

A number of possible causes of contact angle hysteresis can be identified. Usually solid surfaces are not homogeneous on a microscopic scale, and the contact line tends to get trapped at certain positions on the surface. Slow kinetic processes also may involve molecular rearrangements at the surface as well as molecular transport to and from the contact zone.

2.3 Capillary Condensation, Ostwald Ripening, and Nucleation Are Practical Manifestations of Surface Phenomena

The surface tension and surface free energy concepts discussed in Sections 2.1 and 2.2 provide a basis for understanding a number of practically important phenomena. We illusatrate this by three examples: capillary condensation of a liquid on a rough surface, growth of colloidal precipitates through Ostwald ripening, and the nucleation of a new phase under supersaturated conditions.

2.3.1 Surface Energy Effects Can Cause a Liquid to Condense on a Surface Prior to Saturation in the Bulk Phase

Consider a vapor in contact with a (solid) surface. If the corresponding liquid wets the surface, the vapor will have an extra tendency to condense on the surface to form a wetting film. At a given temperature, this film will be formed at a vapor pressure p lower than the bulk saturation pressure p_0. The chemical potential in the vapor is

$$\mu_g = \mu_g^\theta + RT \ln p \qquad (2.3.1)$$

while the chemical potential of the condensed bulk liquid is

$$\mu_i^\theta = \mu_g^\theta + RT \ln p_0 \qquad (2.3.2)$$

The free energy change in forming a liquid film of thickness δ of an area A involves a transfer from vapor to liquid of $\delta A/V_L$ moles, where V_L represents the molar volume of the liquid. A change in surface free energy $(\gamma_{lv} + \gamma_{sl} - \gamma_{sv})A$ also occurs. The total free energy change is

$$\Delta G = \frac{\delta A}{V_L} RT \ln\left(\frac{p_0}{p}\right) + (\gamma_{lv} + \gamma_{sl} - \gamma_{sv})A \qquad (2.3.3)$$

This free energy change is zero when the vapor pressure p is given by

$$RT \ln\left(\frac{p_0}{p}\right) = \frac{V_L}{\delta}(\gamma_{sv} - \gamma_{sl} - \gamma_{lv}) \qquad (2.3.4)$$

where we recognize the spreading coefficient of eq. 2.1.15 in the right-hand side. Clearly, surface condensation occurs only when the liquid wets the surface, because then the right-hand side of eq. 2.3.4 is positive.

In the derivation we have treated the thickness of the film δ as given. In fact, to determine δ, we must go beyond the macroscopic thermodynamic level and discuss intermolecular and surface forces in detail. The background for such a description is presented in Chapter 5.

Even for a liquid that does not wet the surface, condensation at $p < p_0$ can occur if the surface arrangement is more favorable than that of a single planar surface. Consider, for example, two planar parallel surfaces with a gap of thickness h between them and in equilibrium with a bulk vapor phase. Under what conditions does the liquid condense in the gap? In the condensed state, the gap contains hA/V_L moles and the free energy change on condensation is

$$\Delta G = \frac{hA}{V_L} RT \ln\left(\frac{p_0}{p}\right) + 2(\gamma_{sl} - \gamma_{sv})A \qquad (2.3.5)$$

in analogy with eq. 2.3.3. The factor of 2 arises owing to the presence of two interfaces. Condensation occurs at the threshold pressure where $\Delta G = 0$ and

$$RT \ln\left(\frac{p_0}{p}\right) = \frac{2V_L}{h}(\gamma_{sv} - \gamma_{sl}) = \frac{2V_L}{h}\gamma_{lv}\cos\theta \qquad (2.3.6)$$

We have used eq. 2.1.13 to obtain the last equality.

Problem: Calculate the relative humidity at 25°C at which water will condense in a gap of thickness 100 Å on a surface for which (a) the contact angle is small ($<10°$), and (b) for a typical hydrophobic surface with $\theta = 115°$.

Solution:

(a) When $\theta < 10°$ we have $\cos \theta \simeq 1$.

$$\ln(p_0/p) = \frac{2(18 \times 10^{-6}\,\text{m}^3)(72 \times 10^{-3}\,\text{J/m}^2)}{(1 \times 10^{-8}\,\text{m})(2.5 \times 10^3\,\text{J})} = 0.1$$

$p = 0.91p_0$ or 91% relative humidity.

(b) When $\theta = 115$, $\cos \theta = -0.42$, and $p = 1.04p_0$ and water will only condense in the gap when the air is 4% supersaturated with water vapor.

An interesting aspect of eq. 2.3.6 is that for $\theta > 90°$ (so that $\cos \theta < 0$), the vapor will be less likely to condense in the gap than in the bulk. This tendency implies that if we consider the reverse process of a liquid filling the system that approaches the boiling point, the vapor is preferentially formed in the gap. Because the vapor has a much lower density, the contribution from the surface free energy is much larger, counted per molecule in the gap. Vaporization typically occurs at larger critical separation for $\cos \theta < 0$ than for the condensation when $\cos \theta > 0$.

An actual solid surface is normally rough. It may contain cracks, and pores can lead to the interior of the solid. For such a surface in equilibrium with a vapor, the more macroscopic phenomenon of a condensation of a nonwetting liquid may occur in addition to physical and chemical adsorption. The final result of the condensation events depends on general thermodynamic properties, such as vapor pressure, contact angle, and surface tension, as well as on the local geometry of the surface. Clearly, the finer the cracks or pores, or the rougher the surface, the more the liquid is likely to condense.

2.3.2 Surface Free Energies Govern the Growth of Colloidal Particles

Consider a dispersion of colloidal particles that have been formed by precipitating a new phase from solution. The "particles" may be either solid or liquid; they may even take the form of gas bubbles. Their constituent molecules are present as a solute in the liquid at a saturation concentration. Molecules exchange between the particles and the solution so that each particle is in local equilibrium with the solution close to it.

The free energy of a particle i with n_i molecules and area A_i consists of a bulk term, $n_i \mu_p^\theta$, plus a surface term $A_i \gamma_{ps}$, where γ_{ps} stands for the particle–solution surface free energy. The free energy change ΔG_i caused by taking a molecule from the solution at concentration c_i and adding it to particle i, is

$$\Delta G_i = (\mu_p^\theta - \mu_i^\theta) - kT \ln c_i + \gamma_{ps} \Delta A_i \qquad (2.3.7)$$

This free energy change equals zero at local equilibrium.

The area change, ΔA_i, upon adding a molecule to the particle depends on the size of the particle. For spherical particles of radius R_i, the volume is

$$\frac{4\pi}{3} R_i^3 = n_i V_M \tag{2.3.8}$$

where V_M represents the volume per molecule. Thus, increasing n_i by one leads to a change in the radius by

$$\Delta R_i = \frac{V_M}{A_i} \tag{2.3.9}$$

and the corresponding area change is

$$\Delta A_i = \frac{2V_M}{R_i} \tag{2.3.10}$$

This equation shows that the larger the particle, the smaller will be the area change and the smaller the surface energy term in eq. 2.3.7. At local equilibrium, c_i becomes smaller as the particle becomes larger.

In a dispersion containing both large and small particles, the solute concentration is higher close to the small ones, and a diffusional flow of solute moves from the small particles to the large ones. As a result, the small particles decrease in size and dissolve and the large ones grow. This process is called Ostwald ripening. As we will discuss later, Ostwald ripening provides one important mechanism for destabilizing colloidal dispersions.

2.3.3 Surface Free Energies Oppose the Nucleation of a New Phase

It is a common observation that liquids can be heated somewhat above the equilibrium boiling point before vapor bubbles actually form. Similarly, a vapor can be brought to a temperature/pressure below the boiling point before liquid drops actually appear in the bulk phase, and a liquid can remain intact even when cooled to below the freezing point. The time stability of metastable states depends on the kinetics of the different processes leading to the formation of the stable new phase. We say that the new phase must nucleate and that the surface free energy of the new interface makes nucleation a sluggish process.

For example, let us consider the nucleation of a liquid phase from a vapor of pressure p. The free energy change of forming a liquid drop of radius R is

$$\Delta \mathbf{G} = -nkT \ln\left(\frac{p}{p_0}\right) + 4\pi R^2 \gamma$$

$$= -\frac{4}{3}\pi \frac{R^3}{V_M} kT \ln\left(\frac{p}{p_0}\right) + 4\pi R^2 \gamma \tag{2.3.11}$$

in analogy with eq. 2.3.5. When $p > p_0$ the first term on the right-hand side will make $\Delta \mathbf{G} < 0$ for sufficiently large R, but for sufficiently small R, the positive second term will dominate. Consequently, a maximum in $\Delta \mathbf{G}$ occurs at some intermediate value of R. Figure 2.10 illustrates the variation of $\Delta \mathbf{G}$ with R. At the maximum, p is related to the radius by

$$kT \ln \left(\frac{p}{p_0} \right) = \frac{2\gamma V_M}{R_c} \tag{2.3.12}$$

which is called the Kelvin equation where R_c is the Kelvin radius. The pressure p given by this equation is often interpreted as the equilibrium vapor pressure over a surface of curvature $1/R$ and surface tension γ. A better description is that p represents the maximum possible vapor pressure in the presence of drops of radius R_c before the bulk liquid phase is necessarily formed.

If we introduce a drop of radius R_1 into a vapor of a given vapor pressure p, what is the fate of the drop? If R_1 is smaller than the radius R_c, given by the Kelvin equation, we are on the left-hand side of the maximum in Figure 2.10 and vaporizing the liquid gains free energy. On the other hand, if $R_1 > R_c$, condensing more vapor on the drop gains free energy and leads to the formation of a bulk liquid phase.

For a solute in a liquid slightly above its saturation concentration c_s, a similar nucleation problem is at hand. If we were to repeat the derivation of the preceding paragraph, giving the solute concentration c the role of the vapor pressure p, and

Figure 2.10
The free energy of a water droplet in steam as a function of the radius calculated from eq. 2.3.11. The temperature is 0°C and the supersaturation $p/p_0 = 4$. Note that the free energy maximum corresponds to several tens of kT.

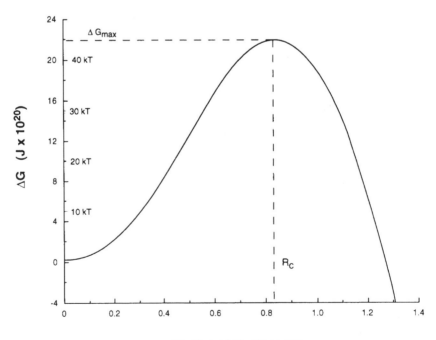

Radius of Nucleus nm

making c_s correspond to p_0, the resulting Kelvin equation would

$$kT \ln\left(\frac{c}{c_s}\right) = \frac{2\gamma V_s}{R_c} \qquad (2.3.13)$$

where γ equals the surface free energy solution-precipitate, and V_s is the solute volume. R_c is the particle radius that by thermodynamic necessity nucleates a bulk solid.

Similar nucleation problems arise when a pure liquid crystallizes or when a liquid is heated above its boiling point. In each case the thermodynamic driving force for forming the new bulk phase must balance the surface free energy of the small nuclei of the new phase. In the formation of vapor bubbles, pressure equilibrium provides an extra complication and the corresponding Kelvin equation contains more terms.

2.3.4 Combining the Kelvin Equation with a Kinetic Association Model Provides an Expression for the Rate of Homogeneous Nucleation

In practice, homogeneous nucleation is a very delicate process. It transforms the system from a homogeneous metastable phase to an equilibrium bulk phase. As an example, consider the condensation of a supersaturated vapor to a liquid. In the preceding section we used a macroscopic thermodynamic view to determine the critical drop size at which spontaneous growth occurs at a given vapor pressure and temperature.

To see how such a critical drop can form, we must adopt a more molecular view of the system. Molecules constantly collide, associate, and disassociate. This means that occasionally an aggregate can grow in size even when macroscopic thermodynamic arguments say that a decrease in size is more favorable.

We can analyze the aggregation process if we assume that aggregates/droplets A_n grow (and shrink) by addition (or subtraction) of only single molecules A, that is,

$$A_{n-1} + A \underset{k_n^-}{\overset{k_n^+}{\rightleftarrows}} A_n \qquad (2.3.14)$$

The ratio of the rate constants k_n^+/k_n^- is related to the equilibrium constant of reaction K_n (eq. 2.3.14)

$$\frac{k_n^+}{k_n^-} = K_n$$

which in turn depends on the free energy of forming an aggregate, n. We assume that we reach the supersaturated state by a rapid temperature jump, $T_2 \rightarrow T_1$, such that at the higher temperature, T_2, only unassociated molecules can be found in the vapor. At the lower temperature, T_1, an association starts that ultimately leads to the formation of a bulk liquid phase.

It is important to realize that in principle a single droplet with a radius $R \gtrsim R_c$ can act as a nucleus for the condensation of an entire system. In an extreme case a macroscopic change of the system could be triggered by a single molecular event!

The rate of forming aggregates with n monomers is

$$\frac{d[A_n]}{dt} = k_n^+[A_{n-1}][A] - k_n^-[A_n] - k_{n+1}^+[A_n][A] + k_{n+1}^-[A_{n+1}] \qquad (2.3.15)$$

and $[A_n](t)$ increases relative to the time of the temperature jump. The crucial step for the nucleation is the formation of the aggregate corresponding to the critical radius

$$R_c = \frac{2\gamma V_M}{kT \ln(p/p_0)} \qquad (2.3.16)$$

with an aggregation number

$$n_c = \frac{4\pi}{3} \frac{R_c^3}{V_M} \qquad (2.3.17)$$

The excess free energy of this aggregate is

$$\Delta G_c = \frac{4\pi}{3} R_c^2 \gamma = \frac{16\pi}{3} \frac{\gamma^3 V_M^2}{[kT \ln(p/p_0)]^2} \qquad (2.3.18)$$

combining eqs. 2.3.11 and 2.3.16.

As a first approximation for n_c in the rate equation (eq. 2.3.15), we can assume that only the first term contributes prior to nucleation, since $[A_n]$ and $[A_{n+1}]$ are negligible quantities. To calculate the number of nuclei $N(n_c)$, we must multiply by the volume of the sample, V, so that

$$\frac{d}{dt} N(n_c) \simeq k_{n_c}^+[A_{n_c-1}][A] \cdot V \qquad (2.3.19)$$

The integral

$$\int_0^{t_0} \frac{dN(n_c)}{dt} dt$$

counts the average number of critical aggregates formed up to time t_0. We can now define a critical nucleation time t_{nuc} as the average time required to form one critical nucleus. Thus,

$$1 = \int_0^{t_{\text{nuc}}} \frac{dN(n_c)}{dt} dt \simeq [A] V k_{n_c}^+ \int_0^{t_{\text{nuc}}} [A_{n_c-1}] dt \qquad (2.3.20)$$

If the critical radius R_c is large, $[A_{n_c-1}]$ will remain very small and t_{nuc} is long. In fact, it may be so long that in practice nucleation does not occur For a somewhat higher degree of supersaturation, t_{nuc} is long but accessible. Nucleation will occur after some time in one or a few places in the sample. For still higher degrees of supersaturation, t_{nuc} will be short compared to the characteristic mixing time in the sample.

In the absence of convection (stirring), diffusion dominates the molecular transport and the characteristic diffusion time becomes

$$t_D = \frac{V^{2/3}}{D} \qquad (2.3.21)$$

where D stands for the diffusion coefficient. When $t_{nuc} \ll t_D$, nucleation occurs throughout the sample with the number of nuclei given by

$$N(\text{nuclei}) \approx V(Dt_{nuc})^{-3/2} \qquad (2.3.22)$$

For a slow nucleation process we can obtain an estimate of t_{nuc} by assuming that aggregates with $n < n_{crit}$ rapidly establish a quasi-equilibrium. When this case applies, $[A_{n_c-1}]$ rapidly approaches a steady state value

$$[A_{n_c-1}] = [A] \exp\left(\frac{-\Delta G_c}{kT}\right) \qquad (2.3.23)$$

with ΔG_c as in eq. 2.3.18. (The difference between $\Delta G(n_c)$ and $\Delta G(n_c - 1)$ is negligible.) We can now solve eq. 2.3.20 to obtain the average nucleation time

$$
\begin{aligned}
t_{nuc} &= \frac{\exp\{(4\pi/3)R_c^2\gamma/kT\}}{V[A]^2 k_{n_c}^+} \\
&= \frac{\exp\{16\pi\gamma^3 V_M^2[3(kT)^3(\ln(p/p_0))^2]\}}{V[A]^2 k_{n_c}^+}
\end{aligned}
\qquad (2.3.24)
$$

We see that within this limit the critical nucleation time does depend on the macroscopic volume V of the sample. The larger the sample, the more likely the single event of forming a nucleus.

The exponent on the right-hand side of eq. 2.3.24 varies dramatically with the degree of supersaturation. An example is water vapor at 20°C where $p_0 = 17.5 \, \text{mm Hg}$. The molecular volume of liquid water is $3.0 \times 10^{-29} \, \text{m}^3$, and the surface tension $\gamma = 72 \, \text{mN/m}$. At 100% supersaturation, that is, $p/p_0 = 2$, the critical radius is

$$R_c(p/p_0 = 2) = 1.6 \, \text{nm}$$

and the exponential term adopts the large value of 6×10^{73}, making the estimated nucleation time exceedingly great.

At higher degrees of supersaturation the value of the exponential factor decreases rapidly. To obtain a quantitative number for t_{nuc}, we should specify the value of the association constant $k_{n_c}^+$. However, the analysis given above is not accurate enough to motivate a detailed estimate of the binary association constant. (In Chapter 8, we will see that this constant should be on the order of 10^{-16} m^3/s.) Instead, we quote the result of a more detailed analysis of the kinetic scheme of eq. 2.3.15 which yields

$$t_{nuc} = \frac{(kT)^2}{V \times V_M p^2} \left(\frac{\pi m}{2\gamma}\right)^{1/2} \exp\{16\pi\gamma^3 V_M^2/[3(kT)^3(\ln p/p_0)^2]\} \qquad (2.3.25)$$

assuming ideal gas behavior of the vapor phase. For water at 20°C this equation reduces to

$$t_{nuc} = \frac{4.4 \times 10^{-25}}{Vp^2} \exp\{85.1/(\ln p/p_0)^2\}(s) \qquad (2.3.26)$$

where sample volume V and partial pressure p should enter in SI units.

Problem: Calculate the nucleation time for water at 20°C ($p_0 = 2.3 \times 10^3$ Pa) when $p/p_0 = 3.3$ and $V = 1$ m^3.
Solution: According to eq. 2.3.26

$$t_{nuc} = \frac{4.4 \times 10^{-25}}{1 \times 3.3^2 \times 2.3^2 \times 10^6} \exp\{85.1/(\ln 3.3)^2\} = 0.04 \text{ s}$$

This implies 25 nuclei per V m^3 per second which is barely an observable nucleation rate.

These calculations show that homogeneous nucleation is a sluggish process, particularly when the surface (interface) free energy is large. When the nucleating phase is a solid, an additional complication occurs that was neglected in this discussion: the structure of a small cluster can be very different from that of the bulk crystal, making the effective γ depend strongly on cluster size.

The calculation suggests that phase changes typically occur far from the equilibrium transition. Sometimes this is indeed true, as you can see in a bottle of soda pop, which only slowly releases CO_2. In most practical situations, however, a new phase is nucleated heterogeneously. In a can of soda pop, for example, gas bubbles do come off at the surface and at the container wall. Furthermore, adding some grains of salt or sugar releases a burst of CO_2. Avoiding heterogeneous nucleation requires extremely careful removal of colloidal impurities. For example, a standard way to bring water down to temperatures at which it nucleates homogeneously to ice is to emulsify it, thus creating a large number of independent compartments. The odd comparment may contain a nucleating impurity, but this ice nucleus will not affect other compartments, except for an Ostwald ripening process.

2.4 Thermodynamic Equations That Include Surface Contributions Provide a Fundamental Basis for Characterizing Behavior of Colloidal Particles

In the preceding sections, we developed equations to describe surface tension and surface free energy, as well as the mechanical behavior of liquid systems. The Gibbs model gives us a workable way to conceptualize small-scale systems in thermodynamic terms.

2.4.1 The Gibbs Model Provides a Powerful Basis for Analyzing Surface Phenomena by Dividing a System into Two Bulk Phases and an Infinitesimally Thin Dividing Surface

The presence of a surface affects virtually all the thermodynamic parameters of a system. Gibbs developed a model for analyzing a system's thermodynamics by dividing the system into three parts: the volumes of the bulk phases V' and V'' plus the surface that separates them. We can apportion any extensive thermodynamic property of a system between these parts and write the energy ascribed to its surface, $\mathbf{U}_\sigma$ as

$$\mathbf{U}_\sigma = \mathbf{U} - u'V' - u''V'' \tag{2.4.1}$$

where $\mathbf{U}$ represents the total energy of the system and u' and u'' are the energies per unit volume in the two phases. Similar surface quantities can be written for the other thermodynamic quantities.

By dividing the total volume of the bulk phases into the precise quantities V' and V'', we construct an imaginary system that has the same thermodynamic properties as the real system, yet separates the two phases by an infinitesimally thin dividing surface. Although in a real system physical and thermodynamic properties change rapidly (but not discontinuously) at the interface, in this imaginary system they remain constant up to the surface. Replacing the real system with the model makes it easier to think of such extensive quantities as energy and volume in discrete segments rather than continuously varying quantities.

Molecular composition also changes across the interface. In a two-phase multicomponent system with volumes V' and V'', we can write, c_i' and c_i'' as the concentration of component i in each phase. Then the number of moles of component i in each phase becomes

$$n_i' = c_i'V' \quad \text{and} \quad n_i'' = c_i''V'' \tag{2.4.2}$$

The amount of component i in the interface is

$$n_i^\sigma = n_i - c_i'V' - c_i''V'' \tag{2.4.3}$$

where n_i is the total number of moles in the whole system. We can define a surface concentration or surface excess for component i as

$$\frac{n_i^\sigma}{A} = \Gamma_i \qquad (2.4.4)$$

Equation 2.4.3 makes it clear that n_i^σ (and thus Γ) can be either positive or negative.

Varying the location of the dividing surface by only a few angstroms can change the magnitude and sign of the value of Γ_i dramatically.

2.4.2 The Gibbs Adsorption Equation Relates Surface Excess to Surface Tension and the Chemical Potential of the Solute

We can develop thermodynamic relations for the surface variables in a way that is analogous to those used for bulk phases. For example, the differential of the surface internal energy is

$$dU^\sigma = T\,dS^\sigma + \gamma\,dA + \sum_i \mu_i(T)\,dn_i^\sigma \qquad (2.4.5)$$

where the bulk work term $p\,dV$ has been replaced with the surface work term $\gamma\,dA$. The same manipulations that lead to the relation $U = TS - pV + \Sigma\,\mu_i n_i$ for the bulk yield (see eq. 1.5.19)

$$U^\sigma = TS^\sigma + \gamma A + \sum_i \mu_i n_i^\sigma \qquad (2.4.6)$$

for the surface quantities. If we then differentiate eq. 2.4.6 and compare it with eq. 2.4.5, we obtain the Gibbs adsorption equation

$$S^\sigma\,dT + A\,d\gamma + \sum n_i^\sigma\,d\mu_i = 0 \qquad (2.4.7)$$

At a constant temperature, this gives the Gibbs adsorption isotherm

$$-d\gamma = \sum_i \frac{n_i^\sigma}{A}\,d\mu_i = \sum_i \Gamma_i\,d\mu_i \qquad (2.4.8)$$

which relates the change in surface tension, $d\gamma$, to changes in the chemical potential, $d\mu_i$, through the surface excess. This thermodynamic equation is generally valid, but in practical applications it is useful to judiciously specify the location of the dividing surface in order to arrive at a conceptual and algebraic simplification of eq. 2.4.8. Rather than keeping the general thermodynamic description, we consider an interface between a vapor and a binary solution so that eq. 2.4.8 reduces to

$$-d\gamma = (\Gamma_1\,d\mu_1 + \Gamma_2\,d\mu_2) \qquad (2.4.9)$$

In many solid systems interfaces between phases are sharply defined and concentration changes abruptly on an angstrom scale in slight contrast to liquid systems, where concentrations vary across an interfacial region that is typically a few molecular diameters thick, that is on the order of a nanometer. Figure 2.11A illustrates this variation by showing how the water concentration changes from 55 M in bulk liquid to 1×10^{-3} M in the vapor phase (20°C). Note that although the concentration of water changes dramatically, its chemical potential remains equal in the two bulk phases and also remains constant across the interface. This situation is analogous to the relation between concentration and chemical potential in external fields discussed in Section 1.5.2.

To determine the surface excess of water, Γ_{H_2O}, in moles per unit area, we first locate in the interface a dividing plane situated at a specific distance (Figure 2.11A), z_0, and define

$$\Delta c(z) = c(z) - c_{liq} \quad \text{for} \quad z < z_0$$
$$\Delta c(z) = c(z) - c_{vap} \quad \text{for} \quad z > z_0 \quad (2.4.10)$$

Then we obtain the surface excess, Γ_{H_2O}, as an integral

$$\Gamma = \int_{-\infty}^{+\infty} \Delta c(z)\, dz \quad (2.4.11)$$

or

$$\Gamma = \int_{z_0}^{+\infty} (c(z) - c_{vap})\, dz + \int_{-\infty}^{z_0} (c(z) - c_{liq})\, dz \quad (2.4.12)$$

across the interfacial region. The integral is finite since $\Delta c = 0$ well into the respective bulk phases. By projecting an interfacial volume onto an interfacial area, we have formulated the concept of a surface concentration or surface excess.

If we choose to locate z_0 as shown in Figure 2.11A, then $\Gamma_{H_2O} = 0$. The area below z_0 is defined by the second integral of eq. 2.4.12. This area is negative because $c(z) < c_{liq}$, while the area

Figure 2.11
Schematic showing how the concentrations of water, $c(z)_{H_2O}$, and SDS, $c(z)_{SDS}$, vary across the vapor–liquid interface. (A) In water the Gibbs dividing surface, z_0, is located so that the surface excess $\Gamma_{H_2O} = 0$ since the two areas are of equal and oposite sign. (B) In SDS this choice of z_0 leads to $\Gamma_{SDS} > 0$ since both areas make positive contributions. (C) A representation of the increase in SDS in interfacial zone due to its surface activity.

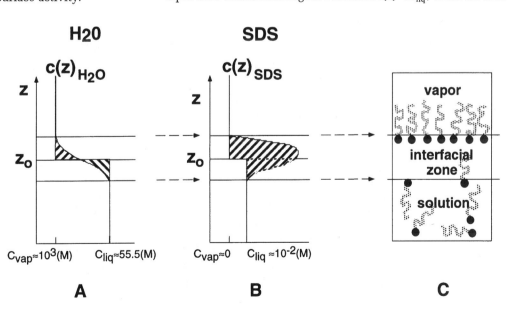

$z > z_0$, defined by the second integral, is positive because $c(z) > c_{vap}$. Since the two areas are equal but opposite in sign, Γ_{H_2O} is zero.

The value of this specific choice for the dividing surface is that when we apply equation 2.4.9 to a binary system, this reduces to

$$d\gamma/d\mu = -\Gamma_2 \qquad (2.4.13)$$

We can relate eq. 2.4.13 to accessible experimental quantities. Differentiating the chemical potential, assuming that it can be expressed in terms of concentration c_2 of the solute, $\mu_2 = \mu_2^\theta + RT \ln c_2$, at constant temperature yields $d\mu_2 = RT \, d \ln c_2$. Combining this with 2.4.13 yields

$$\Gamma_2 = -\frac{1}{RT} \frac{d\gamma}{d \ln c_2} \qquad (2.4.14)$$

For SDS solutions, the surface tension decreases strongly with increasing c_{SDS} from a value of $72 \, mJ/m^2$ for water to $40 \, mJ/m^2$ when $c_{SDS} = 8 \times 10^{-3} \, M$ (see Figure 2.11a). When we increase the concentration, the chemical potential increases, and the rapid decrease of γ for SDS reflects the interfacial adsorption as illustrated in Figure 2.11c. The concentration profile for SDS with the interfacial plane z_0 located so $\Gamma_{H_2O} = 0$ is shown in Figure 2.11b. Both of the integrals in eq. 2.4.12 are positive because c_{liq} and c_{vap} are both smaller than $c(z)$. When the bulk concentration of SDS is $10^{-2} \, M$, the surface concentration corresponds to approximately one molecule per $0.5 \, nm^2$ which in turn gives a volume concentration of around $1.8 \, M$ assuming a thickness of $2.2 \, nm$ for the surface layer. SDS concentrates at the interface in order to decrease contact between the nonpolar hydrocarbon chains and the water.

In NaCl solutions, the surface tension increases from $72 \, mJ/m^2$ in water to 74.4 at $1 \, M$ making Γ_{NaCl} negative. This increase corresponds to a surface concentration that is less than that of the bulk solutions. Ions are repelled from the liquid–vapor interface by the lower dielectric constant of the vapor. Unlike bulk concentrations, surface concentrations can be either positive or negative.

Problem: What is the value of Γ_{H_2O} when the reference place z_0 is chosen so that $\Gamma_{SDS} = 0$?
Solution: To equalize the areas defined by eq. 2.4.12, we must locate z_0 far into the vapor phase. Since we can choose to locate it wherever we want, this location is formally acceptable, but in practice such a choice could easily lead to conceptual confusion. In order to find the value of Γ_{H_2O}, we extrapolate the bulk concentration of water up to the reference plane and observe that Γ_{H_2O} is negative, as illustrated below.

2.4.3 The Langmuir Equation Describes Adsorption at Solid Interfaces Where We Cannot Measure Surface Tension Directly

Solubilization, detergency, and flotation are among the many applications that involve adsorption of surfactant molecules onto solid surfaces. Because changes that occur in surface tension as a result of composition changes at solid interfaces are not experimentally accessible, we cannot characterize the adsorption process in these situations by the Gibbs isotherm. Instead, we describe the behavior of such systems in terms of the Langmuir isotherm.

The Langmuir isotherm is conceptually simple and widely used to represent experimental data. It is most easily derived using a two-dimensional version of the lattice model introduced in Section 1.4. Assume that there are N independent binding sites on a surface in equilibrium with a dilute solution (or low pressure gas) of the adsorbant. Then we can follow eq. 1.4.8 and write the free energy of the surface as

$$\mathbf{G}_s = \mathbf{G}_s^\theta + N_2\mu_s^\theta + RT\{N_2 \ln X_2 + (N - N_2)\ln(1 - X_2)\} \tag{2.4.15}$$

where N_2 stands for the number of adsorbed solute molecules, N is the total number of sites, and $X_2 = N_2/N$. The chemical potential at the surface is thus

$$\left(\frac{\partial \mathbf{G}}{\partial N_2}\right) = \mu_2(s) = \mu_2^\theta(s) + RT\ln\left\{\frac{X_2}{1 - X_2}\right\} \tag{2.4.16}$$

In the bulk solution (b),

$$\mu_2(b) = \mu_2^\theta(b) + RT\ln c_2(b) \tag{2.4.17}$$

where $c_2(b)$ is the concentration of solute in the solution.

At equilibrium, the two chemical potentials are equal and Θ, the fraction of the surface site occupied by the solute molecules, becomes

$$\Theta = X_2 = \frac{Kc_2(b)}{Kc_2(b) + 1} \tag{2.4.18}$$

where K stands for the equilibrium constant,

$$RT\ln K = \{\mu_2^\theta(b) - \mu_2^\theta(s)\}$$

This equation is called the Langmuir adsorption isotherm.

At high solute concentrations or high surface affinities, K-large, $c_2(b)K \gg 1$, and $\Theta \simeq 1$, which means that the surface is saturated. At the other limit, $c_2(b)K \ll 1$ and Θ is proportional to $c_2(b)$.

Assumptions used to derive the Langmuir equation are rather stringent. For example, ideal behavior on the surface implies a homogeneous surface that lacks dislocations or any other structural nonidealities that might induce preferred adsorption. Interactions between adsorbed molecules must be ignored, and only monolayer formation can be included. Although we almost never realize these conditions in practice, the Langmuir isotherm provides a conceptual basis for thinking about surface adsorption as well as a basis for modeling the adsorption process. Furthermore, the equation makes a convenient form for fitting data because it accommodates many situations in which the Langmuir model assumptions do not strictly apply.

2.5 Monolayers Are Two-Dimensional Self-Organizing Systems

We have now discussed the basic principles of surface chemistry needed to understand monolayers. Sections 2.5.1 and 2.5.2 deal with two limiting cases: molecules in equilibrium with those in the bulk solution and with those that are insoluble in bulk.

2.5.1 Monolayers Formed by Soluble Amphiphiles Can Be Characterized by Surface Tension Measurements Using the Gibbs Adsorption Isotherm

Amphiphilic molecules orient themselves at the interface of aqueous solutions to form monolayers, as we saw in the discussion of SDS in Chapter 1. The phase in contact with water can be air, saturated vapor, or an immiscible organic liquid.

We can calculate the free energy of transferring an amphiphile from bulk solution to the air–water interface from

$$\mu_s^\theta - \mu_w^\theta = \boldsymbol{R}T \ln\left(\frac{X_w}{X_s}\right) \tag{2.5.1}$$

assuming ideal behavior in both solution and interface.

In 1917 Langmuir analyzed data for several homologous families of amphiphiles. He assumed that adding surfactant to lower γ enables us to measure the relative values of μ_s directly from the values of X_w that produce equal lowering of the surface tension in each solution.

Langmuir showed that all the results for a series of fatty acids, esters, and alcohols with a varying number of carbon atoms μ_s could be expressed by

$$\mu_s^\theta - \mu_w^\theta = \text{constant} - 2.6n_c \text{ kJ/mol} \tag{2.5.2}$$

where the constant depends only on the head group. An analysis for the transfer of similar compounds from water to the neat phase gives $\mu_s^\theta - \mu_w^\theta = -3.4$ to -3.7 kJ/mol per methylene group (see Section 1.7).

We can evaluate the area per surfactant at the air–water interface, $a_0 = (N_{Av}\Gamma_2)^{-1}$, from the change in surface tension with surfactant concentration by using the Gibbs adsorption equation

$$-d\gamma = \Gamma_2 \, d\mu_2 \qquad (2.5.3)$$

For ionic surfactants like SDS, the chemical potential is

$$\mu_2 = \mu_2^\theta + RT \ln c_{DS^-} + RT \ln c_{Na^+} \qquad (2.5.4)$$

below the CMC, where we can assume, ideal behavior. Thus,

$$-d\gamma = RT\Gamma(d \ln c_{DS^-} + d \ln c_{Na^+}) \qquad (2.5.5)$$

and electroneutrality requires that $c_{DS^-} = c_{Na^+}$ if no electrolyte is added. This in turn gives

$$\frac{d\gamma}{d \ln c_{DS}} = -2RT\Gamma \qquad (2.5.6)$$

Data from the γ versus $\log c$ shown in Figure 1.1 give $a_0 = 52 \, \text{Å}^2$ for SDS using eq. 2.5.6.

If excess electrolyte is present (i.e., sufficient salt to make electrolyte concentration constant, adding SDS only affects the concentration, then eq. 2.5.5 becomes

$$\frac{d\gamma}{d \ln c_{DS^-}} = -RT\Gamma \qquad (2.5.7)$$

and in 0.1 M NaCl, $a_0 = 38 \, \text{Å}^2$ for SDS. The change in a_0 in these two situations reflects a decrease in the electrostatic head group repulsion.

Since an equilibrium exists between the surfactant in solution and at the interface, we cannot vary the interfacial properties independently. For example, decreasing the interfacial area simply causes the surfactant to desorb back into the solution until the chemical potentials of surfactant at the interface and in the solution equalize again.

2.5.2 Monolayers Formed by Insoluble Amphiphiles Behave as Separate Phases and Are More Readily Characterized Using the Langmuir Balance

The usual way to prepare monolayers of insoluble amphiphiles is to dissolve them in a volatile organic solvent and deposit drops of the solution onto the air–water interface. Initially the spreading coefficient is positive, but when the solvent evaporates, a layer of amphiphile forms. This technique permits quantitative deposit of very small amounts of solute on the surface. Since the amphiphiles are insoluble in water and of low volatility, they remained confined to the surface.

We can express the lowering of the surface tension of the interface caused by the presence of adsorbed molecules as

$$\Pi_s = \gamma^o - \gamma \qquad (2.5.8)$$

where γ^o is the surface tension with no amphiphile present and γ its value with adsorbed amphiphile. Thus, Π_s represents the surface pressure needed to prevent the film from spreading.

Using the Langmuir balance, we can directly measure Π_s as a function of vapor–liquid interfacial area. As Figure 2.12 shows, the balance employs a float, which separates the film from the clean solvent surface, placed in a trough made of brass, stainless steel, glass, or Teflon that is typically 2–3 feet long and 1 foot wide. The trough must be filled so that the water stands above its rim. Coating the rim of non-Teflon troughs with paraffin makes it possible to obtain this high contact angle.

A head, located about one-third of the way from one end of the apparatus, supports the float suspension system. The float itself is constructed from a thin strip of Teflon, paraffin-coated mica, or platinum and rides freely on the fluid surface. Thin gold or plastic foil barriers seal gaps between the ends of the float and the trough. These barriers permit the float to move in response to differences in pressure, Π_s, on each side of the float, but prevent the film from escaping. A mirror attached to the stiff wire that connects the float to a torsion wire monitors the float position. Any motion of the float rotates the mirror and displaces the position of a spot of reflected light.

Before the film is deposited, the entire surface can be cleaned by sweeping the movable barriers back and forth across the trough. Amphiphile dissolved in a volatile solvent is then carefully pipetted onto the aqueous surface. After the organic solvent evaporates, Π_s can be measured as a function of A by moving the sweep toward the head.

Langmuir balances have become highly automated. Computer-controlled operations, including surface cleaning, application of amphiphiles, measurement of Π_s versus A, and recleaning the surface for subsequent measurements, make it possible to obtain 50 or more Π_s versus A areas for different materials without human involvement.

To prevent contamination of the water, organic solvent, or any chemicals used in the experiment, including the amphi-

Figure 2.12
A Langmuir film balance: (1) trough, (2) sweep, (3) float, (4) mirror, and (5) main torsion wire. The deflection of the float is measured optically using the mirror. (W.A. Adamson, *Physical Chemistry of Surfaces*, 5th ed., Wiley-Interscience, New York, 1990, p. 113.)

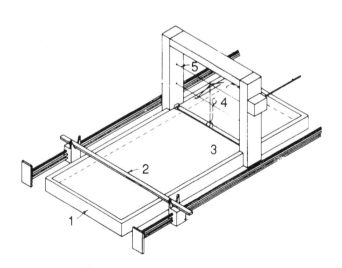

philes, we usually place the Langmuir balance inside an air thermostat. Extreme precautions against contamination are necessary because a human fingerprint contains enough surface-active material to form numerous monolayers in a typical Langmuir balance.

2.6 Π_s Versus a_0 Surface Isotherms for Monolayers Containing Insoluble Amphiphiles Parallel P Versus V Isotherms for Bulk Systems

2.6.1 The Insoluble Monolayer Displays Several Aggregation States

Figure 2.13 displays a schematic of features of pressure–area isotherms observed for monolayers containing a single amphiphile. At very low pressures, all monolayers exhibit gaseous behavior, and as Π_s approaches zero, we can apply a two-dimensional equivalent of the ideal gas law,

$$\Pi_s A = n\boldsymbol{R}T \tag{2.6.1}$$

or in a version more useful for our purposes

$$\Pi_s a_0 = \boldsymbol{k}T \tag{2.6.2}$$

where $a_0 = A/(nN_{Av})$ signifies the area each molecule occupies. In this so-called gaseous region of the isotherm, the hydrocarbon chains lie flat on the surface.

Figure 2.13
Composite two-dimensional pressure (Π_s)–area (a_0) isotherm, which includes a wide assortment of gaslike, liquidlike, and solidlike monolayer phenomena. To permit all features to be included on one set of coordinates, the scale of the figure is not uniform. Π_v is the highest pressure of the gaslike film. Π_c is the collapse pressure of the film. (P.C. Hiemenz, *Principles of Colloid and Surface Chemistry*, 2nd ed., Dekker, New York, 1986, p. 364.)

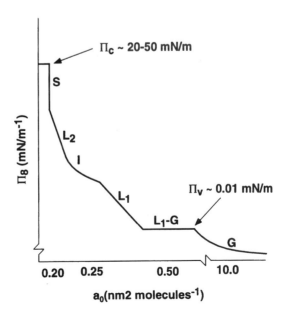

These equations are valid when the dominant contribution to the free energy comes from the entropy of the individual molecule confined to an area A in the same way as for an ideal gas whose only free energy contribution is due to the entropy of a gas molecule confined to a volume V.

Plots of the compressibility factor Z (defined as $a_0 \Pi_s / kT$ in the case of a surface) easily reveal deviations from ideal conditions. Figure 2.14 shows Z plots for a homologous series of carboxylic acids at 25°C. Negative deviations occur at lower pressure, just as they do for analogous three-dimensional plots. Because of increasing attraction between molecules as their size increases, these deviations become more pronounced as the alkyl chain lengthens. At higher pressures, $Z > 1$ reflects an excluded area analogous to the excluded volume in three dimensions.

At the other end of the isotherm, we observe high Π_s, low a_0, and solidlike behavior. In this case, the hydrocarbon chains are oriented vertically and packed together. The film is relatively incompressible. For alcohols and esters, $a_0 \approx 0.19 \, \text{nm}^2$, which corresponds to the close-packed cross section we obtain for bulk, solid hydrocarbons by x-ray scattering. Compressing the solid region further leads to collapse of the film.

A monolayer's transition from gas to solid is more complex than for a corresponding bulk phase. We can observe two liquid phases, L_1 (the liquid expanded state) and L_2 (the liquid condensed state), with a transition region I between them.

This behavior originates as follows. The transition from G to L_1 occurs as it does in a gas–liquid transition with significant compression at constant pressure, Π_v. Generally, the transition

Figure 2.14
Plots of $Z = \Pi_s a_0 / kT$ versus Π_s for n-Alkyl carboxylic acids: (1) C_4, (2) C_5, (3) C_6, (4) C_8, (5) C_{10}, and (6) C_{12}. Ideal behavior corresponds to $Z = 1$. (P.C. Hiemenz, *Principles of Colloid and Surface Chemistry*, 2nd ed., Dekker, New York, 1986, p. 367.)

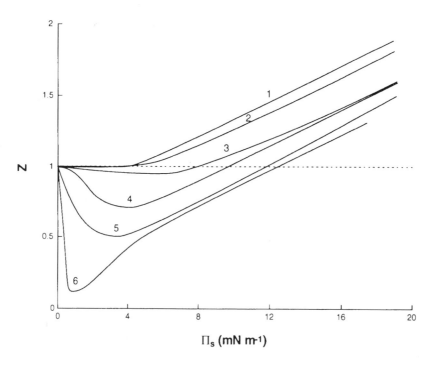

$\Pi_s \ (\text{mN m}^{-1})$

pressure is quite low. For tetradecanol, for example, $\Pi_v = 1.1 \times 10^{-4}$ N/m at 15°C.

For single-chain amphiphiles with saturated unbranched hydrocarbon chains, $a_0(L_1)$ falls in the range 0.3 to 0.5 nm^2. Although this value is considerably less than we might expect if the entire hydrocarbon chain were free to move, it is notably larger than it would be for close packing of the chain. In the liquid expanded state the molecules have self-assembled into a state analogous to the one found in liquid crystalline bilayers. They diffuse rapidly along the surface and there is a substantial conformational disorder in the alkyl chain. As the pressure increases and the film compresses there is a break in the Π_s versus a_0 curve. In this transition region, the isotherm maintains a finite positive slope rather than the horizontal line associated with first-order phase transitions. Although some sort of cooperative interaction clearly takes place the physics underlying the I region has not been explained in a satisfactory way.

Further compression of the monolayer forms the condensed liquid phase, L_2. Both the L_2 and solid regions show lower compressibility than the L_1 region due to strong intermolecular forces. In Chapter 10 we will come back to the complex equilibria occurring in the liquid condensed phase.

2.6.2 Fluorescence Microscopy Can Visualize the Aggregation State of Monolayers Directly

The ability to visualize amphiphilic microstructures deposited on the surface of the Langmuir balance is a tremendous aid in understanding monolayer systems. Fluorescence microscopy can detect molecules at exceedingly low concentration ($\sim 10^{-12}$ mol/L) and permits us to visualize phase transformations.

Images showing the behavior of dimyristoylphosphatidic acid (DMPA) monolayers containing a 1% fluorescence dye L-α-dipalmitoyl-nitrobenzoxadiazol-phosphatidylethanolamine (DP-NBD-PE) provide one example of the microscope's effectiveness. At this concentration, the dye barely affects the diagrams of pressure versus area.

Figure 2.15 shows the fluorescence images corresponding to various $\Pi_s - a_0$ values marked on the isotherms for DMPA monolayers. At Π_c and an area of ~ 80 Å^2 a solidlike phase appears as well as a liquidlike one. The fluorescent probe is not soluble in the solidlike phase, which appears dark in the microscopic view. The pictures show that as the average area per molecule decreases, the dark regions increase. However, macroscopic phase separation does not occur, and the Π_s versus A curve is not perfectly flat.

The solid patches repel one another, resulting in an ordered structure. The out-of-plane components of the molecular dipoles cause the repulsion. These dipolar components also provide a repulsive component to the energy of a single domain, favoring several small ones relative to one large patch. The competition between the electrostatic effect and the line tension

Figure 2.15
(A) Pressure–area isotherm ($T = 19°C$) of DMPA with 1 mol % of the dye DP-NBD-PE and 10^{-4} M NaCl, 5×10^{-5} M ethylenediamine tetraacetic acid in the water subphase.
(B) Corresponding fluorescence micrographs at the marked values on the isotherm. The light areas indicate the fluid phase, the dark areas the gel phase. (C.A. Helm, H. Möhwald, K. Kjaer, and J. Als-Nielsen, *Biophys. J.* **52**, 381 (1987).)

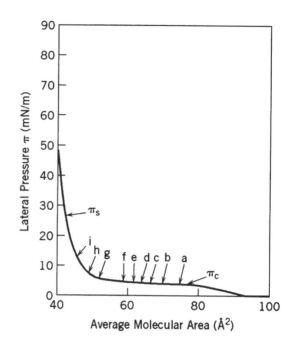

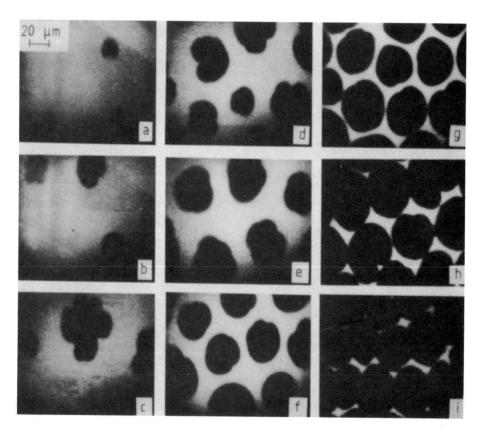

promoting domain growth leads to an optimal domain size typically in the micron range. This competition can lead to a number of fascinating patterns in addition to those shown in Figure 2.15.

2.6.3 The Langmuir–Blodgett Technique Provides a Way to Deposit Monolayers or Multilayers onto Solid Surfaces

By modifying the Langmuir balance, we can deposit monolayers on a solid substrate to create mono- or multilayer films. Modifications include a dipping device to lower or raise the substrate through the monolayer, an automated movable barrier, which moves during the deposition process so as to maintain a control value of Π_s, and a surface pressure sensor, which controls the movable barrier.

Successively dipping a plate in and out of a liquid covered by a monolayer builds up multilayers; a Langmuir–Blodgett film may contain multiple layers as shown in Figure 2.16.

The Langmuir–Blodgett technique provides a way to construct organized systems of molecules one monolayer at a time. Potential applications of the technique include electron beam resists for micro- and macrolithography, insulating films in semiconductor devices, nonlinear optical elements, coatings for communication devices, biosensors, and highly selective membranes for biotechnology. These application opportunities have stimulated considerable research interest in the basic physics and chemistry of the assembly process, synthesis of appropriate amphiphilic molecules, and assessment of the character and durability of the resulting films. A major practical problem in these applications is the long-term stability of the film.

Figure 2.16
Deposition of multiple monolayers by the Langmuir–Blodgett technique. (A. Weissberger and B.W. Rossiter, eds., *Physical Methods of Chemistry*, Vol. 1, Part IIIB, Wiley-Interscience, New York, 1971, p. 595.)

2.7 Scanning Tunneling and Atomic Force Microscopies Permit Imaging of Molecular Structures at Solid Interfaces

The organization and molecular structure of amphiphilic molecules and polymers adsorbed onto solid surfaces plays key roles in determining the properties of many colloidal systems.

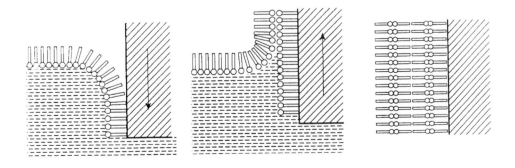

Figure 2.17
Schematic diagram of a scanning tunneling microscope (STM). To begin a measurement, the scanning tip is brought to within 1 nm of the surface by activating a stepping motor (not shown) that lowers the entire piezoelectric tube (coarse adjustment). Increasing the voltage on the piezoelectric tube moves it to a distance (fine adjustment) at which a tunneling current can be detected. Varying the voltage to a different electrodes attached in- and outside the piezo tube make it possible to move the tip in all directions. To create an image, the tip is rastered across the sample surface in the x and y directions while a feedback loop adjusts the voltage to the tube to maintain constant current. Recording the voltages as a function of position creates a three-dimensional image.

Scanning tunneling microscopy (STM) and atomic force microscopy (AFM) can provide three-dimensional images of surfaces that approach atomic or molecular resolution.

Figure 2.17 is a schematic diagram demonstrating the operation of an STM. A sharp metal tip attached to a cylindrical piezoelectric tube is brought to within approximately 1 nm of the surface of a conducting material. When a bias voltage between 2 mV and 2 V is applied, electrons tunnel between the tip and the surface.

Varying the voltage across the piezoelectric tube enables the tip to scan over the surface in the x and y directions, and the tunneling current is measured and recorded. As the tip approaches a high (or low) spot on the surface, the tunneling current increases (or decreases) in response to the change in separation between the tip and the surface. An electronic feedback network adjusts the vertical position of the tip to maintain a constant tunneling current. Plotting a contour map of tip height z versus lateral position x and y produces a three-dimensional image of the scanned surface.

A useful approximation for the tunneling current I_i at constant bias voltage is

$$I_i \sim \exp(-2\kappa_e h) \qquad (2.7.1)$$

where h stands for the separation between the tip and the sample and (κ_e) represents the decay of the length of the wave function of the electron, which depends on the work functions of the tip and the surface. With a typical work function of 4 eV, $(\kappa_e) \approx 10\ nm^{-1}$, the tunneling current changes by an order of magnitude when the separation h varies by 0.2 nm. If the current is kept constant to within 2%, then h remains constant

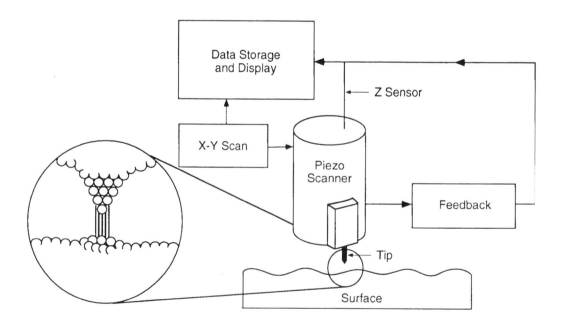

to within 10^{-3} nm. STM resolution approaches 0.2 nm in the lateral dimension and 10^{-3} nm in the vertical direction. When interpreting molecular images from heterogeneous surfaces, the approximation given by eq. 2.7.1 is often inadequate and more sophisticated analysis is required.

The tunneling tip can be made by simply cutting a platinum wire at an oblique angle with a pair of scissors. Because the tunneling current is extremely sensitive to distance, the atom that protrudes the farthest from the others is the one that becomes involved in the tunneling process. It is this property of tunneling that leads to the high resolution of STM images.

Figure 2.18 shows an STM image of a graphite surface in which individual atoms and surface defects are visualized. The cross-sectional profile illustrates changes in surface height arising from individual molecules and defects. Figure 2.19a shows clusters of gold molecules deposited onto graphite by thermal evaporation. Clusters can be seen to follow defects in the graphite surface. Figure 2.19b shows part of a single gold cluster. The cross-sectional profile directly measures individual step heights and widths. STM images of the top of the gold plateaus permit visualization of individual gold atoms. By generating a series of samples with increasing surface coverage, we can trace the growth of a gold film from small isolated clusters to a continuous film and can image surface defects with atomic resolution.

Figure 2.20 shows a schematic of the apparatus used in atomic force microscopy. In AFM measurements we replace the STM tip by one made of diamond or silicon nitride (Si_2N_4) mounted onto a mirrored cantilever spring. When the tip is scanned over the surface, we can detect its vertical movement by reflecting a laser beam off the back of the spring and onto a position-sensitive detector. The signal from the detector goes to

Figure 2.18
STM image of graphite showing individual carbon atoms. The cross profile indicated by the line across the image provides a measure of height versus distance (distances in nm).

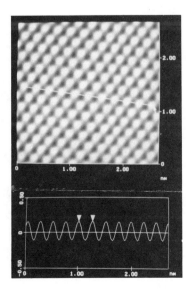

Figure 2.19
(a) STM images of
thermally deposited gold
islands on graphite. Defects
in the graphite surface
dictate the orientation of
the islands. (b) STM image
of part of a gold island.
The vertical distance
between the arrows is
8.4 nm. (c) When the image
is replotted and truncated,
the 0.35 nm steps that
comprise the island
become clearly evident.
The height of the step is the
vertical distance between
the arrows. (L. Stong, D.F.
Evans, and W. Gladfelter,
Langmuir **7**, 442 (1991).)

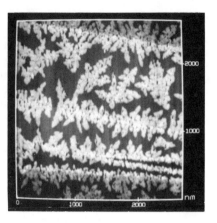

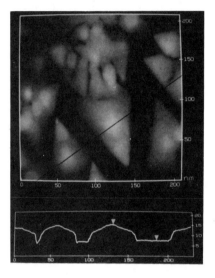

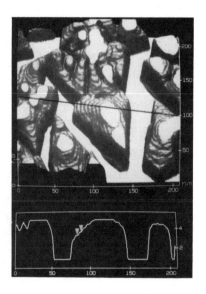

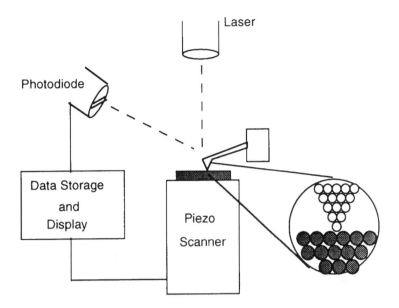

Laser

Photodiode

Data Storage
and
Display

Piezo

Scanner

Figure 2.20
Schematic diagram of an atomic force microscope (AFM). To begin an AFM measurement, the diamond or Si_3N_4 tip attached to a cantilever spring is brought into contact with the sample. Reflecting a laser light off the back of the spring and onto a position-sensitive detector captures any deflection of the tip as it moves across the surface. A feedback loop operated via the piezoelectric tube moves the sample in the vertical direction to maintain a constant force between the tip and the sample. Recording the signal from the feedback loop as a function of position creates a three-dimensional image of the surface.

an electronic feedback network that raises or lowers the sample so as to maintain constant force. A major advantage of the AFM over the STM is that the former can image nonconducting surfaces, although often at lower resolution.

Figure 2.21 shows an atomic force microscope (AFM) image of the individual methyl groups contained in a monolayer of $(C_{20}H_{41})_2N(CH_3)_2$ cationic surfactant adsorbed onto a negatively charged surface.

Lateral force microscopy (LFM) provides a way to differentiate between heterogeneous regions on a surface. As an AFM tip moves across the surface, a friction force is exerted on the tip by the surface. This force in turn generates a torque on the cantilever spring, and the resulting twist of the spring deflects the laser beam horizontally onto the photodetector. By replacing the two-position photodetector with a four-position detector, we can produce AFM and LFM images simultaneously.

We can use the AFM/LFM technique to follow the nucleation and growth of dewetting processes that lead to the rupture of thin liquid films on a solid surface and thus illustrate some of the concepts developed in this chapter. Simple liquids, like water, usually exhibit an extremely short time scale for dewetting as a consequence of the low viscosity of the liquid. The transformation of a continuous film of water into droplets on a greasy window provides an example. However, by heating a thin polystyrene film above its melting point, we can show the dewetting process due to the polymer's high viscosity. First we produce a thin (~ 20 nm), continuous, and smooth polystyrene film by spin-coating a polymer solution onto a silicon wafer and then drying in a vacuum oven to remove the solvent. Heating the polymer film above its glass transition temperature of 105°C initiates the dewetting process. Removing the wafer from the oven solidifies the liquid polymer and quenches the dewetting

Figure 2.21
AFM image showing the individual methyl groups of a monolayer containing $(C_{20}H_{41})_2N(CH_3)_2$ ions adsorbed onto a negatively charged mica surface. The monolayer extends uniformly for micrometers over the mica surface. Images obtained in aqueous solution and in air are very similar for this system. (Y.-H. Tsao, S.X. Yang, D.F. Evans, and H. Wennerström, *Langmuir* **7**, 3154 (1991).)

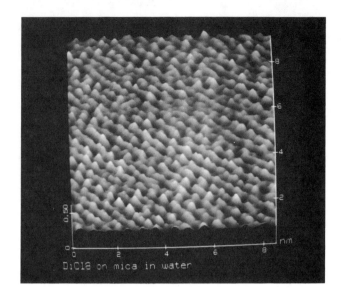

process. By repeating this cycle of heating and quenching, and imaging the same region of the film in the AFM between cycles, we can follow the nucleation and growth processes.

We can identify two nucleation processes:

1. Homogeneous nucleation occurs when surface waves on the liquid transform under the influence of attractive forces into indentations that move down into the film as illustrated in Figure 2.22. When indentations reach

Figure 2.22
AFM section views of a growing indentation in the surface of a 22 nm polystyrene film after (a) 20 minutes and, (b) after 70 minutes. Markers show relative dimensions of the indentations. The indent in (a) is 0.76 μm wide and 4.3 nm deep. The indent in (b) is 1.0 μm wide and 8.6 nm deep. Annealing temperature was 144°.

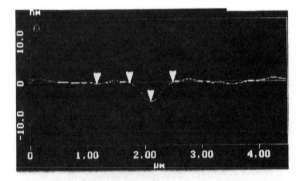

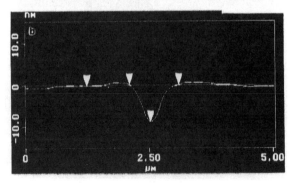

Figure 2.23
AFM image of a 1.3 μm
hole nucleated by a
0.37 μm particle. Vertical
distance between markers
is 22.8 nm.

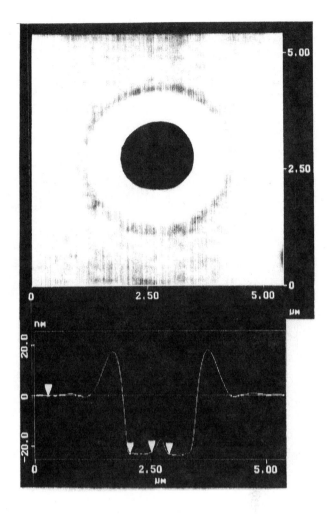

the substrate, they form a hole that grows gradually
outward. AFM images permit us to follow the growth of
indentations, while LFM allows us to ascertain that
the surface of the hole is the SiO_2 substrate.

2. Heterogeneous nucleation occurs when particles initially
located at the air–polymer surface sink into the polymer
film under the influence of attractive forces and gravity
as shown in Figure 2.23. Upon reaching the SiO_2
substrate, the individual particles attach to the silicone
surface and polymer film dewets radically outward from
each attachment site. We can increase the ratio of
heterogeneous to homogeneous nucleation sites by
contaminating the polymer surface with additional
particles before heating. Growth occurs when individual holes in the polymer film expand radially
until material from adjacent holes join together
(Figure 2.24) and eventually grow polyhedral structures
such as those shown in Figure 2.25. Further heating
causes these polyhedra to break up and form drops on the
surface.

Figure 2.24
AFM image of 0.9 μm holes in a 10.5 nm thick polystyrene film. No nucleation sites are visible in the center of the holes.

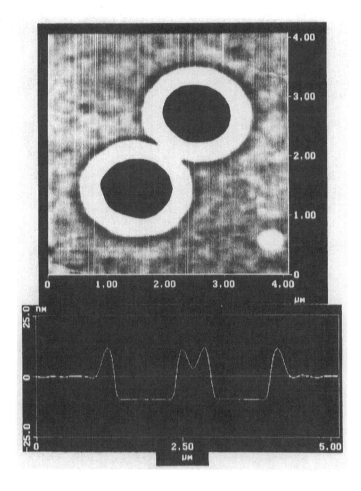

Exercises

2.1. Two liquids, A and B, phase separate when mixed into one phase consisting of 87.5% A and 12.5% B and one phase of 12.5% A and 87.5% B. Use the regular solution model to estimate the interfacial free energy. $T = 278$ K, the interaction parameter w is 6 kJ/mol, and the area per molecule at the interface is 3×10^{-19} m^2 and $z_b = 6$.

2.2. Two soap bubbles blown from the same solution have radii R_1 and R_2. The bubbles are cautiously joined by a thin tube. What is the outcome if one can avoid bursting the bubbles?

2.3. An air bubble of radius 1×10^{-6} m is generated in water with a syringe. The bubble is gently transferred to a solid hydrophobic surface. Calculate the contact angle. Is the volume of the bubble significantly changed upon adsorbing on the surface? Use $\gamma_{H_2Ov} = 72$, $\gamma_{S.H_2O} = 50$, and $\gamma_{Sv} = 20$ mJ/m^2.

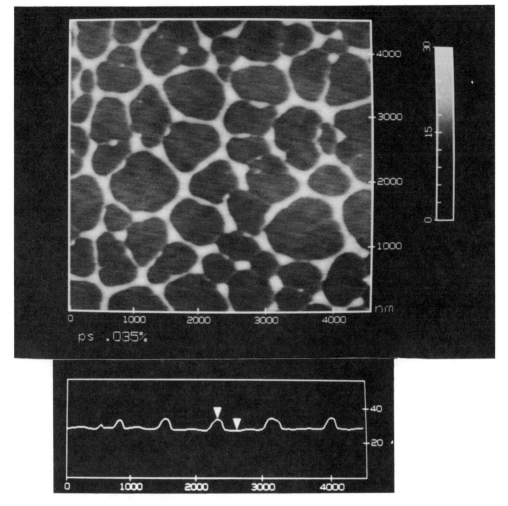

Figure 2.25
AFM image of a dewetted
polystyrene film on silicon
illustrating the formation
of a weblike (Voronoi
tessellation) pattern. Lower
panel shows a linear trace
across the surface showing
the heights of the
polystyrene ridges. (T.G.
Stange, R. Mathew, D.F.
Evans and W.A.
Hendrickson, *Langmuir* **7**,
922 (1992).)

2.4. What is the critical drop size for nucleating liquid water at
$p = 1$ atm., and (a) $T = 99.99°C$, (b) $T = 99°C$, (c)
$T = 90°C$? Between 90 and 100°C the surface tension of
water varies approximately linearly from 60.8 to
58.9 10^{-3} J/m^2.

2.5. You have a metastable water vapor at 20°C and 50 mm Hg.
Estimate the homogeneous nucleation rate per volume.

2.6. The Langmuir model for adsorption isotherms can be
extended to allow for interactions between adsorbed
molecules. Using the same arguments as in the regular
solution theory, an extra term, $N_2 X_2 w_{22}$, contributes
to the free energy. Derive the adsorption isotherm: that is,
X_2 as a function of $c_2(b)$.

2.7. The table shows the measured surface tension of aqueous
solutions of dodecyldimethylammonium chloride at
293 K for (a) no added salt and (b) 0.20 M NaCl solution.

No Salt		0.20 M NaCl	
C (M)	γ (mJ/m^2)	C (M)	γ (mJ/m^2)
1×10^{-5}	72	1×10^{-5}	72
8×10^{-5}	71.5	1.5×10^{-5}	70
3×10^{-4}	71	5×10^{-5}	66
1.0×10^{-3}	68	1.6×10^{-4}	58
2.0×10^{-3}	64	4.0×10^{-4}	52
3.5×10^{-3}	58	9.0×10^{-4}	46
5.6×10^{-3}	52	1.5×10^{-3}	40
9.5×10^{-3}	45	2.5×10^{-3}	35
1.1×10^{-2}	41	3.9×10^{-3}	34
1.4×10^{-2}	37.5	5.0×10^{-3}	34
2.0×10^{-2}	37.5	4.0×10^{-2}	34
4.0×10^{-2}	37.5		

Why does the surface tension go to a constant value at high concentration? Give an answer both with respect to molecular events and in relation to the Gibbs adsorption equation.

2.8. Use the data of Exercise 2.7 to estimate the surface excess and area per molecule at $c = 1 \times 10^{-3}$ M and $c = 1.4 \times 10^{-2}$ M for the salt-free case.

2.9 Use the data of Exercise 2.7 to determine the surface excess and the area per molecule at $c = 1 \times 10^{-3}$ and $c = 2 \times 10^{-2}$ M for the case with $c_{NaCl} = 0.2$ M.

2.10. The table gives the measured surface pressure of a dipalmitoylphosphatidylcholine film at the n-heptane–water (0.1 M NaCl) interface for a range of surface concentrations. Verify that $\Pi_s = kT\Gamma$ at low surface coverage. Determine also the second surface virial coefficient defined analogously to B_2 in eq. 1.5.13 (20°C).

$10^{-17} \times \Gamma$ (m^{-2})	$10^3 \Pi_s$ (J/m^2)
2	1.02
4	2.97
6	5.86
8	10.0
10	14.8
12	21.6

2.11. Taking the surface tension to be 58 mN/m at 100°C, determine the apparent boiling point of water if boiling means the formation of bubbles of vapor 1.15×10^{-6} m in diameter. At 100°C the heat of vaporization of water is 42 kJ/mol and the density is 958 kg/m^3. On the basis of your result, provide a short physical explanation of the "bumping" that sometimes occurs when pure water boils.

2.12. The surface tensions of aqueous solutions of KI have the following values at 20°C.

KI (mol/L)	0.1	0.5	1.0	2.0
Activity, $a_{+/-}$	0.0917	0.444	0.897	1.824
Surface tension, γ (mN/m)	72.59	72.50	71.35	69.95

Using the Gibbs adsorption equation, calculate the surface excess concentration Γ_{KI} in molecules per 1000 Å^2 for a 0.8 molar concentration of KI. Surface tension of pure water at 20°C is $\gamma_0 = 72.83$ mN/m.

2.13. Surface tensions of solutions of sodium dodecyl sulfate (SDS) in 0.1 NaCl at 20°C are as follows:

SDS (mM)	2.0	1.0	0.5	0.2	0.1	0.04	0.02	0.01
γ (mN/m)	37.90	38.26	45.28	53.87	60.04	66.13	69.31	71.44

The surface tension of a 0.1 M NaCl solution is 72.93 mN/m.
(a) Calculate Γ_{SDS} as a function of concentration by fitting the data to the Gibbs adsorption equation.
(b) Use the fit in (a) to calculate the surface pressure versus area curve for a monomolecular film of SDS over a 0.1 M NaCl solution.

2.14. When 5.19×10^{-5} g of palmitic acid $(C_{15}H_{31}COOH)$ in the form of a dilute solution in benzene is spread on the surface of water, it can be compressed to an area of 265 cm^2 when a condensed film is formed. Calculate the area (Å^2) occupied by a single molecule in the closely packed layer.

2.15. H. B. Bull [*J. Am. Chem. Soc.* **65**, 4 (1945)] determined the pressure–area curve for egg albumin using a film balance. The following data were obtained at 25°C.

Π_s (mN/m)	A(m^2/g)
0.200	275
0.300	178
0.500	97
0.600	76

Using a graphical plot, calculate the molecular weight of egg albumin.

2.16. Given the following surface and interfacial tensions, at 20°C, work out answers to (a)–(c).

Interface	γ(mN/m)	Interface	γ(mN/m)
Air–water	72	Mercury–water	375
Air–octyl alcohol	28	Mercury–octyl alcohol	348
Air–hexane	18	Mercury–hexane	378
Air–mercury	476	Water–octyl alcohol	9
		Water–hexane	50

(a) Will octyl alcohol spread at a mercury–water interface?

(b) If it were to spread, how would the polar group in the alcohol be oriented at the water and at the mercury interface? Explain.

(c) Will hexane spread on the water–mercury interface?

2.17. The following results were obtained for the adsorption of carbon dioxide on one gram of charcoal at 0°C at various pressures:

P (mm Hg)	25.1	137.4	416.4	858.6
x (mg)	0.77	1.78	2.26	2.42

(a) Use these data to test the validity of the Langmuir adsorption isotherm.

(b) Calculate the maximum amount of CO_2 adsorbed to form a monomolecular layer.

(c) If the diameter of a carbon dioxide molecule is 0.35 nm, determine the surface area of the charcoal (m^2/kg), assuming the formation of a monomolecular layer.

2.18. In the adsorption of nitrogen on mica at 90 K, the volumes, of gas adsorbed (reduced to standard temperature and pressure) were 1.082 and $1.769 \times 10^{-6} m^3$ for a constant weight of mica at pressures of 5.60×10^{-4} and 5.45×10^{-3} torr, respectively.

(a) Assuming that the Langmuir adsorption isotherm is applicable to this case, calculate the volume of nitrogen (measured at STP) adsorbed by the same weight of mica at a pressure of 1.0×10^{-2} torr.

(b) Calculate the total surface area of mica used in the experiment. The area of nitrogen is 15 Å^2.

2.19. n-Octane is found to spread to a duplex film on an air–water interface, provided some myristic acid is also present in the oil. The duplex film has a surface pressure of 5 mN/m. How many milligrams of myristic acid are present per square centimeter, assuming that it acts entirely as a gaseous film at the oil–water interface?

(a) Surface tension of n-octane: 25 mN/m

(b) Interfacial tension with water: 53 mN/m

(c) Surface tension of water at 25°C; 72 mN/m

(d) Molecular weight of myristic acid: 228 g/mol

Literature

References[a]	Section 2.1: Surface Tension	Section 2.2: Experimental Techniques	Section 2.3: Capillarity and Nucleation	Section: 2.4 Adsorption	Sections 2.5–2.7 Monolayers–Characterization
Adamson & Gast	CHAPTER IV Spreading CHAPTER X Contact angles	CHAPTER II Surface tension CHAPTER XI Adsorption from solution	CHAPTER II Capillarity CHAPTER IX Nucleation	CHAPTER III Gibbs equation and examples CHAPTER VII–VIII Solid surfaces	CHAPTER IV Monolayers: characterization and applications
Davis	CHAPTER 7.1 Interfacial tension	CHAPTER 7.2–7.3 Capillary	CHAPTER 7.4 Interfacial thermo-dynamics		
Hiemenz & Rajago-palan	CHAPTER 6 Surface tension and contact angles			CHAPTER 7 Adsorption from solution and solid–gas adsorption	
Hunter (I)	CHAPTER 5.2 Surface tension	CHAPTER 5.7 Liquids in capillaries	CHAPTER 5.5 5.9 Capillarity, nucleation	CHAPTER 5.3–5.4 Surface thermody-namics and Gibbs adsorption	
Hunter (II)	CHAPTER 5 Surface tension CHAPTER 5.10 Contact angle and melting	CHAPTER 5.12 Determining surface tension and contact angle			
Meyer	CHAPTER 8 Surface tension				
Shaw	CHAPTER 4 Surface tension and spreading CHAPTER 6 Contact angles			CHAPTER 4 Adsorption and orienta-tion at inter-faces	
Tanford				CHAPTER 10 Monolayers	

[a]For complete reference citations, see alphabetic listing, page xxxi.

(Continued)

Literature

References[a]	Section 2.1: Surface Tension	Section 2.2: Experimental Techniques	Section 2.3: Capillarity and Nucleation	Section: 2.4 Adsorption	Sections 2.5–2.7 Monolayers– Characterization
Vold and Vold	CHAPTER 1 Liquid– vapor interface CHAPTER 2 Surface films in liquids CHAPTER 7.1–7.2 Contact angles			CHAPTER 2.2 Gibbs monolayers CHAPTER 4.3–4.5 Adsorption on solid surfaces	CHAPTER 2.4–2.6 Langmuir– Blodgett films

[a]For complete reference citations, see alphabetic listing, page xxxi.

ELECTROSTATIC INTERACTIONS IN COLLOIDAL SYSTEMS

CONCEPT MAP

Intermolecular Interactions

- In principle, we can obtain the interaction energies V_{ij} between two molecules by solving the Schrödinger equation.

- However, it is more useful to write V_{ij} as the sum of five interactions: (1) overlap repulsion, (2) charge transfer, (3) electrical multipole–electrical multipole, (4) electrical multipole–induced multipole, and (5) induced dipole–induced dipole, because we can express some of these interactions in classical rather than quantum mechanical terms.

Electrostatic Intermolecular Interactions in the Vapor Phase

- Interactions between isolated molecules in the vapor phase provide the simplest way to understand the basic concepts. Table 3.1 summarizes the essential equations.

Coulombic Interactions

- Arise when the electrical field emanating from one molecule couples to the existing charge distribution of other molecules.

- Can be attractive or repulsive, depending on charge and orientation of the molecules.

$V_{\text{ion}-\text{ion}}(R) \propto q_1 q_2/R$ and interactions are long-ranged. q is the net charge.

$V_{\text{ion}-\text{dipole}}(R, \theta) \propto q_1 m_2/R^2$ and depends on the orientation and magnitude of the dipole moment m (Figure 3.2).

$V_{\text{dipole}-\text{dipole}}(R, \theta, \phi) \propto m_1 m_2/R^3$ and depends on θ and ϕ, as depicted in Figure 3.3.

- Can account for the hydrogen bond.

Induced Interactions

- Arise when the electrical field emanating from one molecule induces charge polarization in other molecules.

- Are always attractive

$V_{\text{ion}-\text{ind dipole}}(R) \propto -q_1 m_2/R^4$, where α is the polarizability (eq. 3.4.2).

$V_{\text{dipole}-\text{ind dipole}}(R) \propto -m_1^2 \alpha/R^6$.

If $V_{\text{coulombic}} \propto 1/R^n$ then $V_{\text{ind}} \propto 1/R^{2n}$ because one $1/R^n$ term induces the response while the second produces the interaction.

The Dielectric Approximation

- The Poisson equation (3.2.1) is difficult to apply in a condensed phase where multiple interactions occur. However, by averaging over solute-solvent and solvent-solvent interactions transforms Poisson's equation into a useful form (eq. 3.5.13) by introducing the relative dielectric permittivity ε_r.
  In this way one can determine electrostatic solvation energies and solute-solute interaction free energies.

- The Born equation (3.6.2) gives the free energy associated with transferring an ion from one phase to another.

Angle-Averaged Potentials

Interactions involving dipoles are often less than kT, and we must average over all orientations using the Boltzmann distribution to determine their interaction energy. Such averaging yields free energies half as large as the corresponding energies.

Angle-averaged interactions are always attractive for dipoles

$$(V_{\text{dipole}-\text{dipole}}) \propto -\frac{m_1^2 m_2^2}{kTR^6}, \text{ which is called the Keesom interaction}$$

Chemical Origins of Surface Charges

- Virtually all vapor– liquid–solid interfaces acquire charge by dissociation or adsorption of ionic constituents.

- In addition to electrostatic interactions, short-range chemical interactions between ions and surface can lead to pronounced effects. These can be described by a conventional chemical equilibrium when electrostatic contributions are included.

- Many ionic crystals acquire charge because two ionic species possess different equilibrium adsorption affinities for the surface. In the case of Ag^+ and I^- in the AgI colloidal system, covalent interactions influence how the charge on the particles is set by the ratio of c_{Ag}^+/c_I^- concentration in solution.

- The Hofmeister series describes specific counterion binding in many systems.

- Because electrostatic forces are long-ranged, charged interfaces play an important role in many interfacial processes.

Charged Interfaces: The Poisson–Boltzmann Equation

We can describe entropic and energy interactions between charges fixed at a surface and those free to move in a solution by combining the Poisson and Boltzmann equations to yield the Poisson–Boltzmann (PB) equation (eq. 3.8.4).

- The **Gouy–Chapman theory** describes the properties of a charged planar interface by relating the surface charge density σ_0 and the surface potential Φ_0 (eq. 3.8.15). It shows how the potential in solution decays exponentially with distance (eq. 3.8.22, Figure 3.7) with a decay constant given by the Debye length $1/\kappa$ (eq. 3.7.13).

- The **Debye length** is a key parameter that measures how the combination of valency, concentration, and dielectric constant contribute to the screening of interactions between charges in solutions.

- Near a charged surface, counterions concentrate, while coions are repelled forming the **electrical double layer**. The Gouy–Chapman theory permits us to calculate how ionic concentrations vary in the solution (Figure 3.8).

The **Debye–Hückel theory** describes how the potential decreases exponentially with distance around a spherical ion (eq. 3.8.25).

The Graham equation (3.8.26) which follows from the Poisson-Boltzmann equation relates how the concentration of ions close to a charged surface varies with charge density and electrolyte concentration.

Electrostatic Free Energy of a Charged Surface

By separately calculating the entropic and energy contribution to the electrostatic free energy of solution using the PB equation, we gain considerable insight into the properties of charged surfaces (Figure 3.12).

By evaluating the electrostatic lateral pressure associated with a charged interface, we obtain a very useful relation between the change in free energy with surface area and energy per area (eq. 3.9.17).

3

As we saw in Chapter 2, intensive physical properties such as concentrations, energy density, and dielectric polarizability vary over a short distance at the interface of two phases. For example, at the air–water interface, the dielectric constant changes rather abruptly from 78 to 1. The orientation and polarization of interfacial molecules that result from this change constitute a microscopic separation of charge in the localized area.

Preferential adsorption of electrolytes or ionic surfactants produces a charged interface, and polar interactions involving head groups and solvent play a critical role in self-assembled structures, such as monolayers, micelles, bilayers, vesicles, and emulsions. They also dominate self-assembly and stability in other colloidal solutions, charged latex particles, silica particles, clays, and other colloidal systems. For this reason, the study of electrostatic interactions between molecules and particles is a central theme of colloidal science.

Electrostatic interactions play an essential role in many industrial and biological processes. Electrochemical reactions, including the reduction of aluminum, the purification of copper, and the production of Cl_2 gas, all occur at charged interfaces. Operation of most microelectronic solid state devices depends on the charge separation that occurs at the interfaces between materials. In biology, selective adsorption and transport are governed by electrical properties associated with biological membranes or other surfaces. In fact, understanding the behavior of charged interfaces is a major step toward understanding many phenomena that affect our daily lives.

Most textbooks on colloids begin their discussion of electrostatics with a form of the Poisson equation (see Section 3.5) containing the solvent dielectric constant to describe ion–ion interaction. Then other important electrostatic interactions, such as ion–dipole or dipole–dipole interactions, are considered on an ad hoc basis. Although in one sense this approach is appealing in its directness, it glosses over many fundamental issues that are prerequisites for understanding the subject. We begin our discussion with the equations in their most basic form and show how electrostatic interactions emerge in a unified and consistent way. Although this approach is more abstract and requires deeper contemplation on the part of the reader, the insight that emerges makes the effort worthwhile. In another sense it provides a concrete link between molecular and colloidal interactions, one of the main themes of this book.

3.1 Intermolecular Interactions Often Can Be Expressed Conveniently as the Sum of Five Terms

The notion of intermolecular interactions is based on the idea that in a condensed phase we can identify molecular units that remain largely unaffected by their immediate environment. By solving the Schrödinger equation for the isolated molecular unit, we can obtain the intramolecular energies and charge distributions. We can also determine V_{ij}, the interaction between two molecules, by this procedure. In practice, it is more convenient to interpret the total interaction potential energy as the sum of five parts:

$V_{ij} =$ (1) overlap repulsion + (2) charge transfer

$+$ (3) electrical multipole—electrical multipole

$+$ (4) electrical multipole—induced electrical multipole + (5) dispersion

Each of these five terms depends on molecular properties derived from the Schrödinger equation. However, the advantage of organizing V_{ij} in this way is that some of the terms can be expressed more simply in classical rather than quantum mechanical terms.

1. Overlap repulsion occurs when the electron clouds of two closed-shell molecules overlap. In accord with the Pauli exclusion principle, guest electrons are forced into excited state orbitals of the host molecule and produce a strong increase in energy. The resulting repulsion is approximately proportional to the square of the overlap, which increases strongly as the separation between molecules decreases. This effect determines a molecule's size.

We can express the overlap repulsion for atoms or molecules as a power law

$$V(r) \simeq \left(\frac{\sigma}{r}\right)^n \qquad (3.1.1)$$

When $n = 12$, we obtain the repulsive term in the Lennard–Jones potential. When $n = \infty$, we obtain an expression for hard spheres, since for $r > \sigma$ the value of $V(r)$ is effectively zero, whereas for $r < \sigma$ it is infinite.

2. Charge transfer interactions occur when one molecule donates excess electrons to an acceptor molecule with an electron deficiency. This type of interaction occurs when the molecules are in close proximity, but because its effect is relatively unimportant in colloidal systems, we will not consider it further.

3. Electrical multiple–electrical multipole interactions occur between molecules that possess net charges or an asymmetrical distribution of electrons or nuclei. We can describe this term by classical electrostatics.

4. Electrical multipole–induced multipole interactions occur between one molecule with a permanent electrical multipole and a polarizable molecule. This term is semiclassical in the sense described in Section 3.4.

5. The dispersion term can be described only in quantum mechanical terms. It corresponds to a spontaneous dipole—induced dipole interaction. We will postpone discussing it until Chapter 5, where we develop equations for London dispersion attractive interactions.

Overlap and dispersion terms describe properties common to all molecules (both polar and nonpolar). Multipole and induced multipole interactions occur more selectively and are thus potentially more effective in promoting an organized system. They are the focus of this chapter.

Many scientific writers treat the hydrogen bond as a special type of interaction and include an additional term in the expression for V_{ij}. For reasons that will emerge later in this chapter, we will consider this bond to be a special case of electrostatic interaction.

3.2 Multipole Expansion of the Charge Distribution Provides a Convenient Way to Express Electrostatic Interactions Between Molecules

In an electrostatic description, we can represent a molecule by a charge distribution, $\rho(\vec{r})$, set up by its electrons and nuclei. This charge distribution generates an electrostatic potential Φ according to the Poisson equation

$$\varepsilon_0 \nabla^2 \Phi(\vec{r}) = -\rho(\vec{r}) \tag{3.2.1}$$

where $\nabla^2 = \partial^2/\partial x^2 + \partial^2/\partial y^2 + \partial^2/\partial z^2$ is the Laplacian operator and ε_0 represents the dielectric permittivity of a vacuum ($\varepsilon_0 = 8.854.10^{-12}\,C/V \cdot m$). When discussing electrostatic effects, we will consistently use SI units. In older literature cgs units are customary. In the Poisson equation (eq. 3.2.1), a difference by a factor $4\pi\varepsilon_0$ exists between the two systems of units. You should remember this when comparing equations in different texts.

We can formally solve eq. 3.2.1 for a fixed charge distribution $\rho(\bar{r})$ by first noting that it is a linear equation so that the principle of superposition applies. That is, the potential Φ generated by two charge distributions, $\rho_1(\bar{r}) + \rho_2(\bar{r})$, is the sum of $\Phi = \Phi_1 + \Phi_2$ of the potentials generated by $\rho_1(\bar{r})$ and $\rho_2(\bar{r})$ separately. Second, we know that the potential from a point charge q_0 at $\bar{r}_0$ is $\Phi(\bar{r}) = q_0/(4\pi\varepsilon_0|\bar{r} - \bar{r}_0|)$, which consequently is the solution of eq. 3.2.1 when $\rho(\bar{r})$ represents a point charge. We can look at the actual charge distribution, $\rho(\bar{r})$, as a collection of point charges, and, using the principle of superposition, integrate over $\rho(\bar{r})$ to get the potential

$$\Phi(\bar{r}) = \frac{1}{4\pi\varepsilon_0} \int \frac{\rho(\bar{r}')}{|\bar{r} - \bar{r}'|} (d\bar{r}')^3 \qquad (3.2.2)$$

A charge q_2 at $\bar{r}$ will interact with an external potential Φ_1 to yield the energy $V_{12}^{el} = q_2\Phi_1(\bar{r})$. Thus, the charge distribution $\rho_1(\bar{r})$ of one molecule will sense the potential $\Phi_2(r)$ from another molecule to yield the interaction energy

$$V_{12}^{el} = \int \rho_1(\bar{r})\Phi_2(\bar{r})(d\bar{r})^3 = \frac{1}{4\pi\varepsilon_0} \iint \frac{\rho_1(\bar{r})\rho_2(\bar{r}')}{|\bar{r} - \bar{r}'|} (d\bar{r})^3(d\bar{r}')^3$$

$$= \int \rho_2(\bar{r}')\Phi_1(\bar{r}')(d\bar{r}')^3 \qquad (3.2.3)$$

whose coordinates are defined in Figure 3.1. Note also that we could equivalently consider the potential generated by molecule 1 at the position of molecule 2.

This deceptively simple formula contains a multitude of effects. If by solving the Schrödinger equation for the molecules, we determine the charge distributions ρ_1 and ρ_2, we could numerically integrate eq. 3.2.3 to obtain the interaction energy. Even though this gives accurate numbers it doesn't provide a conceptual basis for understanding the qualitative properties of

Figure 3.1
Coordinates used to calculate the electrostatic interaction between two molecules, 1 and 2, with charge distributions ρ_1 and ρ_2, respectively.

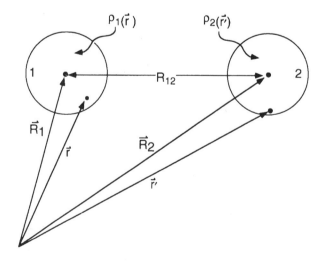

electrostatic intermolecular interactions. With molecules we are accustomed to characterizing their electrical properties through the net charge, the dipole moment and possibly the quadrupole moment. This point of view is based on a mathematical device called the multipole expansion.

Choose a central point $\vec{R}_1$, within ρ_1 and a central point, $\vec{R}_2$ within ρ_2. The denominator in eq. 3.2.3 can be rewritten as

$$|\vec{r} - \vec{r}\,'| = |\vec{R}_2 - \vec{R}_1 + (\vec{r}\,' - \vec{R}_2) - (\vec{r} - \vec{R}_1)| \qquad (3.2.4)$$

When the terms $(\vec{r}\,' - \vec{R}_2)$ and $(\vec{r} - \vec{R}_1)$ are small relative to the intermolecular separation $\vec{R}_{12} = (\vec{R}_2 - \vec{R}_1)$, we can eliminate $\vec{r}\,' - \vec{R}_2$ and $\vec{r} - \vec{R}_1$ from the denominator in eq. 3.2.3 by a series expansion. This expansion is straightforward but somewhat involved mathematically. Our discussion simply states the expressions for the most important terms. The general result can be written as a sum over different multipole terms having a progressively stronger distance and angular dependence

$$V_{12}^{el} = \sum_{m,n=1}^{\infty} V_{12}^{m,n}(R_{12}) = \sum_{m,n=1}^{\infty} C_{m,n} R_{12}^{-m-n+1} \qquad (3.2.5)$$

where the summation indices denote the order of the multipoles. The coefficient $C_{m,n}$ is a product of molecular multipole moments and an angular dependent term. We will consider the several lowest terms of this series and give the appropriate expressions for $C_{m,n}$ rather than deriving them. In Sections

TABLE 3.1 Electrostatic Interaction Energies Between Molecules in the Gas Phase

Value of m,n in eq. 3.2.5	Name	Interaction Energy	Comment
1,1	Coulomb ion–ion	$\dfrac{q_1 q_2}{4\pi\varepsilon_o R_{12}}$	See Figure 3.1
1,2	Ion–dipole	$\dfrac{q_1 m_2 \cos\theta}{4\pi\varepsilon_o R_{12}^2}$	See Figure 3.2
2,2	Dipole–dipole	$\dfrac{-m_1 m_2 f(\theta,\phi)}{4\pi\varepsilon_o R_{12}^3}$	$f(\theta,\phi) = 2\,\cos\,\theta_1\,\cos\,\theta_2 - \sin\,\theta_1\,\sin\,\theta_2\,\cos(\phi_2 - \phi_1)$ See Figure 3.3
—	Ion-induced dipole	$\dfrac{-\alpha q_1^2}{2(4\pi\varepsilon_o)^2}\dfrac{1}{R_{12}^4}$	Attractive α-polarizability
—	Dipole-induced dipole (Debye)	$\dfrac{-m_1^2\alpha(1 + 3\,\cos^2\,\theta)}{2(4\pi\varepsilon_o)^2 R_{12}^6}$	Attractive
2,2	Thermally averaged dipole—dipole (Keesom)	$\dfrac{-2m_1^2 m_2^2}{3(4\pi\varepsilon_o)^2 kT R_{12}^6}$	Attractive $kT > \dfrac{m_1 m_2}{4\pi\varepsilon_o R_{12}^3}$

3.2—3.4, we focus on interactions between molecules that occur in the gas phase; the most important interaction formulas are summarized in Table 3.1. In Section 3.5, we generalize the results to condensed phases.

The lowest order term is obtained for $m = n = 1$ and simply represents the Coulomb interaction

$$V_{\text{ion}-\text{ion}} = \frac{q_1 q_2}{4\pi\varepsilon_0 R_{12}} = \frac{z_1 z_2 e^2}{4\pi\varepsilon_0 R_{12}} \tag{3.2.6}$$

where the total charge $q_i = \int \rho_i\, dr^3$ is the **monopole** moment in this general expansion. We can use eq. 3.2.6 to describe the interaction between two ions where the magnitude and sign of each ionic charge is given by $q = ze$ (e = elementary charge).

Problem: Calculate the Coulomb interaction between a sodium ion and a chloride ion at contact, $R_{12} = 0.276\,\text{nm}$.
Solution:

$$V_{\text{ion-ion}} = \frac{(1)(-1)(1.602 \times 10^{-19}\text{C})^2}{4\pi(8.854 \times 10^{-12}\text{C/Vm})(0.276 \times 10^{-9}\text{m})} = -8.4 \times 10^{-19}\text{J}$$

In the problem above we found that the Coulombic interaction energy between two typical ions at contact was $-8.4 \times 10^{-19}\text{J}$. A comparison with the thermal energy $kT = (1.38 \times 10^{-23}\text{J/K})\,(300\,\text{K}) = 4.1 \times 10^{-21}\text{J}$ at $300\,\text{K}$ shows that interaction per ion pair in a vacuum is around $-200\,kT$. Thus, Coulomb interactions are very strong. In fact, only at separations greater than $56\,\text{nm}$ will the Coulomb interaction be less than kT. In the gas phase, even a small net charge on a dust particle strongly influences its interactions with other particles. Often this effect is important, such as when considering the stability of aerosols.

A noncharged polar molecule has a dipole moment, $\vec{m}_i$, defined by

$$\vec{m}_i = \int \rho_i(\vec{r}) \cdot \vec{r}\, d\vec{r}^3 \tag{3.2.7}$$

In the simple case of ρ equal to two opposite charges, $\pm q$, separated by a distance l, $m = ql$. For two elementary charges, $q = \pm e$, by $l = 0.1\,\text{nm}$, the dipole moment is $m = (1.602 \times 10^{-19}\text{C})(10^{-10}\text{m}) = 1.6 \times 10^{-29}\ \text{C}\cdot\text{m} = 4.8\,D$. The Debye unit D often is used for dipole moments and equals $3.336 \times 10^{-30}\,\text{C}\cdot\text{m}$. Permanent dipole moments occur in asymmetric molecules and thus not in single atoms.

As an example consider the case illustrated in Figure 3.2 where an ion (1) of charge ze ($m_1 = 0$) interacts with a dipole (2) with two charges, $\pm q$, separated by a distance l. We obtain the electrostatic interaction energy by summing the two Coulomb terms for the charge ze interacting with the charges $\pm q$ so that

$$V^{el} = \frac{zeq}{4\pi\varepsilon_0}\left(\frac{1}{R_+} - \frac{1}{R_-}\right)$$

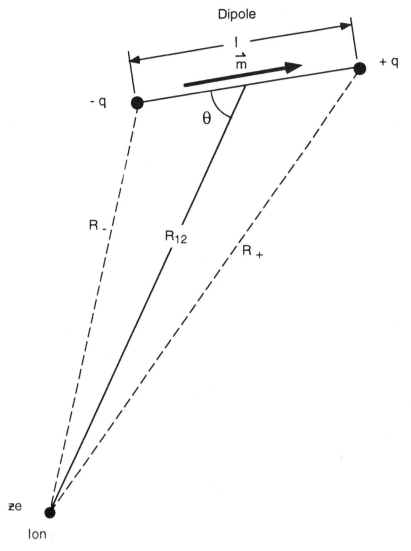

Figure 3.2
Ion–dipole interactions.
The ion–dipole distance
R_{12} is defined by the
distance between the ion
and the center of the dipole
$0.5l$. The interaction angle
is defined as the acute
angle between R_{12} and m.

When $R_{12} \gg l$, this expression can be simplified through use of the cosine theorem

$$R_{\pm}^2 = R_{12}^2 \left[1 \pm \frac{l \cos \theta}{R_{12}} + \left(\frac{l}{2R_{12}} \right)^2 \right]$$

and through a series expansion ($l \cos \theta \ll R_{12}$)

$$R_{\pm} \simeq R_{12} \left(1 \pm \frac{l \cos \theta}{2R_{12}} \right)$$

Using $l \cos \theta \ll R_{12}$ again, the ion–dipole interaction becomes

$$V_{\text{ion}-\text{dipole}}(R_{12}, \theta) \cong -\frac{(ze)(ql) \cos \theta}{4\pi\varepsilon_0 R_{12}^2} \qquad (3.2.8)$$

ELECTROSTATIC INTERACTIONS IN COLLOIDAL SYSTEMS / **109**

If a cation is near the dipole, the maximum attraction occurs when the dipole points away from the ion ($\theta = 0$), while the maximum repulsion occurs when the dipole points toward the ion ($\theta = 180$). The general expression emerging from eq. 3.2.5 with $m = 1$, $n = 2$ and $m = 2$, $n = 1$ is

$$V_{\text{ion}-\text{dipole}} = \frac{1}{4\pi\varepsilon_0 R_{12}^3} (-q_1\vec{m}_2 \cdot \vec{R}_{12} + q_2\vec{m}_1 \cdot \vec{R}_{12}) \quad (3.2.9)$$

Problem: Estimate the Coulombic interaction between a water molecule and a sodium ion using eq. 3.2.8. The dipole moment of a water molecule equals $1.85\,D$, and assume a separation of $0.23\,$nm for a water molecule in contact with a sodium ion.
Solution: The maximum interaction occurs for $\theta = 0$ in eq. 3.2.8 and

$$V_{\text{ion-dipole}}(R_{12}, \theta = 0) = \frac{-(1.602 \times 10^{-19}\,\text{C})(1.85 \times 3.336 \times 10^{-30}\,\text{C}\cdot\text{m})}{4\pi(8.854 \times 10^{-12}\,\text{CV}\cdot\text{m})(0.23 \times 10^{-9}\,\text{m})^2}$$

$$= -1.7 \times 10^{-19}\,\text{J} = -40\,kT \quad (T = 300\,\text{K})$$

Dipole–dipole interactions dominate the electrostatic interaction in polar liquids. They correspond to $m = n = 2$ and are given by

Figure 3.3
Dipole–dipole interactions. The dipole–dipole distance R_{12} represents the distance between the centers of dipoles. Relative orientations of the dipoles are specified by the polar 0_1, θ_2, and azimuthal ϕ_1, and ϕ_2 angles.

$$V_{\text{dipole}-\text{dipole}}(R_{12}, \theta_1, \theta_2, \phi_2 - \phi_1)$$
$$= -\frac{m_1 m_2}{4\pi\varepsilon_0 R_{12}^3} [2\cos\theta_1 \cos\theta_2 - \sin\theta_1 \sin\theta_2 \cos(\phi_2 - \phi_1)]$$

$$(3.2.10)$$

The two dipoles are separated by R_{12} and are oriented with polar angles θ_1, ϕ_1 and θ_2, ϕ_2 with respect to $\vec{R}_{12}$ as illustrated in Figure 3.3. This interaction is strongly orientation dependent. For example, with $\vec{m}_1$ fixed but $\vec{m}_2$ rotated isotropically

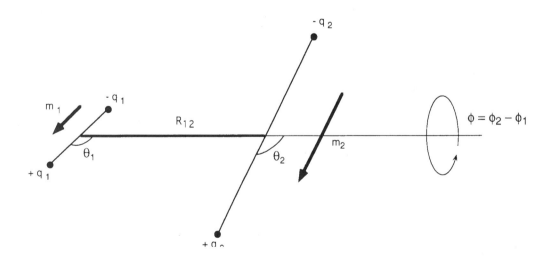

over all orientations, the net effect is zero. In a liquid with only weak orientational correlations, this factor substantially reduces the net averaged interaction. Maximum attraction occurs when two dipoles are aligned parallel to $\vec{m}_{12}$ ($\theta_1 = \theta_2 = 0$). This attraction is given by

$$V(R_{12}, 0, 0, \phi) = -\frac{m_1 m_2}{2\pi\varepsilon_0 R_{12}^3} \qquad (3.2.11)$$

For two equal parallel dipoles with moments of $1\,D$, the interaction energy in vacuum equals kT at either $R_{12} = 0.36\,\mathrm{nm}$ (when the dipoles are in line with $\vec{R}_{12}$) or at $R_{12} = 0.29\,\mathrm{nm}$ (when they are perpendicular to $\vec{R}_{12}$). Since these distances are on the order of molecular separations in liquid and solids, dipole–dipole interactions generally are not strong enough to lead to substantial mutual alignment of polar molecules in the liquid state.

The quadrupole–quadrupole interaction, $m = n = 3$, forms the dominant electrostatic term in liquids formed by molecules having zero dipole moment. Although the quadrupolar term is shorter ranged and more orientation dependent than the dipole–dipole term, as the general properties of the series in eq. 3.2.2 indicate, it is of utmost importance for understanding the properties of the condensed phase for such molecules as CO_2, N_2, and C_6H_6. Figure 3.4 illustrates the charge distributions of the quadrupolar molecules CO_2 and C_6H_6. For example, one major reason for the noticeably greater water solubility of benzene over other hydrocarbons of similar size lies in the slight polarity caused by the high quadrupolar moment of C_6H_6.

Problem: Calculate the quadrupolar moment of CO_2. Assume that the charge distribution can be represented by partial charges on the atom. Take the oxygen charge to be $-0.32e$, and use a bond length of $0.12\,\mathrm{nm}$. The quadrupolar moment Q is given by (for an axially symmetric molecule)

$$Q = \sum_i q_i z_i^2$$

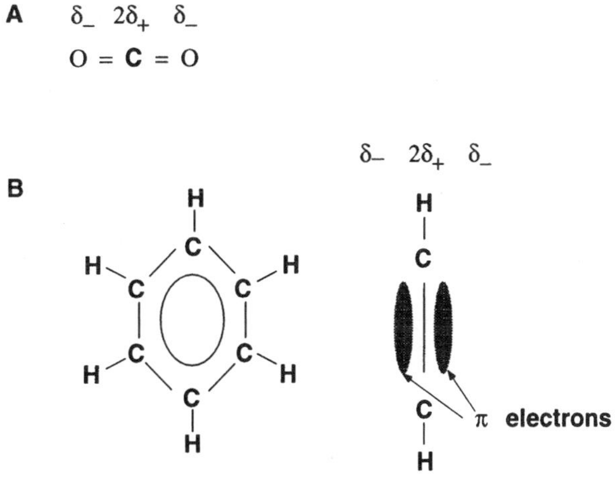

Figure 3.4
Illustration of the charge distributions in carbon dioxide and benzene. For (A) CO_2 the oxygens have a partial negative charge making the central carbon strongly positive. For (B) benzene the π-electrons make the space on both sides of the ring plane negative with the compensating positive nuclear charge localized in the plane. Note that despite the structural differences between the molecules the charge distributions along the symmetry axes are similar.

where z_i denotes the coordinate along the symmetry axis and the sum is over all charges.

Solution: Choose $z = 0$ on the carbon atom (the final do not depend on this choice)

$$Q = -0.32e \times (-0.12)^2 10^{-18}\,\text{m}^2 - 0.32e \times 0.12^2 \times 10^{-18}\,\text{m}$$

$$= 4.6 \times 10^{-21}\,\text{m}^7 \times 1.602 \times 10^{-19}\,\text{C} = 7.4 \times 10^{-40}\,\text{Cm}^2$$

This is a typical size of a molecular quadrupole moment.

The hydrogen bond often induces molecular association in colloidal systems. In most textbooks it is presented as an intermolecular interaction different from the five parts discussed in Section 3.1. In fact, the hydrogen bond is dominated by electrostatics but with the special property that it doesn't quite fit into the multipole expansion scheme given in eq. 3.2.5. To be useful, the multipole expansion for the interaction energy should converge rapidly. Usually we want to include only the first nonzero term and perhaps the second one. If convergence is not reached within two terms, there are good reasons to look for an alternative formulation.

For example, the electron density around a hydrogen atom is low, particularly when it is attached to an electronegative atom such as N, O, or F. Then the overlap repulsion with other atoms is relatively weak and allows for a close approach of ~ 1.5–$2.0\,\text{Å}$ to atoms in other molecules. A strong electrostatic interaction results if these atoms have a high electron density, as do N, O, and F. With such small values of R_{12}, the series expansion in eq. 3.2.5 does not converge rapidly. When higher order terms in the expansion provide substantial contributions to the interaction energy V_{12} it follows that the interaction is of shorter range and is more orientation dependent. The relative weight of the higher order terms change from one type of hydrogen bond to the other so one cannot give a general expression for either the precise form of the distance dependence or of the angular dependence. At large separations it is the dipole–dipole term that dominates, provided the molecules are neutral, but in the vicinity of the most favorable separation this is not true. The term *hydrogen bond* is thus a convenient way to denote a strong short range highly directional electrostatic interaction. The wide application of this term, not least in biological systems, has led to its electrostatic origin being partly forgotten.

In a hydrogen bond there is a donor, that is a hydrogen atom covalently bonded to an electronegative atom, and an acceptor which is another electronegative atom. In liquids like water, hydrazine, and ethylene glycol, the molecule can act as both a donor and an acceptor and hydrogen bonds act to give a strong association and cohesion in such liquids. We discussed some of the consequences of this cohesion in Section 1.7.

We can use the water dimer to illustrate some properties of hydrogen bonds. Figure 3.5 shows the energy of a water

Figure 3.5
(**A**) The interaction
potential between two
water molecules as a
function of the oxygen–
oxygen distance $R_{O..O}$
calculated using a quantum
chemical formalism. The
molecules
are kept at a fixed
orientation corresponding
to that of the lowest energy
configuration. The three
curves show the total
interaction, the
electrostatic interaction
and its dipolar part. At long
range the interaction is
dominantly electrostatic
with other components
coming in close to the
optimal separation of
$R_{O..O} = 2.86$ Å. In the
distance range shown the
dipolar part is less than
50% of the total
electrostatic interaction.
(**B**) The calculated water–
water interaction energy
as a function of the
orientation of the hydrogen
bond donor molecule
illustrating the electrostatic
nature of the hydrogen
bond. The curves show
how the angular
dependence of the total
interaction is completely
determined by the
electrostatic interaction.
The two minima
correspond to the two
hydrogen bond
configurations, while the
dipolar part has its
minimum value at the
maximum between the two
hydrogen bond minima.
Note that the wells in the
interaction around $\alpha = 0$
and $\alpha = 100°$ are separated
by a barrier of a few $\pmb{k}T$
so that the interaction is
strong enough to cause a
net orientation in a
thermally averaged system.
(Courtesy of Gunnar
Karlström.)

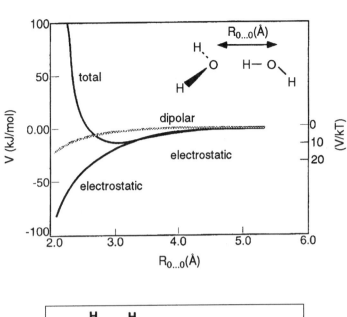

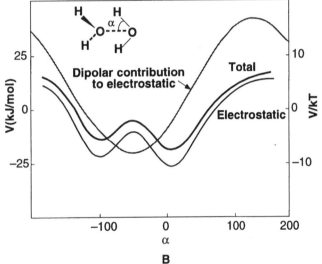

dimer, where the hydrogen atoms, for simplicity, have been
confined to the same plane as the oxygens. The minimum in the
intermolecular interaction (Figure 3.5) occurs for an oxygen–
oxygen distance of 0.286 nm and $V = 4.8 \times 10^{-20}$ J or $V/kT \sim 12$
when $T = 298$ K. Note that the hydrogen atom of the hydrogen
bond falls nearly but not perfectly on the oxygen–oxygen line.
Figure 3.5 shows the strong orientation dependence of the
interaction at the equilibrium separation. This is perfectly
described, except for a constant, by only considering the
charge–charge interaction as in eq. 3.2.3, while the dipolar
component by itself fails to give the qualitatively correct
behavior.

3.3 When Electrostatic Interactions Are Smaller than the Thermal Energy, We Can Use Angle-Averaged Potentials to Evaluate Them and Obtain the Free Energy

Molecules in a liquid or gas tumble through the thermal motion. Interactions involving dipoles and higher multipole moments are angle dependent and the rotation results in an averaging of the interaction. If a molecule adopts all orientations with equal probabilities the average multipole interaction is zero. However, in a thermal average the low energy configurations occur preferentially determined by a Boltzmann weighting factor.

We can calculate the angle-averaged energy component $U(R)$ of an orientation dependent interaction energy from the relationship

$$U(R) = \int V(R, \Omega) e^{-V(R,\Omega)/kT} \frac{d\Omega}{4\pi} \qquad (3.3.1)$$

where the integration is performed over polar, θ, and azimuthal, ϕ, angles; $d\Omega = \sin\theta d\theta d\phi$. When integrated over all orientations, $d\Omega$ gives 4π.

When $V(R, \Omega)$ is less than kT, we can expand the exponential function in eq. 3.3.1

$$e^{-V(R,\Omega)/kT} = 1 - \frac{V(R,\Omega)}{kT} \cdots \qquad (3.3.2)$$

to give

$$U(R) = \int \left(V(R, \Omega) - \frac{V(R, \Omega)^2}{kT} + \cdots \right) \frac{d\Omega}{4\pi} \qquad (3.3.3)$$

By using eq. 3.2.8 for $V_{\text{ion-dipole}}(R_{12}, \theta)$, we can obtain the angle-averaged energy for the charge–dipole interaction

$$U(R) = \int \left(\frac{(ze)m \, \cos\theta}{4\pi\varepsilon_0 R^2} - \left(\frac{(ze)m}{4\pi\varepsilon_0 R^2} \right)^2 \frac{\cos^2\theta}{kT} + \cdots \right) \frac{d\Omega}{4\pi} \qquad (3.3.4)$$

Substituting the spatially averaged values of $\langle \cos\theta \rangle = 0$ and $\langle \cos^2\theta \rangle = 1/3$ gives

$$U(R) \approx -\frac{(ze)^2 m^2}{3(4\pi\varepsilon_0)^2 kTR^4} \quad \text{for} \quad kT > \frac{(ze)m}{4\pi\varepsilon_0 R^2} \qquad (3.3.5)$$

which is always attractive, a general property of second order terms.

The angle-averaged free energy for the dipole–dipole interaction we obtain from eq. 3.2.10 leads to

$$U(R) = -\frac{2m_1^2 m_2^2}{3(4\pi\varepsilon_0)^2 kTR^6} \quad \text{for} \quad kT > \frac{m_1 m_2}{4\pi\varepsilon_0 R^3} \qquad (3.3.6)$$

This Boltzmann-averaged interaction between two permanent dipoles, known as the Keesom interaction, constitutes one of the three important contributions to the van der Waals interactions.

Angle-averaged potentials in eqs. 3.3.5 and 3.3.6 result from a compromise between energy, which favors perfect alignment, and entropy, which strives toward a total randomization of the angular variables. $U(R)$ contains only the energy, and to obtain the free energy representing the effective potential, we must estimate the entropy contribution. Because of the simple T^{-1} dependence in $U(R)$, calculating the entropy contribution becomes a straightforward process.

The Helmholtz free energy, $\mathbf{A}$, is

$$\mathbf{A} = \mathbf{U} - T\mathbf{S} = \mathbf{U} + T\left(\frac{\partial \mathbf{A}}{\partial T}\right), \qquad (3.3.7)$$

which can be transformed to the Gibbs-Helmholtz equation

$$\frac{\partial}{\partial T}(A/T) = -U/T^2 \qquad (3.3.8)$$

If we now integrate this relation using the circumstance that $U = const/T$

$$A/T = -\int_T^\infty const \times T^{-3} dT = const\frac{1}{2T^2} = \frac{U}{2T} \qquad (3.3.9)$$

and thus

$$A = \frac{1}{2}U \qquad (3.3.10)$$

for a system where the energy is proportional to the inverse temperature. Thus, the free energy of the charge–dipole interaction is half the value given in eq. 3.3.5 and the same reasoning applies for the Keesom interaction in eq. 3.3.6.

In the subsequent sections we will encounter more examples of the fact that electrostatic interactions also contain an entropic contribution as soon as some molecular degree of freedom has the possibility to respond thermally to the electrical fields.

3.4 Induced Dipoles Contribute to Electrostatic Interactions

We expanded eq. 3.2.3, describing the interaction between two charge distributions, under the assumption that the interaction did not change the distributions ρ_1 and ρ_2. However, in a real system, one molecule responds internally to the presence of another. The separation between strong intramolecular and weak intermolecular interactions suggests that we can treat this response in terms of a perturbation so that

$$\rho_i(R_{12}) = \rho_i(R_{12} \to \infty) + \delta\rho_i \qquad (3.4.1)$$

where $\delta\rho_i \ll \rho_i$. Induced changes $\delta\rho_i$ lower the interaction energy with the second charge distribution by an amount $\int \Phi_1 \delta\rho_2 dV + \int \Phi_2 \delta\rho_1 dV$. However, setting up $\delta\rho_i$ involves an intramolecular energy penalty. For small perturbations, in which the response is linear, this penalty is simply half of the energy gain, as the following example illustrates.

The two most important induced electrostatic interactions are ion–induced dipole and dipole–induced dipole interactions. Both involve polarization of one molecule by the electrical field emanating from nearby molecules (ions). We define the polarizability α of an atom or molecule by the magnitude of the induced dipole moment m_{ind} acquired in an electrical field E as

$$m_{ind} = \alpha E \qquad (3.4.2)$$

Polarizability is nonzero for all atoms and molecules and plays an important role in determining colloidal forces.

To understand the nature of polarizability in more detail, we can model the interaction between an electrical field and a nonpolar molecule by considering the latter as consisting of two opposite charges $\pm q$ interacting by a spring with spring constant k. At rest they coincide, but an external electric field generated by an adjacent ion or dipole will cause them to separate by a distance X. Thus, the internal energy V_{int} for small deformations is

$$V_{int} = \frac{k}{2}X^2 \qquad (3.4.3)$$

The interaction energy V_{field} between the electric field $\vec{E}$ and the separated charges provides

$$V_{field} = -EqX \qquad (3.4.4)$$

The total interaction energy V_{tot} is $V_{int} + V_{field} = (k/2)X^2 - EqX$.

At equilibrium a balance exists between the external interaction and the restoring spring force when $(\partial V_{tot}/\partial X) = 0$, and we find the equilibrium charge separation X_0 to be

$$X_0 = \frac{Eq}{k} \qquad (3.4.5)$$

Using eq. 3.2.7, the induced moment thus becomes

$$m_{ind} = qX_0 = \frac{Eq^2}{k} \qquad (3.4.6)$$

Comparison with eq. 3.4.2 shows that the polarizability equals

$$\alpha = \frac{q^2}{k} \qquad (3.4.7)$$

so we see that the smaller the internal restoring force (i.e., the smaller the value of k), the larger the polarizability. The total energy is

$$V_{tot} = V_{int} + V_{field} = \frac{k}{2}\frac{E^2 q^2}{k^2} - Eq\frac{Eq}{k}$$

$$= -\frac{E^2 q^2}{2k} = -\alpha\frac{E^2}{2} \qquad (3.4.8)$$

We see that $V_{tot} = \frac{1}{2}V_{field} = -V_{int}$. This factor of 0.5 between V_{tot} and V_{field} results from the linearity of the response, and it also appears in more sophisticated models of the polarizability. Fundamentally, its origin is similar to that of the factor 0.5 in the Keesom interaction discussed in Section 3.3.

Polarizability has the dimensionality of $\varepsilon_0 \times volume$, and its magnitude typically is somewhat less than the molecular volume times ε_0. For example, in water, $\alpha/\varepsilon_0 = 1.86 \times 10^{-29}\,\text{m}^3$. The molecular van der Waals volume is $3.0 \times 10^{-29}\,\text{m}^3$.

With multiatom molecules, the induced dipole moment depends on the orientation of the polarizing field relative to the molecule, and we should express α as a tensor quantity. To simplify our calculations, we can assume that α is isotropic, which is an acceptable approximation for many molecules.

Now we are in a position to calculate $V_{ion-ind\,dipole}$ by combining eq. 3.4.8 and the one for the electric field associated with an ion of charge ze at $\vec{R} = 0$

$$\vec{E}(\vec{R}) = \frac{ze}{4\pi\varepsilon_0}\frac{\vec{R}}{R^3} \qquad (3.4.9)$$

to give

$$V_{ion-ind\,dipole}(R_{12}) = -\frac{\alpha(ze)^2}{2(4\pi\varepsilon_0)^2 R_{12}^4} \qquad (3.4.10)$$

The range is R^{-4}, rather than R^{-2} as in eq. 3.2.8, because one R^{-2} term induces the response, but a second R^{-2} term produces the interaction.

We can obtain dipole–induced dipole interactions by combining eq. 3.4.8, which gives the potential energy of an induced interaction, with the electrical field associated with a permanent dipole. Both the magnitude and the orientation of the field at the polarizable molecule depend on the angle θ between the dipole moment and the vector $\vec{R}$ joining the two species. The magnitude of the field is

$$E(\cos\theta) = \frac{m(3\cos^2\theta + 1)^{1/2}}{4\pi\varepsilon_0 R^3} \qquad (3.4.11)$$

The resulting interaction energy is

$$V_{dipole-ind\,dipole}(R_{12}, \theta) = -\frac{1}{2}\alpha E^2 = -\frac{m^2\alpha(1 + 3\cos^2\theta)}{2(4\pi\varepsilon_0)^2 R_{12}^6} \qquad (3.4.12)$$

The angle-averaged energy follows directly from eqs. 3.3.3, in which the first term dominates, and 3.4.12 since $\langle \cos^2 \theta \rangle = 1/3$ so that

$$V_{\text{dipole-ind dipole}}(R_{12}) = -\frac{m^2\alpha}{(4\pi\varepsilon_0)^2 R_{12}^6} \qquad (3.4.13)$$

This result represents the Debye attractive interaction and constitutes the second of the three inverse sixth-power contributions to the total van der Waals interaction energy between molecules.

Problem: Calculate the dipole–induced dipole interaction between two water molecules at a separation at $0.29\,\text{nm}$ ($m = 1.860$, $\alpha/\varepsilon_0 = 1.86 \times 10^{-29}\,\text{m}^3$).
Solution:

$$V = -\frac{m^2\alpha}{(4\pi\varepsilon_0)^2 R^6} = -\frac{(1.86)^2 \times (3.3 \times 10^{-30})^2 \times 1.86 \times 10^{-29}}{(4\pi)^2 8.85 \times 10^{-12} \times (0.29 \times 10^{-9})^6} = -8.4 \times 10^{-22}\,\text{J} \leftrightarrow -0.2\,kT$$

or $V \sim -0.2\,kT$ which is only a small contribution to the total interaction in Figure 3.5.

Induced interactions are not pairwise additive, and this fact adds an important complication to our calculations. For permanent moments, the interaction V_{123} between three molecules at fixed orientations is simply

$$V_{123} = V_{12} + V_{13} + V_{23} \qquad (3.4.14)$$

or the sum over the individual pairs. But summing pairs does not hold true for induced interactions. For example, a molecule midway between two similar ions does not experience any electrical field, so the contribution from the ion–induced dipole term is really zero. In such a case, if we assume pairwise additivity, as in eq. 3.4.14, and use eq. 3.4.10, we will get a quite different result—one that amounts to an artificial stabilization of the particular molecular configuration.

3.5 Separating Ion–Ion Interactions from Contributions of Dipoles and Higher Multipoles in the Poisson Equation Simplifies Dealing with Condensed Phases

The electrostatic interaction between isolated pairs of molecules shows substantial but manageable complexity, as demonstrated earlier. In a condensed phase the situation becomes much more intricate. Attempts to use the Poisson equation given in eq. 3.2.1 would require evaluating all multipole inter-

Figure 3.6

Illustration of the drop in
electrostatic potential
across a parallel plate
capacitor with a surface
charge density $-\sigma$ and $+\sigma$
at the two plates. (A) In
air the potential varies
linearly in the gap and the
slope $d\Phi/dz = -\varepsilon_o\sigma$ is
determined solely by the
surface charge density σ.
(B) With a liquid medium
in the gap, the molecules
are oriented and polarized
by the field. As illustrated
this occurs uniformly
throughout the gap with
the result that the potential
still varies linearly, but
with a smaller slope so that
$d\Phi/dz = -\sigma/\varepsilon_r\varepsilon_o$.

actions simultaneously and would result in an intractable problem. This section develops a procedure to circumvent this
difficulty, which permits us to rewrite the Poisson equation in
a more convenient form.

Ion–ion interactions are stronger and longer ranged than
all other multipole interaction. As a result, they typically play
a more important role in determining the interactions in electrolyte or charged colloidal solutions. This fact suggests that we
write the interaction between free charges explicitly while
averaging over the electrostatic interactions in the solvent.
Using the Poisson equation to analyze a parallel plate capacitor
gives us the simplest illustration of the principle for how to
perform such an average.

In the capacitor shown in Figure 3.6, an applied potential
drop $\Delta\Phi$ across the gap produces an induced surface charge
density, σ and $-\sigma$, at the surface of the conducting planes
separated by a distance h. The capacitance C is defined by

$$C = \frac{\sigma}{\Delta\Phi} \text{ area} \qquad (3.5.1)$$

We can calculate C in a straightforward manner from the
Poisson equation (eq. 3.2.1), which reduces to $\nabla^2\Phi(\bar{r}) = d^2\Phi/dz^2 = -\rho/\varepsilon_0$ because the potential is constant parallel to the
plates. A formal integration between two arbitrary points, z_1

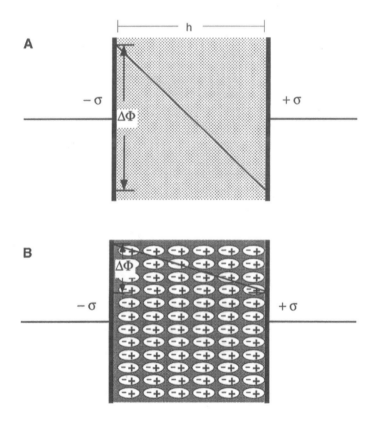

and z_2, gives

$$\left.\frac{d\Phi}{dz}\right|_{z_2} - \left.\frac{d\Phi}{dz}\right|_{z_1} = -\frac{1}{\varepsilon_0} \int_{z_1}^{z_2} \rho \, dz \qquad (3.5.2)$$

If z_2 and z_1 both lie in the gap between the capacitor plates, the charge density ρ equals zero, and the electric field

$$E_z \equiv -\frac{d\Phi}{dz} \quad (\vec{E} = -\vec{\nabla}\Phi) \qquad (3.5.3)$$

is constant. As a result, the potential drop $\Delta\Phi$ is $h(d\Phi/dz)|_{\text{gap}}$. To find the magnitude of the field, we choose z_2 to be inside the positively charged plate and z_1 to be in the gap. Inside the conductor, the field is zero, so that

$$\left.\frac{d\Phi}{dz}\right|_{z_2} - \left.\frac{d\Phi}{dz}\right|_{z_1} = 0 - \left.\frac{d\Phi}{dz}\right|_{\text{gap}} = -\frac{1}{\varepsilon_0} \int_{z_1}^{z_2} \rho \, dz = \frac{\sigma}{\varepsilon_0} \qquad (3.5.4)$$

The capacitance is

$$C = \frac{\sigma}{\left(-h\left.\dfrac{d\Phi}{dz}\right|_{\text{gap}}\right)} \text{area} = \frac{\sigma\varepsilon_0}{h\sigma}\text{area} = \left(\frac{\varepsilon_0}{h}\right)\text{area} \qquad (3.5.5)$$

Equation 3.5.2 represents a specific example of the so-called Gauss's law, which provides a very useful relation between electrical charge and electrical field. In its general form Gauss's law states that the electric field at the surface A of a closed volume V is determined by the enclosed charge; in mathematical terms

$$\varepsilon_0 \int \vec{\nabla}\Phi \cdot \vec{n} \, dA = -\int \rho \, dV \qquad (3.5.6)$$

where $\vec{n}$ signifies a surface normal vector pointing outward. Often we can use this relation to obtain a qualitative understanding (and sometimes also a quantitative solution) to an electrostatic problem. We will return to illustrations of the use of Gauss's law.

We can calculate the energy required to charge the capacitor, i.e. the energy stored, U, by considering a charging process where an infinitesimal amount of charge dq is brought from one plate to the other. As the degree of charging increases, the potential difference between the plates increases; and the differential energy increase is the product of charge transferred, dq, and the potential difference $\Delta\Phi$, which depends on the state of charge density σ'

$$dU = dq\Delta\Phi(\sigma') = d\sigma' \, A \frac{h\sigma'}{\varepsilon_0}$$

By integrating from charge density zero to the final value σ we find

$$U = \frac{Ah}{\varepsilon_0} \int_0^\sigma \sigma' \, d\sigma' = \frac{Ah}{\varepsilon_0} \frac{\sigma^2}{2} = \frac{A\sigma \Delta\Phi}{2}$$

$$= A \frac{\varepsilon_0}{2} \frac{(\Delta\Phi)^2}{h} \tag{3.5.7}$$

In a more general form for an arbitrary globally electroneutral charge distribution

$$U = \frac{1}{2} \int \rho \Phi \, dV = \frac{\varepsilon_0}{2} \int (\nabla\Phi)^2 \, dV \tag{3.5.8}$$

Since the potential is generated by the charge distribution itself there is a factor of one-half entering the expression for the energy to compensate for a doubling counting.

Problem: Calculate the energy of the capacitor by considering a separation of the two plates at constant charge density from zero distance to h. Use equation 3.2.3

$$U = \frac{1}{4\pi\varepsilon_0} \int \frac{\rho_1(r)\rho_2(r')}{|\vec{r} - \vec{r}'|} \, dV dV'$$

Solution: For a capacitor, ρ_1 and ρ_2 are charge distributions σ and $-\sigma$ confined to two parallel planes. For a surface element dA on surface 1 can be described as located at origin

$$\rho_1(0)dV = \sigma dA$$

The distance from this surface element to one, dA', on plate 2 is $(x^2 + y^2 + z^2)^{1/2}$, where z has a fixed value. The energy of brining surface element dA from $z = 0$ to $z = h$ is

$$d(U(h) - U(0)) = \frac{1}{4\pi\varepsilon_0} \sigma \, dA(-\sigma) \int\int_0^\infty \left\{ \frac{1}{(x^2 + y^2 + h^2)^{1/2}} - \frac{1}{(x^2 + y^2)^{1/2}} \right\} dxdy$$

$$= \frac{-\sigma^2}{2\varepsilon_0} dz \left\{ \int_0^\infty \frac{rdr}{(r^2 + h^2)^{1/2}} - \int_0^\infty dr \right\}$$

$$= \frac{-\sigma^2}{2\varepsilon_0} dA \left\{ \Big|_0^\infty [(r^2 + h^2)^{1/2} - r] \right\}$$

$$= \frac{\sigma^2}{2\varepsilon_0} dA \cdot h$$

Integrating over plate 1

$$U(h) - U(0) = \frac{\sigma^2}{2\varepsilon_0} A \cdot h$$

as in eq. 3.5.7.

In this example we express the energy relative to that of zero separation ($z = 0$) between the plates. In eq. 3.2.3 we have, as the reference state, the charges at infinite separation. This is only practical for finitely charged systems.

In a commercial capacitor, a nonconducting (dielectric) medium fills the gap between the plates, as shown in Figure 3.6B. This arrangement increases capacitance because the electrical field polarizes the medium so that the induced charge distribution partly compensates for the surface charges on the conducting plates. In other words, the response of the molecules results in a smaller potential drop $\Delta\Phi$ for a given σ, and thus a higher capacitance.

To formally describe this effect, we have to go back and rewrite the Poisson equation by dividing the total charge distribution ρ into two parts as follows:

$$\rho = \rho_{\text{free}} - \vec{\nabla} \cdot \vec{P} \tag{3.5.9}$$

where $\vec{P}$ is called the polarization vector and is a dipole moment density. The point of separating ρ into two parts is that ρ_{free} describes local net charges, while $\vec{\nabla} \cdot \vec{P}$ describes the effects of dipoles and higher moments. Combining Gauss's law (eq. 3.5.6) and the Poisson relation (eq. 3.2.1), we have

$$\int_V \vec{\nabla} \cdot \vec{P} \, dV = \int_A \vec{P} \cdot \vec{n} \, dA \tag{3.5.10}$$

Thus, the polarization carries an effective net charge only at the surface of the volume. The fundamental Poisson equation is

$$\varepsilon_0 \vec{\nabla} \cdot \vec{\nabla}\Phi = \vec{\nabla} \cdot \vec{P} - \rho_{\text{free}} \tag{3.5.11}$$

Any further progress with this problem requires introducing some approximation that relates polarization to the electrical field. By analogy with the concept of the polarizability of an isolated molecule, we can assume that $\vec{P}$ relates in a linear fashion to the field $\vec{E} = -\vec{\nabla}\Phi$ to yield

$$\vec{P} = \varepsilon_0(\varepsilon_r - 1)\vec{E} \tag{3.5.12}$$

The proportionality factor in this linear relation is chosen so that $\varepsilon_r = 1$ corresponds to no polarization or the vacuum case, and ε_r represents the relative dielectric permittivity, or simply the dielectric constant. For a time-independent field $\vec{E}$, $\varepsilon_r \geqslant 1$.

Combining eq. 3.5.12 with 3.5.11 yields a Poisson equation

$$\varepsilon_0 \varepsilon_r \vec{\nabla}^2 \Phi = -\rho_{\text{free}} \tag{3.5.13}$$

because $\vec{\nabla} \cdot \vec{E} = -\vec{\nabla} \cdot \vec{\nabla}\Phi$. This equation differs from the fundamental Poisson equation only by a factor ε_r. Consequently, the calculation for capacitance in the dielectric case applies as long as we replace ε_0 by $\varepsilon_0 \varepsilon_r$. Thus, the capacitance for the capacitor in Figure 3.6B increases by a factor ε_r and $C = (\varepsilon_0 \varepsilon_r / h \times \text{area})$.

To see the molecular origin of the increased capacitance we note that the field is $E = -\sigma/\varepsilon_0 \varepsilon_r$ and the polarization

$$P = -\varepsilon_0(\varepsilon_r - 1) \frac{\sigma}{\varepsilon_0 \varepsilon_r} = -\left(1 - \frac{1}{\varepsilon_r}\right)\sigma \qquad (3.5.14)$$

using eq. 3.5.12. Note that charge per area is dimensionally equivalent to dipole moment per volume. In the capacitor there is thus a constant polarization of magnitude $(1 - 1/\varepsilon_r)\sigma$ oriented to compensate the charges on the capacitor plates as illustrated in Figure 3.6B. For example liquid water has a dielectric constant of approximately 80 showing that it sets up a compensating polarization amounting to $(1-1/80) \approx 0.9875$ of the charge on the plate.

Problem: Calculate the maximum dipolar polarization in liquid water, given that a water molecule has a dipolar moment of $1.85\,D$.
Solution: A density of $10^3\,\text{kg/m}^3$ implies a volume of $3.0 \times 10^{-29}\,\text{m}^3$ per molecule. In SI units $1.85\,D$ corresponds to $1.85 \times 3.336 \times 10^{-30}\,\text{C}\cdot\text{m}$. For perfectly oriented water molecules the dipolar density and thus the polarization is $P_{\text{max}} = 1.85 \times 3.336 \times 10^{-30}\,\text{C}\cdot\text{m}/3.0 \times 10^{-29}\,\text{m}^3 = 0.21\,\text{C/m}^2$.

Four additional points must be noted to complete the argument in this section. The first is that eq. 3.5.12 treats ε_r as a constant, irrespective of the properties of $\vec{E}$. In fact, our discussion of the capacitor implicitly assumes a dc field. If we apply a slowly alternating field, the dipoles turn as the polarity of the electrodes varies. When the ac frequency increases to a point at which the dipoles no longer can follow the field, the dielectric constant decreases. Therefore, in the limit of high frequency (in which only electronic motions contribute to the polarizability), the dielectric constant equals the square of refractive index of the material.

A second point relates to the gap size h in our discussion of the capacitor. Although the gap is small, it has a macroscopic magnitude. Describing the solvent as a dielectric continuum may be invalid on a microscopic scale; nevertheless, experience has shown that it gives a useful approximation. Therefore, we assume that the Poisson equation holds true on a molecular scale and that ρ_{free} contains the charges that we consider explicitly. The factor ε_r includes the effects of the medium. Under these circumstances, all the equations we derived for the electrostatic interactions between isolated pairs of molecules are still valid in a medium, provided we again replace ε_0 by $\varepsilon_0 \varepsilon_r$.

For example, the Coulomb interaction between two charges q_1 and q_2 in a solvent with a relative dielectric permittivity ε_r is

$$V_{\text{coulomb}}(R_{12}) = \frac{q_1 q_2}{4\pi \varepsilon_0 \varepsilon_r} \frac{1}{R_{12}} \qquad (3.5.15)$$

in analogy with eq. 3.2.6. From a molecular perspective, this simple equation describes a very complicated reality. Thus we can identify four different molecular contributions to $V_{coulomb}$ in eq. 3.5.15. There is a direct ion–ion interaction

$$V = \frac{q_1 q_2}{4\pi\varepsilon_0} \frac{1}{R_{12}}$$

as in eq. 3.2.6. There is also the ion–solvent interaction $\mathbf{U}_{ion\text{-}solvent}$ which has an opposite sign relative to $V_{Coulomb}$, the change in the solvent–solvent interaction $\Delta\mathbf{U}_{solvent-solvent}$, which has the same sign as $V_{coulomb}$; and there is a term $-T\Delta\mathbf{S}_{solvent}$ due to the charge in solvent entropy, mainly from orienting molecular dipoles. For the description to be consistent $\mathbf{U}_{ion-solvent} + \Delta\mathbf{U}_{solvent-solvent} - T\Delta\mathbf{S}_{solvent}$ has to add up to

$$-\frac{q_1 q_2}{4\pi\varepsilon_0} \frac{1}{R_{12}}\left(1 - \frac{1}{\varepsilon_r}\right).$$

The interaction energy of eq. 3.5.15 represents the total work required to bring the charges from infinite separation up to R_{12}. During the process, the solvent molecules adjust to equilibrium. This illustrates the third point; with ε_r present the interaction energies represent free energies because the solvent's response involves both energy and entropy contributions. With ε_r we describe an average over some molecular degrees of freedom in much the same way as do the Keesom interaction of eq. 3.3.6. The relation between the free energy and the internal energy given by

$$\mathbf{A} = \mathbf{U} - T\mathbf{S} = \mathbf{U} + T(\partial\mathbf{A}/\partial T) \tag{3.3.7}$$

requires that we take the temperature dependence of the solvent dielectric constant into account. Thus, if the dielectric constant changes with temperature (as it does for all polar solvents including water, but not appreciably for nonpolar solvents like hydrocarbons), the interaction energy has an entropic component even if the interaction potential does not have an explicit temperature dependence. The entropic contribution for the Coulomb energy (eq. 3.5.15) is

$$-TS = T\left(\frac{\partial\mathbf{A}}{\partial T}\right) = -\frac{q_1 q_2}{4\pi\varepsilon_0 R_{12}}\left(\frac{T}{\varepsilon_r^2}\frac{\partial\varepsilon_r}{\partial T}\right) = -\mathbf{A}\left(\frac{T}{\varepsilon_r}\frac{\partial\varepsilon_r}{\partial T}\right) \tag{3.5.16}$$

As a result, we can obtain the internal energy by multiplying eq. 3.5.16 with the factor $(1 + (T/\varepsilon_r)(\partial\varepsilon_r/\partial T))$. For water at room temperature, $(T/\varepsilon_r)(\partial\varepsilon_r/\partial T) \simeq -1.35$ so that $\mathbf{U} \simeq -0.35\mathbf{A}$.

Problem: Calculate the changes ΔU and $T\Delta S$ in internal energy and entropy when a sodium ion and a chloride ion are brought from a large separation into contact ($R_{NaCl} = 0.276$ nm) in water at 20°C ($\varepsilon_r \simeq 80$).

Solution: The free energy change is

$$\Delta A = \frac{-e^2}{4\pi\varepsilon_0\varepsilon_r}\left[\frac{1}{R_{NaCl}} - \frac{1}{R_\infty}\right]$$

$$= -\frac{(1.6\times10^{-19})^2}{4\pi\times80\times8.85\times10^{-12}}\times\frac{1}{0.279\times10^{-9}} = -1.05\times10^{-20}(\sim-2.6\,kT)$$

or $-6.6\,$kJ/mol. Then

$$\Delta U = -0.35\Delta A = 3.7\times10^{-21}\,J \quad \text{or} \quad 2.3\,\text{kJ/mol}$$

So despite the fact that the chloride ion attracts the sodium ion, the internal energy increases and heat is taken up from the surrounding thermal bath when the ions approach one another.

The entropic part of the free energy is $-T\Delta S = \Delta A - \Delta U = 1.35\Delta A = -1.42\times10^{-20}$ or $-8.9\,$kJ/mol. This large increase in entropy comes from the decrease in water dipolar orientation around the ions as these come closer to each other.

For a general, not neutral charge distibution ρ in a medium the charge–charge interaction energy remains as in eq. 3.5.8

$$U_{el}^\dagger = \frac{1}{2}\int\rho\Phi dV = \frac{1}{2}\varepsilon_0\varepsilon_r\int(\vec{\nabla}\Phi)^2 dV \qquad (3.5.17)$$

with the difference that ε_r enters into the second equality. We have added a dagger superscript to show that the charge–charge interaction is a free energy for a medium where ε_r is temperature dependent. The entropic contribution comes from the orientation of solvent molecules, which is automatically included in ε_r.

3.6 The Poisson Equation Containing Solvent-Averaged Properties Describes the Free Energy of Ion Solvation

The Born model of ion solvation illustrates the use of the continuum dielectric model of a solvent down to molecular length scales. For a spherical ion we can calculate the free energy difference of an ion in the solvent relative to the gas phase by using either of the two expressions in eq. 3.5.17. Although the second expression requires somewhat more computation, it is conceptually clearer.

Using Gauss's law (eq. 3.5.6), we can determine that the field in the radial direction outside a spherically symmetrical charge distribution such as in an atomic ion with net charge q is

$$E_r = \frac{q}{4\pi\varepsilon_0}\frac{1}{r^2} \qquad (3.6.1)$$

in vacuum and a factor $1/\varepsilon_r$ smaller in the solvent. Inside the ion the electrical field is unaffected by the presence of a medium in spherical symmetry. We can see this by using the principle of superposition and consider only the potential due to the medium in the space occupied by the ion. The field across a spherical shell is still, according to Gauss's law, determined by the enclosed charge; but since the medium is absent in the central space, the enclosed charge in this case is zero and that part of the field is zero. Thus, we need to consider the electrical field for $r \geqslant R_{ion}$ only, where R_{ion} represents the radius of the ion

$$U_{vac} = \frac{1}{2} \varepsilon_0 \int E^2 dV = \frac{\varepsilon_0}{2} \int_{R_{ion}}^{\infty} \left(\frac{q}{4\pi\varepsilon_0}\right)^2 \frac{1}{r^4} 4\pi r^2 dr = \frac{1}{8\pi\varepsilon_0} q^2 \frac{1}{R_{ion}} \qquad (3.6.2)$$

We can obtain the free energy in a medium simply by replacing ε_0 with $\varepsilon_0 \varepsilon_r$ as in eq. 3.5.2. The solvation free energy, which is the difference between these expressions, is

$$\mathbf{A}_{solv} = \mathbf{U}_{med}^{\dagger} - U_{vac} = \frac{1}{8\pi\varepsilon_0\varepsilon_r} q^2 \frac{1}{R_{ion}} - \frac{1}{8\pi\varepsilon_0} q^2 \frac{1}{R_{ion}}$$

$$= \frac{-q^2}{8\pi\varepsilon_0} \frac{1}{R_{ion}} \left(\frac{\varepsilon_r - 1}{\varepsilon_r}\right) \qquad (3.6.3)$$

Another way to derive eq. 3.6.3 is by writing $\Phi(\bar{r}) = \Phi_{ion}(\bar{r}) + \Phi_{medium}(\bar{r})$. Using Gauss's law once more, we can determine that for $r < R_{ion}$, the potential from the medium is constant. Therefore, the charge distribution of the ion does not change when it passes from vacuum to medium. As a result, Φ_{ion} is the same in the two cases, and $\mathbf{A}_{solv} = \frac{1}{2}q\Phi_{medium}$ $(r \leqslant R_{ion})$. For $r \geqslant R_{ion}$, $\Phi_{medium} = \Phi - \Phi_{ion} = -(q/4\pi\varepsilon_0)(1 - 1/\varepsilon_r)(1/r)$, and by inserting $r = R_{ion}$ we regain eq. 3.6.3.

Solvation energy in water at room temperature for $q = e = 1.6 \times 10^{-19}$C, $\varepsilon_r = 80$, and $R_{ion} = 2$Å equals -5.7×10^{-19}J $\leftrightarrow -140\,kT$. This calculation demonstrates why we need a polar solvent with a high ε_r to dissolve electrolytes. Changing ε_r from 20 to 80 changes the solvation energy by more than $6kT$ per ion. For a 1:1 electrolyte, this would imply a change in solubility of a factor $(\exp(6))^2 \simeq 1.6 \times 10^5$!

A biological lipid membrane sets up a sheath of low dielectric permittivity $(\varepsilon_r \approx 3)$ between two aqueous regions. Passive diffusion of ions across the membrane is slow because the concentration of ions in an oillike layer is very small. For example, we can use the Born model to estimate the difference in solvation energy for a sodium ion between oil and water, $q^2/(8\pi\varepsilon_0 R_{ion})(79/80 - 2/3) \simeq 44\,kT$, which corresponds to a distribution coefficient $K \simeq \exp(44) \simeq 10^{19}$. This result overestimates the distribution coefficient, but it provides a qualitative explanation of the phenomenon.

Even though the Born model of ion solvation relies on a simplistic model of a solvent, it conveys a great deal of important information. It gives numbers of the right order of magnitude, although the quantitative agreement is on the order of

$\pm 50\%$, depending on the ion and on the choice of R_{ion}. We have emphasized that ion solvation is a long range property. Equation 3.6.3 shows that solvation energy depends on the charge squared and that the polarization effects in the medium are long-ranged. For example, the solvation energy of an ion in a solvent droplet would depend on $1/r_{droplet}$.

3.7 Self Assembly, Ion Adsorption and Surface Titration Play an Important Role in Determining Properties of Charged Interfaces

In practice, colloidal interfaces are charged and one of the themes of this book is to demonstrate how these charges are essential in determining the physical and chemical properties of the colloidal systems in question. There are three basic mechanisms that generate charged interfaces. The first, and most direct, is when the colloid is formed by intrinsically charged units. Examples include: surfactant systems where the self-assembly process generates these charges by collecting the charged amphiphilic molecules into a colloidal unit; polyelectrolytes, where the charge is due to covalently linked ionic groups distributed along the polymer chain; latex particles, where charged groups are covalently attached by sulfonation or carboxylation; clay surfaces where trivalent aluminum atoms replace tetravalent silicon atoms in the oxygen–silicon network generating negatively charged colloidal particles.

A second mechanism for generating charged surfaces arises when titratable groups are attached to the surface. Proteins contain carboxylic and amino groups that make their net charge strongly pH dependent. Polyelectrolytes like polyacrylic acid and polyamines also show this behavior. Silica surfaces contain silanol groups that titrate and a similar situation is found at most metal oxide surfaces.

The third mechanism occurs when an intrinsically uncharged colloidal particle acquires a charge from species adsorbing from solution. A minute contamination of a strongly adsorbing species can generate interfaces containing appreciable charge. Fatty acids, or the corresponding ions, as well as surfactants, adsorb strongly on apolar surfaces and insert into lipid bilayers. Large inorganic ions like I^-, ClO_4^- or SCN^- adsorb to most interfaces. All ionic crystals tend to acquire a surface charge because the ionic species forming the crystal have slightly different affinities to the surface layer.

For most charged surfaces the actual charge is determined by an equilibrium process involving exchange of charge species between the bulk solution and the interface. The final equilibrium state is determined by an interplay between three free energy contributions: the short range chemical interactions at the surface that promote adsorption; the long-range electrostatic interactions that act to prevent the formation of highly charged surfaces; and the entropy of the adsorbing ion, which typically favors desorption and the formation of the charged surface.

We can describe the effect of short range chemical interactions in terms of a conventional chemical equilibrium. For example, an ionic surface group, S^-, can bind a counterion, M^+, (or a proton) to form a neutral species, SM, and establish an equilibrium,

$$S^- + M^+ \rightleftharpoons SM \qquad (3.7.1)$$

The scheme works equally well for S neutral and SM charged, giving us

$$S + M^+ \rightleftharpoons SM^+ \qquad (3.7.2)$$

To write down the proper equilibrium equation for the process described in eq. 3.7.1, we must take into account two additional complications that normally do not arise in bulk solution, where the equilibrium can be written as $K_B = X_{SM}/X_S X_M$. First, association between the ionic species changes the surface charge density and as a result changes the electrostatic ion–ion interaction. Second, since we assume that both S^- and SM (or S and SM^+) are confined to the surface, we must consider the entropy of mixing contribution to the free energy with some care.

Bearing these factors in mind, we can write the chemical potential of a surface group as

$$\mu_S = \mu_S^{\theta}(\text{surface}) + \mu_S(\text{entropy of mixing}) + \mu_S^{el} \qquad (3.7.3)$$

If we assume a random (ideal) mixing of S and SM groups, the entropic term becomes

$$\mu_S(\text{entropy of mixing}) = kT \ln X_S \qquad (3.7.4)$$

where $X_S = n_S/(n_S + n_{SM})$ represents the fraction of charged groups, S, relative to the total number of S groups, that is, the surface mole fraction. Similarly, for the neutral SM group with mole fraction X_{SM} we have

$$\mu_{SM} = \mu_{SM}^{\theta}(\text{surface}) + kT \ln X_{SM} \qquad (3.7.5)$$

The electrostatic contribution, μ_S^{el}, to the chemical potential of the charged surface group represents the free energy required to bring a charge $z_S e$ from the bulk to the surface. This amounts to

$$\mu_S^{el} = z_S e \Phi_0 \qquad (3.7.6)$$

where Φ_0 is called the surface potential and is the difference in electrical potential between surface and bulk, where the potential Φ is set to zero by convention.

The chemical potential of the ion M^+ is uniform and assuming ideal solution behavior

$$\mu_M = \mu_M^\theta(\text{solution}) + kT \ln c_M^* \tag{3.7.7}$$

By requiring that chemical potentials are equal for the left- and right-hand sides of the scheme in eq. 3.7.1—that is, $\mu_S + \mu_M = \mu_{SM}$—we obtain the equilibrium equation

$$\frac{X_{SM}}{X_S c_M} = K_S \exp\left(\frac{-z_S e\Phi_0}{kT}\right) \tag{3.7.8}$$

which explicitly brings out the electrostatic contribution to the equilibrium. The surface equilibrium constant is related to the standard chemical potentials through

$$kT \ln K_S = \mu_M^\theta(\text{solution}) + \mu_S^\theta(\text{surface}) - \mu_{SM}^\theta(\text{surface}) \tag{3.7.9}$$

In this expression, K_S differs from the corresponding equilibrium constant K_B, for the association in the bulk solvent, because of two factors. First, K_B includes a substantial contribution from the long range electrostatic interaction between S^- and M^+. The surface potential Φ_0 includes this contribution, however, so it is absent in K_S. Second, confining S^- and SM to a standard state of a surface environment leads to additional differences between K_B and K_S. The solvation energy at the surface, particularly for the charged species (S^-), can be substantially smaller than it is in the bulk if we assume an aqueous or polar solvent. Nevertheless, K_S and K_B usually do not differ by more than an order of magnitude, since the two effects tend to cancel in part.

Surface group titration provides the clearest example of a substantial nonelectrostatic contribution to the binding between an ion and a surface. In surface titration, a proton binds to a carboxylate ion, an amine, or a silanol group with a covalent bond. A titration curve, for this interaction showing degree of ionization versus pH, will not have the same sigmoid shape as observed in the bulk. Because of the electrostatic interactions, increasing the number of charged groups becomes progressively more difficult the higher the surface charge density.

Equation 3.7.8 expresses the effect of these interactions in the factor $\exp(-z_i e\Phi_0/kT)$. We can calculate the expected titration behavior at a surface if we combine eq. 3.7.8 with an independent estimate that gives the relation between surface charge density and surface potential. We show in the following section of this chapter how this is accomplished.

It is also possible to measure the surface potential of a colloidal particle through the electrophoretic mobility, for example, as we discuss in detail in Chapter 8. As an example, consider the protein enzyme ribonuclease. Like all proteins it has both carboxylic and amino groups, whose degree of protonation change with pH. Figure 3.7 shows the surface potential in the range of pH from 4.0 to 11.0 measured at 25°C and in a

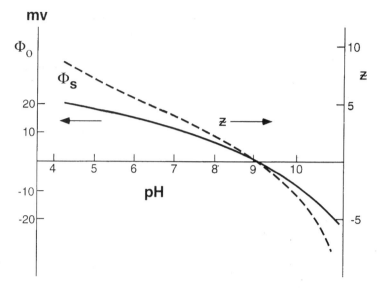

Figure 3.7
The measured surface
potential Φ_0 and charge ze,
from electrophoresis, of
the protein ribonuclease as
a function of pH in 50 mM
KNO_3. At the isoelectric
point just above pH 9 the
protein is net neutral, that
is, carries on average the
same number of positive
and negative charges. The
pH dependence of the
surface potential is due to
the titration of COO^- and
$-NH_2$ groups. (From
Norde *et al.*) (Data from
Norde, Proteins at
Interfaces, The Agricultural
University, Wageningen,
The Netherlands, 1976).

50 mM KNO_3 solution. Ribonuclease is a small protein (molecular weight 12.600 Daltons) and the net charge varies from approximately $+7e$ to $-6e$ over this pH range. The surface potential goes through zero at a pH just above 9. This is called the isoelectric point or the point of zero charge. The ribonuclease has an unusually high fraction of amino groups so that the isoelectric point is at a higher pH than for most proteins. From a functional point of view this is certainly correlated with the fact that ribonuclease has as its substrate the negatively charged ribonucleic acid, RNA.

It is a general observation that colloidal interfaces are more often negatively, rather than positively, charged. For example, it is striking that in the living cell all membranes and organelles carry a negative charge. There are several reasons why interfaces more easily acquire a negative than a positive charge. One of them is that anions tend to bind unspecifically to interfaces.

A general sequence called the Hofmeister series seems to describe ion-binding effects in a large number of systems. In order of decreasing surface affinity, the sequence for more common anions is

$$I^- > ClO_4^-, NO_3^- > Br^- > Cl^- > OH^-, F^-, SO_4^{2-}$$

Although the effect of the ions is sometimes ascribed to their influence on water structure, their relative affinity to interfaces seems to offer a more appropriate explanation.

As a consequence of specific ion binding effects, the surface charge density, σ, and surface potential, Φ_0, are difficult to control in any particular system. In principle, these properties vary with conditions, which we have to remember when we consider the electrostatic contribution to the interaction between surfaces.

3.8 The Poisson–Boltzmann Equation Can Be Used to Calculate the Ion Distribution in Solution

3.8.1 The Gouy–Chapman Theory Relates Surface Charge Density to Surface Potential and Ion Distribution Outside a Planar Surface

Interactions that occur between charges fixed at the surface and those that are free in solution play an important role in colloidal systems. Charged surfaces are characterized by a surface charge density σ and a potential Φ_0. The adjacent solution contains electrolyte and is characterized by the bulk concentration c_{io}, ion valency z_i, and the solvent dielectric constant, ε_r. We want to determine the relationship between σ and Φ_0 and also how the potential and distribution of ions in the solutions varies with distance from the charged interface.

To do this, we solve the Poisson equations. In the region of the solution, this equation takes the form

$$\varepsilon_0 \varepsilon_r \vec{\nabla}^2 \Phi = -\rho_{(\text{free ions})} \tag{3.5.13}$$

where the solution's charge distribution is expressed as

$$\rho_{(\text{free ions})} = e \sum_i z_i c_i^*(\vec{r}) \tag{3.8.1}$$

The asterisk on c indicates that the concentration is measured in molecules per cubic meter rather than in moles per liter, and $c_i^*(\vec{r})$ represents the local concentration of ions. Because the ions in the solution are free to respond to the electrical fields, the solution's charge distribution, ρ, is not independently known. In addition to the electrostatic interaction energy, we must also consider the entropy associated with the solution's ion distribution. The electrostatic interaction favors an ordered and very localized ion arrangement, but entropic factors strive to generate a random uniform distribution of ions.

In an external potential, this compromise between energy and entropy results in a Boltzmann distribution as discussed in Section 1.5.3

$$c_i^*(\vec{r}) = c_{io}^* \exp\left(\frac{-z_i e \Phi}{kT}\right) \tag{3.8.2}$$

where z_i stands for the ion valency and c_{io}^* is the concentration of ion species i at the reference point where $\Phi = 0$, which we usually choose to be the bulk. It now appears straightforward to combine eqs. 3.8.1, 3.8.2 and 3.5.13 to give

$$\rho_{(\text{free ions})} \simeq e \sum_i z_i c_{io}^* \exp(-z_i e \Phi / kT) \tag{3.8.3}$$

and

$$\varepsilon_o \varepsilon_r \nabla^2 \Phi = -e \sum_i z_i c_{io}^* \exp(-z_i e\Phi/kT) \qquad (3.8.4)$$

where the latter is the **Poisson–Boltzmann equation** describing the ion distribution in an electrolyte solution outside a charged interface or particle.

There is, in fact, a fundamental approximation involved in the step connecting eq. 3.8.2 to eq. 3.8.4. The former equation is strictly valid only when the potential is generated by an external charge, while in the Poisson–Boltzmann equation the potential is also generated by the charge of the ions whose distribution we want to determine.

In a real electrolyte solution outside a charged surface the ions and constantly change position. In each moment there is a certain position of all ions and we could, in principle, solve the Poisson equation for that particular charge configuration. In the next moment ions have changed place slightly and we can solve for the new spatial variation of the potential. If we follow the system in this way over a molecularly long time we get a distribution of values for the potential at each position in space. The average value of this potential we call the mean potential, and this is the relevant potential that enters the Poisson–Boltzmann equation. Thus, in deriving the Poisson–Boltzmann equation we have made a so-called mean field approximation replacing true ion distributions by their distribution in the mean, and the electrostatic potential by its mean value. This is a very useful and convenient approximation, but there are circumstances of practical importance where it gives quantitatively incorrect results and it can even fail qualitatively.

We can illustrate the properties of the ion distribution around a single colloidal particle by choosing as our model an infinite uniformly charged planar surface exposed on one side to an electrolyte solution, as shown in Figure 3.8. Because the surface extends uniformly in the x and y directions, only variations in the z direction are important. As a consequence of this lateral symmetry, the Poisson–Boltzmann equation reduces to an ordinary differential equation

$$\frac{d^2\Phi}{dz^2} = -\frac{e}{\varepsilon_o \varepsilon_r} \sum_i z_i c_{io}^* \exp\left(\frac{-z_i e\Phi}{kT}\right) \qquad (3.8.5)$$

To obtain a unique solution, we need two boundary conditions plus a reference point for the potential. Electroneutrality requires that the surface charges be fully neutralized by ions in the solution, and at sufficiently large distances from the surface

$$\left.\frac{d\Phi}{dz}\right|_{z \to \infty} = 0 \qquad (3.8.6)$$

which follows from Gauss's law. Letting $\Phi = 0$ at large z is a

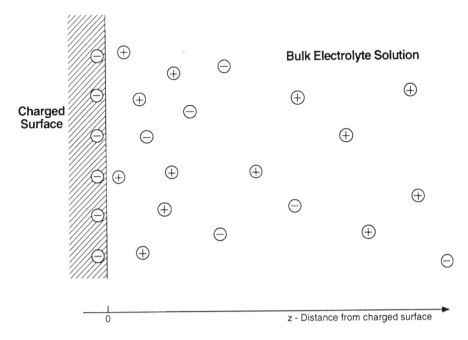

Bulk Electrolyte Solution

Charged
Surface

|
0   z - Distance from charged surface

Figure 3.8
Model used to derive the
Gouy–Chapman theory
from the Poisson–
Boltzmann equation. A
negatively charged wall
that is infinite in the x or y
direction has associated
with it a charge density σ
and a potential Φ. Valency
and concentration of
electrolyte and the
dielectric permittivity of
the solvent determine the
decrease in potential in the
z direction away from the
wall ($\Phi = \Phi_0$ at $z = 0$, $\Phi - 0$
as $z \to \infty$). The ions are
treated as point charges.

natural choice because then c_{io}^* simply represents the bulk
electrolyte concentration. Now we can apply the second boundary condition at the surface, where the situation is similar to the
capacitor discussed in Section 3.5, and the boundary condition
is in fact the same:

$$\left.\frac{d\Phi}{dz}\right|_{z=0} = \frac{-\sigma}{\varepsilon_o \varepsilon_r} \tag{3.8.7}$$

This equation follows from Gauss's law if the field is zero for
$z < 0$. No charged species are allowed for $z < 0$, and because the
system is electroneutral, $d\Phi/dz = 0$ for $z < 0$, provided no external field is applied. A more elaborate description would be
necessary if the surface were so porous that ions could penetrate into the surface material.

Solving a nonlinear second order differential equation
generally causes difficulties, but in this case a little trick can
help us. By using the identity

$$\frac{d}{dz}\left(\frac{df}{dz}\right)^2 = 2\,\frac{d^2f}{dz^2}\frac{df}{dz} \tag{3.8.8}$$

we can transform eq. 3.8.5 into the integrable form

$$\frac{d}{dz}\left(\frac{d\Phi}{dz}\right)^2 = -\frac{2e}{\varepsilon_o \varepsilon_r}\left(\frac{d\Phi}{dz}\right)\sum_i z_i c_{io}^* \exp\left(\frac{-z_i e\Phi}{kT}\right)$$

$$= \frac{2kT}{\varepsilon_o \varepsilon_r}\sum_i c_{io}^* \frac{d}{dz}\exp\left(\frac{-z_i e\Phi}{kT}\right) \tag{3.8.9}$$

Then we choose the limits in the integration to be z and ∞ so that

$$\left(\frac{d\Phi}{dz}\right)^2 = \frac{2kT}{\varepsilon_0\varepsilon_r} \sum_i c_{io}^* \left\{\exp\left(\frac{-z_i e\Phi}{kT}\right) - 1\right\} \quad (3.8.10)$$

When we take the square root of eq. 3.8.10, the result depends on the sign of the surface charge density σ. If we choose a negative surface and specialize to a symmetric $(Z:Z)$ electrolyte solution (such as NaCl, $z = 1$, or MgSO$_4$, $z = 2$) at concentration c_o^*, then

$$\frac{d\Phi}{dz} = -\left(\frac{8kT c_o^*}{\varepsilon_0\varepsilon_r}\right)^{1/2} \sinh\left(\frac{ze\Phi}{2kT}\right) \quad (3.8.11)$$

where $\sinh(x) \equiv \{\exp(x) - \exp(-x)\}/2$ stands for the hyperbolic sine function. The advantage of specializing to a single symmetrical electrolyte is that the right-hand side of eq. 3.8.10 contains an even square.

The nonlinear first-order differential equation in eq. 3.8.11 can be integrated to

$$\Phi(z) = \frac{2kT}{ze} \ln\left\{\frac{1 + \Gamma_0\,\exp(-\kappa z)}{1 - \Gamma_0\,\exp(-\kappa z)}\right\} \quad (3.8.12)$$

Here the **Debye screening length**, $1/\kappa$, given by

$$\frac{1}{\kappa} = \left(\frac{\varepsilon_0\varepsilon_r kT}{\sum_i (z_i e)^2 c_{io}^*}\right)^{1/2} \quad (3.8.13a)$$

emerges naturally. It is a central concept for the description of electrostatic interactions in electrolyte solutions and we will encounter it repeatedly throughout this book. For most applications one considers an aqueous medium at 25°C and for future reference

$$\frac{1}{\kappa} = 3.04 \times 10^{-10}/C_0^{1/2} \text{ (m)} \quad (3.8.13b)$$

for a 1:1 electrolyte at a molar concentration C_0. For a 0.1 M 1:1 electrolyte solution $1/\kappa$ is 0.96 nm and for a 1 mM solution $1/\kappa = 9.6$ nm. These lengths are of a colloidal size and they set the length scale for the range of electrostatic interactions. The second parameter Γ_0 in eq. 3.8.12 is related to the potential at the charge surface, that is, the surface potential, Φ_0, through

$$\Gamma_0 = \frac{\exp(ze\Phi_0/2kT) - 1}{\exp(ze\Phi_0/2kT) + 1} = \tanh\left(\frac{ze\Phi_0}{4kT}\right) \quad (3.8.14)$$

In Chapters 5 and 8 it will turn out that this parameter is important in determining the interaction between surfaces. However, for now we merely note that as $\Phi_0 \to 0$, Γ_0 approaches zero, while as $\Phi_0 \to \infty$, Γ_0 approaches a limiting value of unity.

We can obtain a relation between the surface charge density, σ, and the surface potential by combining eqs. 3.8.7 and 3.8.11, namely,

$$\sigma = (8kTc_o^* \varepsilon_o \varepsilon_r)^{1/2} \sinh \left(\frac{ze\Phi_0}{2kT} \right) \tag{3.8.15}$$

The quantity $ze\phi/kT$ is often encountered in describing charged colloidal systems. At 25°C, $e\Phi/kT = 1$ when $\Phi = 25.7\,\text{mV}$. Keeping this number in mind provides an easy way to estimate the ratio V/kT in charged systems and also to evaluate the exponentials. As $|\sigma|$ increases, $|\Phi_0|$ increases, and vice versa. Thus, we obtain the desired relationships between σ and Φ_0 and can see how Φ varies with electrolyte concentration and distance from the charged surface.

Problem: Determine the surface charge density for a charged interface with a surface potential of $-75\,\text{mV}$ in an aqueous NaCl solution of 0.15 M at 25°C.

Solution: According to eq. 3.8.15

$$\sigma = (8kTc_o^* \varepsilon_o \varepsilon_r)^{1/2} \sinh \left(\frac{ze\Phi_0}{2kT} \right)$$

$c_o = 0.15\,\text{M} \to c_o^* = 0.15 \times 6.023 \times 10^{26}\,\text{m}^{-3}$

$\varepsilon_r = 78.5, \ \varepsilon_o = 8.85 \times 10^{-12}\,\text{C/Vm}$

$z = 1, \ kT = 4.1 \times 10^{-21} \to (8kT\varepsilon_0\varepsilon_r)^{1/2} = 4.77 \times 10^{-15}$

$\to \sigma = (0.15 \times 6.023 \times 10^{26})^{1/2} \cdot 4.77 \cdot 10^{-15}$

$$\times \frac{1}{2} \left\{ \exp \left(\frac{1.6 \times 10^{-19} \times -75 \times 10^{-3}}{2 \times 4.1 \times 10^{-21}} \right) - \exp \left(\frac{-1.6 \times 10^{-19} \times -75 \times 10^{-3}}{2 \times 4.1 \times 10^{-21}} \right) \right\}$$

$= 4.5 \times 10^{-2}(-2.04) = -0.09\,\text{C/m}^2$

which corresponds to one negative charge per $180\,\text{Å}^2$.

The charged surface and the neutralizing diffuse layer of counterions are said to form an **electrical double layer**. The thickness of the diffuse layer is of the order $1/\kappa$, the Debye screening length. Viewed this way, the arrangement of charges resembles that in a parallel plate capacitor discussed in Section 3.5, with a gap size of $1/\kappa$.

We can obtain the differential capacitance per unit area of the electrical double layer by differentiating eq. (3.8.15) with respect to Φ_0, which yields

$$C = \left(\frac{2z^2 e^2 c_o^* \varepsilon_o \varepsilon_r}{kT} \right)^{1/2} \cosh \left(\frac{ze\Phi_0}{2kT} \right) \tag{3.8.16}$$

As Figure 3.9 illustrates, the capacitance increases with increasing potential and electrolyte concentration.

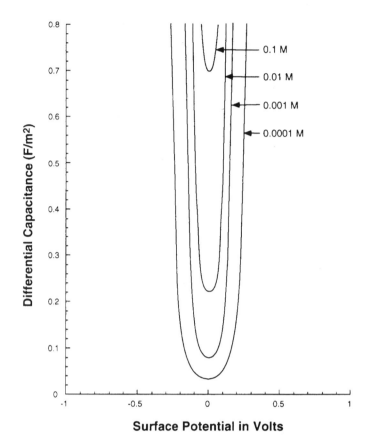

Figure 3.9
Differential capacitance of the Gouy–Chapman layer as calculated from eq. 3.7.16 using different concentrations of a 1:1 electrolyte in water. (H. R. Kruyt, ed., *Colloid Science*, Vol. 1, Chap. 4, Elsevier, New York, 1952, p. 135.)

Differential Capacitance (F/m²) — vertical axis with markings 0, 0.1, 0.2, 0.3, 0.4, 0.5, 0.6, 0.7, 0.8

Surface Potential in Volts — horizontal axis with markings -1, -0.5, 0, 0.5, 1

Curve labels: 0.1 M, 0.01 M, 0.001 M, 0.0001 M

3.8.2 Linearizing the Poisson–Boltzmann Equation Leads to Exponentially Decaying Potentials and the Debye–Hückel Theory

We can gain considerable insight into the relationships between σ, Φ, and κ by first examining the limiting forms of eq. 3.8.12, for the potential profile, and eq. 3.8.15, for the surface potential, in weakly charged or highly screened (large κ) systems. In the limit of low potential, we can linearize the Poisson–Boltzmann equation by writing the right-hand side of eq. 3.8.4 as

$$e \sum_i z_i c_{io}^* \exp\left(\frac{-z_i e\Phi}{kT}\right) \simeq e \sum_i z_i c_{io}^* \left(\frac{1 - z_i e\Phi}{kT + \cdots}\right) = -\varepsilon_o \varepsilon_r \kappa^2 \Phi \qquad (3.8.17)$$

where the first term vanishes because of electroneutrality in the bulk case. Then eq. 3.8.5 simplifies to

$$\frac{d^2\Phi}{dz^2} = \kappa^2 \Phi \qquad (3.8.18)$$

which upon integration using the boundary conditions given in eq. 3.8.6 gives

$$\Phi(z) = \Phi_0 \exp(-\kappa z) \qquad (3.8.19)$$

From the boundary condition eq. 3.8.7, the surface charge density is proportional to the surface potential

$$\sigma = \varepsilon_o \varepsilon_r \kappa \Phi_0 \qquad (3.8.20)$$

The ratio of the charge density to the surface potential yields a capacitance per unit area

$$\frac{C}{\text{area}} = \frac{\sigma}{\Phi_0} = \frac{\varepsilon_o \varepsilon_r}{1/\kappa} \qquad (3.8.21)$$

which also emerges by expanding eq. 3.8.16 assuming $ze\Phi_0/kT < 1$. By analogy with eq. 3.5.5, $1/\kappa$ equals the distance h separating the two parallel capacitor plates. Thus, we can picture the counterions that neutralize the charge on the surface to behave on the average as though they were located at a distance of $1/\kappa$ away from it. As the concentration of ions and consequently of κ increases, the potential decays more rapidly and the diffuse layer moves closer to the charged surface.

To obtain eqs. 3.8.19 and 3.8.20, we assumed that the linearization in eq. 3.8.17 holds for all z values. This condition is a rather restrictive. However, in actuality the exponential decay of the potential in eq. 3.8.19 always applies sufficiently far from the surface where the potential approaches zero. We can see this is so by making a series expansion of the solution in eq. 3.8.12. When $\Gamma_0 \exp(-\kappa z) \ll 1$ (which occurs for a sufficiently large z),

$$\Phi(z) = \frac{4kT}{ze} \Gamma_0 \exp(-\kappa z) \qquad (3.8.22)$$

where Γ_0 is defined in eq. 3.8.14.

Problem: Expand Γ_0 assuming $ze\Phi_0/kT < 1$ to show that eqs. 3.8.22 and 3.8.19 are consistent.
Solution:

$$\Gamma_0 = \frac{\exp(ze\Phi_0/2kT) - 1}{\exp(ze\Phi_0/2kT) + 1} \simeq \frac{1 + ze\Phi_0/2kT - 1}{1 + ze\Phi_0/2kT + 1} \simeq \frac{ze\Phi_0}{4kT}$$

$$\Phi(z) \doteq \frac{4kT}{ze} \frac{ze\Phi_0}{4kT} \exp(-\kappa z) = \Phi_0 \exp(-\kappa z)$$

It is only for a few special cases that the nonlinear Poisson–Boltzmann equation has a closed form solution. However, when it can be linearized the mathematical problems reduce substantially and analytical solutions are obtained for a range of conditions and geometries. Of special practical importance is the solution for a spherically symmetrical case. For example, it forms the basis of the Debye–Hückel theory of electrolyte solutions. This theory focuses attention on one particular central ion j and calculates the distribution of the other ions around it. We assume that the central ion with radius R_{ion} is of the same nature as the free ions and that we can describe the interactions by the linearized Poisson–Boltzmann equation. Because the distribution of other ions around the central ion is spherically symmetrical, we transform the Laplace operator of eq. 3.8.4 to spherical coordinates and combine with eq. 3.8.17 to

$$\frac{1}{r^2}\left(\frac{d}{dr}\left(r^2\frac{d\Phi}{dr}\right)\right) = \kappa^2\Phi \tag{3.8.23}$$

Here Φ is the potential in a coordinate system fixed at the central ion as it moves through solution.

Integration of eq. 3.8.23 yields

$$\Phi = \frac{A\,\exp(-\kappa r)}{r} + \frac{B\,\exp(\kappa r)}{r} \tag{3.8.24}$$

Combined with the boundary conditions $\Phi \to 0$ as $r \to \infty$ and $d\Phi/dr = -z_j e/(4\pi\varepsilon\varepsilon_o r^2)$ when $r = R_{ion}$ (eq. 3.6.1), this gives

$$\Phi = \frac{z_j e}{4\pi\varepsilon_o\varepsilon_r}\frac{\exp[-\kappa(r - R_{ion})]}{r(1 + \kappa R_{ion})} \tag{3.8.25}$$

which is the Debye–Hückel expression for the potential. Thus, in the electrolyte solution the potential decreases exponentially with distance from the central ion rather than the slow $1/r$ decay for the bare ion. The ion cloud screens the potential and effectively makes the electrostatic interactions more short-ranged.

As a result of these calculations, a picture emerges of a central ion surrounded by its ionic atmosphere located at a distance of $1/\kappa$ from the ion's surface. We can extend this idea to the description of weakly charged particles in general and we will return to it in later chapters.

3.8.3 The Gouy–Chapman Theory Provides Insight into Ion Distribution near Charged Surfaces

Using the physical insight gained from the linearized equations, we now delve into a more thorough discussion of the complete Gouy–Chapman eqs. 3.8.12 and 3.8.15. Figure 3.10 displays how potential varies as a function of distance for two limiting cases of constant potential and constant charge density. Recalling eq. 3.8.7, which established the limiting slope of Φ as proportional to the negative of the charge density, we can see from the figures that with a constant charge density, the initial slopes are identical, but Φ_0 decreases upon addition of more electrolyte. At constant potential, the slope becomes larger— that is, σ increases with added electrolyte.

We can use the data shown in Figure 3.10, along with the Boltzmann expression given in eq. 3.8.2, to calculate the local concentration of counter- and coions as a function of bulk concentration and distance (see Figure 3.11). The charge density is the difference between the concentration of these two species. It yields σ when integrated out from the surface, showing that the system is electroneutral on a length scale on the order of $1/\kappa$. Figure 3.11 also indicates how the charge distribution ρ concentrates more toward the surface when the electrolyte concentration (or ion valency) increases.

In practical applications we sometimes want to know the ion concentration close to the surface. This concentration is relevant when a chemical process, such as in micellar catalysis, ion adsorption, or surface group titration, occurs at the surface.

We can obtain a simple expression for the total ion surface concentration by combining eq. 3.8.10 evaluated at $z = 0$ with the boundary condition of eq. 3.8.7 so that

$$\left(\frac{-\sigma}{\varepsilon_o \varepsilon_r}\right)^2 = \frac{2kT}{\varepsilon_o \varepsilon_r} \sum_i c_{io}^* \{\exp(-z_i e\Phi_o/kT) - 1\}$$

$$= \frac{2kT}{\varepsilon_o \varepsilon_k} \sum_i \{c_i^*(0) - c_{io}^*\} \qquad (3.8.26)$$

Figure 3.10
The change in potential as a function of distance and electrolyte concentration at (A) constant surface potential Φ_0 and (B) constant surface charge density σ is calculated using eqs. 3.8.12, 3.8.14–3.8.15. At constant σ, the initial slopes α and β of $d\Phi_0/dz|_{z-0} = -\sigma/\varepsilon_o \varepsilon_r$ (eq. (3.8.7)) are identical, but Φ_0 decreases upon addition of electrolyte. At constant Φ_0, adding electrolyte causes the surface charge density to increase. (H. van Olphen, *Introduction to Clay Colloid Chemistry*, Wiley-Interscience, New York, 1963, p. 34.)

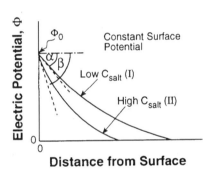

A

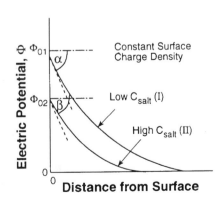

B

Figure 3.11
(a) Charge distribution in the Gouy–Chapman diffuse double layer at constant surface potential. The curves are calculated using the Poisson–Boltzmann relation (eq. 3.7.2) and the potential as a function of distance displayed in Figure 3.7a. The distance D and D' where the local concentrations (c_+ and c_-) begin to depart from the bulk value c_0 decrease with increasing electrolyte. Local increase in concentration of cations at the negatively charged surface is shown by curves $A'D'$ and AD, and the local decrease in concentration of anions by curves $C'D$ and CD. The areas of $A'D'C'$ and ADC are proportional to the net charge in the solution and thus to the charge density on the surface ($A'D'C' > ADC$). (b) Charge distribution at constant surface charge density: the general features are similar to those in (a) except that the areas $A'C'D'$ and ACD are equal. (H. van Olphen, *Introduction to Clay Colloid Chemistry*, Wiley-Interscience, New York, 1963, p. 34.)

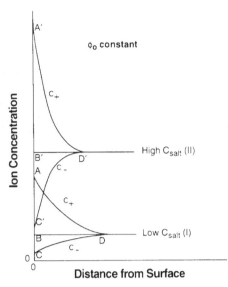

Or, rearranged to yield the sum of all surface concentrations $c_i(0)$ including both counter- and coions,

$$\sum_i c_i^*(0) = \frac{\sigma^2}{2kT\varepsilon_o\varepsilon_r} + \sum_i c_{io}^* = \frac{\sigma^2}{2kT\varepsilon_o\varepsilon_r} + \frac{\Pi_{osm}}{kT} \quad (3.8.26)$$

where the second equality comes from the expression $\sum c_{io}^* = \Pi_{osm}/kT$, which follows from the definition of osmotic pressure in an ideal solution (cf. eq. 1.5.11). Sometimes called the **Graham equation**, the relation expressed in eq. 3.8.26 was validated recently in an exact statistical mechanical model, and in that context, it is called the contact value theorem.

In the limit of high charge densities, the term $\sigma^2/2kT\,\varepsilon_o\varepsilon_r$ dominates the right-hand side of eq. 3.8.26. The surface potential is also high, so few coions reach the surface, and in $\Sigma c_i^*(0)$ the counterion terms dominate. For example, with $\sigma = 0.1\,C/m^2$ or one unit charge per $160\,\text{Å}^2$, $\sigma^2/(2kT\varepsilon_o\varepsilon_r) = 2.9\,M$ in water at room temperature. In this regime, neither the valency of the ion nor the bulk ionic concentrations ($\lesssim 0.5\,M$) appreciably affect the surface concentration for fixed σ. For an aqueous system in which $T\varepsilon_r$ is approximately constant, the surface concentration also is insensitive to temperature changes. These predictions are counterintuitive but well verified by experiments.

In fact, for highly charged surfaces a large fraction of the neutralizing counterions reside in a close vicinity of the charged surface by purely electrostatic forces. Above we found that the counterion concentration was $2.9\,M$ in the solution in direct contact with the surface. Consider a slice of the solution of a thickness, $\Delta \simeq 3\,\text{Å}$, of the size of a solvent molecule. If we neglect the decrease in counterion concentration over this slice the total surface charge density in the layer is for monovalent ions

$$\sigma_{counterions}(\Delta) = 2.9 \times 10^3 N_{Av} \times \Delta \times e$$

$$= 2.9 \times 6.023 \times 10^{26} \times 3 \times 10^{-10} \times 1.6 \times 10^{-19} = 0.084\,C/m^2$$

which amounts to 84% of the surface charge. Although this represents an overestimate, the calculation shows that a majority of the counterions for highly charged surfaces ($|\sigma| \gtrsim 0.05$ C/m^2) reside in close molecular contact with the surface.

Problem: Recalculate $\sigma_{counterions}(\Delta)$ to approximately take the decay in counterion concentration into account, using the known field at the surface.

Solution: Assume the surface to be negatively charged

$$\sigma_{counterions}(\Delta) = e \int_0^\Delta c_{counterions}(z)dz$$

$$= ec_o \int_0^\Delta \exp(-e\Phi/kT)dz = ec_o \exp(-e\Phi_0/kT) \int_0^\Delta \exp\left[\frac{e}{kT}(\Phi_0 - \Phi)\right]dz$$

$$\simeq ec(0) \int_0^\Delta \exp\left[-\frac{e}{kT}z\left.\frac{d\Phi}{dz}\right|_0\right]dZ$$

$$= ec(0) \int_0^\Delta \exp\left(\frac{e\sigma}{kT\varepsilon_o\varepsilon_r}z\right)dz \simeq ec(0) \int_0^\Delta \left(1 + \frac{e\sigma z}{kT\varepsilon_o\varepsilon_r}\right)dz$$

$$= ec(0)\Delta\left(1 + \frac{e\sigma\Delta}{2kT\varepsilon_o\varepsilon_r}\right) = 0.084(1 - 0.76) \simeq 0.02\,C/m^2$$

Since we use the field at the surface we overestimate the drop in potential and thus underestimate $\sigma_{counterion}$. An exact integration of the Gouy–Chapman solution yields

$$\sigma_{counterion} = 0.048\,C/m^2$$

for the particular choice of parameters.

The high concentration of counterions close to the surface raises some questions on the validity of the approximation leading to the Poisson–Boltzmann equation. At this close contact between counterions and surface ions and between the counterions themselves, short range "chemical" interactions are likely to play a role. In practice this turns up as changes in a dependence of the properties of the colloidal system on the specific nature of the counterion. It is clear that these ion specific effects are larger for positively than for negatively charged surfaces; which is to say that anionic counterions show more ion specificity than do cationic ones. Furthermore, it becomes difficult to determine a surface charge density experimentally since the counterions are so closely associated with the surfaces. That these surfaces in practice are also molecularly rough makes the problem even more complex.

One way of dealing with these difficulties is to say that the solution layer closest to a charged surface has properties so different from the bulk that it should be treated as a separate entity. This device was introduced in the 1930s by the German electrochemist Stern and the surface layer is commonly referred to as the Stern layer, whose properties are specified by a number of empirical parameters. It is the opinion of the authors of this book that the Stern layer concept is an intellectual cul de sac for the description of electrostatics in colloidal systems. One reason for this point of view is that from modern spectroscopic measurements we know molecular properties are not dramatically changed for a liquid close to a charged surface. As long as the surface charge is generated by molecular charges, and not through electrons of a metal, counterion concentrations may be high but remain physically reasonable.

An important aspect of charged surfaces is an interesting competition between counterions of different valency due to purely electrostatic factors. By a direct application of the Boltzmann distribution, the surface concentration of an ion i, charge $z_i e$, that is attracted to the surface electrostatically is

$$c_i^*(0) = c_0^* \exp\left(\frac{-z_i e \Phi_0}{kT}\right) \qquad (3.8.2)$$

From example, we can see how negatively charged clays or latex resins can act as ion exchangers to soften water by removing divalent cations like calcium ions. Highly charged ions in solution preferentially concentrate in the electrical double layer adjacent to the surface. If the clay or resins are preloaded in the sodium form, exchange of Ca^{2+} for Na^+ occurs. This ionic selectivity is enhanced when $e\Phi_0/kT \geqslant 1$ and it becomes much less pronounced as Φ_0 becomes small.

Problem: Calculate the ratio between Ca^{2+} and Na^+ surface concentrations relative to the same quantity in the bulk given a surface potential of $-100\,mV$ at $T = 298\,K$.

Solution:

$$\frac{c_{Ca}(0)}{c_{Na}(0)} = \frac{c_{Ca}(\text{bulk})}{c_{Na}(\text{bulk})} \exp(-e\Phi_0/kT) = \frac{c_{Ca}(\text{bulk})}{c_{Na}(\text{bulk})} \exp\left(\frac{+1.6 \times 10^{-19} \times 10^{-1}}{4.1 \times 10^{-21}}\right)$$

$$= \frac{c_{Ca}(\text{bulk})}{c_{Na}(\text{bulk})} \times 50$$

There is thus an enhancement of the relative Ca^{2+} concentration of a factor of 50. Such a charged surface has really the ability to "fish" Ca^{2+} ions from the solution when the electrolyte concentration is low so that the surface potential is high.

3.9 The Electrostatic Free Energy is Composed of One Contribution from the Direct Charge–Charge Interaction and One Due to the Entropy of the Nonuniform Distribution of Ions in Solution

3.9.1 There are Several Equivalent Expressions for the Electrostatic Free Energy

We have, for the Gouy–Chapman and the Debye–Hückel cases, calculated the electrostatic potential and the distribution of ions outside a charged unit. With this information we can also determine the electrostatic contributions to thermodynamic properties. The fundamental quantity of interest is the free energy, which, as we emphasized in Chapter 1 can be analyzed most effectively in terms of the energetic and entropic contributions.

$$\mathbf{A}_{el} = \mathbf{U}_{el} - T\mathbf{S}_{el} \tag{3.9.1}$$

The energy part (also including the entropy of the solvent) derives from the direct interaction between charges, modified by the presence of the solvent. We considered this effect in Section 3.5 and derived

$$\mathbf{U}_{el}^{\dagger} = \frac{1}{2} \int \rho\Phi \, dV = \frac{1}{2}\varepsilon_r\varepsilon_o \int (\vec{\nabla}\Phi)^2 \, dV \tag{3.5.17}$$

As discussed in connection with the Boltzmann distribution of eq. 3.8.2, an entropy term also is associated with the inhomogeneous ion distribution in the presence of an electrical field. Apart from the effect of the field, ions mix ideally with the solvent; therefore, the entropy of the mixing term has the same value locally as for an ideal solution theory (see Section 1.4). For a small volume element dV containing dn_S^* solvent molecules and dn_i^* molecules of ion species i (see eq. 1.4.8)

$$
\begin{aligned}
d\mathbf{S} &= -k\left(\ln X_S\, dn_S^* + \sum_i \ln X_i\, dn_i^*\right) \\
&\simeq -k\left(\sum_i \left(-c_i^* + c_i^* \ln c_i^*\right) + \sum_i c_i^* \ln v_S\right) dV
\end{aligned} \qquad (3.9.2)
$$

In the second equality we have assumed that the solvent is in excess so that the series expansion becomes possible:

$$
\ln X_S = \ln\left(1 - \sum X_i\right) \simeq -\sum X_i \qquad (3.9.3)
$$

We also use $dn_S^* \simeq dV/v_S$, $dn_i^* \simeq c_i^* dV$ and $X_i \simeq v_S c_i^*$ (where v_S is the solvent molecular volume). Here the subscript s denotes the solvent.

The last term in eq. 3.9.2 stands for a constant that can be included in the standard chemical potential. Integrating eq. 3.9.2 leads to an expression for the entropy of mixing of the total system. To explicitly obtain the entropy due to the in-homogeneous ion distribution close to a charged surface, we must subtract the entropy of the corresponding homogeneous ideal system (see eq. 1.4.8). Then the electrostatic contribution to the entropy becomes

$$
\mathbf{S}_{el} = \mathbf{S} - \mathbf{S}_{ideal} = -k\sum_i \left\{\int c_i^* \left(\ln c_i^* - 1\right) dV - n_i^* \left(\ln\left(\frac{n_i^*}{V}\right) - 1\right)\right\} \qquad (3.9.4)
$$

This formula simply reflects the excess entropy of mixing due to the inhomogeneous ion distribution, assuming local ideal mixing. The second term on the right-hand side represents the ideal entropy of mixing in the absence of surface charges.

One advantage of using eq. 3.9.1 to calculate $\mathbf{A}_{el}$ is that one needs to know the potential and ion distribution in the actual system only. In the alternative ways of determining $\mathbf{A}_{el}$ one performs an integration through a range of states. The basis for these methods is a fundamental statistical mechanics relation

$$
\mathbf{A}(\alpha) = \mathbf{A}(\text{ref}) + \int_{\alpha(\text{ref})}^{\alpha} \frac{\mathbf{U}(\alpha')}{\alpha'}\, d\alpha' \qquad (3.9.5)
$$

where α is some parameter entering the expression for the energy.

For charged systems the natural choice for α is a scaled charge $\alpha = q' = \lambda q$ $(0 \leqslant \lambda \leqslant 1)$. We say that the free energy is determined by the work of a charging process. In any colloidal system there are several kinds of charged species and when applying eq. 3.9.5 we have the choice of charging them simultaneously or sequentially. In the former case

$$\mathbf{U}^\dagger(q') = \frac{1}{2} \int \rho(q')\Phi(q')dV \qquad (3.9.6)$$

in analogy with eq. 3.5.17.

If, on the other hand, we consider a single charged species i

$$\mathbf{U}^\dagger(q') = q_i\Phi(\vec{r}_i, q') \qquad (3.9.7)$$

Here $\Phi(\vec{r}_i, q)$ is the potential from the other charges at the position of charge i. There is no double counting of the ion–ion interaction and hence the factor 1/2 is not present.

3.9.2 In the Debye–Hückel Theory the Electrostatic Contribution to the Chemical Potential of an Ion is Obtained by a Charging Process

In Section 3.8.2 we solved the linearized Poisson–Boltzmann equation for a spherically symmetrical case. We can use this result to calculate the electrostatic contribution to the chemical potentials of simple ions in solution. To accomplish this we fix the attention on a single ion of charge $z_i e$. The potential outside the ion is

$$\Phi(r) = \frac{z_i e}{4\pi\varepsilon_o\varepsilon_r} \frac{\exp[-\kappa(r - R_{\text{ion}})]}{r(1 + \kappa R_{\text{ion}})} \qquad (3.8.25)$$

which is composed of one contribution

$$\Phi_{\text{ion}}(r) = \frac{z_i e}{4\pi\varepsilon_o\varepsilon_r} \frac{1}{r} \qquad (3.9.8)$$

from the ion itself plus one generated by the surrounded ion cloud

$$\Phi_{\text{ion cloud}} = \Phi - \Phi_{\text{ion}} \qquad (3.9.9)$$

In an application of eq. 3.9.5 we now consider the charging of the single central ion from zero to $z_i e$ while keeping the other charged ions intact. This process amounts to adding a single new ion to the solution. The free energy change then, by definition, corresponds to the electrostatic contribution to the chemical potential of this charged species

$$\mu_{\text{el}}^i = \int_0^1 \lambda z_i e \, \Phi_{\text{ion cloud}}(R_{\text{ion}}) d\lambda$$

$$= \frac{(z_i e)^2}{4\pi\varepsilon_o\varepsilon_r} \frac{1}{R_{\text{ion}}} \left\{ \frac{1}{1 + \kappa R_{\text{ion}}} - 1 \right\} \int_0^1 \lambda \, d\lambda$$

$$= -\frac{\kappa(z_i e)^2}{8\pi\varepsilon_o\varepsilon_r} \frac{1}{(1 + \kappa R_{\text{ion}})} \qquad (3.9.10)$$

In relevant applications the Debye screening length κ^{-1} is large relative to ion sizes so we can neglect the term κR_{ion} in the denominator leaving

$$\mu_i^{el}(\text{Debye–Huckel}) = -\frac{(z_i e)^2 \kappa}{8\pi\varepsilon_o\varepsilon_r} \qquad (3.9.11)$$

which is the renowned Debye–Hückel ion activity correction term.

3.9.3 The Electrostatic Free Energy of a Planar Charged Surface can be Calculated in Closed Form

For a planar uniform charged surface the Gouy–Chapman theory provides expressions for the electrostatic potential and the ion distribution. However, the expressions for these are fairly complex and it is not evident that one can analytically integrate these further by either eq. 3.9.1 or 3.9.5 to obtain the free energy. Let us thus first look at the simpler case when the linearization approximation can be used. Then there is a simple relation between surface charge density and surface potential

$$\sigma = \varepsilon_o\varepsilon_r\kappa\Phi_0 \qquad (3.8.20)$$

which makes the integration in a charging process straightforward. Let the surface charge density grow from zero to σ with the surface exposed to an inert electrolyte solution. The free energy change then represents the free energy cost of generating a charged surface and

$$A_{el}/\text{area} = \int_0^1 \sigma(\lambda)\Phi_0(\lambda)/\lambda \, d\lambda = \frac{\sigma^2}{\varepsilon_o\varepsilon_r\kappa}\int_0^1 \lambda \, d\lambda = \frac{\sigma^2}{2\varepsilon_o\varepsilon_r\kappa} = \frac{1}{2}\sigma\Phi_0 \qquad (3.9.12)$$

In the general nonlinear case we have the more complex relation between σ and Φ_0 given in eq. 3.8.15. With some difficulty one can perform the integration as in the eq. 3.9.12 above. However, let us use another route to the final result and start from eq. 3.9.1. For a single planar surface we can use the Gouy–Chapman solution for the potential profile. For a 1:1 electrolyte, using eq. 3.5.14,

$$\frac{U_{el}^\dagger}{\text{area}} = \frac{\varepsilon_o\varepsilon_r}{2}\int_0^\infty \left(\frac{d\Phi}{dz}\right)^2 dz = \frac{\varepsilon_o\varepsilon_r}{2}\int_{\Phi_0}^0 \frac{d\Phi}{dz}\, d\Phi$$

$$= -(2kT\, c_o^*\varepsilon_o\varepsilon_r)^{1/2}\int_0^{\Phi_0} \sinh\left(\frac{e\Phi}{2kT}\right) d\Phi$$

$$= \frac{kT\sigma}{e}\frac{1}{s}[(s^2 + 1)^{1/2} - 1] \qquad (3.9.13)$$

where we introduce the dimensionless parameter

$$s \equiv \sigma(8kT\, c_o\varepsilon_o\varepsilon_r)^{-1/2} = \sinh\left(\frac{e\Phi_0}{2kT}\right) \qquad (3.9.14)$$

In eq. 3.9.13 the first equality follows from eq. 3.5.17, the second results from a change of variable, $z \to \Phi$, the third is a consequence of eq. 3.8.11, while the fourth involves an integration of the sinh function combined with the definition in eq. 3.9.14.

For $s \gg 1$ in eq. 3.9.13, which occurs for high charge densities, the electrostatic energy is proportional to σ. If N unit charges generate the surface charge, the energy $\mathbf{U}^\dagger$ per charge is simply kT. We might intuitively expect the energy per charge to increase with increasing charge density, but this does not happen for high charge densities.

In a similar but slightly more involved computation, we can determine the entropic contribution to be

$$-\frac{T\mathbf{S}_{el}}{\text{area}} = \frac{2kT\sigma}{e} \left\{ \ln[s + (s^2 + 1)^{1/2}] - \frac{3}{2s} [(s^2 + 1)^{1/2} - 1] \right\} \tag{3.9.15}$$

Thus, the total electrostatic free energy is

$$\frac{\mathbf{A}_{el}}{\text{area}} = \frac{2kT\sigma}{e} \left\{ \ln[s + (s^2 + 1)^{1/2}] + \frac{1}{s} - \frac{(s^2 + 1)^{1/2}}{s} \right\} \tag{3.9.16}$$

Figure 3.12
Free energy eq. 3.9.16, energy, eq. 3.9.13, and entropy, eq. 3.9.15, associated with forming a charged surface as a function of parameter s (eq. 3.9.14).

Expressions for the different free energy terms are surprisingly simple to calculate algebraically, considering the involved expressions needed to obtain the potential and concentration profiles. Figure 3.12 plots these different terms as a function of s. In the linear regime of the Gouy–Chapman solution, where s is small, the energy term $\mathbf{U}_{el}^\dagger$ equals the entropy term $-T\mathbf{S}_{el}$. For large s, however, $\mathbf{U}_{el}^\dagger$ tends to be a constant value of one kT per

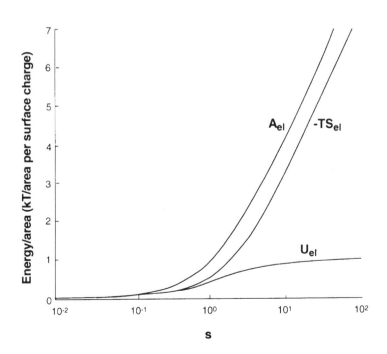

charge while $-TS_{el}$ increases. In this nonlinear regime, a change in s through a change in salt concentration c_o^* leaves the ion–ion interaction $\mathbf{U}_{el}^{\dagger}$ unchanged, despite a substantial change in the entropy of the ion distribution. What does this mean?

According to the Graham equation, eq. 3.8.26, the ion concentration close to the surface scarcely changes when the bulk concentration changes. Entropy increases, upon setting up the inhomogeneous distribution, because the ratio between the surface and the bulk concentration changes. Usually, the decrease in the free energy of the electrical double layer when salt is added is attributed to a screening effect on $\mathbf{U}_{el}^{\dagger}$. This explanation is reasonable in the linear regime with small s; for high s values, however, the effect is an entropic one, resulting from a change in the reference concentration in the bulk. Equation 3.9.5 represents the true thermodynamic internal energy only when ε_r is independent of temperature. As discussed in Section 3.5, a factor of

$$\left(1 + (T/\varepsilon_r)\left(\frac{\partial \varepsilon_r}{\partial T}\right)\right)$$

is used to multiply $\mathbf{U}_{el}$ of eq. 3.9.5 to obtain the true thermodynamic internal energy.

Clearly, electrostatic forces alone would not cause ionic surfactants to self-assemble. At the aggregate surface, an attractive interaction, due mainly to hydrophobic forces, balances the repulsion from the electrostatic interactions. The free energy of the electrical double layer also contributes to the surface pressure for charged monolayers and bilayers. We can calculate this electrostatic lateral pressure $P_{el(lat)}$ from the derivative $\partial A_{el}/\partial(\text{area})$ at constant charge. Because $\sigma \times \text{area}/e$ represents the number of charges, the area derivative of eq. 3.9.8 is

$$P_{el(lat)} = -\frac{\partial \mathbf{A}_{el}}{\partial(\text{area})} = -\frac{2kT\sigma}{e} \times \text{area} \, \frac{\partial s}{\partial(\text{area})} \frac{\partial}{\partial s} \left\{ \ln[s + (s^2 + 1)^{1/2}] + \frac{1}{s} - (s^2 + 1)^{1/2}/s \right\}$$

$$= 2kT\frac{\sigma}{e}\frac{1}{s}[(s^2 + 1)^{1/2} - 1] = \frac{2\mathbf{U}_{el}^{\dagger}}{\text{area}} \qquad (3.9.17)$$

The second-to-last equality requires some algebra, noting that $\partial s/\partial(\text{area}) = -s/\text{area}$ and the last equality follows from eq. 3.9.13.

In the nonlinear regime of the Gouy–Chapman solution, $s \gg 1$, $\mathbf{U}_{el}^{\dagger} = kT$ per charge so that the electrostatic component of the surface pressure is constant and equals $2kT/\text{area}$. Equation 3.9.17 expresses one in a series of simple relations involving electrostatic free energies, and its validity goes far beyond the Gouy–Chapman case. By permitting us to obtain analytical expressions like these for the different physical quantities, the Gouy–Chapman theory contributes a great deal to our conceptual understanding of the electrical double layer.

Exercises

3.1. In a cube of side a, equal charges q are located at all eight corners. Calculate the total electrostatic interaction energy. Compare this with the electrostatic energy of a uniformly charged sphere of charge $8q$ and a radius of $\sqrt{3}/2a$ such that it touches the corners of the cube.

3.2. A simple model to represent the charge distribution in a water molecule is to consider charges q_0, q_H at the positions of the atoms. Since the molecule is net neutral, $q_{H1} = q_{H2} = -0.5q_0$. What is the value of q_0 consistent with a dipole moment 1.87 D? The bond length is 0.096 nm and the bond angle H   H is 104°.

3.3. Calculate the interaction between a water molecule and a spherical anion A for both the hydrogen-bonded

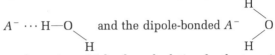

configurations. Make the calculation for the two $A^- \cdots$ O distances of 0.25 and 0.30 nm. Use the electrostatic model of Exercise 3.2 for the water molecule.

3.4. Calculate the electrostatic interaction energy between two water molecules in a hydrogen bond conformation and an oxygen–oxygen distance of 0.28 nm. Assume a planar complex.

$$\text{H} \diagdown \text{O} \ldots \text{H---O} \diagdown$$

Use the model of Exercise 3.2. Compare the interaction with the optimal dipole–dipole interaction of the same oxygen–oxygen separation.

3.5. The Keesom interaction energy for rotating dipoles is obtained from a series expansion, which is not useful when the dipole–dipole interaction can reach a value of kT ($T = 300$K). At what magnitude of the dipole does this happen at room temperature and at a separation between the two equivalent dipoles of 0.30 nm?

3.6. Calculate the hydrogen-bond-acceptor dipole–donor-induced dipole interaction for two water molecules in the hydrogen bond configuration of Exercise 3.4. Assume that polarizabilities and dipole moments on the oxygen atoms.

3.7. The polarizability of a free Cl^- ion can be estimated to 5.0×10^{-39} C·m²/V. What is the polarization interaction in an Na^+Cl^- pair and in a linear $Na^+Cl^-Na^+$ complex? Use a distance of 0.28 nm and neglect the polarization of Na^+.

3.8. Calculate the force per unit area acting between the plates of a planar capacitor when (a) the surface charge is kept constant and (b) the potential difference is kept constant. (Include the electrical work needed to keep the potential constant.)

3.9. Show that the electrostatic potential outside an infinite uniformly charged cylinder varies logarithmically in the radial direction. Determine the potential function when there are q charges per unit length along the cylinder axis.

3.10. The table gives the hydration free energies and the ionic radii determined from crystal data for a number of cations. Explain the trends and comment on the quantitative accuracy of the Born model (T = 298K).

Ion	R_{ion} (nm)	ΔG (hydr) (kJ/mol)
Li^+	0.068	$-$ 505
Na^+	0.098	$-$ 406
K^+	0.133	$-$ 332
Mg^{2+}	0.065	-1895
Cr^{3+}	0.069	-4209

3.11. Calculate the Debye screening length κ^{-1} at 298K for the aqueous solutions: (a) 0.1 M NaCl, (b) 10 mM NaCl, (c) 1 mM NaCl, (d) 0.1 M $MgSO_4$, and (e) *pure* water, assuming $K_w = 10^{-14}$. Discuss what the relevant κ value is in a 28 mM solution of sodium dodecyl sulfate.

3.12. Consider an infinitely charged cylinder of radius R_c as in Exercise 3.9. Let it be surrounded by an electrolyte solution of concentration c_0, which might be very small. Calculate the counterion, charge ze, concentration profile, assuming that the electrostatic potential is due solely to the charged cylinder. Integrate the counterion concentration between $r = R_c$ and $r \rightarrow \infty$. Does a consistent solution exist for all charge densities? Discuss the result.

3.13. A bilayer of amphiphilic molecules separates two aqueous electrolyte solutions. When a charge is brought from one of the solutions to the other, the bilayer will act as a capacitor separating two conducting media. What is the capacitance per unit area if the bilayer is 4 nm thick and has a relative dielectric permittivity of $\varepsilon_r = 3$?

3.14. An amphiphilic bilayer in an aqueous solution can close upon itself to form a spherical shell, a visicle. Calculate the resulting potential difference between inside and outside when 10 potassium ions have been transported across the bilayer. Assume excess electrolyte in the water and a vesicle radius of 20 nm. Use the bilayer characteristics of Exercise 3.13.

3.15. In Exercises 2.7, 2.8, and 2.9 of Chapter 2, we found that the area per molecule at the air–water interface of an aqueous solution of dodecyldimethylammonium chloride was 43 Å^2 for pure surfactant and 41.5 Å^2 when 0.20 M NaCl had been added. These values refer to conditions at the CMC. Calculate the concentration of Cl$^-$ at the charged surface in the two cases, assuming that the amphiphile is fully protonated.

3.16. Dodecyldimethylammonium chloride has a proton that can titrate to form the uncharged primary amine. The pK_α

is around 10.6. At a bulk concentration of 15 mM the pH is approximately 6.2. Calculate the surface charge density at the air–water interface. The area per molecule is 43 Å^2.

3.17. The surface potential of an AgI surface is determined by the activity (concentration) of Ag^+ and I^- ions. Assume that $\Phi_0 = 75$ mV and is unaffected by the addition of electrolyte.
  (a) Calculate the charge density and the Debye length $1/\kappa$ for a solution containing either KNO_3 or $MgSO_4$ at concentrations of 10^{-4} and 10^{-2} M.
  (b) Sketch how Φ decreases as a function of distance from the AgI surface and show how it is related to σ.
  (c) If the bulk solution contains equal concentrations of NO_3^- and SO_4^{2-} (10^{-3} M), estimate the relative concentration of the two anions at the surface.

3.18. (a) Calculate the potential and electrical field associations with an isolated potassium ion in water at distances of 10 and 30 Å from the ion.
  (b) Calculate the potential arising from a potassium ion in a 10^{-3} M KCl solution at distances of 10 and 30 Å from the ion.

3.19. Ionic micelles can be modeled as spheres with their charged head groups on the external surface with counterions in the adjacent solution.
  (a) Use Gauss's law to derive the capacitance for a spherical capacitor consisting of two concentric spheres of radii R_1 and R_2.
  (b) Assuming that SDS micelles have a radius of 20 Å and an effective net charge of -12, calculate the capacitance in a 10^{-4} and a 10^{-2} M NaCl solution, assuming that the outer sphere is located a distance corresponding to the Debye length $1/\kappa$.
  (c) Under what conditions does the expression for a spherical capacitor approach that for a parallel capacitor?

Literature

References[a]	Section 3.1–3.4: Electrostatics: Vapor Phase Ions, Dipoles-Induced Dipoles	Sections 3.5–3.6: Electrostatics: Condensed Phase	Section 3.7 Energetics of Charged Interfaces	Section 3.8–3.9 Charged Interfaces, Gouy–Chapman Theory
Adamson & Gast				CHAPTER V (1–4) Electrical double layers
Hiemenz & Rajagopulan			CHAPTER 12.10	CHAPTER 11.1–11.6
Hunter (I)				CHAPTER 7 Charged Interfaces
Hunter (II)		CHAPTER 2.2 Dielectric materials and electrical fields	CHAPTER 6.9–6.12 Specific ion effects	CHAPTER 6.1–6.6 Gouy–Chapman
Israelachvili	CHAPTER 3–5 Ions, dipoles, and induced dipoles CHAPTER 7 General discussion of pair potential CHAPTER 8.1–8.4 Hydrogen bonding			CHAPTER 12.1–12.5 Gouy–Chapman
Meyer	CHAPTER 4 Long-Range Attractive Forces			CHAPTER 5 Double Layers
Shaw				CHAPTER 7 Double layer
Vold and Vold				CHAPTER 6.5 Electrical interfaces

[a]For complete reference citations, see alphabetic listing, page xxxi.

4

STRUCTURE
AND PROPERTIES
OF MICELLES

CONCEPT MAP

Micellar Cooperative Association Models

- Micelles are the simplest and most thoroughly characterized self-organizing structures. A stepwise association process, $S + S_{n-1} \rightleftarrows S_n$ (eq. 4.1.1), describes their aggregation in general. However, because it is almost impossible to specify all the K_n equilibrium steps, approximate models are used.

  The **isodesmic model** assumes that K_n is independent of n and describes association of some dyes in water. However, it does not capture the cooperativity associated with amphiphilic aggregation.

  The **phase separation model** approximates aggregation as a phase separation process in which the activity of the monomer remains constant above the CMC. It captures the start mechanism of aggregation but not the stop mechanism.

  The **closed-association model** assumes that one aggregation number N dominates and yields $\Delta \mathbf{G} = \mathbf{R}T \ln$ CMC (eq. 4.1.11), thus relating the free energy of micellization to the measured CMC. This model captures both the stop and start cooperative features associated with aggregation processes.

- Examination of eq. 4.1.11 shows that as N becomes larger, the CMC becomes more sharply defined. As $N \to \infty$, the phase separation model becomes a true representation of the system.

- The association model for ionic micelles (eq. 4.1.15) involves an empirical counterion binding parameter, and expressions involving the PB equation are more useful. Examination of the thermodynamics of micellization shows that above the CMC, the monomer's concentration increases very slowly (eq. 4.1.18) and it also provides a way to evaluate ΔH_{mic}, ΔS_{mic}, and the change in CMC with temperature (eq. 4.1.21).

Characterization of Micelles

Micelles are characterized by:

- **CMC**—many physical properties undergo abrupt change at the CMC, including (1) surface tension, which also provides a sensitive measure of surfactant purity, (2) conductance, and (3) surfactant ion activity as measured by specific ion electrodes.

- Aggregation number, N, which can be measured by either light-scattering or fluorescence quencher techniques.

- **Time scales for dynamic processes**, obtained by relaxation measurements, which yield two characteristic relaxation times: τ_1 (10^{-6} to 10^{-3} s) measures the rate of exchange of monomer between micelles, while τ_2 (10^{-3} to 1 s) measures the rate of micelle formation and disintegration. Because micelles must pass through a kinetic bottleneck as they disintegrate τ_2 is longer than τ_1.

- **Diffusion of micellar components**, which involves a complex interplay between monomers, counterions, and micelles.

Quantitative Analysis of Micellar Solutions

We can interpret variations in CMCs by writing $\Delta G_{mic} = \Delta G(HP) + \Delta G(contact) + \Delta G(packing) + \Delta G(HG)$, where

- $\Delta G(HP)$ represents the free energy associated with transferring the hydrocarbon chain out of water and into the oillike interior of the micelle, $\Delta G = -(n_c - 1)3.0-9.6$ kJ/mol (eq. 4.3.2).

- $\Delta\mathbf{G}$(contact) represents the surface free energy attributed to solvent—hydrocarbon contact in the intact micelle. It is proportional to area, A_{mic} and takes the form $\Delta\mathbf{G} = \gamma' A_{\mathrm{mic}}/N$ (eq. 4.3.3).

- $\Delta\mathbf{G}$(packing) represents the positive contribution associated with confining the hydrocarbon chain to the micelle's interior given by $\Delta\mathbf{G} = \text{constant} - \gamma'' A_{\mathrm{mic}}/N$ (eq. 4.3.4).

- $\Delta\mathbf{G}$(HG) represents the positive contribution associated with head group interactions, including electrostatic as well as head group conformation effects.

Ionic Micelles

Increasing the hydrocarbon chain length or adding salt lowers the CMC and increases N, while raising the temperature increases the CMC and lowers N.

Solution of the PB equation provides quantitative expressions for $\Delta\mathbf{G}$(HG) $= \Delta\mathbf{G}_{\mathrm{el}}$ that account for these experimental observations.

Figure 4.12 compares the calculated and observed changes in CMC with chain length and added salt.

Table 4.5 compares the calculated and observed changes in CMC with chain length, salt, and temperature. Figure 4.13 shows how $\Delta\mathbf{G}$(HP), $\Delta\mathbf{H}$(HP), and $\Delta\mathbf{S}$(HP) vary with temperature.

In polar nonaqueous solvents like hydrazine and ethylammonium nitrate, $\Delta\mathbf{G}$(HP) for micellization decreases in accord with the observations for $\Delta\mathbf{G}$ for gas solubilities given in Figure 1.11.

Nonionic Micelles

Nonionic micelles with zwitterionic or nonionic head groups have substantially lower CMCs than corresponding ionic surfactants (Table 1.1).

Added salt has relatively little effect on either the CMC or N. Heating causes N to increase until phase separation occurs at the cloud point temperature.

Nonionic micelles are more difficult to model than ionic micelles because head group interactions are dominated by steric and osmotic effects.

Nonspherical Micelles

Surfactants that form spherical micelles ($N_s = 0.3$) clearly illustrate the start (well defined CMCs), stop (adding more surfactant lead to more micelles not larger ones), and well defined properties ($R_{mic} \propto l_c$) as described in Section 1.2.

Many surfactants associate to form rods, polymer-like threads or branched infinite aggregates near the CMC or at high concentrations. Aggregate growth can be promoted

- In nonionic systems by increasing the temperature which diminishes favorable head group–solvent interactions.

- In ionic systems by adding a cosurfactant with a small head group, adding electrolyte which reduces head group repulsion, or adding an amphiphilic counterion.

Applications

Many biological and industrial applications involve mixed micelles containing:

- Similar surfactants with different chain length where the relative concentration of surfactants in the micelles changes in concentration and can be understood in terms of a mixing model (Figure 4.15).

- Mixed ionic and nonionic head groups where adding an ionic surfactant to a nonionic micellar system raises the cloud point, or adding a nonionic surfactant to an ionic system decreases the CMC.

- Hydrocarbons and fluorocarbons which form two distinct micelles because of the mutual insolubility of these two chains.

Examples involving surfactants include digestion of fats by bile salt micellar solubilization, dissolution mechanisms of fats and oils by aqueous surfactants, and micellar catalysis leading to increased reaction rates.

4

Micelles are the simplest of all amphiphilic self-organizing structures. They also are the best understood, because many experimental techniques have been used to characterize them. Critical micelle concentrations (CMC) occur at experimentally accessible concentrations, and we can study micelles in dilute solution, in which interaggregate interactions can be ignored. Because they display rather narrow size distributions, we often can characterize micelles by a single aggregation number. All these factors make them amenable to theoretical analysis, and quantitative reationships are available. The insights gained from studying micellar systems provide the basis for understanding the more complex bilayers and emulsions as well as self organization in general.

4.1 Micelle Formation Is a Cooperative Association Process

4.1.1 Several Models Usefully Describe Micellar Aggregation

When surfactants associate into micelles, they form a liquidlike aggregate. Because no obvious mechanism leads to a specific aggregation number, it is natural to describe the association in terms of a stepwise addition of a monomer, S, to the aggregate, S_{n-1}, as in

$$S + S_{n-1} \rightleftarrows S_n \qquad (4.1.1)$$

If we neglect additional interactions between aggregates and between monomers, we can write the equilibrium given in eq. 4.1.1 as

$$K_n = \frac{[S_n]}{[S][S_{n-1}]} \qquad (4.1.2)$$

157

Equation 4.1.2 provides a general description of any step-wise association process in dilute solution. In situations that involve aggregation numbers n of order 100, we must specify a rather intractable number of equilibrium constants K_n. For this reason, we can obtain a more useful qualitative sense of the characteristic features of different association patterns by considering some simplified models.

The isodesmic model, which we will consider first, assumes that K_n is independent of n. In this case we can show that regardless of either the total concentration or of K, $[S]K < 1$. The aggregate distribution function

$$f(n) = \frac{[S_n]}{\sum\limits_{n=1}^{\infty} [S_n]} \qquad (4.1.3)$$

decays exponentially with $[S_1] > [S_2] > [S_n]$.

In this model, aggregation is a continuous process that does not show the abrupt onset in a narrow concentration range that typifies micelle formation. The isodesmic model describes the association of some dyes in aqueous solution quite well, but it is less successful as a description of the formation of micelles because the model does not predict a CMC. Its basic shortcoming lies in making K_n independent of n and thus depriving the process of cooperativity.

Micelle formation has several features in common with the formation of a separate liquid phase. This fact provides a basis for a second model in which micelles formally constitute a separate phase. In terms of the association described in eq. 4.1.1, the phase separation model assumes that aggregates with large n dominates all others except the monomer. This assumption implies strong cooperativity because, once aggregation has started, it becomes more and more favorable to add another monomer until a large aggregation number is reached. This signifies a cooperative process. In the pseudoseparate phase, the surfactant possesses a certain chemical potential $\mu^\theta(\text{micelle})$ in the aggregate. When

$$\mu^\theta(\text{micelle}) = \mu^\theta(\text{solvent}) + RT\ln[S] \qquad (4.1.4)$$

monomers and aggregates coexist in equilibrium and $[S]$ is the CMC (neglecting dimers and oligomers). The standard free energy of micelle formation $\Delta G^\theta_{\text{mic}}$ represents the free energy difference between a monomer in the micelle and the standard chemical potential in dilute solution and

$$\Delta G^\theta_{\text{mic}} \equiv \mu^\theta(\text{micelle}) - \mu^\theta(\text{solvent}) = RT\ln\text{CMC} \qquad (4.1.5)$$

Equation 4.1.5 provides a useful approximation for obtaining $\Delta G^\theta_{\text{mic}}$, and the phase separation model captures several, but not all, essential features of micelle formation. Although it describes the start mechanism of the self-assembly process as discussed in Chapter 1, it does not describe the stop mechanism.

The third model, the closed-association model, describes both start and stop features. It assumes that one aggregation number N dominates. With only monomers and N-aggregates,

$$NS \rightleftharpoons S_N$$

$$K_N = \frac{[S_N]}{[S]^N} \tag{4.1.6}$$

Figure 4.1
Critical micelle concentration. As the aggregation number N increases, the fraction of added surfactant that goes to the micelle, $d\{N\{S_N\}\}/d[S]_T$, varies more and more steeply with total concentration $[S]_T$. In the limiting case in which the aggregation number becomes infinite, the transition becomes a step function that unambiguously defines the critical micelle concentration CMC, while small micellar aggregation numbers lead to less well-defined values of CMC. (R.J. Hunter, *Foundations of Colloid Science*, Oxford University Press, Oxford, 1989, p. 576.)

The total surfactant concentration expressed in terms of moles of monomer is

$$[S]_T = N[S_N] + [S] = NK_N[S]^N + [S] \tag{4.1.7}$$

K_N relates to the other equilibrium constants of eq. 4.1.2 by $K_N = \Pi_2^N K_n$. Using eqs. 4.1.6 and 4.1.7, we can obtain a straightforward solution for the derivative $\partial\{N[S_N]\}/\partial[S]_T$, describing what fraction of added surfactants enters into an aggregate.

Figure 4.1 shows three curves with varying values of N. The larger the N value, the more abruptly the derivative $\partial\{N[S_N]\}/\partial[S]_T$ changes from a low concentration value of zero to the high concentration value of unity. When $N \rightarrow \infty$, we regain the results of the phase separation model with a discontinuity in the derivative at the CMC.

Experiments identify the CMC as the concentration at which added surfactant preferentially starts to enter into aggregate. A good measure of this concentration point is where an addd monomer is as likely to enter a micelle as to remain in

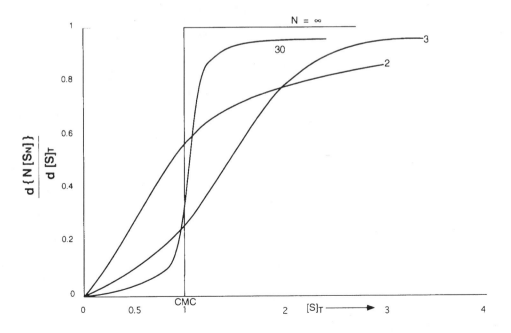

solution so that

$$\frac{\partial\{N[S_N]\}}{\partial[S]_T}\bigg|_{\text{CMC}} = \frac{\partial[S]}{\partial[S]_T}\bigg|_{\text{CMC}} = 0.5 \qquad (4.1.8)$$

Solving for the inverse of eq. 4.1.8, $\partial[S]_T/\partial[S] = 2$ in eq. 4.1.7, we find

$$[S]_{\text{CMC}}^{N-1} = (N^2 K_N)^{-1} \qquad (4.1.9)$$

where $[S]_{\text{CMC}}$ is the monomer concentration at the CMC.

The critical micelle concentration refers to the total surfactant concentration, and if we combine eqs. 4.1.7 and 4.1.9, we find

$$\text{CMC} = [S]_{\text{CMC}}(1 + N^{-1}) = (N^2 K_N)^{-1/(N-1)}(1 + N^{-1}) \quad (4.1.10)$$

From the first equality we see that the amount of micellized surfactant is $[S]_{\text{CMC}}/N$, which becomes more negligible as the value of N increases.

We can relate this expression for the CMC to the one obtained in eq. 4.1.5 for the phase separation model. Noting that $\Delta G^\theta = -RT \ln K_N$ and K_N refers to the association of N monomers to a micelle,

$$-RT \ln K_N = \Delta \mathbf{G}^\theta = N(\mu^\theta(\text{micelle}) - \mu^\theta(\text{solvent})) = N\Delta \mathbf{G}_{\text{mic}}^\theta$$

Combining eq. 4.1.9 with eq. 4.1.10 and neglecting the term N^{-1} in the second equation gives

$$RT \ln \text{CMC} \simeq RT \ln[S]_{\text{CMC}} = -\frac{RT}{(N-1)} \ln(K_N N^2) = \frac{1}{N-1}(N\Delta \mathbf{G}_{\text{mic}}^\theta - 2RT \ln N) \simeq \Delta \mathbf{G}_{\text{mic}}^\theta \quad (4.1.11)$$

In the last equality, we have introducded two approximations whose validity depends on a large value of N. First, we replaced $N - 1$ by N in the denominator; and second, we neglected the logarithmic term in N relative to the term linear in N. Equation 4.1.11 establishes a link between the phase separation and closed association models, but due to the approximations requiring a large N value, we risk introducing sizable errors if we use it for systems with small ($N \lesssim 30$) aggregation numbers.

Typical micellar aggregation numbers lie in the range $N = 30$–100, which ensures that N is large enough to make the CMC a reasonably well-defined quantity in the closed-association model. The relation in eqs. 4.1.5 and 4.1.11 between $\ln \text{CMC}$ and $\Delta \mathbf{G}_{\text{mic}}^\theta$ is also a useful approximation and we will return to it. The closed-association model, however, has a seemingly unphysical feature; for aggregation numbers $n < N$, cooperativity is so strong that aggregates S_n ($n < N$) occur in negligible amounts, and cooperativity is broken for $n > N$ so that aggregates S_n ($n > N$) do not occur either. The result cannot be correct in the strict sense, but the ability of the closed-association model to capture many of the observed properties of micellar systems shows that a qualitative change in cooperativity of the association occurs around the optimal aggregation number.

If we analyze what happens when an apolar liquid reaches its solubility limit in water, we can appreciate the particular features of micelle aggregation. For a small droplet, we can calculate its free energy as the sum of a bulk term $n\mu^{\theta}(\infty)$, plus a surface term γA_n. When a monomer in solution chemical potential μ(solution), is added to a droplet of n monomers to form one of $n + 1$, the change in free energy is

$$\Delta G(n \rightarrow n + 1) = (n + 1)\mu^{\theta}(\infty) + \gamma A_{n+1} - n\mu^{\theta}(\infty) - \mu(\text{solution}) - \gamma A_n$$

$$= \mu^{\theta}(\infty) - \mu(\text{solution}) - \gamma \frac{\partial A}{\partial n}$$

$$= \mu^{\theta}(\infty) - \mu(\text{solution}) + \frac{2\gamma V_s}{R_n} \tag{4.1.12}$$

where R_n represents the radius of the spherical drop and V_s the volume per molecule.

The important feature of eq. 4.1.12 is that the last term decreases with increasing n, since R_n increases. This fact implies that adding a monomer becomes more and more favorable as the aggregate enlarges (ΔG decreases with increasing n). Cooperativity of this kind is precisely what drives a drop to grow indefinitely and form a separate phase.

In the isodesmic model, the standard free energy change that results from adding a monomer is simply $\mu^{\theta}(\infty) - \mu^{\theta}(\text{solvent}) = RT \ln K$ independent of n. Thus, as we have stated, no cooperativity exists in that model. If n is large, the behavior of micellar systems approximates the isodesmic model, but for small n, the systems behave in the same way as the drop. The micelle's molecular organization offers a structural explanation for this transition in behavior. Because head groups are located at the surface and tails are located by preference in the interior, the effective γ value decreases with increasing aggregate size. Repulsive interactions between the head groups then limit aggregate growth.

Because of these competing effects, most micellar systems show a rather narrow size distribution. Table 4.1 illustrates this

TABLE 4.1 The CMC, the Mean Aggregation Number N, and the Standard Deviation σ of the Micelle Size Distribution for a Series of Sodium Alkyl Sulfates, Determined from Kinetic Measurements (See Section 4.2.3)

Surfactant	Temperature (°C)	CMC(M)	N	σ
NaC_6SO_4	25	0.42	17	6
NaC_7SO_4	25	0.22	22	10
$NaC_{12}SO_4$	25	8.2×10^{-3}	64	13
$NaC_{14}SO_4$	40	2.05×10^{-3}	80	16.5

From E.A.G. Aniansson et al., *J. Phys. Chem.* **80**, 905 (1976).

for a number of surfactant systems. Typically, the width of the distribution is on the order of 20–30% of the mean aggregation number. In Figure 4.2 we show the results of a detailed calculation using the stepwise association model with an n-dependence of the equilibrium constant calculated as described in Section 4.3. The figure illustrates how a peak in the distribution function around $n = 55$ suddenly appears in the vicinity of the CMC.

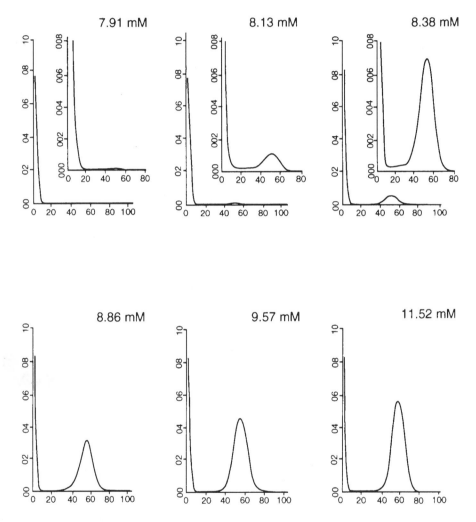

Figure 4.2
These calculated micelle size distributions illustrate how surfactant is distributed among different aggregates with aggregation, that is, $N[S_n]/[S]_T$ versus N. The CMC is estimated to be 8.3 mM, as it is for SDS, but the calculations show that micellar aggregates exist in very low concentrations—even below the formal CMC. As the total concentration is increased from $[S]_T = 7.91$ mM to 11.52 mM, the mean aggregation of the micellar aggregates increases slightly from 48 to 56. Concentrations for aggregates with aggregation numbers in the range $N = 10$—30 are notably small. Above $[S]_T = 8.5$ mM, the micelle size distribution can be well described by a Gaussian with a calculated standard deviation of approximately 7. (G. Gunnarsson, Ph.D. thesis, University of Lund, Lund, Sweden, 1981.)

So far, we have limited our discussion to uncharged surfactants, but we can generalize eq. 4.1.6 to include ionic surfactants by writing an equilibrium between surfactant monomers, S^-, counterions, C^+, and micelles, S_N, as

$$(N - P)C^+ + NS^- \rightleftarrows S_N^{-P} \qquad (4.1.13)$$

for which

$$K_N = \frac{[S_N^{-P}]}{[S^-]^N[C^+]^{N-P}} \qquad (4.1.14)$$

or

$$\Delta \mathbf{G}^\theta_{\text{mic}} = \frac{\Delta \mathbf{G}^\theta}{N} = -\frac{\mathbf{R}T}{N}\ln[S_N^{-P}] + \mathbf{R}T\ln[S^-] + \mathbf{R}T\left(1 - \frac{P}{N}\right)\ln[C^+] \qquad (4.1.15)$$

When N lies in the range 50–100, the $\ln[S_n^{-P}]/N$ term becomes small and consequently negligible by an argument analogous to the one made in eq. 4.1.11. If no salt is added, we can replace both $[S^-]$ and $[C^+]$ with C_{CMC} in the second and third terms to give

$$\Delta \mathbf{G}^\theta_{\text{mic}} = (2 - \beta)\mathbf{R}T\ln C_{\text{CMC}} \qquad (4.1.16)$$

where $\beta = P/N$ is the degree of dissociation of the micelle or $1 - \beta$ is the degree of counterion binding. For a completely ionized micelle, $\beta = 1$ and for a neutral micelle, $\beta = 0$. Thus, this equilibrium model pictures the micelle as a charged entity consisting of N surfactant molecules and $(N - P)$ counterions with a net charge of $-P$.

The notion that counterions bind to a charged surfactant aggregate has played a major historical role in the description of micellar systems. Counterions drawn into the regions of the charged head groups reduce the repulsive electrostatic interactions between them, and this is the heuristic physical basis for the model of counterion binding. The idea provides a good qualitative picture of the phenomenon, but it is not amenable to quantitative analysis. As we will show in Section 4.3, a quantitative alternative to eq. 4.1.15 emerges naturally from the Poisson–Boltzmann equation.

4.1.2 Thermodynamics of Micelle Formation Provide Useful Relationships Between Free Energies and Surfactant Chemical Potentials and Explicit Relations for Enthalpy and Entropy

Free Energy and Chemical Potentials

In the preceding section we analyzed the relationship between stepwise aggregation of amphiphiles and related the corresponding equilibrium constants to standard free energies. We saw that as a consequence of the cooperativity of the micelle formation process, aggregates start to form at a relatively well-defined concentration—that is, the CMC. Cooperative association also affects the concentration dependence of the surfactant chemical potential μ_S.

At equilibrium, μ_S is uniform throughout the system, so we can obtain a value for it by focusing on the free monomer in solution. Neglecting activity coefficient corrections, which are too small to be of consequence for the basic conclusions, we write

$$\mu_S = \mu_S^\theta(\text{solvent}) + \boldsymbol{RT}\ln[S] \qquad (4.1.17)$$

Below the CMC, $[S] \simeq [S]_T$ and the surfactant chemical potential increases logarithmically with concentration as it normally does for any dilute solution. Far above the CMC, where $[S]_T \gg [S]$, we find from eq. 4.1.7 that

$$[S] \simeq [S]_T^{1/N}(NK_N)^{-1/N} \qquad (4.1.18)$$

Substituted into eq. 4.1.17 this gives

$$\mu_S = \mu_S^\theta(\text{solvent}) + \frac{\boldsymbol{RT}}{N}\ln[S]_T - \frac{\boldsymbol{RT}}{N}\ln(NK_N) \qquad (4.1.19)$$

This equation shows how little the surfactant chemical potential varies with total concentration within this concentration range. To obtain the same change in μ_S above the CMC that doubling the concentration produces below the CMC, we need to increase $[S]_T$ by 2^N. For typical aggregation numbers, this is simply impossible. Thus, the surfactant chemical potential remains pratically constant above the CMC up to very high values of the total concentration.

The weak concentration dependence of μ_S above the CMC has important consequences in applications of surfactants. A surface or a bulk phase in equilibrium with a surfactant solution registers the chemical potential, and at the CMC the surfactant already reaches its maximum influence. For example, in cleaning processes, a detergent realizes its maximum effect at the CMC. Adding more detergent only creates a reservoir that can compensate for adsorption losses.

An even more subtle mechanism operates in ionic surfactants, and we can illustrate this mechanism qualitatively by using the association model of eq. 4.1.13. When N-charged monomers, S^-, and $N - P$ counterions, C^+, associate to the micelles, the surfactant chemical potential carries contributions from both ions and

$$\mu_{SC} = \mu_S^- + \mu_C^+ = \mu_S^\theta(\text{solvent}) + \boldsymbol{RT}\ln[S^-] + \mu_{C+}^\theta(\text{solvent}) + \boldsymbol{RT}\ln[C^+]$$

$$= \mu_{SC}^\theta(\text{solvent}) + \boldsymbol{RT}\ln[S^-][C^+] \qquad (4.1.20)$$

As with nonionic surfactants, μ_{SC} is virtually independent of the total concentration above the CMC, which means that the product $[S^-][C^+]$ remains constant. Because a fraction of the counterions does not associate with the micelle, $[C^+]$ increases with total surfactant concentration. Paradoxically, for the product $[S^-][C^+]$ to remain constant, the free charged monomer concentration $[S^-]$ must decrease with increasing concentration, as does the chemical potential μ_{S^-} of the surfactant ion.

One consequence of the decreasing μ_{S^-} is that adsorption on positively charged surfaces actually decreases with increasing surfactant concentration above the CMC. As a result, maximum adsorption should occur close to the CMC.

Enthalpy and Entropy

The relative contributions of enthalpy determine the temperature dependence of the CMC. We can see this effect most easily if we use the approximate equation

$$\Delta \mathbf{G}^{\theta}_{\text{mic}} = \boldsymbol{R}T \ln \text{CMC} \tag{4.1.5}$$

which combined with the Gibbs–Helmholtz equation provide an expression for the enthalpy of micelle formation,

$$\Delta \mathbf{H}^{\theta}_{\text{mic}} = -\boldsymbol{R}T^2 \frac{\partial}{\partial T} \ln \text{CMC} \tag{4.1.21}$$

For an ionic surfactant there is an additional factor $(2 - \beta)$, as in eq. 4.1.16.

4.2 We Can Measure Critical Micelle Concentrations, Aggregation Numbers, and Characteristic Lifetimes by a Number of Methods

Almost every technique available to physical science has been used to study micellar properties. Rather than surveying the literature, this section cites key experimental methods that provide insightful information on the properties of micellar solutions. We will focus first on spherical micelles, then broaden the discussion to include elongated structures.

4.2.1 We Can Determine CMCs by Surface Tension, Conductance, and Surfactant Ion Electrode Measurements

As illustrated in Figure 4.3, many physical properties of surfactant solutions undergo an abrupt change at the CMC. Surface tension and conductance are two simple and widely used methods for measuring these changes, but surfactant ion electrode measurements also provide useful information on aggregation behavior.

Figure 4.3
Effect of surfactant
concentration on various
physical properties. Over a
fairly narrow surfactant
concentration range,
virtually all physical
solution properties display
a pronounced change in
slope that can be used to
identify a CMC. Variations
in CMC remind us that
micellization is a
cooperative process that
occurs over a finite
concentration range, in
contast to an abrupt phase
transition. (B. Lindman
and H. Wennerström,
*Current Topics in
Chemistry*, Springer-
Verlag, Berlin, 1980, p. 93.)

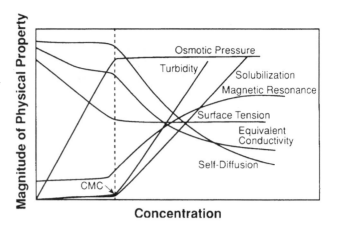

Surface Tension

Surface tension γ decreases from 72 mN/m for pure water to an almost constant value beyond the CMC. It might be tempting to conclude that the surface is saturated by the surfactant at the CMC. However, the proper explanation of the constant γ value above the CMC follows by combining the basic form of the Gibbs adsorption isotherm (eq. 2.4.10) and eq. 4.1.19 for the surfactant chemical potential. Above the CMC, the chemical potential of the surfactant scarcely changes and as a result conditions do not change at the surface. The limiting value of the surface tension ranges from 40 to 30 mN/m depending on the surfactant. The intersection resulting from linear extrapolation of the two straight lines shown in Figure 1.1 provides a measure of CMC value.

Measurements of surface tension constitutes an extremely sensitive test for surfactant purity. For example, a 0.01% impurity of 1-dodecanol, $CH_3(CH_2)_{10}CH_2OH$, in SDS results in curve B shown in Figure 4.4. Although both compounds adsorb at the air—water interface, alcohol concentrates preferentially at the interface because it is more surface active. Since γ(alcohol) $< \gamma$(SDS), the surface tension curve falls below that observed for pure SDS. In the bulk phase, alcohol initiates micelle formation at a value below the CMC for the pure surfactant. The aggregation process also extends over a much broader concentration range. Above the CMC, alcohol desorbs from the interface as more surfactant is added and solubilizes back into the micelles as the micellar concentration increases. The result is a gradual increase in γ. A small amount of surface impurity, therefore, leads to a pronounced dip in the surface tension curve, such as we find in Figure 4.4.

The slope of the γ versus $\ln C_T$ below the CMC provides a way to determine the interfacial area per surfactant at the air—water interface, as discussed in Chapter 2.

Conductance

Change in electrical conductance has been widely used to determine the CMC of ionic surfactants. We can express con-

Figure 4.4
Effect of dodecanol on the surface tension of solutions of SDS: curve A, pure SDS; curve B, unpurified SDS; curve C, pure SDS + 0.5% dodecanol of the amount of SDS.
(P. Elworthy and K. Mysels, *J. Colloid Interface Sci.* **21**, 339 (1966).)

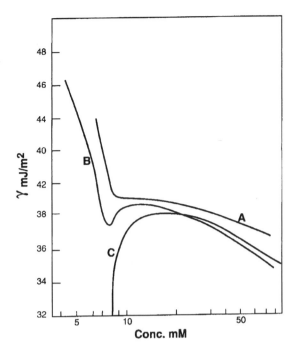

ductance data in one of two ways: as specific conductance, $\bar{\kappa}$,

$$\bar{\kappa} = \frac{k}{R} \qquad (4.2.1)$$

where k is the cell constant and R the measured resistance; or as equivalent conductance

$$\Lambda = \frac{10^3 \bar{\kappa}}{c} \qquad (4.2.2)$$

which gives the conductance per mole of electrolyte.

When we plot conductance data for SDS using eqs. 4.2.1 and 4.2.2, the marked breakpoint that occurs in $\Lambda(c)$ provides a measure of the CMC. This value differs somewhat from the CMC obtained from surface tension measurements. This discrepancy originates in the fact that micellization is not an abrupt phase transition. The association process sets in over a finite concentration range, and to characterize this range by a single number, such as the CMC, reflects a somewhat arbitrary decision. For a typical micelle-forming substance like SDS, an uncertainty on the order of a few percentage units in the CMC will occur. Its extent depends on the procedure used in the determination. Several attempts have been made to define the CMC precisely, but we believe it is better to accept the reality of a gradual aggregate growth and leave the CMC undefined with respect to a few percentage units.

Surfactant Ion Electrodes

Ion-selective electrodes provide a direct measure of surfactant monomer and counterion chemical potentials. Thus, we can use them to determine patterns of surfactant aggregation, including premicellar dimerization.

We can measure the change in electromotive forces (emf) as a function of concentration using the Nernst equation

$$E = E^{o'} - (\mu - \mu^0)/F \simeq E^{o'} - \frac{RT}{F}\ln c \qquad (4.2.3)$$

where $E^{o'}$ represents the standard state emf, and F the Faraday constant.

Electrodes can be arranged in two ways. If we connect the surfactant and counterion electrodes in series, the resulting emf cell directly determines the combined chemical potentials. Alternatively, if we operate the ion electrodes individually against a common reference electrode, we can measure μ_+ and μ_- separately.

Figure 4.5 shows the aggregation pattern observed with dyes, bile salts, and long-chain surfactants. The formation of aggregates causes the measured emf to deviate from the one expected in a simple electrolyte solution.

1. Orange II is the sodium salt of an anionic dye. In such dyes, electron resonance spreads the negative charge over the entire multiring planar molecule. Consequently, the molecules do not possess well-defined, separate hydrophilic or hydrophobic regions, and the dyes aggregate in a stepwise manner that can be described by

Figure 4.5
Emf measurements demonstrate differing aggregation patterns. The Nernstian slope corresponds to ideal behavior of free ions in solution and provides a useful comparison in understanding amphiphilic aggregation patterns. Plots of log(anion activity) versus log(concentration) for Orange II show a deviation from the Nernstian slope with increasing concentration resulting from association to form dimers, trimers, and so on. SDS shows a distinct break at the CMC and a surfactant activity that decreases above that concentration in accord with the prediction of eq. 4.1.20). The taurodeoxycholate displays behavior consistent with an aggregation to small aggregates occurring over an extended concentration range. (K.M. Kale, E.L. Cussler, and D.F. Evans, *J. Phys. Chem.* **84**, 593 (1980).)

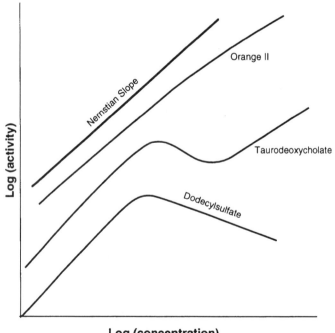

the isodesmic model (K_n = constant). The compound's aggregation behavior is shown in Figure 4.5 by the increasing departure from the Nernst slope with increasing concentration. Analysis with the isodesmic model gives a value of $K = 1200 \, (M)^{-1}$ for Orange II.

2. Sodium taurodeoxycholate is a bile salt possessing a hydroxylated steroid ring structure with hydrophobic and hydrophilic sides. Such compounds initially aggregate to form small micelles with aggregation numbers ranging from 4 to 10. These aggregation numbers grow with increasing concentration. As shown in Figure 4.5, the slope is almost one in dilute solution, but with increasing concentration, the curve displays a maximum, then a minimum, and finally returns to the original slope.

3. SDS shows the typical aggregation pattern of a surfactant with a well-defined CMC. As Figure 4.5 indicates, in dilute solution the surfactant's slope is Nernstian with a well-defined break at its micellization point. Above this concentration, the activity decreases in accord with the discussion in Section 4.1.2.

The emf data for decyltrimethylammonium bromide illustrates measurements employing both ion electrodes in series and operating individually against a common reference electrode. When we plot the combined $(\mu - \mu^0)_{C_{10}N(Me)_3Br}/RT$ against $\log(C_{10}N(Me)_3Br$ below the CMC, the results show a straight line with a slope of unity. Above the CMC, the data produces a straight line with a slight positive slope.

When the ion-selective electrodes operate separately against a reference electrode, the plots of chemical potential in Figure 4.6 yield a slope of 0.98 below the CMC. This nearly

Figure 4.6
Activity of decyltrimethylammonium and bromide ions. Two types of information are obtainable by emf measurements on ionic surfactant solutions. When we use surfactant and bromide connected in series, we determine the chemical potential of the neutral surfactant, which increases minutely above the CMC. With a common reference electrode, we can determine the single-ion chemical potential of surfactant and counterion, which decrease and increase respectively. (K.M. Kale, E.L. Cussler, and D.F. Evans, *J. Phys. Chem.* **84**, 59 (1980).)

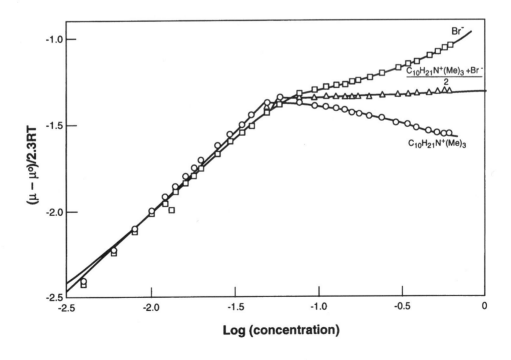

ideal response indicates little or no premicellar aggregation. However, after the sharp change in slope at the CMC, the chemical potential of the DTA$^+$ ion decreases and that of the bromide increases. The difference between the line we obtain by extrapolating the premicellar slope and the bromide slope above the CMC measures the decrease of bromide ion chemical potential as a consequence of its association with the charged micelles.

4.2.2 Micellar Aggregation Numbers Can Be Measured Most Simply by Light Scattering or with Fluorescent Probes

Knowledge of aggregation numbers greatly facilitates our understanding of micellar structure and our ability to make quantitative calculations. Just as we can obtain CMCs by several measurement techniques, we can determine aggregation numbers by a range of methods. We discuss two simple, accurate, and generally applicable methods: light scattering and fluorescence.

Light Scattering

Scattering methods, such as light scattering, x-ray scattering, and neutron scattering, are generally useful for studying structure in colloidal systems. We postpone a broader discussion of these techniques to Chapters 7 and 9. Here we concentrate on light-scattering studies of micellar systems as ways to determine aggregation numbers.

The wavelength of visible light is two orders of magnitude larger than the radii of typical spherical micelles. A light beam hitting a micellar solution will see the micelles as small objects of refractive index n_{mic}, more or less randomly distributed in a solvent of refractive index n_{solv}. Provided $n_{mic} \neq n_{solv}$, the refractive index averaged over a volume of size λ^3 will vary in accordance with the local concentrations of micelles. Variations in n will cause the light beam to scatter.

The intensity of the scattered light I_s depends on four factors so that

$$I_s(\text{sample}) - I_s(\text{solvent}) = \text{instrumental constant} \times \text{angular factor}$$
$$\times \text{optical constant} \times \text{concentration fluctuation factor}$$

(4.2.4)

The instrumental constant is proportional to the intensity I_0 of the incoming light beam and inversely proportional to the square of the distance r_s between the sample and the detector.

The optical constant K_0 depends on the wavelength λ, the solution's index of refraction, n, and its variation with concentration dn/dc. The expression for K_0 is

$$K_0 = 2\pi^2 \frac{n^2}{\lambda^4} \left(\frac{dn}{dc}\right)^2 \bigg/ N_{Av}$$

(4.2.5)

which can be measured using a refractometer. Note that the scattering intensity depends strongly on the wavelength.

We can most easily understand the dependence of scattering intensity on θ, the angle between the primary beam and the scattered beam, by considering the unpolarized incoming light as composed of two perpendicularly polarized components. For small scattering angles, θ, the polarized components contribute equally to the scattering. However, for light scattered perpendicular to the primary beam ($\theta = 90°$), only the component polarized perpendicular to the direction of the scattered beam can contribute, because no polarized component occurs parallel to the direction of propagation. The explicit form of the angular factor $(1 + \cos^2\theta)$ follows from this argument.

From a chemical perspective, the concentration fluctuation term is the most interesting factor in eq. 4.2.4 because it contains the desired information on the properties of the sample. The free energy cost in setting up an inhomogeneity in the micelle concentration is proportional to the derivative of the surfactant chemical potential with respect to concentration, $\partial\mu_s/\partial c$. Using the Gibbs—Duhem relation, $n_{(solvent)} d\mu_{(solvent)} = -n_s d\mu_s$ (eq. 1.5.20), giving

$$\frac{d\mu_s}{dc} \sim -\frac{1}{V_{solv} \cdot c} \frac{d\mu_{(solvent)}}{dc} = \frac{1}{c} \frac{d\Pi_{osm}}{dc} \qquad (4.2.6)$$

Here Π_{osm} represents the osmotic pressure and $(d\Pi_{osm}/dc)^{-1}$ is called the osmotic compressibility.

The smaller $d\mu_s/dc$, the smaller the free energy cost of generating a concentration fluctuation and the larger the scattering intensity. An explicit derivation based on Einstein's fluctuation theory shows that the concentration fluctuation factor in eq. 4.2.4 is

$$\boldsymbol{R}Tc\left(\frac{d\Pi_{osm}}{dc}\right)^{-1} \qquad (4.2.7)$$

We have now specified all the factors in eq. 4.2.4. Separating out the factors that depend on the experimental setup makes reporting data more convenient. We can do this by defining a Rayleigh ratio:

$$R_\theta = \frac{r_s^2}{1 + \cos^2\theta} \frac{I_s}{I_0} \qquad (4.2.8)$$

In terms of the Rayleigh ratio, the scattering caused by the presence of the micelles becomes

$$\Delta R_\theta = R_\theta(\text{solution}) - R_\theta(\text{solvent}) \qquad (4.2.9)$$

or

$$= \frac{2\pi^2 n^2}{\lambda^4}\left(\frac{dn}{dc}\right)^2 \boldsymbol{R}Tc\left(\frac{d\Pi_{osm}}{dc}\right)^{-1} \ (\text{m}^{-1}) \qquad (4.2.10)$$

If we know λ, n, and dn/dc, the light-scattering experiment measures ΔR_θ to give the thermodynamic quantity $c^{-1}(d\Pi_{osm}/dc)$. Since the molecular weight is usually unknown, we rewrite the concentration in mass per volume, c_m.

To obtain the molecular weight and thus the micelle aggregation number, we need a model for the concentration dependence of Π_{osm}. With well-defined particles, we can use a virial expansion of the osmotic pressure as in eq. 1.5.13. The molar concentration c is related to the mass concentration through the molecular weight M; $c_s = 10^3 c_m / M$. Thus,

$$\Pi_{osm} = 10^3 RT c_m \left(\frac{1}{M} + 10^3 \frac{B_2 c_m}{M^2} + \cdots \right) \qquad (4.2.11)$$

and

$$\frac{d\Pi_{osm}}{dc_m} = RT \left(\frac{10^3}{M} + 2B_2' c_m + \cdots \right) \qquad (4.2.12)$$

where $B_2' = 10^6 B_2/M^2$. By extrapolating $d\Pi_{osm}/dc_m$ toward $c_m = 0$, we can determine the molecular weight M from the intercept and the virial coefficient B_2 from the slope.

For micellar systems, we cannot use eq. 4.2.12 because as the concentration c decreases below the CMC, the change in $d\Pi_{osm}/dc$ causes a drastic change in scattering intensity. A reasonable approximate procedure formally includes the monomers in the solvent and replaces c_m by $c_m - CMC_m$ (counted as mass concentration). Then we can modify eq. 4.2.12 to

$$\frac{d\Pi_{osm}}{dc_m} = 10^3 \frac{RT}{M} + 2RTB_2'(c_m - CMC_m) + \cdots \qquad (4.2.13)$$

which we can calculate by extrapolating to $c_m = CMC_m$.

By substituting eqs. 4.2.5 and 4.2.13 into eq. 4.2.10, we find

$$\Delta R_\theta = \frac{K_0(c_m - CMC_m)}{10^3/M + 2B_2'(c_m - CMC_m)} \qquad (4.2.14)$$

which upon rearrangement gives

$$10^{-3} \frac{K_0(c_m - CMC_m)}{\Delta R_\theta} = \frac{1}{M} + 2 \times 10^{-3} B_2'(c_m - CMC_m) \qquad (4.2.15)$$

In Figures 4.7 and 4.8, we illustrate the procedure for determining the aggregation number of tetradecyltrimethylammonium bromide micelles. The concentration dependence of the scattering intensity is shown in Figure 4.7. Around $c_m = 1 \, \text{kg m}^{-3}$, a sharp increase in I_s signals the onset of micelle formation, but the scattering intensity is perceptible even below the CMC. Such a scattering could be caused by small amounts of high molecular weight impurities in water. In fact, removing impurities from the system constitutes a major experimental

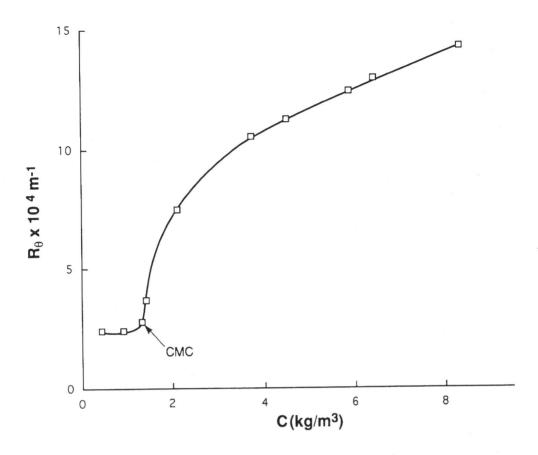

Figure 4.7
Light-scattering data for tetradecyltrimethyl-ammonium bromide, $C_{14}N(Me)_3Br$, as a function of surfactant concentration at 25°C. Below the CMC, the Rayleigh ratio R_θ is almost constant. Above the CMC, R_θ reflects the increase of concentration fluctuations when N molecules associate into a single micelle. (A.H. Beesley, Ph.D. thesis, University of Minnesota (1990).)

difficulty. In Figure 4.8 the data have been replotted using eq. 4.2.15 as $(c_m - CMC_m)K_0/\Delta R_\theta$ versus $c_m - CMC_m$ to obtain a straight line, where the intercept is the inverse molecular weight of the micelle and the slope yields the virial coefficient.

Fluorescence Measurements

An apolar fluorescent molecule dissolved into a micellar solution can be solubilized in the micelles. If the concentration of micelles exceeds that of fluorescent probes, then the solution will contain micelles of two types—those with probes and those without. If the fluorescent probe concentration is sufficiently low, each tagged micelle will contain only one probe. By this procedure, we obtain a known concentration of tagged micelles. If we know the total surfactant concentration, the CMC, and can determine the fraction of tagged micelles by some means, we can calculate the total concentration of micelles and thus the average aggregation number.

When the sample is irradiated by constant low intensity light, we can measure a steady state fluorescence intensity I_0. We now introduce another apolar component (a fluorescence quencher) into the micellar solution. If the excited fluorescent

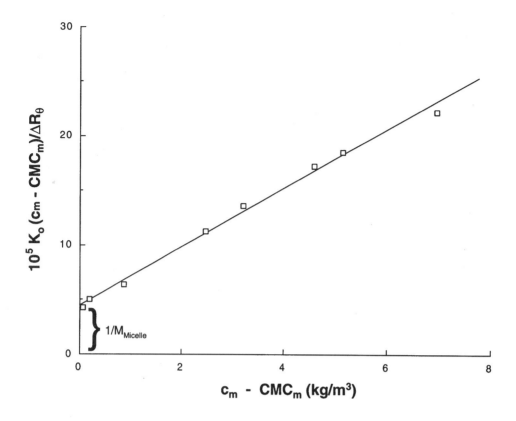

Figure 4.8
Plot of the light scattering data of the previous figure that yields the molecular weight of the micelle as the inverse of the intercept and the second virial coefficient from the slope. Intercept is $4.3 \times 10^{-5}\,(\text{kg}^{-1})$, $M_{\text{mic}} = 2.3 \times 10^{4}$ and with a monomer molecular weight of 336 the aggregation number $N = 70$. The slope is $2.2 \times 10^{-5}\,\text{m}^{3}/\text{kg}^{2}$ and $B_2 = 5.4\,\text{m}^{3}$ per mole. The micelles have a radius of 1.9 nm and occupy a volume $0.018\,\text{m}^{3}$ per mole. The value of B_2 shows that micelles repel, $B_2 > 0$, and excludes a volume three hundred times their own size by electrostatic interactions. (A.H. Beesley, Ph.D. thesis, University of Minnesota (1990).)

molecules come sufficiently close to the quencher, a rapid non-radiative relaxation of the fluorescent state occurs and the probe molecules do not produce any fluorescence. As we successively increase the quencher concentration c_q, the intensity $I(c_q)$, decreases to zero. If we add a large excess of quencher, therefore, we no longer observe fluorescence.

We can assume that both the fluorescent and quencher molecules will be confined within a single micelle during the lifetime of a fluorescent state (~ 10—$100\,\text{ns}$). Furthermore, if the micelle is small enough, a quencher can meet and deactivate an excited probe during this time. Thus, we observe steady state fluorescence only from micelles containing probes but not quenchers. Assuming that probes and quenchers both are distributed randomly between micelles, we can calculate the probability of finding a micelle that contains the probe but lacks the quencher.

If the concentration of micelles equals c_M, the average number of quenchers per micelle, $\bar{q}$, is

$$\bar{q} = \frac{c_q}{c_M} \tag{4.2.16}$$

A random placing of m objects (quenchers) in n boxes (micelles) results in a Poisson distribution for large m and n. This distribution implies that the probability $P(p)$ of having p objects in an arbitrary box is

$$P(p) = \bar{q}^p \frac{\exp(-\bar{q})}{p!} \qquad (4.2.17)$$

where $\bar{q} = m/n = c_q/c_M$. In particular, $P(0)$ signifies the probability of finding an empty micelle (box)

$$P(0) = \exp(-\bar{q}) \qquad (4.2.18)$$

since $0! = 1$. The steady state fluorescence at a quencher concentration c_q is thus

$$I(c_q) = I_0 \exp\left(\frac{-c_q}{c_M}\right) \qquad (4.2.19)$$

provided the probes and quenchers are distributed independently. Here I_0 is the fluorescence intensity in the absence of quenchers.

The slope of the plot of $\ln\{I(c_q)/I_0\}$ versus c_q gives the concentration of micelles c_M. From it we can easily calculate the average aggregation number N as

$$N = \frac{\{[S]_{\text{tot}} - \text{CMC}\}}{c_M} \qquad (4.2.20)$$

This way of determining N illustrates how we can take advantage of the special organization of a micellar solution in which micelles dispersed in a solvent with very different solvation properties.

Although the steady state fluorescence method of determining aggregation numbers is both experimentally and conceptually simple, it works only for small aggregation numbers. Time-resolved fluorescence provides a related but more sophisticated method. In this case, we apply a short ($\lesssim 1\,\text{ns}$) light pulse and measure the time response of the fluorescence. With a fluorescent probe, but no quencher, the fluorescence intensity I decays exponentially in an ideal system

$$I(t) = I(0) \exp\left(\frac{-t}{\tau_0}\right) \qquad (4.2.21)$$

where τ_0 is the fluorescence lifetime.

When quencher molecules are added, the fluorescence intensity decays more rapidly because of the additional quenching process. Figure 4.9 illustrates how the decay curve changes as the quencher concentration increases. With quencher present, the curve clearly shows a nonexponential decay of the fluorescence. To analyze these curves quantitatively, we again assume a Poisson distribution of quenchers among the micelles. In addition, we choose a small probe concentration relative to the micelle concentration, c_M.

The quenching is not instantaneous but is characterized by a rate constant, k_q. Thus, the rate of quenching increases

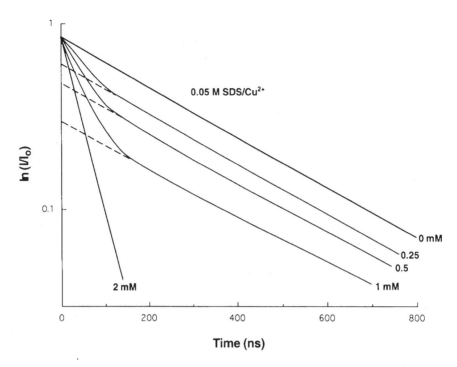

Figure 4.9
Fluorescence quenching data used to calculate the aggregation number of micelles. Plot of relative fluorescence intensity, $\ln(I/I_0)$, for pyrene in 0.05 M SDS versus time as a function of added quencher, cupric ion. In the absence of quencher, the slope of the line measures the lifetime of the fluorescent state of pyrene in the micelles. As quencher is added we observe a double decay curve. Rapid decay is a consequence of quenching the excited pyrene by Cu^{2+}. Analysis of the data using eq. 4.2.19 provides the average aggregation number. (F. Grieser and R. Tauschtreml, *J. Am. Chem. Soc.* **102**, 7258 (1980).)

with the number of quenchers in the micelle. A detailed kinetic derivation shows that the fluorescence intensity should decay according to the equation

$$\ln\left[\frac{I(t)}{I(0)}\right] = -t/\tau_0 + \bar{q}[\exp(-k_q t) - 1] \qquad (4.2.22)$$

where $\bar{q}$ stands for the average number of quenchers per micelle as in eq. 4.2.16.

Equation 4.2.22 contains three parameters: fluorescence lifetime τ_0, which can be determined for the system without quencher; the quenching rate constant, which should be independent of quencher concentration; and the average number of quenchers $\bar{q}$, which is proportional to the quencher concentration, $\bar{q} = c_q/c_m$. By using a computer to fit the three parameters to a set of curves as in Figure 4.9, we obtain an accurate estimate of the micelle concentration c_M. A substantial advantage of this method in comparison to steady state fluorescence is that we obtain a much more detailed check on whether the theory can reproduce the experimental data. If it cannot, some of the assumptions must be invalid.

We can quickly determine c_M by focusing on the decay curves at long times $t \gg 1/k_q$. Then we can neglect the term $\exp(-k_q t)$ in eq. 4.2.22 and reduce it to

$$\ln\left[\frac{I(t)}{I(0)}\right] = \frac{-t}{\tau_0} - \bar{q} \qquad (4.2.23)$$

By taking the difference between $\ln[I(t)/I(0)]$ for no quencher ($\bar{q} = 0$) and a curve with a finite quencher concentration at equivalent long time, we can obtain $\bar{q}$ directly. Then we can calculate the aggregation number as we did previously.

4.2.3 Kinetic Experiments Provide Valuable Insight into the Time Scales of Dynamic Processes in Micellar Solutions

Micellar formation, dissolution, and size redistribution involve multistep kinetic processes described by the following series of elementary steps:

1. Monomer exchange between aggregate and solution (cf. eq. 4.1.1)

$$S + S_{n-1} \underset{k_n^-}{\overset{k_n^+}{\rightleftarrows}} S_n \qquad (4.2.24)$$

The ratio between the two rate constants is fixed by the equilibrium constant of eq. 4.1.2, which ensures that the forward and backward rates are equal at equilibrium.

$$\frac{k_n^+}{k_n^- = K_n} \qquad (4.2.25)$$

2. Aggregate fusion and fission expressed as

$$S_n + S_m \rightleftarrows S_{m+n} \qquad (4.2.26)$$

resulting in a change in the size and number of micelles.

3. Change in counterions responsible for the change in charge on an ionic micelle

$$S_n C_{n-P}^{p-} \rightleftarrows S_n C_{n-P-1}^{(P+1)} + C^+ \qquad (4.2.27)$$

Counterions are associated to the micelles by long range electrostatic forces and do not experience any kinetic barriers. Thus, the diffusion-controlled redistribution of counterions takes place in the electrostatic field set up by the micelles and occurs on a very short time scale.

We cannot use classical kinetic experiments to determine the rates of micelle formation because the process occurs so rapidly. Instead, we must use fast relaxation procedures, such as temperature jump, pressure jump, or ultrasonic absorption. The idea is to take a system that is initially at equilibrium, subject it to a small but rapid change in an intensive parameter (e.g., T or p), and then follow the relaxation to the new equilibrium state.

To complete a T-jump experiment, we place the solution in a spectrometric cell containing two large electrodes. A huge capacitor charged up to 30 kV is shorted across the electrodes; discharged electrical energy causes the temperature of the solution to increase by $\approx 10°C$ in a microsecond. Any process with a ΔH will respond to the temperature change at its characteristic rate, which we can follow using changes in the spectrum of appropriately chosen dye additives.

Different fast relaxation methods sample different time scales. Therefore, to characterize a process with multiple equilibria, each with its characteristic relaxation time, we must use several measuring techniques.

As shown in Table 4.2, measurements on a variety of surfactant solutions yield a biexponential relaxation process with two well-defined times: τ_1, which typically falls in the range 10^{-6}–10^{-3} s, and τ_2, in the range 10^{-3}–1 s. We will see that the cooperative nature of the micelle formation process causes these two relaxation times.

It has been shown that experimental observations were fully consistent with the kinetic scheme in eq. 4.2.24, in which micelles change size by addition or removal of one monomer at a time. In a general association/dissociation of aggregates, as in eq. 4.2.26, the kinetic behavior should be more complex.

A micellar system responds to a T- or a p-jump perturbation in two steps. First it relaxes rapidly to a conditional equilibrium state under the constraint that the number of micelles remain constant. To understand the kinetics involved in this process, we need to think about the micellar system in terms of multiple equilibria and an aggregate size distribution. The perturbation causes a small shift in the optimal size distribution. Monomers leave some micelles and associate to others, restoring an optimal size distribution. However, such processes leave the number of micelles unchanged.

Detailed analysis of the fast relaxation process results in the expression

$$\tau_1^{-1} = \frac{k^-}{\sigma^2} + k^- \frac{(c - \text{CMC})}{(\text{CMC} \cdot N)} \qquad (4.2.28)$$

Here N is the most probable aggregation number and σ stands for the standard deviation of the micelle size distribution function (cf. Figure 4.2).

TABLE 4.2 Relaxation Times τ_1 and τ_2 for Some Sodium Alkyl Sulfates

Surfactant	Temperature (°C)	Concentration (M)	τ_1 (μs)	τ_2 (ms)
$\text{NaC}_{16}\text{SO}_4$	30	1×10^{-3}	760	350
$\text{NaC}_{14}\text{SO}_4$	25	2.1×10^{-3}	320	41
	30	2.1×10^{-3}	245	19
	35	2.1×10^{-3}	155	7
	25	3×10^{-3}	125	34
$\text{NaC}_{12}\text{SO}_4$	20	1×10^{-2}	15	1.8
	20	5×10^{-2}		50

From E.A.G. Aniansson et al., *J. Phys. Chem.* **80**, 905 (1976).

Experiments confirm the predicted inverse linear dependence between the relaxation time τ_1 and the surfactant concentration c. We can determine k^- and σ in a τ_1 versus c—CMC plot if we know independently the aggregation number N.

After the rapid relaxation caused by monomer exchange between existing micelles, a slower process takes place that involves a change in the number of micelles. For a particular micelle to dissolve—if it starts from an aggregation number that is close to optimal—it must pass through a stage involving very unfavorable aggregation numbers analogously with the nucleation process discussed in Section 2.3.4. The result is a kinetic bottleneck consisting in the formation of the least probable aggregation numbers between N and unity (cf. Figure 4.2). A detailed analysis yields

$$\tau_2^{-1} = \frac{N^2}{CMC} \frac{1}{R_1} \left(1 + \sigma^2 \frac{c - CMC}{CMC} \frac{1}{N}\right)^{-1} \qquad (4.2.29)$$

where the equilibrium and the kinetic properties of micelles with the least probable aggregation numbers determine the "resistance" R_1. Through the concentration dependence of τ_2, we can get an independent measure of σ^2/N.

Table 4.3 summarizes the parameters obtained from an analysis of kinetic data for a series of sodium alkyl sulfates. Note that the dissociation constant, k^-, changes by a factor of 1400 in going from the C_6 to the C_{14} compound, while the association constant, k^+ changes only by a factor of 7.

In fact, k^+ is close to the diffusion-limited rate. (See Section 8.3.1.) Because the micelle is essentially a liquidlike drop, the monomer association reaction involves no slow structural reorganization or bond formation. The sole barrier felt by a monomer entering the micelle arises from long range electrostatic repulsions due to the micellar charge. With longer alkyl chains, the CMC is lower and the electrolyte concentration, consisting of monomer surfactants and counterions, is lower. The charge of micelle surface is less screened and slows down

TABLE 4.3 Kinetic Parameters of Association and Dissociation of Alkyl Sulfates from Their Micelles

Surfactant	N	CMC (M)	k^- (s^{-1})	k^+ ($M^{-1}s^{-1}$)
NaC_6SO_4	17	0.42	1.32×10^9	3.2×10^9
NaC_7SO_4	22	0.22	7.3×10^8	3.3×10^9
NaC_8SO_4	27	0.13	1.0×10^8	7.7×10^9
NaC_9SO_4	33	6×10^{-2}	1.4×10^8	2.3×10^9
$NaC_{11}SO_4$	52	1.6×10^{-2}	4×10^7	2.6×10^9
$NaC_{12}SO_4$	64	8.2×10^{-3}	1×10^7	1.2×10^9
$NaC_{14}SO_4$	80	2.05×10^{-3}	9.6×10^3	4.7×10^8

From E.A.G. Aniansson et al., *J. Phys. Chem.* **80**, 905 (1976).

the association rate for long-chain surfactants relative to short-chain ones.

For the most probable aggregation numbers, $N - 1$, and N, we have at equilibrium $k^+[S_1][S_{N-1}] = k^-[S_N]$. Since the monomer concentration equals the CMC, this reduces to $k^+\text{CMC} \simeq k^-$. Thus, a weakly decreasing k^+ and a strongly decreasing CMC, due to the increasing hydrocarbon chain length, causes k^- to decrease even more strongly.

We can learn several things from the analysis of the micellar kinetics. Micelles actually grow in a stepwise manner, as shown in the scheme in eq. 4.2.24, and two aggregates rarely associate into one big aggregate. (For micelles that grow to rods, this conclusion might not be valid.) Because few kinetic barriers to micellar growth exist on the molecular scale, most processes can be seen as diffusion controlled if we correct for the effects of long range electrostatic interactions.

An additional useful insight is that the cooperativity of micelle formation causes the slow relaxation time, τ_2. This process exists only because intermediate aggregation numbers are unlikely. The existence of the fast relaxation time demonstrates the existence of a distribution of aggregation numbers, but the experimental values reflect a moderately narrow distribution of aggregate sizes. For typical ionic micelles, the standard deviation, σ, in the distribution is of order 10.

4.2.4 Dynamics of Solutes Dissolved in Micelles Provide a Measure of the Time Scales for Solubilization Processes

A micelle's hydrocarbon core provides a small oillike reservoir that can be used to solubilize nonpolar molecules in an aqueous environment. This characteristic of micellar solutions finds wide applications, as discussed in Section 4.4. For this reason, it is important to understand the dynamics of the solubilization processes.

We can obtain exchange rate information by measuring residence times of suitable probes. Phosphorescence lifetimes are long enough for this purpose, and Table 4.4 gives some

TABLE 4.4 Residence Times in Micelles

| Probe Molecule | Residence Times (μs) | | Solubility in Water (M) |
	SDS (micelles)	$C_{16}TAB$ (micelles)	
Anthracene	59	303	2.2×10^{-7}
Pyrene	243	588	6×10^{-7}
Biphenyl	10	62	4.1×10^{-5}
Naphthalene	4	13	2.2×10^{-4}
Benzene	0.23	1.3	2.3×10^{-2}

From M. Almgren, F. Grieser, and J.K. Thomas, *J. Am. Chem. Soc.* **101**, 279 (1979).

typical residence times. As we might expect, they generally decrease with the probe molecule's increasing solubility in water. In fact, the association rate is diffusion controlled, just as it is with surfactant monomers. We can attribute the variations observed in residence times directly to differences in the distribution coefficients of the probe molecule between the aqueous phase and the micelle. On the time scales required for most industrial applications these exchange processes will result in uniformly distributed solubilizates.

4.2.5 Diffusion Plays an Important Role in Virtually All Micellar Processes

Diffusion in surfactant solutions is important in many diverse processes, including digestion, growth and dissolution of cholesterol gallstones, drug solubilization, cleaning of fabric, and degreasing of surfaces. There are three ways to experimentally probe the diffusion properties in micellar systems:

1. Diffusion is a macroscopic concentration gradient in which local concentrations are usually measured by some optical method. Such an experiment yields the **mutual-diffusion coefficient**.

2. Tagged particle diffusion in an otherwise homogeneous system. We can tag a particle by using radiactive tracers or, more elegantly, by using nuclear magnetic resonance techniques with pulsed field gradients. The observable parameter in this case is the **self-diffusion coefficient**.

3. Dynamic light scattering, which uses temporal fluctuations in the scattered light to sudy the translational motion of scattering objects. From such measurements, we also can extract the mutual-diffusion coefficient.

Figure 4.10 shows the measured self-diffusion coefficient of all components in a micellar solution of decylammonium dichloroacetate with added trace amount of tetramethylsilane (TMS). A noticeable break in the curves around 25 mM signals a CMC at this concentration. We can use the diffusion data to separate the contribution to the diffusion coefficient from free and micellized species with the formula:

$$D_{obs} = \frac{c^{mic}}{c} D^{mic} + \frac{c^{free}}{c} D^{free} \qquad (4.2.30)$$

where c stands for the total concentration. Figure 4.11 shows the deduced values for c^{mic} and c^{free} for both surfactant and counterion. The free monomer concentration decreases above the CMC, yet the degree of counterion binding remains constant at 0.80 up to concentration 10 times the CMC.

The micelle self-diffusion coefficient, which is measured through the self-diffusion of the completely solubilized TMS in Figure 4.10, decreases strongly with increasing concentration. Strong intermicellar repulsion, mainly of electrostatic origin, causes this decrease.

Figure 4.10
Self-diffusion coefficients of all the molecular species in a micellar solution of decylammonium dichloroacetate determined by NMR. The self-diffusion of the solvent (water), counterion ($CHCl_2COO^-$), surfactant ion ($C_{10}H_{21}NH_3^+$), and a hydrophobic solubilizate (tetramethyl silane, TMS) varies with total surfactant concentration. The micelle determines the diffusion of the TMS. (B. Lindman and P. Stilbs, in *Proceedings of Enrico Fermi School of Physics*, C.V. Degiorgio and M. Corti, eds., North-Holland Physics Publishing, Amsterdam, 1985, p. 94.)

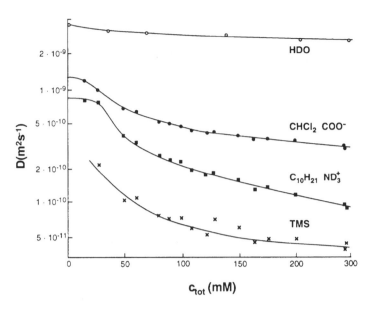

The mutual-diffuson coefficient D_{mut} has a qualitatively different concentration dependence. Except at high salt concentrations, where most of the electrostatic effects are screened out, D_{mut} increases with increasing concentration. In a concentration gradient, however, the repulsive interactions tend to increase diffusion as the micelles actively push one another toward regions of low concentrtation. Electrostatic interactions are responsible for an additional and more subtle effect: small counterions diffuse intrinsically more rapidly than the micelles, but due to the electroneutrality condition the counterions have

Figure 4.11
Concentration of various micellar species in a decylammonium dichloroacetate solution. Using the data given in Figure 4.10, we can calculate the concentration of micellarized amphiphile ions, $c_{+\text{mic}}$, ■, free amphiphilic ions $c_{+.\text{free}}$, □; bound counterions $c_{-\text{mic}}$, ●; and free counterions $c_{-.\text{free}}$, by ○, plotted as a function of c_{tot}, β is the degree of dissociation (cf. eq. 4.1.16). (P. Stilbs and B. Lindman, *J. Phys. Chem.* **85**, 2587 (1981).)

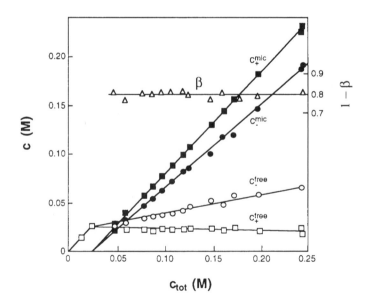

to move with the micelles, and vice versa, over macroscopic distances. The observed D_{mut} is then a weighted average over micelles and counterions. The latter pull the former along and the observed D_{mut} is larger than the self-diffusion coefficient D^{mic}.

4.3 The Properties of Many Micellar Solutions Can Be Analyzed Quantitatively

In Chapter 1 we made the general case that hydrocarbon insolubility in polar solvents drives amphiphilic aggregation and head group interactions oppose it. We also argued that a well-defined CMC exemplifies the start part of the start–stop process that characterizes aggregation, and the formation of well-defined finite species illustrates the stop. In Section 4.1 we obtained relationships between the CMC and the free energy. By bringing these ideas together, we can show quantitatively how these factors control aggregation processes in micelles.

Experimentally, we can easily access the standard free energy of micelle formation ΔG^{θ}_{mic} by its relation to the CMC through eq. 4.1.5. To interpret the variations in the CMC with external conditions, we can decompose ΔG^{θ}_{mic} into a number of components as

$$\Delta G^{\theta}_{mic} = \Delta G(HP) + \Delta G(contact) + \Delta G(packing) + \Delta G(HG) \qquad (4.3.1)$$

We can obtain the hydrophobic free energy contribution, $\Delta G(HP)$, for transferring the hydrocarbon chain out of the solvent into the oillike interior of the micelle from the corresponding values for pure hydrocarbons (as summarized in Table 1.6). For a transfer from water to hydrocarbon.

$$\Delta G(HP) = -(n_c - 1)3.0 - 9.6 \, kJ/mol \qquad (4.3.2)$$

when n_c equals the number of carbon atoms in the alkyl chain, 3.0 kJ/mol is the value for each methylene group, and 9.6 kJ/m that for the methyl group. The presence of double bond makes $\Delta G(HP)$ somewhat less negative in eq. 4.3.2.

The free energy $\Delta G(HP)$ represents a complete transfer out of the solvent. In reality, considerable solvent exposure exists at the micelle surface. Counted per molecule, the solvent contact decreases with increasing micelle size, and this decrease is the source of the cooperativity in the micelle formation process.

We can estimate the exposed hydrocarbon area of an SDS spherical micelle by using the length and volume values of alkyl chains cited in Section 1.3. With 12 carbons in the chain, we can estimate the radius of the hydrocarbon core as ~ 1.6 nm using eq. 1.3.3, where we assume that the radius matches the extended chain length. Using eq. 1.3.2 to estimate the volume of a single chain, we can calculate that we can pack approximately 60 chains in the volume $4\pi 1.6^3/3$. The measured aggregation number of SDS is, in fact, close to 60 (see Table 4.1).

The sulfate head group has a larger cross-sectional area than an extended chain. From the S—O bond length of 0.175 nm and an oxygen van der Waals radius of approximately 0.15 nm, we can estimate the cross-sectional area to be just over 0.3 nm^2. In Section 2.5, we saw that SDS molecules could pack with an area of 0.38 nm^2 at the air–water interface.

On the average, the sulfate atom at the micellar surface can be found about 0.3 nm outside the hydrocarbon core. At this location the total area is $4\pi \times 0.19^2$ and the area per monomer is then ~ 0.75 nm^2. An exposed area of 0.45 nm^2 remains at $r = 1.9$ nm and a somewhat smaller one projected down to the hydrocarbon core at $r = 1.6$ nm. This calculation demonstrates that large hydrocarbon–water contact area exists in a spherical micellar aggregate.

We can assume that the positive free energy $\Delta \mathbf{G}$(contact) attributed to the solvent hydrocarbon contact in the intact micelle is proportional to the area A_{mic} of the aggregate, so that

$$\Delta \mathbf{G}(\text{contact}) = \gamma' \frac{A_{\text{mic}}}{N} \qquad (4.3.3)$$

In this equation, $\Delta \mathbf{G}$(contact) takes the form of a surface free energy expression. The coefficient γ' does not represent the micelle's surface tension, since eq. 4.3.3 expresses only one of several free energy components.

As a surfactant molecule enters a micellar aggregate, its head group is more or less fixed to the micellar surface. This factor restricts the hydrocarbon chain's conformational freedom. On the average, the chain becomes somewhat straighter within the aggregate than in bulk liquid as a result of the combined effects of the micelle–solvent (which the chain does not penetrate) and constraints from neighboring chains. These conformational effects represent a positive free energy contribution, $\Delta \mathbf{G}$(packing), that is difficult to analyze quantitatively. Here we make the assumption that

$$\Delta \mathbf{G}(\text{packing}) = \text{constant} - \frac{\gamma'' A_{\text{mic}}}{N} \qquad (4.3.4)$$

since the smaller the head group area, the smaller the chain's conformational freedom. We also must realize that if the aggregate radius exceeds the estimated length of a chain, $\Delta \mathbf{G}$(packing) becomes large and positive.

Head group interactions $\Delta \mathbf{G}$(HG) make the most significant positive contribution to $\Delta \mathbf{G}_{\text{mic}}^{\theta}$. However, we must consider the different types of head groups separately when discussing the $\Delta \mathbf{G}$(HG) term.

For nonionic surfactants with an oligopolyethylene oxide head group, the micelle surface region is analogous to a polymer solution, while for noncharged but strongly polar head groups, such as those in a zwitterionic surfactant, strong dipolar forces manifest themselves at the micelle surface. We postpone a discussion of these systems until Section 4.3.5.

Ionic surfactants are the most studied. Fortunately, they also are the most amenable to quantitative analysis of head group effects. As the micelle is formed, an electrical double layer is generated. We can estimate the free energy of this double layer using the Poisson–Boltzmann equation described in Chapter 3.

4.3.1 The Poisson–Boltzmann Equation Describes Head Group Interactions in Ionic Micelles

Association of ionic surfactants into aggregates generates a highly charged surface, and the electrostatic interactions make a large positive contribution to the free energy of micelle formation. Thus, changes in the factors that influence the electrostatic free energy strongly affect the micelle formation process.

We can summarize some key experimental observations that are fully or partly due to simple electrostatic effects

- For a homologous series of surfactants, the CMC decreases with the increasing number of carbons in the alkyl chain. However, the decrease is much more pronounced for uncharged surfactants than for ionic ones.

- Addition of electrolyte decreases the CMC and raises the aggregation number.

- Increasing the surfactant concentration results in aggregate growth and leads ultimately to the transformation from one microstructure to another.

- Changes in the ionic surfactant head group can lead to substantial changes in the CMC for a constant apolar chain.

- Normally changes between different atomic counterions of the same valency produce only a small effect, but increase in counterion valency leads to a substantial decrease in the CMC. However, for cationic surfactants there is a clear counterion specific variation of the CMC.

We can model these trends quantitatively by using three different approximations derived from the Poisson–Boltzmann (PB) equation to calculate $\Delta G(HG) = G_{el}$.

Approach 1: Approximation of the Micellar Surface as a Plane

In Chapter 3 we showed that we can calculate the electrostatic free energy for a planar surface analytically as in eq. 3.9.16. The free energy per monovalent surface charge is

$$\mathbf{A}_{el} = \mathbf{G}_{el} = 2kT \left\{ \ln[s + (s^2 + 1)^{1/2}] - \frac{1}{s}[(s^2 + 1)^{1/2} - 1] \right\} \tag{4.3.5}$$

where the dimensionless charge density s is

$$s = \sigma(8kTc_o^* \varepsilon_r \varepsilon_o)^{-1/2} \tag{3.9.4}$$

for a 1:1 electrolyte at concentration c_o.

We can calculate the charge density at the micelle surface from

$$\sigma = \frac{Q_{mic}}{4\pi R_{mic}^2} \tag{4.3.6}$$

where Q_{mic} equals the aggregation number and R_{mic} can be taken as the length of an extended surfactant molecule, including its head group.

The relation between the free energy and the optimum size of an ionic micelle can be expressed quantitatively in terms of the electrical energy. We can obtain an expression for the optimum size by minimizing the free energy

$$\frac{\partial \Delta \mathbf{G}_{mic}^\theta}{\partial R_{mic}} = \frac{\partial \Delta \mathbf{G}_{mic}^\theta}{\partial A_{mic}} = 0 \tag{4.3.7}$$

where A_{mic} stands for the micellar area. The derivative

$$\frac{\partial}{\partial A_{mic}}(\Delta\mathbf{G}(HP) + \Delta\mathbf{G}(packing) + \Delta\mathbf{G}(contact)) = \gamma' - \gamma'' \equiv \gamma \tag{4.3.8}$$

is easy to evaluate. The electrostatic part is more difficult, but the result

$$\frac{\partial \mathbf{G}_{el}}{\partial A_{mic}} = -\frac{2\mathbf{U}_{el}}{A_{mic}} \tag{4.3.9}$$

which we developed in Chapter 3 (eq. 3.9.17) for planar surfaces, is equally valid in spherical geometry, which we discuss next. Thus, combining eqs. 4.3.7, 4.3.8, and 3.9.17 gives us the relation

$$\gamma = \frac{2\mathbf{U}_{el}}{A_{mic}} \tag{4.3.10}$$

at the optimal micelle size. By calculating $\mathbf{U}_{el}$ using known aggregation numbers, we can determine the unknown parameter γ.

To evaluate s in eq. 3.9.14, we require values of the electrolyte concentration c_0. The total electrolyte concentration includes contributions from both added electrolyte and unmicellized surfactant, that is,

$$c_0 = c_{salt} + CMC$$

Because the CMC depends on c_{salt}, we must employ an iterative procedure in the calculations.

Approach 2: Exact Relation to PB Equation Using the Linearized Form

In spherical symmetry, which is the relevant case for micelles, we can solve the PB equation only as in the linearized Debye–Hückel theory. Then the expression for the electrostatic free energy of a micelle can be calculated in the same way as for the Born model of ion solvation discussed in Section 3.6. In analogy to eq. 3.8.25, the potential on the micelle charges at $r = R_{mic}$ is

$$\Phi(R_{mic}) = \frac{Q_{mic}}{4\pi\varepsilon_r\varepsilon_o}\frac{1}{R_{mic}}\frac{1}{1+\kappa R_{mic}} \tag{4.3.11}$$

The free energy can be calculated most simply from a charging procedure

$$G_{el} = \int_0^{Q_{mic}} \phi(Q)\,dQ = \frac{Q_{mic}^2}{8\pi\varepsilon_r\varepsilon_o}\frac{1}{R_{mic}}\cdot\frac{1}{1+\kappa R_{mic}} \tag{4.3.12}$$

When the Debye screening length κ^{-1} goes to infinity, eq. 4.3.12 reduces to the Born solvation energy described in eq. 3.6.3. The Debye—Hückel linearization is limited to small potentials $(ze\Phi < kT)$ and applies only for micellar systems at higher salt contents.

Approach 3: Approximate Analytical Solution to the PB Equation in Spherical Geometry

The nonlinear PB equation that describes the distribution of ions about a sphere in a 1:1 electrolyte is

$$\nabla^2\Phi = \frac{1}{r^2}\left(\frac{d}{dr}\left(r^2\frac{d\Phi}{dr}\right)\right) = \frac{2ec_0^*}{\varepsilon_r\varepsilon_o}\sinh\left(\frac{e\Phi}{kT}\right) \tag{4.3.13}$$

This equation has no analytical solution, but after tedious algebra, we can obtain a series expansion in terms of $(\kappa R_{mic})^{-1}$ valid for $\kappa R_{mic} \gtrsim 1$.

$$G_{el} = 2kT\left\{\left[\ln(s+[1+s^2]^{1/2})+\frac{1-[1+s^2]^{1/2}}{s}\right]-\frac{2}{s\kappa R_{mic}}\ln\left(\frac{1+[1+s^2]^{1/2}}{2}\right)+\cdots\right\} \tag{4.3.14}$$

per charged amphiphile. Here s has the same definition as in eq. 3.9.14.

Comparing this equation with eq. 4.3.5, we see that the first term on the right-hand side of eq. 4.3.14 is the electrostatic free energy of a planar double layer and the second term is an explicit correction due to the curvature of the surface. The curvature term is negative and shows that the electrical free energy is smaller on the (convex) curved surface because in addition to the screening effects, the electrical field decays out into the solution on the curved surface for geometrical reasons.

We can also evaluate U_{el} to determine γ from eq. 4.3.10. We have $U_{el} = G_{el} - T(\partial G_{el}/\partial T)$ (cf. eq. 3.3.7), which can be used to evaluate the curvature-dependent part of U_{el}. Equation 3.8.5 gives the planar U_{el}. When evaluating the derivative, we use the relations $\partial s/\partial T = -s/(2T)$ and $\partial(1/\kappa)/\partial T = 1/\kappa \cdot 1/(2T)$, which follow from eqs. 3.7.13 and 3.9.14. Combining eqs. 4.3.10, 3.9.13, and 4.3.14, we find

$$\gamma = 2kT\frac{\sigma}{e}\left\{\frac{1}{s}[(s^2+1)^{1/2}-1] - \frac{2}{\kappa R_{mic}}\frac{s}{1+s^2+(1+s^2)^{1/2}} + \frac{4}{s\kappa R_{mic}}\ln\left(\frac{1+(1+s^2)^{1/2}}{2}\right)\right\} \quad (4.3.15)$$

The analytical solution, valid for $\kappa R_{mic} \gtrsim 1$, is called the dressed micelle model. It yields equations that appear somewhat algebraically complex, but they are readily evaluated on an electronic calculator using as input parameter the micellar aggregation number N, the micellar radius R_{mic}, and the Debye screening length κ^{-1}.

If we need to calculate the electrostatic free energy for a range of electrolyte concentrations and micellar radii, including cases when $\kappa R < 1$, we ultimately must solve the PB equation numerically. Today this is straightforward. However, in a sense, numerical results are less informative because the dependence on the different parameters appears through explicit numbers rather than in analytical formulae.

4.3.2 Variations in the CMC Caused by Electrostatic Effects Are Well Predicted by the Poisson—Boltzmann Equation

Figure 4.12 shows how ln(CMC) varies with alkyl chain length and sodium chloride concentration for a series of sodium alkyl sulfates. To understand the dependence on the NaCl concentration, we assume that the micelle itself remains unchanged by the addition of salt. According to eq. 4.3.1, only the head group term $\Delta G(HG)$ changes, since the electrostatic interactions are sensitive to the presence of salt. Using eq. 4.1.5, the change $\Delta\Delta G_{mic}^{\theta}$ in the free energy of micelle formation is

$$\Delta\Delta G_{mic}^{\theta} \equiv \Delta G_{mic}^{\theta}(c_{salt}) - \Delta G_{mic}^{\theta}(c_{salt} = 0)$$

$$= G_{el}(c_{salt}) - G_{el}(c_{salt} = 0)$$

$$= RT\{\ln[CMC(c_{salt})] - \ln[CMC(c_{salt} = 0)]\} \quad (4.3.16)$$

The difference induced by adding electrolyte $\Delta\Delta G_{mic}^{\theta}$ can be calculated, with varying accuracy, using any of the methods for calculating electrostatic free energies already described. The only difficulty is that the total electrolyte concentration, $c_0 = c_{salt} + CMC$, must be determined iteratively. Figure 4.12 shows

Figure 4.12
Effect of hydrocarbon chain length and added salt on the CMC for sodium alkyl sulfates. As the chain length increases for a given head group, ΔG(HG) (eq. 4.3.1) becomes increasingly negative, leading to a lower CMC. Adding salt diminishes the head group repulsions, $\mathbf{G}$(HG), which also lowers the CMC. This effect is more pronounced for the longer chain surfaces because the total electrolyte concentration involves contributions from free surfactant and from salt. The solid lines calculated using eq. 4.3.17 show that theory quantitatively accounts for the variation of CMC. (G. Gunnarsson, B. Jonsson, and H. Wennerstrom, *J. Phys. Chem.* **84**, 3114 (1980).)

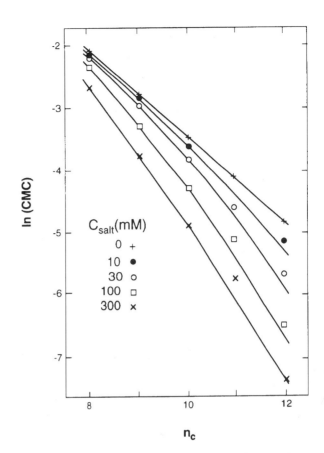

that the values calculated through a numerical solution of the PB equation reproduce the experimental CMC values very well.

Figure 4.12 also shows how the CMC varies with chain length. For example, the chain length dependence clearly is strongest at the highest salt concentration. How should that be interpreted? Going back to eq. 4.3.1 for $\Delta \mathbf{G}^{\theta}_{\text{mic}}$, we see that in changing the chain length—that is, n_c—we affect several terms of the right-hand side. However, if the micelle radius is that of the fully stretched chain, the exposed area per molecule remains essentially constant in a homologous series. Thus, we can ignore the changes in the contact term, $\Delta \mathbf{G}$(contact), and $\Delta \mathbf{G}$(packing). The two important terms in eq. 4.3.1 are the hydrophobic contribution $\Delta \mathbf{G}$(HP) and the electrostatic term $\Delta \mathbf{G}$(HG) = $\mathbf{G}_{\text{el}}$.

The chain length dependence of $\Delta \mathbf{G}$(HP) is given in eq. 4.3.2, and we can obtain the change $\Delta \Delta \mathbf{G}^{\theta}_{\text{mic}}(n'_c, n_c)$ in the standard free energy of micelle formation in going from n_c to n_c carbons in the chain as

$$\Delta \Delta \mathbf{G}^{\theta}_{\text{mic}}(n'_c, n_c) = -3.0 \times 10^3 (n'_c - n_c) + \mathbf{G}_{\text{el}}(n'_c) - \mathbf{G}_{\text{el}}(n_c)$$

$$= \mathbf{R}T \ln \left[\frac{\text{CMC}(n'_c)}{\text{CMC}(n_c)} \right] \qquad (4.3.17)$$

TABLE 4.5 Quantitative Analysis of the Free Energy Contributions in the Formation of Ionic Micelles: Calculations Based on the Dressed Micelle Model

Surfactant Property	CMC (mM)	N	a_0 (Å^2)	$RT \ln CMC^a$ (kJ/mol)	G_{el} (kJ/mol)	γ (J/m$^2 \times 10^3$)	$\gamma a_0 N_{Av}$ (kJ/mol)	$\Delta G(HP)$ (kJ/mol)
SDS (M) c_{salt}								
0	8.3	56	62	−22.0	12.8	16.0	6.0	−40.8
0.01	5.7	64	60	−22.9	11.7	15.8	5.7	−40.3
0.03	3.1	71	58	−24.4	10.4	15.2	5.3	−40.2
0.10	1.47	93	53	−26.3	8.8	15.2	4.8	−39.9
T (°C) $C_{14}N(Me)_3Br$								
25	3.8	70	64	−23.7	14.0	15.8	6.1	−43.9
76	6.7	55	70	−26.2	14.7	16.6	7.0	−47.9
114	13.5	35	81	−26.8	13.3	15.1	7.4	−47.5
166	39	20	100	−26.5	10.9	12.5	7.5	−44.9

From D.F. Evans, Langmuir **4**, 3 (1988).
^{a}CMC expressed in terms of mole fractions.

where G_{el} should be expressed in joules per mole of surfactant. The lines in Figure 4.12 shows that the numerical calculations provide a good representation of the experimental data points. Although a clear deviation exists for the $n_c = 11$ surfactant, it probably represents an experimental artifact resulting from the use of an impure surfactant for this particular chain length.

The slope of the ln(CMC) versus n_c curve is higher at higher salt concentrations as a result of the effect of the $G_{el}(n_c)$ terms in eq. 4.3.17. If we neglect the electrostatic effects, the CMC decreases with a factor of ~ 3.2 for each additional CH_2 group in the chain. However, the electrostatic terms in eq. 4.3.17 lower the magnitude of $\Delta\Delta G^\theta_{mic}$. In the absence of added salt, only the free surfactant and the counterions act as electrolyte, so the lower the CMC, the lower the electrolyte concentration and the more repulsive the head group interactions.

At high salt concentrations (such that $c_{salt} \gg$ CMC), the electrolyte concentration at the respective CMCs does not depend on n_c. The term $G_{el}(n'_c) - G_{el}(n_c)$ is then essentially zero, except for the small dependence on R_{mic} that is apparent from eq. 4.3.14.

In Table 4.5 we illustrate how the dressed micelle model (eqs. 4.3.14 and 4.3.15) can quantitatively explain the changes in aggregation number accompanying changes in salt concentration and temperature. The table shows how the aggregation number N—and thus the radius and the area per surfactant—adjusts when the electrostatic head group repulsion changes as a result of changes in total electrolyte concentration. The value of γ, as calculated from eq. 4.3.15, stays remarkably constant at around 15×10^{-3} J/m^2. This fact strongly supports the assumptions inherent in eqs. 4.3.3 and 4.3.4.

4.3.3 The Contribution of the Solvophobic Free Energy ΔG(HP) Decreases when Micelles Form in Nonaqueous Solvents

Micelles also can be formed in a number of polar solvents in addition to water. The decomposition of ΔG^θ in eq. 4.3.1 remains valid if we interpret the term $\Delta G(HP)$ as the free energy of transferring hydrocarbon from the solvent to a hydrocarbon environment. However, the coefficients in the empirical formula of eq. 4.3.2 change with a change in solvent. In nonaqueous solvents, the dependence of CMC and N on surfactant chain length and salt parallels the dependence we observe in water. In hydrazine, the free energy of transferring a CH_2 out of the solvent and into the micelle is 3.0 kJ/mol or slightly lower than the value of 3.2 kJ/mol observed in water. Plots of ln(CMC) versus log(CMC + c(NaCl)) for SDS give almost identical slopes in water and hydrazine. Analysis with the Poisson—Boltzmann equation (Table 4.6) gives $\Delta G(HP)$ that are about 4 kJ/mol lower than the values observed in water. Based on the discussion of gas solubilities given in Chapter 1, we would expect comparable $\Delta G(HP)$ values in water and hydrazine.

Micelles in ethylammonium nitrate (EAN), an 11 mol/L fused salt, are not amenable to analysis by the Poisson–Boltzmann equation. Instead, we can use the phase separation model to obtain a transfer free energy of 1.76 kJ/mol for a CH_2 group. The CMCs in EAN are about a factor of 10 higher than those in water, and the aggregation numbers are about a factor of 4 smaller. These micellar properties are consistent with the differences in the gas solubility behavior discussed in Chapter 1.

4.3.4 Enthalpy and Entropy of Micellization Change Much More Rapidly with Temperature than the Free Energy

As an example of how the properties of spherical ionic micelles change with temperature, we will focus first on tetradecyltrimethyl ammonium bromide ($C_{14}N(Me)_3NBr$) because its CMCs

TABLE 4.6 Comparison of Thermodynamics of Micelle Formation and Hydrocarbon Transfer in Water and Hydrazine

	Surfactant	T (°C)	$-\Delta G(HP)$ (kJ/mol)	$-\Delta H(HP)$ (kJ/mol)	$\Delta S(HP)$ (J mol^{-1}K^{-1})
H_2O	$C_{14}N(Me)_3Br$	25	43.9	13	104
		95	46.4	52	10
		166	44.3	65	−50
H_4N_2	$C_{12}N(Me)_3Br$	35	30.0	46	−54
	$C_{12}OSO_3Na$	35	36.0	59	−75

From D.F. Evans, *Langmuir* **4**, 3 (1988).

and aggregation numbers are available from 25 to 166°C in water. Across this temperature range, $C_{14}N(Me)_3NBr$'s CMC increases by a factor of 8 and its aggregation number decreases substantially from 70 to 20. However, the total free energy of micellization, which includes all the contributions given in eq. 4.3.1, decreases by 2.8 kJ/mol in going from 25°C to 166°C (Table 4.5). Values of ΔH^θ and ΔS^θ calculated from eqs. 4.1.21 and 1.4.1 are given in Table 4.6.

Because the values of ΔG^θ, ΔH^θ, and ΔS^θ reflect all the contributions listed in eq. 4.3.1, they do not provide penetrating insights into micellization processes. If we use the PB equation, however, we can dissect ΔG^θ_{mic} into $\Delta G(HP)$ and $\Delta G(HG)$. Figure 4.13 shows $\Delta G(HP)$ obtained from the dressed micelle model equation and the corresponding values of $\Delta H(HP)$ and $\Delta S(HP)$ from the Gibbs–Helmholtz equation. The solvophobic driving force $\Delta G(HP)$ is almost independent of temperature from 25°C to 166°C, but it displays a small minimum near 90°C (similar to the behavior of simple hydrocarbons in water). Large changes in enthalpy and entropy compensate one another, leaving the free energy almost invariant.

These changes in $\Delta G(HP)$, $\Delta H(HP)$, and $\Delta S(HP)$ parallel those for transfer of nonpolar solutes from H_2O to oil discussed in Section 1.7. (Note the sign convention used in Table 1.4.) Since micellization involves the transfer of hydrocarbon chains from water to the oillike interior of a micelle, the similarities between Tables 4.6 and 1.4 are not surprising.

Figure 4.13
Thermodynamics of transferring the hydrocarbon chain of the tetradecyltrimethyl-ammonium bromide ion from water to the interior of a micelle as a function of temperature. By combining the change in CMC with temperature and the dressed micelle model, large changes in ΔH and $T\Delta S$ become evident. However, ΔH and $T\Delta S$ compensate each other in such a way that ΔG remains almost constant with temperature. (D.F. Evans, *Langmuir* 4, 3 (1988).)

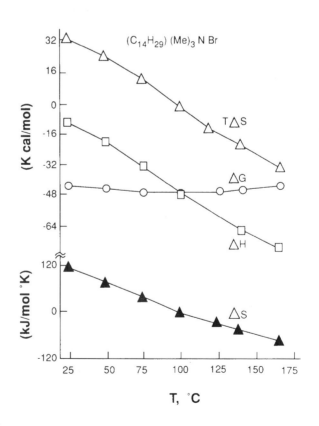

The values for $C_{12}N(Me)_3NBr$ and $C_{12}OSO_3Na$ in hydrazine at 35°C (Table 4.6) resemble those we find for $C_{14}N(Me)_3NBr$ and other surfactants in water at high temperatures where hydrophobic hydration has largely disappeared. Large and negative entropies reflect the unfavorable entropy change associated with transferring a hydrocarbon from solution into the confines of a small micelle. The enthalpic self-organization that drives micellization reflects the solvophobicity of both water and hydrazine.

4.3.5 Uncharged Surfactants Have Much Lower CMCs than Ionic Surfactants

Above we have carefully analyzed the head group interactions with ionic surfactants and demonstrated that they are dominated by long range ion–ion interactions as illustrated by the sensitivity of the CMC to electrolyte concentration. Noncharged surfactants with zwitterionic dipolar or nonionic head groups have substantially lower CMCs than the corresponding ionic surfactants (see Table 4.7); for example, CMC(SDS) = 8 × 10^{-3} M, CMC($C_{12}E_8$) = 7.1 × 10^{-5} M. Clearly, the free energy cost of bringing the polar heads together is much smaller when the head group is uncharged than when it is charged. Yet if some head group—head group repulsion did not exist, the aggregation process would proceed to form a separate phase. Furthermore, the CMC depends more strongly on the alkyl chain length. The head group repulsion varies substantially in magnitude between different uncharged surfactants. Comparing

TABLE 4.7 CMC Values for Some Noncharged Surfactants

Surfactant[a]	T(°C)	CMC
$C_{10}E_8$	25	1.0×10^{-3}
$C_{11}E_6$	25	3.0×10^{-4}
$C_{12}E_8$	15	9.7×10^{-5}
$C_{12}E_8$	25	7.1×10^{-5}
$C_{12}E_8$	40	5.8×10^{-5}
$C_{14}E_8$	25	9.0×10^{-6}
$C_{12}E_6$	25	6.8×10^{-5}
α-d-Octylglucoside	25	1.0×10^{-2}
β-d-Octylglucoside	25	2.5×10^{-2}
β-d-Decylglucoside	25	2.2×10^{-3}
β-d-Dodecylglucoside	25	1.9×10^{-4}
Sucrose monolaurate	20	1.85×10^{-4}
Dodecyl betaine	25	1.4×10^{-3}

[a] $C_nE_m = CH_3(CH_2)_{n-1}(OC_2H_4)_mOH$.
From P. Mukerjee and K.J. Mysels, *CMC of Aqueous Surfactant Systems*, National Standards Reference Data Service, U.S. National Bureau of Standards, 1971.

$C_{12}E_8$, dodecyl glycoside and dodecylbetain we see in Table 4.7 that their CMC values span a factor of 20.

In contrast, adding two CH_2 groups lowers the CMC by approximately a factor of 10 irrespective of the chemical nature of the head group. Using eq. 4.1.11 to relate the CMC to $\Delta \mathbf{G}^0_{mic}$ and assuming that increasing the alkyl chain will not alter the head group repulsion, we find that the hydrophobic energy increases by

$$\tfrac{1}{2}RT \ln 10 = 2.8 \, \text{kJ/mol} \qquad (4.3.18)$$

per CH_2 group. This value is of the same magnitude but lower than the values around 3.5 kJ/mol quoted in Table 1.6 for the transfer to bulk phases, but somewhat larger than the value 2.6 kJ/mol given in eq. 2.5.2 that is valid for transfer to monolayers. Now let us analyze the role that head group interactions play for sugar, zwitterionic, and nonionic ethyleneoxide surfactants.

For a long time biochemists have used sugar surfactants like octylglucoside for solubilizing membrane proteins. In plants and bacteria, sugars constitute the major class of lipid head groups (see Chapter 6). Recently, sugar surfactants also have received great interest for use in commercial detergents due to their biodegradability properties. Experimentally, we find that the CMC (and aggregation properties in general) depend on structural details of the head group. For example, Table 4.7 shows that there is a factor of 2.3 difference between the CMCs of α- and β-octylglucosides. On the other hand, the CMCs and aggregate structures are insensitive to changes in temperature and salt concentration as well as largely unaffected by changes in surfactant concentration. We can rationalize this behavior by analyzing the nature of head group—head group interactions in some molecular detail.

A mono- or disaccharide head group consists of one or two stiff rings of considerable molecular size but little internal flexibility. The rings are covered with OH groups that interact favorably with water (or other strongly polar solvents) but very unfavorably with alkanes. Two adjacent head groups containing sugar rings cannot achieve a strong attractive interaction because the stiffness of the ring prevents an optimal arrangement of the atoms in the head group. Since water molecules are small, they can form hydrogen bonds in a more optimal way. This is the cause of the high solubility of sugar in water. When the sugar ring is fixed to an alkyl chain anchored in an aggregate, the sugar—sugar effective intermolecular interaction—both laterally along the surface and transversely between surfaces—appears to consist merely of a short range repulsion that prevents direct sugar–sugar contact. In building up an aggregate, this repulsion implies the existence of a rather well-defined optimal head group area, a_0, for each particular sugar head group. The area a_0 is quite insensitive to external variables such as temperature, salt content, and surfactant concentration. However, the area the head group occupies does depend on how the group is attached to the alkyl chain.

In zwitterionic surfactants, the head group contains a cationic center, which is normally a quaternary ammonium or primary amine, and an anionic center, which could be a $-COO^-$ (as in betaines), $-OPO_3^-$ (as in phospholipids), or $-OSO_3^-$ and $-SO_3^-$ (as in some commercially used surfactants). Despite the local ionic nature of the head group, the CMCs and aggregate sizs are insensitive to moderate (<0.5 M) electrolyte concentrations in the aqueous medium. However, zwitterionic systems are more sensitive than sugar surfactants to changes in temperature and in surfactant concentration. For many zwitterionic micellar systems we can observe a separation into a more dilute and a more concentrated phase if we lower the temperature. The critical point typically occurs at high surfactant concentrations (15–30% w/w). Due to the strongly polar character of zwitterionic head groups, we should expect electrostatic interactions to play a major role for both intra- and interaggregate interactions. However, these interactions are of short range and their effects are qualitatively different from the ones discussed above with ionic surfactants. Because zwitterionic head groups are particularly important for biological amphiphiles forming bilayers, we postpone a detailed analysis of them to Chapter 6.

Nonionic surfactants with the head group consisting of a chain $-(OC_2H_4)_mOH$ of m ethylene oxide units find much use in commercial detergents. By a lucky coincidence, such surfactants also show intriguing self-assembly properties. Surfactants, $CH_3(CH_2)_{n-1}(OC_2H_4)_mOH$, abbreviated C_nE_m, with normal alkyl chains have been most useful for scientific studies. In this section, we discuss the basic properties of micellar solutions, but in Chapters 10 and 11 we will return to a more general description of the rich self-assembly phenomena. Micellar solutions are characterized by:

1. Marked decrease in the CMC with increasing temperature.

2. Increase in the aggregate size with increasing temperature.

3. Separation into a micelle-rich and a micelle-poor phase with *increasing* temperature, as illustrated in Figure 4.14.

4. A critical point, often called the cloud point, that occurs at low concentrations; for example, $w_{crit} = 0.023$ in Figure 4.14.

5. Aggregates that are sensitive to the presence of electrolyte at moderate to high (>0.1 M) concentrations. This effect is ion specific.

For C_nE_m surfactants, the polar group, $-(OC_2H_4)_nOH$, has approximately the same physical size as the alkyl chain part. The ethylene oxide moities only interact relatively weakly with water, but when $n \geqslant 3$, the total hydrophilicity can still become large enough to ensure surfactant behavior. Having such a long and flexible head group also results in a very important free energy contribution due to the entropy of the head group conformations. The smaller the area per head group, the more densely packed are the ethylene oxide groups and the fewer the possible conformations. Such a constraint results in an effective lateral repulsion between head groups that increases

Figure 4.14
Effect of temperature on the aqueous phase behavior of a nonionic surfactant, $C_{10}H_{21}(OC_2H_4)_4OH$. With increasing temperature, nonionic micellar solutions phase separate to form micellar-rich and micellar-lean phases. The cloud point curve at which this phase transformation occurs is indicated. The composition is measured as weight percent surfactant. (J.C. Lang and R.D. Morgan, *J. Chem. Phys.* **73**, 5852 (1980).)

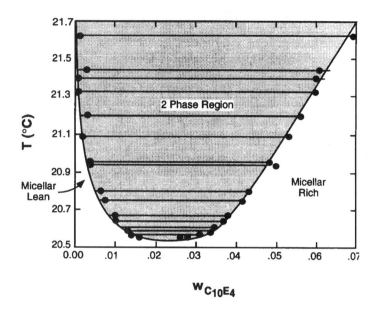

gradually with decreasing head group area. Clearly, the more ethylene oxide groups there are, the stronger the repulsion. For C_{12} chains ($n = 12$), we can obtain spherical micelles for $m = 5$ and up, but the tendency for micellar growth is larger the lower the m value. Due to the entropic nature of the effective head group repulsion, we expect micellar properties to depend strongly on temperature, but the observed temperature dependence is anomalous since phase separation occurs upon heating. This behavior is opposite to what we derived for the regular solution model and also to what we find for zwitterionic surfactants.

We can obtain an initial understanding of the cause of this anomalous temperature effect by noting that a similar phenomenon occurs for solutions of pure polyethylene oxide polymer in water. In this case we observe a separation into dilute and concentrated polymer phases at temperatures above 100°C, showing that water becomes a less good solvent for the ethylene oxide groups as temperature increases. This effect parallels a general feature of dipolar or hydrogen bond electrostatic interactions in water that we briefly discussed in Section 3.3. However, an additional important effect is operating for ethylene oxide chains. The polarity of an ethylene oxide chain varies substantially with its conformation. A detailed analysis reveals few polar conformations of low energy while less polar ones are more abundant but of somewhat higher energy. An increase in temperature leads to an increased population of the less polar configuration, resulting in a lower affinity for water.

For a micellar aggregate formed by surfactants of the C_nE_m type, an increase in temperature primarily leads to a desire to exclude some water from the palisade layer of ethylene oxide units in one of two ways. Either the area per head group is reduced, or head group regions of two micelles overlap so that ethylene oxide groups from other aggregates replace water. In practice, we observe both effects, and the reduction of head group area leads to an increase of the surfactant number N_s and, thus, aggregate growth. Favorable overlap between the head group regions of two micelles provides a mechanism for an attractive micelle–micelle interaction that ultimately triggers phase separation.

For all three types of surfactants, the sugar, the zwitterionic, and the nonionic, we possess a qualitative understanding but not a quantitative description of the interactions between the head groups and their consequences. Accurately accounting for the interplay between intermolecular organizing forces and entropic randomizing effects in the interfacial transition zone between the aqueous medium and the apolar interior of the aggregate poses a substantial challenge. Despite many unsolved issues, studies of micellar systems have become one of the most efficient routes to a detailed understanding of molecular events that occur at a polar–apolar interface.

4.3.6 Micelles Can Grow in Size to Short Rods, Long Polymer-Like Threads, and Even Branched Infinite Aggregates

So far we have focused our discussion of micellar properties on spherical aggregates. We have pointed out that we can characterize a micelle forming surfactant not only by the cooperativity of the formation but also by the mechanism that stops aggregate growth to a macroscopic size and results in the formation of a separate phase. For most typical surfactants the stop process emanating from head group–head group repulsion is so strong that the dominant aggregate is a spherical micelle. However, in many cases, surfactant association leads to the formation of large aggregates either in the vicinity of the CMC or at higher concentrations. The principle to use in achieving aggregate growth is to reduce head group–head group repulsion. How to accomplish this goal depends on the particular system.

A general strategy is to add a cosurfactant with a small head group, like a long-chain alcohol, but this method is difficult to implement. The problem in this case is that we can easily reduce the head group repulsion too much and cause a transition to a lamellar phase. For an ionic surfactant like SDS we can induce growth by introducing high concentrations of electrolyte, effectively screening the long range electrostatic interactions. Furthermore, the longer the apolar chain the less salt is needed to induce growth. The larger the aggregate radius the smaller is the change in curvature for a sphere-to-rod transition. This implies a smaller change in electrostatic energy according to eq. 4.3.14 which can be used to analyze the electrostatic effects quantitatively.

A more specific way to reduce ionic head group repulsion is to add an amphiphilic counterion in stoichiometric concentrations. A much-studied system of this type is $C_{16}N(Me)_3Br$ plus a Na-salicylate. The aromatic ring of the salicylate enters the micelle, and we observe extensive aggregate growth even at low concentrations. Such solutions show spectacular rheological properties and can be used to manipulate the flow properties of aqueous systems. The solution contains long thread-like aggregates that behave much like a polymer.

In the growth process

$$S + S_{n-1} \rightleftarrows S_n \qquad (4.1.1)$$

with n larger than that of the spherical micelle ($n \simeq 60$), adding a monomer simply adds to the length of a rod. To a good approximation, change in standard free energy is independent of n since the new surfactant enters a cylindrical portion of the aggregate for all these n values. Consequently, we can apply the isodesmic aggregation model. As discussed in Section 4.1, this model leads to an exponentially decaying distribution function. However, the decay constant depends on the total concentration of surfactant $[S]_T$ so that the mean aggregate size increases with $[S]_T^{1/2}$. (See also Exercise 4.2.)

Nonionic surfactants of the $C_m E_n$ type display yet another type of behavior with respect to aggregate growth. Here we reduce head group–head group interactions by increasing the temperature. As discussed above, higher temperature leads to liquid–liquid phase separation, but aggregate growth occurs as well. At the onset of this growth, the aggregation behavior is similar to that found for the SDS and CTAB systems already discussed, but the long thread-like aggregates that we observe in those systems do not appear when we increase the surfactant concentration. Instead, branching seems to occur, leading to the formation of aggregates of infinite extension. In this way, we can explain how a system transforms continuously from a dilute micellar solution to a reversed micellar solution when surfactant is present in excess.

4.4 Micellar Solutions Play a Key Role in Many Industrial and Biological Processes

4.4.1 Commercial Detergents Contain a Mixture of Surfactants

When we mix two micelle-forming surfactants, the normal outcome in a mixed system is that micellar aggregation also occurs. Commercial detergents always contain a mixture of surfactants. One reason for this is that impure substances are cheaper to manufacture; but a more interesting reason is that mixing surfactants often improves performance. To have a background for understanding this phenomenon, let us analyze some properties of mixed systems by considering a binary surfactant mixture. The surfactants can differ by: (1) length of the chains, (2) the chemical nature of the apolar chains, and (3) the nature of the head groups.

Consider first a mixture of two surfactants that differ in

the length of their alkyl chains but have the same head group. Because the molecules are very similar, we can (to a good approximation) assume that they mix ideally in the micellar aggregate. However, in the monomer state the surfactant with the longer chain has a higher chemical potential at equal concentrations. As we increase the concentration of the surfactants, the one with the longer chain will have the highest tendency to form micelle. We can calculate the resulting CMC using the phase separation model. For the two surfactants, index 1 and 2, the chemical potential in the micellar pseudophase is

$$\mu_i(\text{mic}) = \mu_i^\theta(\text{mic}) + RT \ln X_i \qquad (4.4.1)$$

where X_i is the mole fraction of surfactant i in the micelle so that $X_1 + X_2 = 1$. In solution we have

$$\mu_i(\text{aq}) = \mu_i^\theta(\text{aq}) + RT \ln c_i \qquad (4.4.2)$$

Before any micelles have formed,

$$c_i = \alpha_i c_{\text{tot}} \qquad (4.4.3)$$

where α_i is the fraction of surfactant i in the original mixture and $\alpha_1 + \alpha_2 = 1$. From the pure systems we know that equilibrium between the micellar pseudophase and the solution exists at the CMC so that

$$\mu_i(\text{mic}) = \mu_i^\theta(\text{aq}) + RT \ln \text{CMC}_i \qquad (4.4.4)$$

where CMC_i is the CMC of component i. Equating the chemical potentials in the micellar and aqueous phases

$$\mu_i(\text{mic}) = \mu_i(\text{aq}) \qquad i = 1, 2 \qquad (4.4.5)$$

we can solve for CMC_{tot} where micelles first appear and the composition in the micellar aggregates. Combining eqs. 4.4.1 to 4.4.5, we find

$$\text{CMC}_{\text{tot}} = \left\{ \frac{\alpha}{\text{CMC}_1} + \frac{(1-\alpha)}{\text{CMC}_2} \right\}^{-1} \qquad (4.4.6)$$

and

$$X_1 = \alpha \{ \alpha + (1-\alpha)\text{CMC}_1/\text{CMC}_2 \}^{-1} \qquad (4.4.7)$$

For example, if $\alpha = 0.5$ and $\text{CMC}_1 = 0.1 \cdot \text{CMC}_2$, then $\text{CMC}_{\text{tot}} = \text{CMC}_1 \cdot 20/11$ and $X_1 = 10/11$. In this case, the CMC is reached just below the point where component 1 reaches its CMC, irrespective of the presence of component 2. Likewise, the micellar aggregate contains surfactant 1 as the dominant component. As further surfactant is added, component 1 continues to enter the micelles preferentially so that $c_2(\text{aq})$ increases more with c_{tot} than $c_1(\text{aq})$. For $c_{\text{tot}} \gg \text{CMC}_2, \text{CMC}_1$, the composition of the micellar aggregates must reflect the composition of the mixture so that $X = \alpha$. In this concentration range

$$c_i(\text{aq}) = \text{CMC}_i \cdot \alpha \qquad (4.4.8)$$

Figure 4.15 illustrates how the monomer concentration varies with the total concentration for a particular case. Clearly, the micelle formation process is much more drawn out in this example than for a pure surfactant, and the properties of the solution change less dramatically at the onset of micelle formation. Furthermore, the monomer concentration of the longer chain surfactant peaks around the CMC and then slowly approaches the asymptotic value αCMC_1. Since we are using the phase separation model, we cannot directly predict aggregate sizes. However, the micelles have a high content of component 1 just above the CMC, and having a longer chain should produce larger micelles. Thus, we also see that the micelles that are formed first are larger than those formed at higher concentrations. This effect is rather subtle, and in cases where component 1 forms large aggregates, such as bilayers, by itself we can observe large aggregates at low concentrations which dissolve into small micelles as the concentration increases. The same peak in monomer concentration explains the dip in the surface tension of curve B of Figure 4.4 for an SDS solution contaminated with dodecanol.

Fluorocarbon and hydrocarbon surfactants both form micelles. What happens when we mix such surfactants with the same polar head? Hydrocarbons and flurocarbons do not mix readily, so we definitely have a nonideal mixing in the micelles. We can apply the regular solution model to describe this nonideality. If we use the phase separation model again, we can derive how the CMC is changed in the mixture. Depending on the value of the interaction parameter, w, the regular solution model predicts mixing $w < 2\boldsymbol{R}\boldsymbol{T}$ or phase separation $w > 2\boldsymbol{R}\boldsymbol{T}$.

Figure 4.15
The calculated monomer concentrations as a function of the total concentration of an equimolar mixture of two surfactants 1 and 2. For the pure components we have $\text{CMC}_1 = 1 \times 10^{-3}\,\text{M}$ and $\text{CMC}_2 = 1 \times 10^{-2}\,\text{M}$. For $C_{\text{tot}} < 1.8 \times 10^{-3}\,\text{M}$ there are only monomers in the solution and $C_1 = C_2$. At $C_{\text{tot}} = 1.8 \times 10^{-3}\,\text{M}$ micelles start to form that preferentially contain surfactant 1, whose monomer concentration actually decreases with increasing total concentration for $C_{\text{tot}} > 1.8 \times 10^{-3}\,\text{M}$.

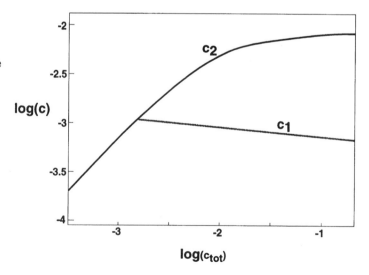

In the latter case, we should expect two populations of micelles, one rich in hydrocarbon and one rich in fluorocarbon tails. Although the phase separation model gives qualitatively reasonable predictions, it has a serious shortcoming in this application. Typical micellar aggregation numbers are on the order of 50 to 100 and the micelles cannot be considered as macroscopic systems. The Stirling approximation (see Section 1.4) does not apply, and there will be distribution of micellar compositions. If the interaction parameter, w, is clearly larger than $2RT$, this distribution function has two peaks, while if $w \lesssim 2RT$, the average composition also is the one most likely to occur in the micelles.

Mixing surfactants with different head groups can induce a multitude of effects. If we add a noncharged surfactant to an ionic surfactant system, we can reduce the electrostatic repulsion and the CMC will decrease below the value we derived for ideal mixing. We can estimate the change in electrostatic free energy on inserting one noncharged monomer into an ionic micelle. The monomer reduces the average charge density and thus lowers the free energy. From Figure 3.12, we see that for typical micellar charges ($s \simeq 10$) the electrostatic free-energy-per-surface-charge is around $4kT$. Thus replacing a charged with an uncharged surfactant results in a substantial extra stabilization of the micelles.

We can obtain an even more dramatic effect by mixing anionic and cationic surfactants to create so-called catanionic systems. An oppositely charged surfactant is twice as effective in reducing the charge density as a noncharged one. In these mixtures, we can easily produce aggregate growth; and bilayers, vesicles, or crystals form in the vicinity of equimolar mixtures.

We get another perspective of electrostatic interactions by considering the reverse situation: adding an ionic surfactant to a noncharged system. Whether this addition increases or decreases head group repulsion will depend on the details of the short range interaction between the two types of head groups. However, it is clear that even slightly charging the micelles increases long range repulsion. For example, introducing 1 to 2 SDS molecules into the nonionic micelles of the $C_m E_n$ type will push the cloud point illustrated in Figure 4.14 to a noticeably higher temperature.

In this section below we give several examples of the role that micelles play in biological and industrial processes. Four properties of micellar solutions play key roles in application design:

1. Micelles possess liquidlike hydrocarbon interiors that can solubilize nonpolar groups in aqueous solution.

2. The micelle–water interface covers a large area. For example a 0.05 M SDS solution contains $1.5 \times 10^4 \, \text{m}^2$ per liter of solution.

3. Micellar dynamic processes are rapid.

4. Transport processes involving diffusion in ionic surfactants may be complex.

4.4.2 Digestion of Fats Requires Solubilization by Bile Salt Micelles

Digestion of fats provides a major source of energy for basic metabolic processes. Figure 4.15 shows how enterohepatic circulation transport bile salts, which act as the body's detergent, through the digestive tract. This process illustrates several interesting features of solubilization and the dynamic behavior of micellar solutions.

The liver synthesizes bile salts from cholesterol and excretes them into the common bile duct as mixed micelles containing lecithin and cholesterol. During periods of fasting, bile is concentrated by a factor of 10 to become a 30 wt% solids solution stored in the gallbladder.

Figure 4.16 illustrates the seven steps of the fat digestion process:

Step 1. Fats from the stomach—in the form of triglycerides with hydrocarbon chains ($n = 16$–18)—are squirted into the duodenum, the upper part of the small intestine, and emulsified by mechanical aggregation.

Step 2. The presence of fats in the duodenum stimulates release of pancreatic enzymes that hydrolyze the triglycerides to fatty acids and 2-monoglycerides. Simultaneously the gallbladder expels bile into the intestine.

Step 3. Because fatty acids are insoluble in water at low pH, they combine with the bile salts to form mixed micelles containing fatty acids and 2-monoglyceride lipids. Bile salts are seldom absent, since without them the body cannot digest fats with hydrocarbon chains longer than eight carbons.

Step 4. Diffusion and convection transport the mixed micelles to the walls of the intestine.

Step 5. The lipid adsorb into the bilayer membranes of the epithelial cells and diffuse across the intestinal wall, where enzymes catalytically reesterify the fatty acids and 2-monoglycerides to triglycerides. Bile salts, which are more polar, are not adsorbed in this process.

Step 6. Bile salts continue to shuttle hydrolyzed lipids to the intestinal wall as food moves down the intestine.

Step 7. In the ileum, the last segment of the small intestine, bile salts are adsorbed by an active transport and flow to the liver via the portal vein. In the liver they are separated from the blood and returned to the common bile duct. If food remains to be digested, bile salts flow back into the intestine. Otherwise they are stored in the gallbladder.

About 3 g of bile salt circulates throughout the human body's enterohepatic system 10 times a day, with a loss of only half a gram from this effective pool of 30 g. The recycling process of bile salts operates, therefore, with an efficiency of 98%.

Concentrated detergents disrupt membranes. So how does the gallbladder tolerate concentrated bile salt solutions? The

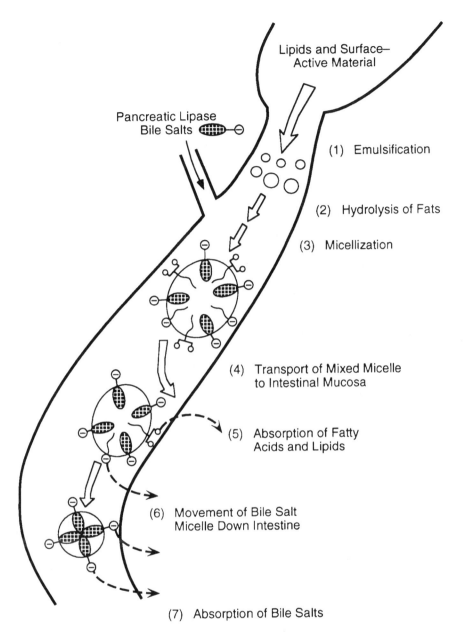

Lipids and Surface–
Active Material

Pancreatic Lipase
Bile Salts

(1) Emulsification

(2) Hydrolysis of Fats

(3) Micellization

(4) Transport of Mixed Micelle
to Intestinal Mucosa

(5) Absorption of Fatty
Acids and Lipids

(6) Movement of Bile Salt
Micelle Down Intestine

(7) Absorption of Bile Salts

Figure 4.16
Schematic illustration of
the sequential steps in the
solubilization and
digestion of fats in the
stomach and upper
intestine.

body's trick is to lower the activity of the bile salts by forming
mixed micelles containing lecithin and cholesterol. Often bile
becomes supersaturated with cholesterol, which then precipitates out to form cholesterol gallstones, a disease that affects 8%
of the adult population. Removal of the gallbladder constitutes
the most common form of major surgery in Western countries.
Since gallstones can be dissolved under some conditions by
feeding the patient selected bile salts, the micellar dissolution
mechanism is interesting also from a medical standpoint.

4.4.3 Solubilization in Micellar Solutions Involves a Complex Combination of Solution Flow and Surface Chemical Kinetics

Solubilization of nonpolar materials by surfactants plays an important role in many industrial and biological processes. Experiments in which fluid flow is defined unambiguously are necessary to explain such mechanisms. Because dissolution is a simpler process in aqueous than in detergent solutions, we will consider solubilization in water first.

Solubilization in Water

Figure 4.17 shows the dissolution of substances into water as a two-step process: a surface reaction followed by diffusion of the solute away from the surface and into bulk water. We can express the dissolution rate J (kg/m$^2 \cdot$ s) as

$$J = K(c_s - c) \tag{4.4.9}$$

where K represents the overall mass transfer coeffiient, c_s the saturation solubility of the compound, and c the actual concentration at any given time.

Each step in the dissolution process possesses a characteristic velocity and also (by analogy with two electrical resistors in series) a characteristic resistance, $1/k_i$. The largest resistance dominates the overall dissolution rate. Therefore, we can describe overall resistance $1/K$ as

$$\frac{1}{K} = \frac{1}{k_1} + \frac{1}{k_1} \tag{4.4.10}$$

where k_1 and k_2 are identified with surface reaction and diffusion. Fluid flow affects the second term in this equation.

A spinning disk is a particularly useful device for studying dissolution under flow conditions. Because radial flow across a spinning disk results in the same dissolution rate at the disk's center as near its edges, the fluid flow is well defined. The solute can be cast or pressed into the disk, and we

Figure 4.17
Two-step kinetic sequence for dissolution of solute molecules from a solid substrate into an adjacent solution. The first step involves a surface reaction characterized by a rate constant k_1. The second step involves diffusion of the solute away from the substrate into solution with a rate constant k_2, which depends on fluid velocity.

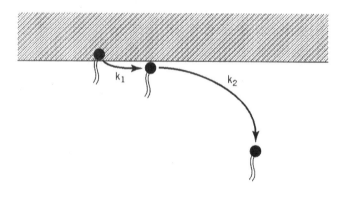

determine the dissolution rate by following the appearance of the materials in the solution through chemical analysis or by using a radioactively tagged solute. All the experiments are performed varying the spinning rate.

Figure 4.18 gives dissolution rates for two substances — benzoic acid and aspirin — as a function of the square root of the fluid velocity, which is written most conveniently in a dimensionless form as the Reynolds number, Re. For a spinning disk this number is

$$Re = \frac{R^2 \omega \rho}{\eta} \tag{4.4.11}$$

where R is the disk's radius, ω the angular velocity, ρ and η the solution's density and viscosity, respectively.

The initial surface reaction step for aspirin is rate controlling because its dissolution rate is independent of Re. With benzoic acid, however, dissolution increases linearly with $(Re)^{1/2}$, implying that $k_2 > k_1$. If the Reynolds number for a spinning disk is larger than 100 but less than 20,000, we can correlate dissolution rate and fluid flow as

$$\frac{k_2 R}{D} = 0.62 \left(\frac{\omega R^2 \rho}{\eta} \right)^{1/2} \left(\frac{\eta}{\rho D} \right)^{1/2} \tag{4.4.12}$$

or

$$Sh = 0.62(Re)^{1/2}(Sc)^{1/3} \tag{4.4.13}$$

The dimensionless groups Sh, Sc, and Re measure ratios of the

Figure 4.18
Dissolution rates for two substances — benzoic acid and aspirin — as a function of the square root of the fluid velocity, which is written most conveniently in a dimensionless form as the Reynolds number.

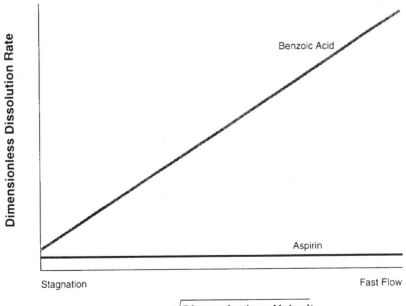

Dimensionless Dissolution Rate

Benzoic Acid

Aspirin

Stagnation

Fast Flow

√Dimensionless Velocity

important physical processes:

Sherwood number Sh: convection mass transfer velocity/ diffusion velocity

Schmidt number Sc: diffusivity of momentum/diffusivity of mass

Reynolds number Re: inertial forces/viscous forces.

The slope of the measured curve for benzoic acid exactly equals $0.62(Sc)^{1/3}$.

Using the spinning disk technique enables us to follow the rate at which SDS dissolves fatty acids. We employ a disk containing radioactively tagged acids and determine the rate at which they appear in solution. The important variables in this experiment are fluid flow, SDS concentration, and temperature. Restricting the discussion to $Re > 100$, we observe that in general the dissolution rate increases linearly with surfactants concentration above the CMC but shows a complex dependence on fluid flow. Initially the dissolution rate increases linearly with $Re^{1/2}$ like the dissolution rate for benzoic acid in water. However, at high Re, the dissolution rate becomes independent of flow like the dissolution of aspirin in water.

The dependence of dissolution rate on surfactant concentration and fluid flow can be modeled as a five-step process (see Figure 4.19) parallel to the Langmuir–Hinshelwood model for gas–solid catalysis. First, surfactant molecules diffuse through the bulk solution to the surface. Second, these molecules are absorbed onto the surface. Third, they react to form a mixed micelle containing both surfactant and the fatty acid. Fourth,

Figure 4.19
Schematic mechanism for solubilizing fatty acids by a micellar solution. Steps 1 and 5, which involve diffusion of surfactant or surfactant plus fatty acids, depend on fluid flow. Steps 2, 3, and 4 involve surface reactions and are independent of flow. The kinetic data suggest that the rate-determining steps are steps 4 and 5, involving mixed micelles. (A.F. Chan, D.F. Evans, and E.L. Cussler, *AIChE J.* **22**, 1106 (1976).)

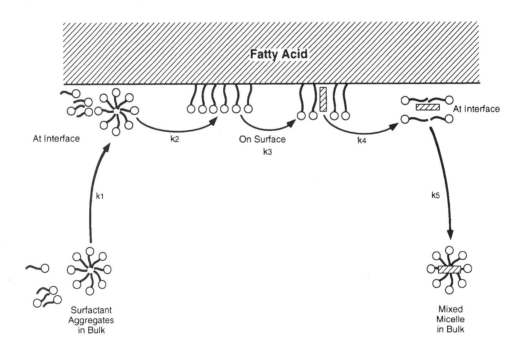

the mixed micelle is desorbed, and fifth, it diffuses away. Fluid velocity affects the diffusion of surfactant in steps 1 and 5, but not in steps 2, 3, or 4.

The complex dependence of solubilization rate on Re requires that the rate-controlling steps involve some combination of steps 1 and 5 and 2, 3, or 4. If we write out the kinetic expressions for various combinations, we find that only the expression involving steps 4 and 5 as rate-controlling factors gives predictions in accord with observations.

In this case, the rate equation takes the form

$$\frac{d[c]}{dt} = \frac{V_{\max}[c_s]}{K_{\max} + [c_s]} \tag{4.4.14}$$

where $d[c]/dt$ is the dissolution rate and c_s is the solubility of the fatty acid in the SDS solution, which varies linearly with surfactant concentration above the CMC. This equation is similar in form to the Langmuir equation (eq. 2.4.18).

We can express eq. 4.4.14 in a more convenient form for analyzing data as

$$\left[\frac{dc}{dt}\right]^{-1} = \frac{1}{V_{\max}} + \frac{K_{\max}}{V_{\max}} \frac{1}{[c_s]} \tag{4.4.15}$$

V_m and $K_{\max}$ involve a complex combination of rate and equilibrium constants. For our purposes it is sufficient to note that

$$V_{\max} = \text{constant} \quad \text{and} \quad \frac{K_{\max}}{V_{\max}} = \frac{A}{k_4} + \frac{B}{k_5} \tag{4.4.16}$$

where k_5 should vary with $\text{Re}^{1/2}$.

We can use the following tests to verify the predictions we obtain from these equations.

1. The reciprocal of the solubilization rate $[dc/dt]^{-1}$ should vary linearly with the reciprocal of the solubility $[1/c_s]$, which is itself linearly related to the SDS concentration. Figure 4.20 shows results of this kind for the dissolution of stearic acid.

2. The plot intercept should be independent of the flow, as shown in Figure 4.21.

3. A plot of the slopes of the lines in Figure 4.21 versus $\text{Re}^{-1/2}$ should yield a straight line because it involves k_5, the diffusion of the mixed micelle away from the surface of the disk.

4. The intercpt in Figure 4.21 should be positive, since it equals A/k_4, as observed with steric and palmitic acids. However, with lauric acid, the intercept is negative and signals a different solubilization mechanism, which we will describe in the discussion of the temperature dependence of solubilization.

Why are steps 4 and 5 the rate-controlling steps? We know that the two diffusion steps (1 and 5) depend on properties of

Figure 4.20
Dissolution of stearic acid, $CH_3(CH_2)_{16}COOH$, shows how the inverse solubilization rate varies linearly with the reciprocal of the solubility as predicted by eq. 4.4.7. (A.F. Chan, D.F. Evans, and E.L. Cussler, *AIChE J.* **22**, 1006 (1976).)

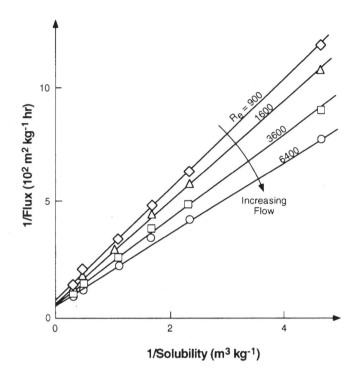

Figure 4.21
Variation of solubilization rate with flow. The slopes of the lines shown in Figure 4.20 for steric acid, $CH_3(CH_2)_{16}COOH$, along with similar data for palmitic acid, $CH_3(CH_2)_{14}COOH$, and lauric acid, $CH_3(CH_2)_{12}COOH$, are plotted versus the inverse square root of the Reynolds number. According to eq. 4.4.16, the intercepts of the plot should be positive and are proportional to the rate constant k_4. (A.F. Chan, D.F. Evans, and E.L. Cussler, *AIChE J.* **22**, 1006 (1976).)

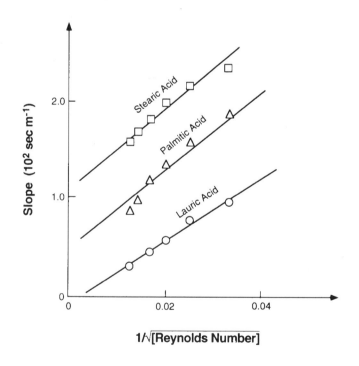

the solution, but not on properties of the fatty acid surface. Step 1 depends on the diffusion of surfactants to the surface, step 5 on the diffusion of fatty acid and surfactant away from the surface. Owing to the low aqueous solubility of the fatty acid, this step probably involves a mixed micelle. Monomer diffusion is faster than the diffusion of a mixed micelle.

We can apply analogous reasoning to the adsorption and desorption processes in steps 4 and 5. Adsorption can occur via surfactant monomers or micelles, but desorption probably involves a large and multiply charged mixed aggregate. Experiments show that surface reactions that take place in step 3 occur more rapidly than desorption, although solubilization measurements do not elucidate the nature of the chemical species involved in this difference. The temperature measurement described in the next section also gives us some information on this topic.

Temperature Dependence of Solubilization in Micellar Solutions

Figure 4.22 shows dissolution rates for palmitic and stearic acids and their eutectic mixtures at two different Reynolds numbers. At a characteristic temperature, which depends on the substrate and surfactant concentration, dissolution increases dramatically. For example, if we use a polarizing microsope to examine fatty acid crystals in contact with an SDS solution, we observe liquid crystals forming at the temperature that produces rapid solubilization. We can define this point as the penetration temperature, and it occurs in water about 10°C below the melting point of the anhydrous lipid or amphiphile.

Figure 4.22
Variation of solubilization rate of plamitic acid, stearic acid, and their eutectic mixture in 4% SDS solutions as plotted versus temperature. Above the penetration temperature, the dissolution rate increases dramatically because the solubilization mechanism changes. Fatty acid liquid crystals also form, and large fatty acid aggregates are dispersed in solution. Below this temperature, the mechanism shown in Figure 4.19 is followed. Results are shown for two Reynolds numbers, 6400 and 1600. (J.A. Scheiwitz, A.F. Chan, E.L. Cussler, and D.F. Evans, *J. Colloid and Interface Sci.* **84**, 47 (1981).)

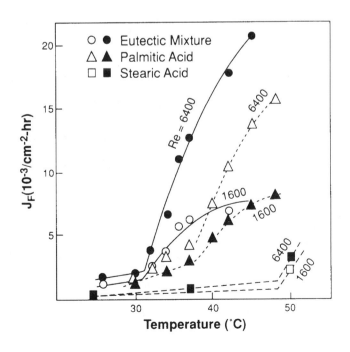

Below the penetration temperature, solubilization occurs by slow dissolution until the mixture reaches an isotropic state. Above the penetration temperature, liquid crystals form at the interface. A stronger dependence on fluid flow under these circumstances may be caused by liquid crystalline material shearing off the surface and subsequently reequilibrating in the bulk solution. Because lauric acid's penetration temperature (15°C) lies below the temperature of the dissolution measurements displayed in Figure 4.22, fluid flow may account for its negative intercept.

4.4.4 Micellar Catalysis Exploits the Large Surface Areas Associated with Micelles and Also Illustrates the Graham Equation

In micellar catalysis we use the special properties of micellar solutions to facilitate chemical reactions. Typically, one of the reactants (e.g., an ion) is highly polar, while the other is apolar and sparingly soluble in water. Under such circumstances, finding a suitable solvent for performing the reaction can be difficult.

In a micellar solution, however, the apolar reactant can be solubilized in the micelles while the polar one is dissolved in the aqueous region. The two reactants can meet at the micellar surface, where the chemical transformation occurs.

An ester hydrolysis by acid can serve to illustrate the principle:

$$RC\overset{\overset{\displaystyle O}{\|}}{}-O-R'+H_2O \xrightarrow[k^2]{H^+} RC\overset{\overset{\displaystyle O}{\|}}{}-OH+R'OH \tag{4.4.17}$$

What type of surfactant should be used to catalyze this reaction? To maximize the local concentration of both ester and H^+ in the same region, we can use an anionic surfactant because the H^+ ions will be attracted to the micellar surface electrostatically. The ester is solubilized with the relatively polar ester group localized at the micellar surface.

Figure 4.23 shows how the rate constant varies with surfactant concentration. A dramatic increase in the rate constant occurs at the CMC, with an enhancement factor of approximately 50. This increase illustrates one of the principles behind enzymatic catalysis. Simply by positioning the two reactants in close proximity, we can substantially enhance the rate of a bimolecular reaction.

A feature of Figure 4.23 is that we observe the maximum rate of ester hydrolysis around the CMC, and adding more surfactant leads to a decrease in the rate. At first sight this rate decrease seems counterintuitive, but it can be readily explained by using the Graham equation (3.8.26).

Surface charge density determines the counterion concentration at the highly charged surface. Assuming no chemical preference, with a CMC of 8 mM and a pH of 6 (implying a concentration of H^+ of 1×10^{-6} M), the ratio of surface concentration of Na^+ to H^+ is 8×10^3.

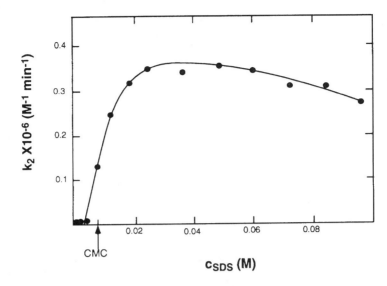

Figure 4.23
Measured second-order
rate constant k_2 for the
acidic hydrolysis of methyl
orthobenzoate in an
aqueous solution at 25°C as
a function of the
concentration of SDS
(CMC = 8 mH). (R.B.
Dunlap and E.H. Cordes, *J.
Am. Chem. Soc.* **90**, 4395
(1968).)

   The micelle has approximately one charge per 0.75 nm². According to the Graham equation, this yields a total counterion surface concentration of 9.3 M. We can obtain the concentration of H^+ at the surface as

$$[H^+]_{surface} = \frac{1}{8 \times 10^3} \times 9.3 = 1.2\,\text{mM} \qquad (4.4.18)$$

which is 300 times higher than the bulk concentration.

   As more surfactant is added, the concentration of the Na^+ counterion increases. It becomes more and more difficult for the H^+ to compete, and its surface concentration gradually decreases, leading to a decrease in the overall rate constant. The dramatic changes in the rate constant shown in Figure 4.22 are not caused by some change in local chemistry; they are due simply to changes in the concentrations of H^+ at the location of the ester.

Exercises

4.1. Calculate the equilibrium constant K_N in the closed association model for a system with (a) CMC = 1.0 M and $N = 10$ and (b) CMC = 8.0 mM and $N = 60$. Estimate the error in the approximation used in eq. 4.1.11 to relate the CMC and $\Delta G^{\theta}_{\text{mic}}$. Use eq. 4.1.8 to identify the CMC.

4.2. In the isodesmic model of stepwise aggregation

$$[S_n] = [S][S_{n-1}]K$$

(a) Show that the total surfactant concentration is

$$S_{\text{tot}} = \sum_{n=1}^{\infty} n[S_n] = \frac{[S]}{\{1 - [S]K\}^2}$$

Hint: Use the formulas for a geometric series repeatedly.

(b) Show that in the limit of $K[S]_{\text{tot}} \gg 1$ it follows that

$$K[S] \simeq 1 - (KS_{\text{tot}})^{-1/2}$$

(c) Then show that in this limit the size distribution function is decaying exponentially and $[S_n] \simeq K^{-1} \exp[-n(KS_{\text{tot}})^{-1/2}]$.

4.3. We can combine the closed-association model and the isodesmic model by stipulating that for the equilibrium $NS \rightleftharpoons S_N$ there is an equilibrium constant K_N, while $S_{n+1} + S \rightleftharpoons S_n$ with equilibrium constant K for $n > N$. Show that

$$S_{\text{tot}} = [S] + \frac{NK_N[S]^N(1 - [S]K + [S]K/N)}{(1 - [S]K)^2}$$

4.4. In Exercise 2.7 of Chapter 2 we gave the concentration dependence of the surface tension for dodecyldimethylammonium chloride solutions with no salt and with 0.20 M NaCl. Determine the CMCs.

4.5. Use the Langmuir adsorption isotherm model to calculate the adsorption of SDS to a solid surface. However, assume the closed association model for SDS in the bulk. The adsorption equilibrium constant is $K = 1.2 \times 10^2 \, \text{M}^{-1}$.

4.6. In pressure-jump experiments on micellar solutions of sodium hexadecyl sulfate (SHS) at 45°C one observes a fast relaxation process with a time constant that varies with concentration as follows

Concentration SHS (mM)	τ_1 (μs)	Concentration SHS (mM)	τ_1 (μs)
0.58	807	1.0	365
0.66	665	1.1	325
0.70	603	2.0	114
0.90	430	5.0	38

The CMC is 0.52 mM and the aggregation number $N = 100$. Determine the rate constant k_N^- for a monomer leaving the micelle. Estimate also the standard deviation σ of the micelle size distribution. What is the association rate constant k_N^+? Calculte the average lifetime of a monomer in the micelle.

4.7. The aggregate size of potassium dodecanoate micelles in an aqueous solution containing 27% (w/w) dodecanoate was determined using time-resolved fluorescence. The molar ratio of dodecanoate to fluorescent pyrene was $93 \times 10^3 : 1$. The fluorescence decayed exponentially with a time constant $\tau_0 = 200 \, \text{ns}$. In a second sample a quencher (benzophenone) was added at a 65:1 molar ratio of dodecanoate to quencher. The fluorescence at long times also decayed exponentially in this sample with the same time constant $\tau_0 = 200 \, \text{ns}$, while the decay was faster at short times. Extrapolating the long time behavior back to $t = 0$ resulted in an apparent zero time intensity of 0.33 times the value in absence of the quencher. Calculate the mean aggregate size.

4.8. The table gives the measured CMC of an SDS aqueous solution as a function of the concentration of sodium chloride in the solution. The table also gives the measured mean aggregation numbers.

C_{NaCl} (M)	CMC (mM)	N
0.00	8.1	58
0.01	5.7	64
0.03	3.1	71
0.10	1.5	93
0.30	0.71	123

Calculate for all five systems:
(a) The volume of the hydrocarbon core.
(b) The effective radius of the hydrocarbon core, assuming spherical geometry.
(c) The cross-sectional area per chain at the aggregate surface (spherical geometry).
(d) The area per charge, assuming it is located 0.3 nm outside the hydrocarbon core.

4.9. Use the data of Exercise 4.8 to calculate the electrostatic free energy per monomer. What is the effective γ value?

4.10. Estimate the CMC of hexadecyl sulfate at 45°C given that the CMC of sodium dodecyl sulfate is 8.9 mM at this temperature.

4.11. Light-scattering measurements on tetradecyltrimethylammonium bromide solutions at 25°C give the following Rayleigh ratios, R:

H₂O $K_0 = 3.7 \times 10^{-5}$ m²/kg²		0.05 M KBr $K_0 = 3.7 \times 10^{-5}$ m²/kg²	
$10^3 c$ (g/mL)	$10^4 R$ m^{-1}	$10^3 c$ (g/mL)	$10^4 R$ m^{-1}
0.406	2.318	0.0275	2.292
0.910	2.347	0.0556	2.332
1.298	2.788	0.0857	2.194
1.426	3.702	0.173	2.332
2.116	7.428	0.291	3.634
3.764	10.665	0.436	5.599
4.534	11.270	0.829	11.195
5.962	12.500	1.406	18.527
6.526	12.982	1.917	25.270
8.408	14.331	2.498	31.671
		3.010	37.410

Data from A.H. Beesley, Ph.D. thesis, University of Minnesota (1990).

(a) Using these data, verify that in water the CMC is 1.2×10^{-3} g/mL and $R_{CMC} = 2.3 \times 10^{-4}$ m^{-1} and in 0.05 M KBr the CMC is 0.202×10^{-3} g/mL and $R_{CMC} = 2.3 \times 10^{-4}$ m^{-1}.

(b) Calculate the aggregation number and second virial coefficient. How does the value of B compare to the estimated volume of a micelle?

(c) Explain qualitatively why N and B change in the way they do upon addition of salt.

4.12. The table gives CMCs and aggregation numbers for the nonionic surfactant dodecyldimethylamine oxide $(n - C_{12}H_{25}(Me)_2N^+O^-)$.

$T(°C)$	10^3 CMC (g/100 mL)	N
1	2.84	77
27	2.10	76
40	1.83	78
50	1.75	73

Data from *J. Phys. Chem.* **66**, 295 (1962).

(a) Calculate the area per head group, assuming that the micellar radii equal the length of the fully extended hydrocarbon chain.

(b) Using the equilibrium model, calculate the values of ΔG, ΔH, and ΔS for micellization at 25°C.

(c) Estimate the values for ΔG(HG) at 25 and 40°C, using the value of ΔG(HP) from eq. 4.3.2. (To a good approximation and over this narrow temperature range, you can assume that ΔG(HP) is almost constant.)

4.13. Upon addition of HCl, dodecyldimethylamine oxide (DDAO) is protonated and forms an ionic surfactant, $C_{12}H_{25}(CH_3)_2, N^+OHCl^-$ (DDHAC). The properties of the ionic and nonionic micelles are compared in the table.

Surfactant	Added Salt	CMC (g/0.1 L)	N
DDAO	0	0.048	76
DDAO	0.2 M NaCl	0.034	78
DDHAC	10^{-3} M HCl	0.16	76
DDHAC	10^{-2} M HCl	0.15	77
DDHAC	0.1 NaCl/10^{-3} HCl	0.041	98
DDHAC	0.2 NaCl/10^{-3} HCl	0.029	117

Data from *J. Phys. Chem.* **66**, 295 (1962). The original values of CMC and N for DDHAC were given in terms of the molecular weight of DDAO; the values have been lowered by a factor of 0.86 in this table.

Discuss the different trends emerging from the table.

4.14. The table gives CMCs and aggregation numbers for dodecyltrimethylammonium bromide as a function of temperature.

T	$10^2\,CMC$ (M)	$T(°C)$	N
5	1.55	20	65
15	1.53	30	61
25	1.47	40.2	54
40	1.63	49.5	51
55	1.87		

CMC data from T. Ingram and M.N. Jones, *Trans. Faraday Soc.* **65**, 297 (1964); *N* data from R. Malliaris, J. LeMoigne, J. Sturm, and R. Zana, *J. Phys. Chem.* **89**, 2709 (1985).

(a) Using the equilibrium model, calculate ΔG_{mic}, ΔH_{mic}, and ΔS_{mic} for micellization at 15 and 40°C.

(b) Using the dressed micelle model, calculate the contributions of $\Delta G(HG) + \Delta G(contact)$ as a function of temperature, and using eq. 4.3.1, estimate $\Delta G(HP)$.

(c) From the temperature dependences of $\Delta G(HG) + \Delta G(contact)$ and $\Delta G(HP)$ calculate the corresponding values of ΔH and ΔS and discuss in terms of the values calculated in (a).

Literature

References[a]	Section 4.1: Micelle Formation	Section 4.2: Experimental Techniques	Section 4.3–4.4: Quantitative Analysis of Micellarization and Applications
Adamson & Gast	CHAPTER 8.5 Micelles		
Hiemenz & Rajagopalen	CHAPTER 8 Colloidal structures in surfactant solution	CHAPTER 2 Diffusion CHAPTER 5 Light scattering	
Hunter (II)	CHAPTER 10.1–10.4 CMC and aggregation process	CHAPTER 2.1 Diffusion CHAPTER 2.3 Light scattering CHAPTER 10.6 Fluorescence CHAPTER 10.7 Micelle kinetics	CHAPTER 10.5 Electrostatic effects
Israelachvili	CHAPTER 16.5–16.7 Micelles and other aggregates		
Kruyt	CHAPTER III Light scattering	CHAPTER I.4 Diffusion and Brownian motion	
Meyers	CHAPTER 15 Micelles		
Shaw	CHAPTER 4 Surfactant solutions	CHAPTER 3 Light scattering	
Vold and Vold	CHAPTER 18 Micelles		
Tanford	CHAPTERS 6–10 Micelles		

[a]For complete reference citations, see alphabetic listing, page xxxi.

5

FORCES IN COLLOIDAL SYSTEMS

CONCEPT MAP

Forces in Colloidal Systems

Describing colloidal interaction forces at a molecular level remains an unsolved challenge. Because the distances involved in colloidal systems are often large compared to molecular sizes, we can often average over molecular structures. At the end of this chapter, we consider interaction forces in which solvent molecularity plays an important role.

For parallel surfaces, we usually can write the interaction force per area, $F/\text{area} = \Pi_{osm}$ (eq. 5.1.2), in terms of the osmotic pressure Π_{osm} for the solvent between the surfaces.

Electrostatic Double-Layer Forces

Interaction between a charged and a neutral surface (Figure 5.1) with counterions only illustrates many features of electrostatic forces. Separate evaluation of the energetic and entropic contributions leads to complex expressions (eqs. 5.1.7 and 5.1.11).

- Combining ΔU and ΔS leads to a simple expression for the free energy, derivative $F/A = \Pi = kT\Sigma_i c_i^*(h)$ (eq. 5.1.12), which contains only a contribution from ΔS. $\Sigma c_i^*(h)$ is the ion concentration at the neutral wall, which can be evaluated (eq. 5.1.17) from the PB equation.

- When the two surfaces move together, counterions are forced into a smaller space, decreasing their entropy and giving rise to a repulsive interaction force.
  Two surfaces with identical charges (Figure 5.3) repel each other when their double layers overlap. The variation in potential between the plates is symmetrical. The force is evaluated by solving the PB equation for $\Pi = kT\Sigma_i c_i^*$ (midplane). The origin of the repulsive force is entropic.

- A useful approximate solution, $F/A \propto \Gamma_0^2 \exp(-\kappa h)$ (eq. 5.1.31), shows that (1) the force decays exponentially with separation h; (2) rates of decay depend on the Debye length κ^{-1}; and (3) surface potential enters as $\Gamma_0 = \tanh(ze\Phi_0/4kT)$, whereas $\Phi_0 \to 0$, $\Gamma_0 \to ze\Phi_0/4kT$ and as $\Phi_0 \to \infty$, $\Gamma_0 \to 1$ (i.e., the force becomes independent of Φ_0 at high potentials).

- **Two surfaces with equal signs but different magnitudes** always repel each other. When **two surfaces bearing opposite signs** move together, a long range attraction can change into a repulsion at shorter separations.

London Dispersion Forces: Induced Dipole–Induced Dipole Interactions

Motion of electrons around one nucleus creates an instantaneous dipole that generates a field at a second atom, which in turn induces an instantaneous dipole. Quantum mechanics shows that this interaction leads to an attractive force $V_{12}^{\text{disp}}(R) = -C_{12}/R_{12}^6$ (eq. 5.2.1).

The coefficient C_{12} depends on atomic polarizabilities and ionization potentials (eq. 5.2.3).

London dispersin, Keesom, and Debye forces combine to constitute the van der Waals attractive force.

London Dispersion Forces Between Colloidal Particles

Two macroscopic plates separated by a slit of width h (Figure 5.6) serve as a model.

Assuming pairwise additivity, we can add all interactions between plates (separated by a vacuum) to obtain $U_{12}/\text{area} = -H_{12}/12\pi h^2$ (eq. 5.2.7), where $H_{12} = \pi^2 C_{12}\rho_1\rho_2$ is the Hamaker constant; ρ represents the density of material. This summation transforms a short range interaction between atoms ($1/R^6$) into a long range interaction ($1/h_2$).

Bringing two identical surfaces into contact measures the work of cohesion and thus relates dispersion energies to surface tension, as illustrated by eq. 5.2.12.

When solvent fills the slit, we can obtain an effective Hamaker constant by considering the interaction energy between plates of media 1 and 3 separated by a liquid 2 (Figure 5.10):

- For identical plates, $H_{121} \cong \{\sqrt{H_{11}} - \sqrt{H_{22}}\}^2$, where H_{11} and H_{22} are the values for individual materials. This interaction is always attractive, but greatly diminished compared to homogeneous interactions in a vacuum.

- For dissimilar plates, $H_{123} \cong \{\sqrt{H_{22}} - \sqrt{H_{11}}\}\{\sqrt{H_{22}} - \sqrt{H_{33}}\}$, resulting in attractive or repulsive interactions, depending on the magnitudes of the individual Hamaker constants.

The Lifshitz theory, which is based on a continuum electrodynamics point of view, provides an alternative and more general way to calculate attractive interactions.

The Derjaguin Approximation

The Derjaguin equation relates in a general way the force $F_{12}(h)$ between curved surfaces to the interaction energy $U_{12}(h)$ between flat surfaces (see Figure 5.11), provided the radius of curvature R is larger than the range of the interactions.

- The relations between $F_{12}(h)$ and $U_{12}(h)$ for a sphere and a plate (eq. 5.2.31), two crossed cylinders (eq. 5.2.32), and two spheres (eq. 5.2.33) involve products of the radii of curvature, R_1 and R_2.

Attractive Forces Generated by Correlated Electrostatic Interactions

Examining interactions between ionic or dipolar groups at a level beyond the mean field approximation shows that correlations can generate attractive forces.

- Lateral inhomogeneities generated in the distribution of ions located between two similarly charged surfaces induce the formation of instantaneous fields. With divalent ions or highly charged surfaces these fields generate an attractive force which can overwhelm repulsive double layer forces (Figure 5.12).

- When two surfaces containing dipolar groups approach, correlations lead to attractive forces. The thermally averaged force varies with a distance dependence of h^{-5} (eq. 5.3.1); while for a rigid lattice the force decays exponentially (eq. 5.3.2).

- When a surface containing a nonuniform charge distribution is formed, for example, by adsorbed micelles or polyelectrolytes or surface phase domains, an electrical field is generated that extends into the adjacent solution. This field can induce a correlated response from a second surface giving rise to an attractive force.

Solid Surfaces – Forces Arising from Density Variations

When two surfaces approach, their molecular components play an increasingly important role in determining interaction forces.

- Surface induced molecular variations in solvent density lead to an oscillatory force when two surfaces approach (eq. 5.4.1 and Figures 5.18 and 5.19).

- Capillary phase separation occurs when two surfaces approach and induce a transition to a phase with lower surface energy as illustrated in Figure 5.20. Such transitions generate attractive forces (eqs. 5.4.5 and 5.4.11).

- When a solute does not absorb on a surface, it becomes excluded between two closely separated surfaces. The resulting difference in osmotic pressure generates an attractive force (eq. 5.4.14 and Figure 5.24).

- Solute adsorption leads to an on-average repulsive contribution to the interactions as surfaces approach. The magnitude of the force depends on whether desorption occurs as the surface moves into contact.

Fluid-Like Surfaces – Forces Arising from Entropic Interactions

With liquid interfaces, entropic effects provide insight into the origin of interaction forces.

- When two surfaces covered with polymer approach, the polymer tails reaching out into solution impede the configurational freedom of the polymers attached to the other surface. This reduction in configurational entropy generates a repulsive force which can overwhelm either van der Waals or double layer forces.

- When phospholipid bilayers approach, a short-ranged repulsive force (Figure 5.25) is observed at 2–3 nm that increases exponentially with decreasing distance of approach.

- Flexible bilayers undergo undulations which are damped out when bilayers approach. This creates a repulsive force of entropic origin which varies as h^{-3} (eq. 5.5.1).

- It is conjectured that repulsive forces between similar surfaces always have an entropic origin.

The Hydrophobic Effect

- Because of its technological and biological importance, water is the most well-characterized solvent.

- In the liquid state, water possesses a unique three-dimensional structure, and the interplay between structural effects and solvophobicity that drives aggregation is complex and confusing.

- We can understand the origin of this confusion by considering:

  1. The change in solubility (free energy, Figure 5.26) of hydrocarbons in water with decreasing temperature initially decreases, goes through a minimum near 20°C, and then actually increases. This argues that as water's structure increases with decreasing temperature, the solubility of hydrocarbons increases.
  2. Analysis of the changes in ΔH and ΔS upon transferring nonpolar molecules into water (Table 5.1) suggests the opposite conclusion, namely that water's structural effects drive molecules out, thereby lowering solubility.

- These two conflicting interpretations can be reconciled, and comparison of solubilities in water and hydrazine provides one way of understanding some of the complexities of aqueous solutions.

- Careful attention should be exercised to distinguish between the terms *hydrophobic hydration*, *hydrophobic interaction*, and *hydrophobic effect*.

Hydrodynamic Interactions

When particles move past one another, hydrodynamic interactions create flow in the surrounding liquid, which in turn decreases diffusion coefficients (eq. 5.7.5).

When a particle moves in a medium, fluid flow generates a hydrodynamic repulsive force, which decreases the velocity of approach as well as separation. This is called the lubrication force.

5

Many practical applications of colloid chemistry revolve around the problem of controlling the forces between colloidal particles, between particles and surfaces, or between two surfaces. Consequently, colloid scientists have devoted considerable experimental and theoretical effort to resolving this problem. Forces between semimacroscopic objects result from a complex interplay between many interacting molecules or atoms both in the particles and in the media separating them. Thus, the total force depends on intermolecular forces of all the types discussed in the introduction to Chapter 3.

The interplay within a colloidal system is so intricate that describing it in full molecular detail constitutes a major challenge that has not yet been resolved. Most distances involved in the colloidal domain are large compared to molecular sizes, however, and this fact allows us to partly average over the local molecular structure. If we execute such a scheme in a suitable way, forces of different types between colloidal particles emerge naturally. In this chapter we will discuss these forces separately and then return to the problem of the effects of their combined action in subsequent chapters.

The force F between two surfaces or particles at a separation h and a constant external pressure is determined by the change in Gibbs free energy $\mathbf{G}$ with separation

$$F = -\left(\frac{\partial \mathbf{G}}{\partial h}\right)_T = -\left(\frac{\partial \mathbf{H}}{\partial h}\right)_T + T\left(\frac{\partial \mathbf{S}}{\partial h}\right)_T \qquad (5.1.1)$$

The sign convention is such that the repulsive force is positive. As discussed already in Chapter 3 we usually consider condensed phases to be incompressible, in which case the force in eq. 5.1.1 could equally well be related to the derivative of Helmholtz free energy and we can then replace $\mathbf{H}$ by $\mathbf{U}$ in the right-hand side of the equation.

We will pay particular attention to the case of two parallel planar surfaces. In this case, our main interest is the force per unit area, which can be obtained by dividing all terms in eq. 5.1.1 by the area. For this case F/area is sometimes termed the *disjoining pressure*, but we will not use this terminology further.

If the two surfaces are separated by vacuum, or more realistically by a low density gas, we can interpret eq. 5.1.1 as giving the experimentally relevant force directly. However, in most applications the surfaces are separated by a liquid medium. As the two surfaces approach each other, the liquid must leave the region between the surfaces. For most systems, we can assume that the liquid is practically incompressible. Then we can rewrite eq. 5.1.1 for planar parallel surfaces as

$$\frac{F}{\text{area}} = -\frac{1}{\text{area}}\left(\frac{\partial \mathbf{G}}{\partial h}\right)_T = -\left(\frac{\partial \mathbf{G}}{\partial V}\right)_T$$

$$= -\frac{1}{V_s}\left(\frac{\partial \mathbf{G}}{\partial n_s}\right)_T = \Pi_{\text{osm}} \qquad (5.1.2)$$

where we see that the force per unit area is simply the osmotic pressure for the solvent between the surfaces.

The force in eq. 5.1.2 is the force experienced when the liquid is expelled upon approaching the surfaces to a reservoir of pure liquid at the same temperature. If it ends up in a different state, transferring the pure solvent to a solution, vapor, or whatever the final state is makes an extra contribution to the change in free energy. However, the key quantity to calculate to obtain the net force is

$$\left(\frac{\partial \mathbf{G}}{\partial h}\right)_T$$

in eq. 5.1.2 or equivalently

$$\left(\frac{\partial \mathbf{A}}{\partial h}\right)_T$$

In calculating this derivative we will make an effort to always directly connect the intermolecular interactions that combine into this thermodynamic quantity. Thus, in Section 5.1 we express the electrostatic free energy primarily in terms of ion–ion interactions and ion entropies. Similarly, in Section 5.2 we start out by considering the dispersion interaction between molecules and find the force between bulk phases in terms of a pairwise additive molecular contribution. The alternative would be to characterize the media in terms of their frequency-dependent dielectric properties. This approach yields a more general (but also more abstract) description whose beauty we can appreciate more readily after mastering the material presented in Section 5.2.

5.1 Electrostatic Double-Layer Forces Are Long-Ranged

5.1.1 A Repulsive Electrostatic Force Exists Between a Charged and a Neutral Surface

When two charged surfaces approach each other in an electrolyte medium, an effective interaction becomes appreciable on a length scale determined by the Debye screening length, $1/\kappa$. In effect, the counterion distributions, or as it is usually expressed, the electrical double layers, start to overlap. To better understand the physical mechanisms behind such interactions, we will first consider the somewhat simpler case of two parallel surfaces, one charged, the other neutral, as illustrated in Figure 5.1. This case is important in practical terms, but it also provides important insight into the mechanism of the interaction of charged surfaces in general.

When the neutral surface approaches a charged surface, the associated counterions in the intervening solution have less and less available space in which to move. This spatial constraint implies a decrease in the entropy of the ion distribution

Figure 5.1
A negatively charged and a neutral surface separated by a distance h in an electrolyte solution. The surface charge density is σ, the solvent dielectric constant ε_r The mean concentrations, $\bar{c}_{\pm}$, of cations (counterions) and anions (coions) are related by $\bar{c}_{+} = \bar{c}_{-} + \sigma/h$. Two surfaces moved together encounter a repulsive force of a magnitude given by eq. 5.1.12.

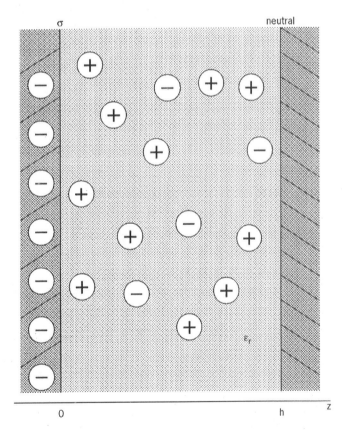

and, as we will see, results in a repulsive force. To determine the force, we must consider the free energy **A** as a function of distance. In Section 3.9, we showed how to calculate the electrostatic part $\mathbf{A}_{el}$ for a single charged surface exposed to an electrolyte solution using the Gouy–Chapman theory. Clearly the same method could be applied for the case of a charged and a neutral wall. However, the solution to the Poisson–Boltzmann (PB) equation is not quite the same. Instead of eq. 3.8.6 stating that the electrical field is zero far from the charged surface, we now have a case in which the field must be zero at the neutral wall due to electroneutrality so that

$$\left.\frac{d\Phi}{dz}\right|_h = 0 \tag{5.1.3}$$

For convenience we also chose the potential to be zero at the same point

$$\Phi(h) = 0 \tag{5.1.4}$$

In principle, we could now proceed as in Sections 3.8 and 3.9 by first solving the Poisson–Boltzmann equation and then calculating the electrostatic free energy using eqs. 3.9.1, 3.5.17, and 3.9.4. In fact, it turns out to be more informative to consider the free energy $\mathbf{A}_{el}$ without first solving the PB equation.

As with the Gouy–Chapman solution, we gain more insight by considering the energy and entropy contributions to the free energy separately. Using eq. 3.5.17, the electrostatic free energy is

$$\mathbf{U}_{el} = \frac{1}{2} \int\limits_{+\infty}^{+\infty} \int\limits_{0}^{+\infty} \rho\Phi \, d\vec{r}^3$$

$$= \frac{1}{8\pi\varepsilon_r\varepsilon_0} \int\limits_{-\infty}^{+\infty}\!\!\!\int\limits_{0}^{h} \frac{\rho(\vec{r})\rho(\vec{r}')}{|\vec{r}-\vec{r}'|} \, d\vec{r}^3 \, d\vec{r}'^3 \tag{3.5.14}$$

where in the second equality the potential hs been written in terms of the charge distribution generating it

$$\Phi(\vec{r}) = \frac{1}{4\pi\varepsilon_r\varepsilon_0} \int \frac{\rho(\vec{r}')}{|\vec{r}-\vec{r}'|} \, d\vec{r}'^3 \tag{3.2.2}$$

To obtain the contribution from the energy to the force, we take the derivative of $\mathbf{U}_{el}$ with respect to h. Now $\mathbf{U}_{el}$ depends on h in two different ways: first, through the dependence of the integration limits, where we can use the general relation

$$\frac{d}{dx} \int_a^x f(x')dx' = f(x) \tag{5.1.5}$$

and second, through the dependence on h of the charge distribution $\rho(\vec{r})$. Since the right-hand side of eq. 3.5.17 is symmetrical with respect to the interchange $\vec{r} \leftrightarrow \vec{r}\,'$, each type of derivative gives rise to two identical terms and

$$\frac{\partial \mathbf{U}_{el}}{\partial h} = \frac{1}{8\pi\varepsilon_r\varepsilon_o} \left[\int\!\!\int\!\!\int\limits_{-\infty}^{+\infty}\!\!\int_0^h \left\{ \frac{\rho(h)\rho(\vec{r}\,')}{|\vec{r}_h - \vec{r}\,'|} \, dx\,dy\,d\vec{r}\,'^3 + \frac{\rho(\vec{r})\rho(h)'}{|\vec{r} - \vec{r}_h|} \, dx'\,dy'\,d\vec{r}^3 \right\} \right.$$
$$\left. + \int\!\!\int\!\!\int\limits_{-\infty}^{+\infty}\!\!\int_0^h \left\{ \frac{\dfrac{\partial\rho(\vec{r})}{\partial h}\rho(\vec{r}\,')}{|\vec{r} - \vec{r}\,'|} + \frac{\rho(\vec{r})\dfrac{\partial\rho(\vec{r}\,')}{\partial h}}{|\vec{r} - \vec{r}\,'|} \right\} d\vec{r}^3\,d\vec{r}\,'^3 \right] \tag{5.1.6}$$

where $\vec{r}_h = (x, y, h)$ represents a position on the neutral surface.

The expression for $\partial \mathbf{U}_{el}/\partial h$ can be simplified by using eq. 3.2.2 in the reverse direction going from a charge distribution to a potential. Remembering that the labeling of the integration variable is immaterial, we find

$$\frac{\partial \mathbf{U}_{el}}{\partial h} = \rho(h)\Phi(h)\cdot\text{area} + \int\!\!\int\limits_{-\infty}^{+\infty}\int_0^h \frac{\partial\rho}{\partial h}\Phi\,d\vec{r}^3$$

$$= \text{area}\int_0^h \frac{\partial\rho}{\partial h}\Phi\,dz \tag{5.1.7}$$

where the last equality follows from the choice $\Phi(h) = 0$ in eq. 5.1.4.

As we will see later in a specific example, the contribution $\partial \mathbf{U}_{el}/\partial h$ is positive and provides an attractive component to the force. We might expect this result because pushing the counterions toward the charged wall should decrease the Coulomb energy, since the average distance between positive and negative ions decreases.

An important contribution to the force also arises from the change in ion distribution when we push with the neutral wall. We must consider both the electrostatic and the ideal part of this entropy so that according to eq. 3.9.4

$$-T\mathbf{S} = -T(\mathbf{S}_{el} + \mathbf{S}_{ideal}) = kT\sum_i \int\!\!\int\limits_{-\infty}^{+\infty}\int_0^h c_i^*(z)(\ln c_i^*(z) - 1)\,d\vec{r}^3$$

$$= \text{area}\,kT\sum_i \int_0^h c_i^*(\ln c_i^* - 1)\,dz \tag{3.9.4}$$

where the asterisk on c denotes the expression of the ion concentration in terms of number of ions per volume. Taking the derivative of $\mathbf{S}$ with respect to h creates a dependence both through the limits of integration and through the h-dependence of the ion distribution so that

$$-T\frac{\partial\mathbf{S}}{\partial h} = \text{area}\cdot kT\sum_i \left\{ c_i^*(h)\ln c_i^*(h) - c_i^*(h) + \int_0^h \frac{\partial c_i^*}{\partial h}\ln c_i^*\,dz + \int_0^h \left(\frac{\partial c_i^*}{\partial h} - \frac{\partial c_i^*}{\partial h} \right)dz \right\} \tag{5.1.8}$$

where the last integral is clearly zero. We can eliminate the $\ln c_i^*$ terms by using the Boltzmann distribution of eq. 3.8.2. In logarithmic form, this distribution reads

$$\ln c_i^*(h) = \ln c_{io}^* - \frac{z_i e \Phi(h)}{kT} \qquad (3.8.2)$$

so that

$$-T\frac{\partial \mathbf{S}}{\partial h} = \text{area} \left\{ -\Phi(h) \sum_i z_i e c_i^*(h) - \int_0^h \Phi \sum_i z_i e \frac{\partial c_i^*}{\partial h} \, dz \right.$$
$$\left. + kT \sum_i \left\{ \ln c_{io}^* \left[c_i^*(h) + \int_0^h \frac{\partial c_i^*}{\partial h} \, dz \right] - c_i^*(h) \right\} \right\} \qquad (5.1.9)$$

The term $\sum z_i e c_i^*$ represents the charge distribution ρ, which thus enters the first two terms. Furthermore,

$$\text{area} \left[c_i^*(h) + \int_0^h \frac{\partial c_i^*}{\partial h} \, dz \right] = \text{area} \frac{\partial}{\partial h} \int_0^h c_i^* dz = \frac{\partial}{\partial h} n_i = 0 \qquad (5.1.10)$$

because the total number of ions n_i of species i is not allowed to change. If we again use the combination $\Phi(h) = 0$, we are left with

$$-T\frac{\partial \mathbf{S}}{\partial h} = -\text{area} \left\{ \int_0^h \frac{\partial \rho}{\partial h} \Phi \, dz + kT \sum c_i^*(h) \right\} \qquad (5.1.11)$$

Combining eqs. 5.1.1, 5.1.2, 5.1.4, and 5.1.11, we find the remarkably simple result

$$\frac{F}{\text{area}} = \Pi_{\text{osm}} = kT \sum_i c_i^*(h) \qquad (5.1.12)$$

The term $kT \sum c_i^*(h)$ can be interpreted as the osmotic pressure resulting from the ion concentration at the neutral surface. Because concentrations are always positive, the force is also positive, which implies a repulsive interaction.

We can understand the source of the repulsive force as follows. The entropy in the counterion distribution explains why a diffuse double layer initially forms at a charged surface. Binding all counterions to the surface minimizes the energy. On the other hand, releasing the ions lowers the free energy because it results in an increase in entropy. However, counterions cannot leave the surface entirely because such high energy is needed to separate the charges completely. Therefore, the nonuniform ion distribution outside the charged surface represents an optimal compromise between these two competing effects.

When a confining neutral surface approaches a charged one, the ion distribution between them must rearrange in the smaller space that is available, so the entropy of ions decreases. A decrease in energy partly compensates for this change. For the ions at the neutral wall, interaction with the charged wall is screened totally by the intermediary ions, since $(d\Phi/dz) = 0$ at $z = h$, and only the entropy term counts in the calculation. We can see this exemplified in eqs. 5.1.4 through 5.1.12, in which all the cancellations that occur in the derivation reflect this balance between energy and entropy.

5.1.2 We Can Solve the Poisson–Boltzmann Equation When Only Counterions Are Present Outside the Charged Surface

To actually determine the force between two surfaces, we still must solve the Poisson–Boltzmann equation to obtain the concentrations $c_i^*(h)$ at the neutral surface. In general, this solution is quite complex, but if only counterions are involved, we can employ a simple analytical solution of

$$\nabla^2\Phi = \frac{ze}{\varepsilon_r\varepsilon_o}\, c^*(h)\exp\left(\frac{-ze\Phi}{kT}\right) \tag{5.1.13}$$

where $c^*(h)$ signifies the concentration of the ionic species at the neutral wall. Electroneutrality requires that

$$\int_0^h c^*(z)dz = \frac{-\sigma}{(ze)} \tag{5.1.14}$$

where σ is the surface charge density. We can solve the Poisson–Boltzmann equation for this special case in much the same way that we solved the Gouy–Chapman case in Section 3.8. The counterion distribution comes out as

$$c^*(z) = 2kT\varepsilon_r\varepsilon_o \frac{s^2}{(zeh)^2} \frac{1}{\cos^2\{s(1 - z/h)\}} \tag{5.1.15}$$

In this equation s represents a dimensionless parameter analogous to s in eq. 3.9.14, but here the boundary conditions imply

$$s\tan(s) = -\frac{ze\sigma h}{2kT\varepsilon_r\varepsilon_o} \tag{5.1.16}$$

Because $\tan(s)$ diverges at $\pi/2$, s can take only values within the range $0 \leqslant s < \pi/2$.

We obtain the force from $c^*(h)$ according to eqs. 5.1.12 and 5.1.15

$$\frac{F}{\text{area}} = \frac{2(kT)^2\varepsilon_r\varepsilon_o}{(ze)^2}\left(\frac{s}{h}\right)^2 \tag{5.1.17}$$

For high charge densities or large h, s approaches an asymptotic value of $\pi/2$ in eq. 5.1.16, and in this nonlinear regime of the Poisson–Boltzmann equation, we find the simple expression for the force

$$\frac{F}{\text{area}} = \frac{\pi^2 (kT)^2 \varepsilon_r \varepsilon_o}{2(ze)^2} \frac{1}{h^2} \qquad (5.1.18)$$

with the somewhat surprising feature that the force is independent of the surface charge density. We will see that this feature is typical of highly charged surfaces. In the limit of small s, $(\tan(s) \simeq s)$ the concentration entering eq. 5.1.12 simply represents the average concentration, $\bar{c} = -\sigma/(zeh)$, so the force shifts from an h^{-1} dependence at short separations to an h^{-2} dependence at longer distances. This effect is still very long-ranged.

As a practical example, let us calculate the thickness of the aqueous film on a charged surface exposed to humid air. This surface might be clean glass or one with an adsorbed ionic surfactant layer.

With a relative humidity of p/p_0 in the air, some water will condense on the surface layer and *free* the (monovalent) counterions to some extent. At equilibrium, the water chemical potential in the film, which is related to Π_{osm} (i.e., F/area), is the same as in air, and

$$V_{H_2O} \Pi_{osm} = -kT \ln \frac{p}{p_0} \qquad (5.1.19)$$

where V_{H_2O} is the molecular volume of liquid water. Unless h is very small, the highly charged surface allows us to use eq. 5.1.18, noting that the force per unit area equals the osmotic pressure,

$$h_{eq} = \left[\frac{V_{H_2O} kT \varepsilon_r \varepsilon_o}{2z^2 \ln(p_o/p)} \right]^{1/2} \frac{\pi}{e} \qquad (5.1.20)$$

if we assume that double-layer effects dominate. At 99% relative humidity, $h_{eq} = 1.3\,\text{nm}$ but increases strongly as p approaches saturation.

As an alternative to eq. 5.1.12 we could have used the ion concentration at the charged wall to calculate the force with equal success, provided we corrected for the effects resulting from the nonzero electrical field. Equation 3.8.15 — which remains valid for the two-wall case if we choose the appropriate reference point — demonstrates this use. Thus,

$$\left(\frac{d\Phi}{dz}\right)^2 = \frac{2kT}{\varepsilon_r \varepsilon_o} \sum_i c_i^*(h) \left\{ \exp\left(\frac{-z_i e\Phi}{kT}\right) - 1 \right\} \qquad (5.1.21)$$

According to eq. 5.1.3, the potential and the field are zero at $z = h$. The field $-d\Phi/dz$ at the surface simply relates to the surface charge density through the boundary condition 3.8.7. If we use eq. 5.1.12, which relates $c_i(h)$ to the force, and evaluate eq. 5.1.21 at $z = 0$ — that is, at the charged surface — we find

$$\frac{F}{\text{area}} = \Pi_{\text{osm}} = kT \sum_i c_i^*(0) - \frac{\sigma^2}{2\varepsilon_r \varepsilon_o} \qquad (5.1.22)$$

as in eq. 3.8.26. This equation remains valid irrespective of the charges on the two surfaces, a fact that proves useful in the following section.

5.1.3 Ion Concentration at the Midplane Determines the Force Between Two Identically Charged Surfaces

When two equally charged surfaces approach each other in a liquid medium, ther diffuse double layers overlap and a repulsive force results. This repulsion sets in at longer separations than are observed when only one neutral surface is involved. We now consider two identical charged surfaces, as shown in Figure 5.2. The boundary conditions of the Poisson–Boltzmann equation are the same ones that we used for the Gouy–Chapman solution at the charged surfaces in eq. 3.8.7, but because of the symmetrical situation, the field at the midplane must be zero,

Figure 5.2
Two interacting parallel negatively charged surfaces separated by a distance h in an electrolyte solution experience a repulsive force (eq. 5.1.28) as they are pushed together.

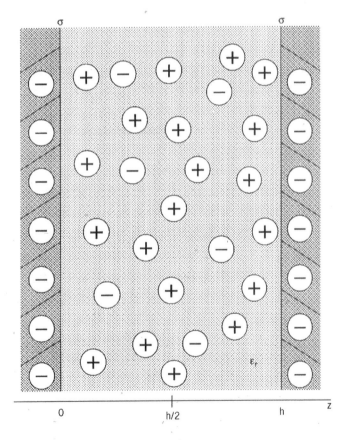

Figure 5.3
Superposition
approximation for two
similarly charged surfaces.
If the charge distributions
ρ_1 and ρ_2 outside the
respective surfaces are
assumed to be unaffected
by the interaction between
the surfaces, the resulting
charge distribution is
$\rho = \rho_1 + \rho_2$. By the
linearity of the Poisson
equation (see Chapter 3),
the resulting potential Φ
represents the sum
$\Phi = \Phi_1 + \Phi_2$ of the
potentials Φ_1 and Φ_2
outside the unperturbed
surfaces. Then the
potential Φ_d stands for the
sum Φ_1 $(z = h/2) + \Phi_2$
$(z = h/2)$.

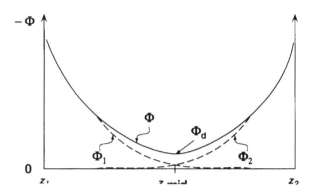

as can be seen by inspecting Figure 5.3. This fact allows us to replace the boundary condition at one of the surfaces by a condition at the midplane ($z = h/2$ in Figure 5.2) to give

$$\left.\frac{d\Phi}{dz}\right|_{z_{\text{mid}}} = 0, \quad \Phi(z_{\text{mid}}) = \Phi_d \tag{5.1.23}$$

The reason for fixing the potential at a nonzero value will become apparent in the following paragraphs.

The Poisson–Boltzmann equation for one half of the gap shown in Figure 5.2 is the same as for the system in Figure 5.1, which we discussed in Section 5.1.1. In that situation, electrical neutrality between the charged wall and the neutral wall occurred of necessity; if two similar double layers overlap, however, symmetry brings about neutrality relative to the midplane. Whatever the physical cause of the boundary conditions, the solution of the Poisson–Boltzmann equation is the same: The ion distribution is identical and so is the free energy. Thus, the derivation of eq. 5.1.12 remains valid and

$$\Pi_{\text{osm}} = kT \sum_i c_i^* (\text{midplane}) \tag{5.1.24}$$

We can explain the repulsion between the two surfaces in eq. 5.1.19 with the same mechanistic reasoning we used to explain the charged and neutral surfaces in Section 5.1.1. When the two double layers start to overlap, ion concentrations increase, lowering the entropy. A decrease in energy offsets some of the resulting increase in free energy, but at the midplane— where no fields act on the ions—only the entropy effect is significant.

Equation 5.1.24 applies directly to the swelling of both lamellar liquid crystals and some clays with a basic lamellar structure. As long as the double-layer forces dominate, the charged surfaces in a lamellar liquid crystal repel each other, and the liquid crystal swells indefinitely in pure water. In practice, although we usually do observe a strong swelling, other factors also play a role in the process.

As an illustration, we can consider a clay with a platelike structure that has swollen in pure water. How does the thickness of the water layer vary as we squeeze the clay mechanically? For a pure water medium, the only ions present are the counterions, and we can use eq. 5.1.15 to calculate their distribution. If we assume that the surface is highly charged, we can use eq. 5.1.18 to describe the osmotic pressure. Because this pressure balances the applied pressure p_{ext}, we can obtain the water layer thickness h from

$$h = \left(\frac{2\varepsilon_r \varepsilon_o}{z^2 p_{ext}}\right)^{1/2} \left(\frac{kT\pi}{e}\right) \tag{5.1.25}$$

noting that h in Figure 5.1 corresponds to $h/2$ in Figure 5.2.

Equations 5.1.20 and 5.1.25 illustrate how clays can swell under humid conditions yet release water under mechanical pressure—an effect familiar to most children and to farmers who try to drive tractors on clay soil on a damp autumn day. In a dry atmosphere, clay retains water, and this property makes clay a beneficial ingredient in agricultural soil. Equation 5.1.20 describes the swelling quantitatively, if we reinterpret h_{eq} as $h_{eq}/2$.

5.1.4 The Bulk Solution Often Provides a Suitable Reference for the Potential

The solvent and the ions between the surfaces of colloidal particles in a medium directly contact the bulk solution. As the surfaces are pushed toward each other, solvent is expelled from the gap. Equation 5.1.24 gives the local osmotic pressure. In the bulk the osmotic pressure is $kT\Sigma_i c_{io}^*$, where c_{io}^* equals the bulk concentration of ion i. The net force acting between the surfaces is then

$$\frac{F}{area} = \Pi_{osm}(\text{midplane}) - \Pi_{osm}(\text{bulk})$$

$$= kT\left\{\sum_i c_i^*(\text{midplane}) - \sum_i c_{io}^*\right\} \tag{5.1.26}$$

In addition, the bulk sets the chemical potential of the ionic species so that it serves as a natural reference point for the electrostatic potential. This explains why we chose the value $\Phi(h/2) = \Phi_d \neq 0$ in eq. 5.1.23. From the Boltzmann distribution, we can relate the ion concentration at the midplane c_i^* (midplane) to the bulk concentration c_{io}^* through

$$c_i^*(\text{midplane}) = c_{io}^* \exp\left(\frac{-z_i e \Phi_d}{kT}\right) \tag{5.1.27}$$

Thus, we find

$$\frac{F}{area} = kT \sum_i c_{io}^* \left[\exp\left(\frac{-z_i e \Phi_d}{kT}\right) - 1\right] \tag{5.1.28}$$

Equation 5.1.28 constitutes a central equation for the force between two equal charged surfaces in equilibrium with a bulk solution. This equation has been derived in many different ways. The derivation chosen here has the advantage of explicitly describing the force as a derivative of the free energy and showing that it is composed of an electrostatic interaction energy and an entropy of the counterion distribution. In this way we gain insight into the molecular origin of the force.

One virtue of having such physical insight is that it brings out some shortcomings in the theory. In Section 5.1.1, we considered the close analogy between the cases of charged and neutral walls. The discussion of the two identically charged walls in Section 5.1.2 shows that in the latter case the two halves do not interact directly. This identity between the system shown in Figure 5.1 and the halves of Figure 5.2 is a consequence of using the mean field approximation of the Poisson–Boltzmann equation. In reality, instantaneous electrical fields go from one half to the other in Figure 5.2, since the lateral charge distribution is inhomogeneous at each particular moment in time. Only the average distribution is laterally homogeneous.

In systems with strong interactions between individual ions (such as those with divalent ions or for low ε_r), these instantaneous fields are large enough to permit the development of substantial correlations between the ions in the two halves. We postpone a detailed discussion of this effect until Section 5.3.

To use eq. 5.1.28 to obtain an expression for the force in terms of known quantities, we still need to solve the Poisson–Boltzmann equation. In the general case with several ionic species present, it is best to use numerical methods. However, the limiting case that afford analytical approximations are of considerable interest. When two double layers start to overlap, the potential Φ_d is displaced only slightly from zero, since $\Phi_d \to 0$ as $h \to \infty$. Thus, for weakly overlapping double layers, we can expand the exponent in eq. 5.1.28 in a power series. The bulk term cancels the first term, and the second term cancels as a result of electroneutrality. The first nonzero contribution, therefore, comes from the third term in the expansion, and

$$\frac{F}{\text{area}} \simeq \frac{1}{2} \sum_i \frac{c_{io}^*(ze)^2 \Phi_d^2}{kT}$$

$$= \frac{\varepsilon_r \varepsilon_o \kappa^2 \Phi_d^2}{2} \tag{5.1.29}$$

In this weak overlap limit, we can estimate Φ_d by taking advantage of the fact that the potential profile differs only slightly from the Gouy–Chapman solution for a single surface. We can assume that the potential at the midplane Φ_d simply equals the sum of the Gouy–Chapman potentials from the two surfaces, which follows from the superposition principle applied to the Poisson equations. (See Figure 5.3.) Using the linearized form of the Gouy–Chapman solution as given in eq. 3.8.19, we find

$$\Phi_d\left(\frac{h}{2}\right) = 2\,\frac{4kT}{ze}\,\Gamma_0\,\exp\left(\frac{-\kappa h}{2}\right) \qquad (5.1.30)$$

valid for a symmetrical $z{:}z$ electrolyte.

As two double layers start to overlap, they experience a repulsive force

$$\frac{F}{\text{area}} \simeq 32(kT)^2\,\frac{\varepsilon_r \varepsilon_o}{z^2 e^2}\,\kappa^2 \Gamma_0^2\,\exp(-\kappa h)$$

$$= 64kTc_o^*\Gamma_0^2\,\exp(-\kappa h) \qquad (5.1.31)$$

where c_o^* represents the bulk electrolyte concentration and $\Gamma_0 = \tanh(ze\Phi_o/4kT)$, as in eq. 3.8.14.

Equation 5.1.31 reveals several important facts. At long range, the double-layer force decays exponentially with a decay constant equal to the Debye screening length. The force is sensitive to the electrolyte concentration through the factor c_o^* and more significantly through κ. For a $z{:}z$ electrolyte, κ^{-1} is proportional to z, and so divalent ions weaken the force substantially compared to monovalent ions.

The nature of the charged surface enters the calculation through the quantity Γ_0. From eq. 3.8.14 we see that for low surface potentials ($ze\Phi_o/kT \ll 1$)

$$\Gamma_0 \simeq \frac{1}{4}\left(\frac{ze}{kT}\right)\Phi_o \qquad (3.8.14)$$

while Γ_0 approaches a constant value of unity for high surface potentials. Thus, irrespective of the charge at the surface, an upper limit to the force exists at a given separation and κ, as was also found in eq. 5.1.18.

Equation 5.1.31 can describe the double-layer force only when it is so weak that the linearization approximations apply. Within this limit, the two double layers are perturbed only slightly relative to infinite separation, and possible chemical rearrangements at the surface are negligible. The situation becomes more complex when the surfaces interact more strongly. Then we must not only solve the Poisson–Boltzmann equation numerically except in the limit of short separations of highly charged surfaces the coions are completely expelled from the region between the surfaces. Then we have the case of counterions only and we can use eqs. 5.1.17 or 5.1.18 for the force.

However, we face a problem with boundary conditions. Normally we do not know whether the ionization of surface groups or adsorption of ions generates the charge at the surfaces. As two surfaces approach each other, these surface groups can titrate or the ions can desorb or adsorb and thus affect the force. Measurements of the force in the strong interaction regime can provide a method for studying the chemical processes at the surface.

Limiting cases exist for modeling these surface effects. For example, if we assume that the charge is constant, we totally neglect the surface charge rearrangement effects and overestimate the force at shorter range. At the other extreme, if we assume an infinitely adjustable electrical surface so that the electrostatic potential remains constant, we obtain a lower limit to the force. The reality is usually somewhere between constant charge and constant potential. We presented a more detailed model for this possibility in Section 3.7. When we apply it to the force problem, we call it the charge regulation model.

5.1.5 Two Surfaces with Equal Signs but Different Magnitudes of Charge Always Repel Each Other

So far, our concentration on the standard case of two planar and equally charged parallel walls has allowed us to develop a reasonably simple quantitative theoretical description of the double-layer interaction. As a rule, reality is more complex than our idealized examples. For instance, particle surfaces are not always planar, and in Section 5.2.4 we discuss the complications that arise from different surface geometries. Another complication relates to so-called heterocoagulation, in which two dissimilar surfaces or particles can come together. For the sake of clarity, we will maintain the parallel planar configuration in our discussion of this complication.

For an unsymmetrical system, the ion concentration at the midplane no longer constitutes a relevant factor. The boundary conditions relate to the two charged surfaces with

$$\frac{d\Phi}{dz} = -\frac{\sigma_1}{\varepsilon_r \varepsilon_o} \quad \text{at } z_1 \tag{5.1.32}$$

and

$$\frac{d\Phi}{dz} = \frac{\sigma_2}{\varepsilon_r \varepsilon_o} \quad \text{at } z_2 \tag{5.1.33}$$

where z_1 and z_2 $(z_1 < z_2)$ are the locations of the two charged surfaces and σ_1 and σ_2 the respective surface charge densities. We can separate the two cases in which σ_1 and σ_2 bear the same sign and the opposite sign.

For charged surfaces of the same sign, the double layers still overlap. At some positions z_0 between the surfaces, the field is zero, and this plane represents a dividing surface for the double layers as it did in the symmetrical case. (See Figure 5.4). In this plane, only the osmotic effect operates and

$$\frac{F}{\text{area}} = kT \sum_i (c_i^*(z_0) - c_{io}^*) \tag{5.1.34}$$

Figure 5.4
Schematic illustration of the charge distribution, ρ, the electrostatic potential Φ, and its derivative $d\Phi/dz$ between two negatively charged surfaces of unequal surface charge densities, $|\sigma_1| > |\sigma_2|$. At $z = z_0$ the derivative $d\Phi/dz$ is zero, while $|\Phi|$ and ρ have their minimal values. The force is obtained from eq. 5.1.29 with $\Phi_d = \Phi(z_0)$.

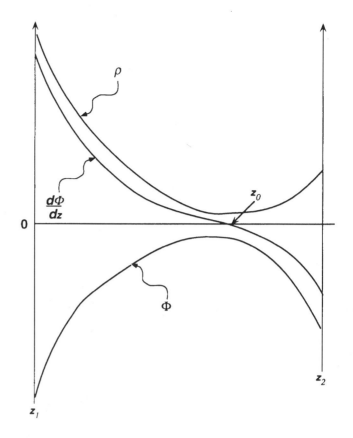

in analogy with eq. 5.1.26. In general, to find z_0 and $c_i^*(z_0)$, we must solve the Poisson–Boltzmann equation.

For weakly overlapping double layers, eq. 5.1.29 still applies if we interpret Φ_d as the potential at z_0. At z_0 we can view the electrical field, which is zero by definition, as being composed of two compensating parts. In the region of weak overlap, these two fields approximately equal those of the unperturbed fields from each surface separately. Using eq. 3.8.22, we can obtain z_0 from

$$\Gamma_{01} \exp\{-\kappa(z_0 - z_1)\} = \Gamma_{02} \exp\{-\kappa(z_2 - z_0)\} \quad (5.1.35)$$

which leads to

$$z_0 = \frac{1}{2}\kappa^{-1}\ln\left(\frac{\Gamma_{01}}{\Gamma_{02}}\right) + \frac{1}{2}(z_1 + z_2) \quad (5.1.36)$$

For these formulas to apply, the charge density on the less charged wall must be large enough that $|\sigma_2| > \varepsilon_r\varepsilon_o|\Phi_{01}| \exp\{-\kappa(z_2 - z_1)\}$. Otherwise we locate z_0 outside the gap ($z_0 > z_2$), which shows that the weak overlap approximation is not applicable.

Using the same approximations that led from eq. 5.1.29 to eq. 5.1.31 for the symmetrical case, eq. 5.1.36 for z_0 results in

$$\frac{F}{\text{area}} = 64kTc_o^*\Gamma_{01}\Gamma_{02}\exp(-\kappa h) \qquad (5.1.37)$$

which simply replaces Γ_0^2 in eq. 5.1.31 by the geometrical mean, $\Gamma_{01}\Gamma_{02}$, in eq. 5.1.37.

From eq. 5.1.34 we can see that the force between the two surfaces is always repulsive. Calculations exist for the constant potential boundary condition that results in attraction in some regime. A closer look at them reveals that attraction occurs only when the charge density on one of the surfaces actually changes sign.

5.1.6 As Surfaces Bearing Opposite Signs Move Closer Together, Long Range Electrostatic Attraction Changes to Repulsion

An attractive force at long range results when σ_1 and σ_2 carry opposite signs. One surface can act as a giant counterion for the other, and the original counterions are released into the bulk solution with a concomitant increase in entropy. In this case, the solution of the Poisson–Boltzmann equation follows from the same boundary conditions we used with surfaces of the same sign of σ in eq. 5.1.32.

However, because no point with zero field exists in the intervening solution, we must determine the force in a different way. Formula 5.1.22 remains valid, so if we know the ion concentration at the wall and the surface charge density, the force follows from

$$\frac{F}{\text{area}} = -\frac{\sigma^2}{2\varepsilon_r\varepsilon_o} + kT\sum_i \{c_i^*(\text{wall}) - c_{io}^*\} \qquad (5.1.38)$$

Because of two surfaces partly neutralize each other, counterions leave the gap between the surfaces and initially $c_i^*(\text{wall})$ will decrease. If $\sigma_2 = -\sigma_1$, the attraction will be sustained up to contact. At short separations, $h \ll \kappa^{-1}$, the bulk term cancels the wall concentration in eq. 5.1.38 and the force becomes $-\sigma^2/(2\varepsilon_r\varepsilon_o)$, which simply represents the force between two planar capacitor plates.

When $|\sigma_1| \neq |\sigma_2|$, some counterions must remain in the gap. These counterions concentrate when the two surfaces come close and ultimately produce a repulsive force. This effect shows that if electrostatic forces dominate, we should expect heterocoagulation of oppositely charged particles of constant charge to result in a complex where the particles do not reach molecular contact.

5.2 Van der Waals Forces Comprise Quantum Mechanical Dispersion, Electrostatic Keesom, and Debye Forces

5.2.1 An Attractive Dispersion Force of Quantum Mechanical Origin Operates Between Any Two Molecules

Attractive interactions between any pair of molecules arise from the correlation between the electrons motions in the two molecules. Even if an atom (or molecule) has no permanent electric moment, electrons circle the nuclei and create instantaneous dipoles. These dipoles generate a field at the second atom (or molecule) whose electrons respond instantaneously (or more precisely at the speed of light). Because Fritz London first analyzed this mechanism of intermolecular interactions, it often is called the London dispersion interaction. Another name, which refers to the mechanistic description presented above, is the induced dipole-induced dipole interaction, described in Chapter 3.

The dispersion interaction emerges naturally from a quantum mechanical description of the interaction between two molecules. It is mechanistically analogous to the classical thermally averaged Keesom interaction. The difference is that for the dispersion interaction the correlation is between electrons that are described quantum mechanically, rather than the thermally excited molecular rotation. Neglecting molecular anisotropies, we can write the interaction potential $V_{12}^{\text{disp}}(R)$ in a quantum mechanical formalism as

$$
V_{12}^{\text{disp}}(R) = -\frac{1}{24(\pi\varepsilon_o)^2} \frac{1}{R_{12}^6} \sum_{n,k} \frac{|\langle n|\vec{m}|0\rangle_1|^2 |\langle k|\vec{m}|0\rangle_2|^2}{(E_1^n - E_1^o) + (E_2^k - E_2^o)}
$$

$$
\equiv \frac{-C_{12}}{R_{12}^6} \tag{5.2.1}
$$

where $\langle n|\vec{m}|0\rangle_1$ represents the so-called transition dipole moment going from the ground state $|0\rangle$ energy E_1^o to an excited state $|n\rangle$ energy E_1^n in molecule 1. The transition dipole moment gives the amplitude of the optical transition between the two states. The quantities $\langle k|\vec{m}|0\rangle_2$, E_2^o and E_2^k represent the corresponding quantities for molecule 2.

In eq. 5.2.1 the R^{-6} dependence has the same origin as the dipole–induced dipole interaction in eq. 5.2.1. The first R^{-3} dipolar factor reflects the strength of the electrical field and the second reflects the interaction between the spontaneous and the induced dipole. Because E_1^o and E_2^o are the ground states, the denominators in eq. 5.2.1 are always positive. As a result, each term in the sum is positive so that the total dispersion interaction is attractive.

The origin of the name *dispersion interaction* is not obvious, but it stems from the fact that the molecular quantities entering eq. 5.2.1 are precisely those that determines the optical properties of the molecules. For example, the frequency-dependent isotropic polarizability $\alpha(\omega)$ at the frequency ω is given by

$$\alpha(\omega) = \frac{2}{3} \sum_n \frac{(E^n - E^o)|\langle n|\bar{m}|0\rangle|^2}{(E^n - E^o)^2 - \hbar^2\omega^2} \tag{5.2.2}$$

This equation contains the same quantities as eq. 5.2.1, but in a different combination ($h = 2\pi\hbar$ = Planck's constant). This polarizability directly relates to the frequency dependence of the refractive index and thus to the dispersion of light.

London derived and approximate relation for the dispersion interaction in terms of the polarizabilities at zero frequency $\alpha(0)$ and the ionization potentials I of the molecules, that is, the energy required to ionize the molecule,

$$V_{12}^{\text{disp}}(R) \simeq -\frac{3}{32(\pi\varepsilon_o)^2} \frac{I_1 I_2}{I_1 + I_2} \frac{\alpha_1(0)\alpha_2(0)}{R_{12}^6} \tag{5.2.3}$$

Dispersion interaction, which is a main cause of cohesion is condensed states (i.e., solids and liquids), does not vary strongly from molecular pair to molecular pair. Because the dispersion interaction has limited selectivity, electrostatic interactions are more likely to cause phase separation and self-assembly in molecular liquid systems. However, some general trends determine the magnitude of the dispersion term. An increase in electrons per unit volume in the molecules increases V_{12} by adding terms to the sum in eq. 5.2.1 and slightly lower excitation energies.

Another important feature of the dispersion interaction that emerges from its quantum mechanical derivation is that it is pairwise additive to a good approximation. Even in a condensed phase, we can obtain the total dispersion energy with reasonable accuracy by simply summing over all molecular pairs. The fact that virtually all molecules are in their electronic ground states at ambient temperatures simplifies this operation. Thus, the total dispersion interaction is independent of temperature in a condensed phase except for the indirect sensitivity to density changes affecting the average separation R_{12}.

5.2.2 We Can Calculate the Dispersion Interaction Between Two Colloidal Particles by Summing Over the Molecules on a Pairwise Basis

If we assume pairwise additivity, we can determine the dispersion force between two colloidal particles by summing over all pairs of molecules, one molecule on each particle. The basic model of a slit separating two surfaces, as in Figure 5.5, still proves useful. We can assume that a low density gas, which has negligible influence on the interaction, fills the slit.

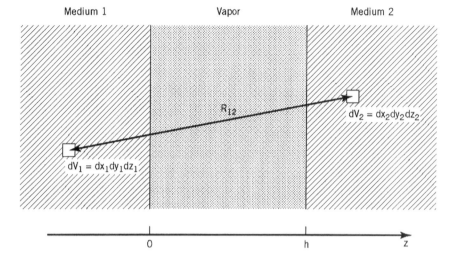

Medium 1   Vapor   Medium 2

Figure 5.5
Calculation of the London dispersion interaction between two half-spaces separated by a distance h across a vapor. An interaction proportional to R_{12}^{-6} exists between the two volume elements, dV_1 and dV_2. The total interaction is obtained by integrating this interaction over the two half-spaces, respectively.

Using eq. 5.2.1, the contribution from volume elements $dV_1 = dx_1 dy_1 dz_1$ and $dV_2 = dx_2 dy_2 dz_2$ (one on either side) to the total interaction energy $\mathbf{U}_{12}$ between the two half-spaces 1 and 2 is

$$dU_{12} = -\frac{C_{12}\rho_1\rho_2 dV_1 dV_2}{\{(x_2 - x_1)^2 + (y_2 - y_1)^2 + (z_2 - z_1)^2\}^3} \quad (5.2.4)$$

where ρ is the density. The total interaction energy follows by direct integration

$$\mathbf{U}_{12} = -\frac{H_{12}}{\pi^2} \iiint\limits_{-\infty}^{+\infty} \int_{-\infty}^{0} \int_{h}^{\infty} \frac{dx_1 dx_2 dy_1 dy_2 dz_1 dz_2}{\{(x_2 - x_1)^2 + (y_2 - y_1)^2 + (z_2 - z_1)^2\}^3} \quad (5.2.5)$$

where we combine $\pi^2 C_{12}\rho_1\rho_{12}$ to form the Hamaker constant H_{12}, a constant that plays a central role in the description of van der Waals forces between particles. The Hamaker constant has the dimension of an energy, and it depends both on the strength of the intermolecular interaction through C_{12} and on the density. Clearly, we expect a weaker interaction between gaseous media than liquid bulk media of the same compounds. As derived here, H_{12} is determined solely by the dispersion interaction. Later we will see that the total Hamaker constant also includes other contributions.

We can solve the integral in e.g. 5.2.5 by first substituting variables $x = (x_2 - x_1)$, $x' = (x_2 + x_1)$, $y = (y_2 - y_1)$, $y' = y_2 + y_1$. This substitution brings out an expected proportionality between $\mathbf{U}_{12}$ and the area because only the relative position of the two volume elements in the x, y direction matters and

$$\frac{\mathbf{U}_{12}}{\text{area}} = -\frac{H_{12}}{\pi^2} \iint_{-\infty}^{+\infty} \int_{-\infty}^{0} \int_{h}^{\infty} \frac{dx\, dy\, dz_1 dz_2}{\{(x^2 + y^2 + (z_2 - z_1)^2\}^3} \quad (5.2.6)$$

The right-hand side can be integrated most simply by using cylindrical coordinates such that $r^2 = x^2 + y^2$ and

$$\frac{U_{12}}{\text{area}} = -\frac{H_{12}}{\pi^2} \int_0^\infty \int_{-\infty}^0 \int_h^\infty \frac{2\pi r \, dr \, dz_1 dz_2}{\{r^2 + (z_2 - z_1)^2\}^3}$$

$$= -\frac{H_{12}}{2\pi} \int_{-\infty}^0 \int_h^\infty \frac{dz_1 dz_2}{(z_1 - z_2)^4}$$

$$= -\frac{H_{12}}{6\pi} \int_h^\infty \frac{dz_2}{z_2^3} = -\frac{H_{12}}{12\pi} \frac{1}{h^2} \qquad (5.2.7)$$

By integrating the short range R^{-6} dispersion term in four directions, we have transformed this interaction into a long range h^{-2} dependence. (The exponent changes by one for each direction over which the integration is performed.)

At first sight this result is somewhat surprising. We know that in the bulk phase the dominant contribution to the energy comes from nearest neighbors as a result of the R^{-6} dependence of the intermolecular interaction. For a narrow slit with h of the order of a molecular diameter, the dominant contribution to U_{12} comes from the surface layer–surface layer interaction. However, the larger the slit, the less dominant this contribution becomes, so that the total interaction varies less strongly with the separation, h.

A similar mechanism operates for a single molecule (2) outside the surface of a bulk phase (1). As in eq. 5.2.7, we can obtain the interaction potential $V_{12}(h)$ by omitting the integration over dz_2 and dividing out the irrelevant density factor ρ_2, so that

$$V_{12}(h) = -\frac{H_{12}/\rho_2}{6\pi} \frac{1}{h^3} \qquad (5.2.8)$$

This expression closely resembles the one describing the force F between the two half-planes

$$F_{12} = -\frac{dU_{12}}{dh} = -\text{area } H_{12} \frac{1}{6\pi} \frac{1}{h^3} \qquad (5.2.9)$$

where we have retained the convention that attractive forces are negative. Such analogies between interaction energies and forces can be very useful, as we will demonstrate further in Section 5.2.4.

For a liquid held together only by dispersion forces, the surface tension γ clearly relates to the dispersion interaction. To see this quantitatively, we can write the total energy $\mathbf{U}$ of the system in Figure 5.5 as the sum of a bulk term $n_1 u_1$ (bulk) $+ n_2 u_2$ (bulk), in which n_i represents the number of molecules in phase i, plus a surface term, $\gamma_1 A + \gamma_2 A$, and an interaction term to give

$$\mathbf{U} = n_1 u_1(\text{bulk}) + n_2 u_2(\text{bulk}) + (\gamma_1 + \gamma_2)A - \frac{A H_{12}}{12\pi} \frac{1}{h^2} \qquad (5.2.10)$$

At contact between the two phases, this equation should reduce to $\mathbf{U} = n_1 u_1 + n_2 u_2 + \gamma_{12} A$ where γ_{12} equals the interfacial tension ($\gamma_{12} = 0$ if $1 = 2$). A distance of closest approach, δ, exists between the half-planes, and by inserting $h_{min} = \delta$, we find

$$\gamma_1 + \gamma_2 = \frac{H_{12}}{12\pi} \frac{1}{\delta^2} + \gamma_{12} \qquad (5.2.11)$$

When medium 1 = medium 2, $\gamma_{12} = 0$ and the surface tension is

$$\gamma_1 = \frac{H_{11}}{24\pi} \frac{1}{\delta^2} \qquad (5.2.12)$$

This distance δ should be on the order of an atomic van der Waals radius. As an example, we can consider a hydrocarbon whose CH_2 group has a van der Waals radius of approximately $2.0\,\text{Å}$. The Hamaker constant for a hydrocarbon is around $H \simeq 5 \times 10^{-20}\,\text{J}$. Inserting these values into eq. 5.2.12 leads to an estimate of the surface tension of a hydrocarbon of $\sim 17\,\text{mJ/}$ m^2, which is in reasonable agreement with the experimental value of approximately $26\,\text{mJ/m}^2$.

A liquid film formed in a vapor phase (such as in a soap bubble) shown in Figure 5.6, provides a situation complementary to the one illustrated in Figure 5.5. In this case, we can obtain the energy of the liquid by integrating eq. 5.2.3 over the film. However, because of the complementary geometries in Figures 5.5 and 5.6, we also can use eq. 5.2.8 to solve the problem algebraically. To illustrate this point, we can subdivide a bulk phase into three parts (1, 2, and 1') as in Figure 5.7. The total energy is

$$\mathbf{U} = (n_1 + n_2 + n_1')u(\text{bulk})$$
$$= \mathbf{U}_{11} + \mathbf{U}_{12} + \mathbf{U}_{22} + \mathbf{U}_{1'1'} + \mathbf{U}_{1'2} + \mathbf{U}_{11'} \qquad (5.2.13)$$

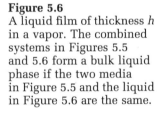

Figure 5.6
A liquid film of thickness h in a vapor. The combined systems in Figures 5.5 and 5.6 form a bulk liquid phase if the two media in Figure 5.5 and the liquid in Figure 5.6 are the same.

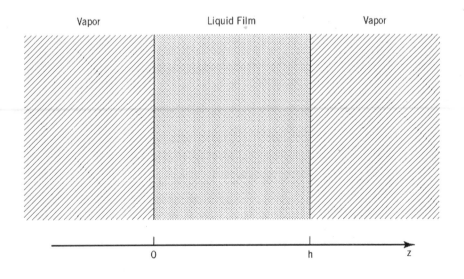

Vapor   Liquid Film   Vapor

0   h   z

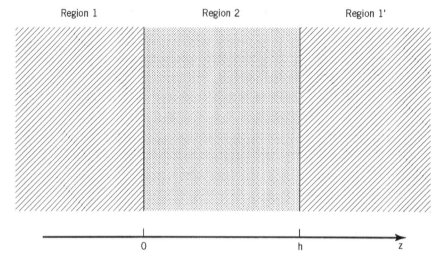

Figure 5.7
Decomposition of a bulk
phase into three regions, to
calculate the energy of
the liquid film in Figure
5.6. Region 2 corresponds
to the liquid film.

The quantity U_{22} represents the integration of eq. 5.2.4 over the film itself, and it constitutes the only contribution to the energy of the film in Figure 5.6. The point is now that we can use the result eq. 5.2.10 to obtain the other unknown quantities in eq. 5.2.13. In eq. 5.2.10 we determined the energy of two half-spaces separated by a distance h. If 1 and 2 in this equation represent the same material, we could relabel them 1 and 1'; then U in eq. 5.2.10 corresponds to $U_{11} + U_{1'1'} + U_{11}$, in eq. 5.2.13. We can obtain the remaining equation needed to calculate U_{22} by noting that

$$U_{11} + U_{12} + U_{22} = U_{11} + n_2\, u(\text{bulk}) \qquad (5.2.14)$$

because both region 1 and region 1 plus 2 represent a bulk phase with one free surface. By combining eqs. 5.2.13, 5.2.10, and 5.2.14, we find the energy of the film is

$$U_{22}(h) = n_2\, u(\text{bulk}) + 2\gamma A - \frac{H}{12\pi}\frac{1}{h^2}A \qquad (5.2.15)$$

since $U_{12} = U_{1'2}$ by symmetry.

The film's energy, therefore, is higher than that of a bulk solution (see eq. 5.2.12). We might have expected this result, but because the h-dependent term is negative, the tendency of the dispersion force to thin the film is perhaps not quite so obvious.

As an application of eq. 5.2.15, we can consider a soap film drawn from an aqueous solution with varying salt content. The charges on the soap molecules create a repulsive double-layer force that is balanced by an attractive force whose main component stems from the dispersion interaction. Because of the balancing effect of these forces, the film can reach a meta-stable equilibrium at some distance h, where the net force is zero. By changing the salt concentration c_o in the bulk, we can vary the double-layer force. Figure 5.8 shows an experiment

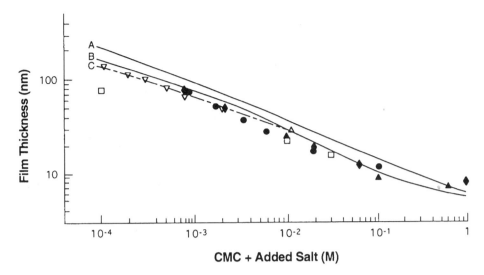

Figure 5.8
Measured thickness of soap films as a function of the electrolyte concentration in the aqueous phase in contact with the film; symbols represent measurements of different researchers. The lines are a theoretical fit, considering that the repulsive double-layer interaction, with surface potential Φ_0, is balanced by a combination of a van der Waals interaction $H = 5 \times 10^{-20}$ J and a hydrostatic pressure P_h. Curve A, $\Phi_0 = \infty$, $P_h = 25.5$ N/m^2; curve B, $\Phi_0 = 65$ mV, $P_h = 0$; curve C, $\Phi_0 = 65$ mV, $P_h = 25.5$ N/m^2. (K. Lyklema and K. J. Mysels, *J. Am. Chem. Soc.* **87**, 2539 (1965).)

plot of h versus c_o. It also confirms the prediction of how the force varies with h, which we can evaluate by using eq. 5.1.28 and 5.2.15 ($H = 3.7 \times 10^{-20}$ J, the Hamaker constant for water at $T = 300$ K). A priori theory does not provide a quantitatively accurate prediction, but it clearly explains the qualitative trends. By adjusting Φ_{do} or H, we can get quantitative agreement for a large range of values of h and c_o'.

5.2.3 The Presence of a Medium Between Two Interacting Particles Modifies the Magnitude of the Hamaker Constant

In most applications of dispersion forces to colloidal problems, interaction between two particles occurs across a liquid medium; an example such as the soap film in Figure 5.6 is really an exception. In an inhomogeneous situation (particle 1–medium–particle 2), dispersion interactions occur between all the constituents. They substantially reduce the effective interaction relative to the case with a vapor gap shown in Figure 5.5. The effective dispersion force may even become repulsive rather than attractive, for reasons that will become clear later.

In keeping with our general approach, we can consider two half-planes of media 1 and 3 separated by a liquid 2 as shown in Figure 5.9. By analogy with eq. 5.2.13, the total energy becomes a sum of six terms

$$\mathbf{U}_{123}(h) = \mathbf{U}_{11} + \mathbf{U}_{33} + \mathbf{U}_{13}(h) + \mathbf{U}_{22}(h) + \mathbf{U}_{12}(h) + \mathbf{U}_{23}(h) \tag{5.2.16}$$

The sum of the first three terms follows from eq. 5.2.10 by a suitable change of indices. The fourth term, $\mathbf{U}_{22}$, is the same as in eq. 5.2.15. Finally, we can obtain $\mathbf{U}_{12}(h)$ by combining eqs.

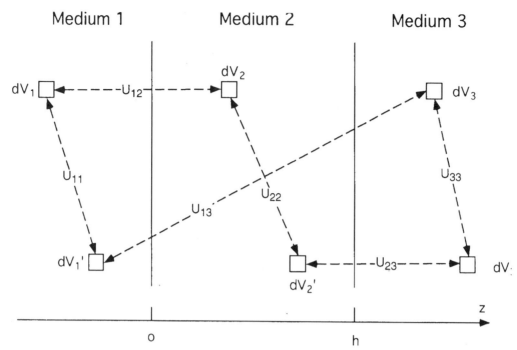

Figure 5.9
Interaction of bulk phases 1 and 3 across a film of phase 2 with thickness h. The interaction energy $\mathbf{U}_{123}(h)$ is obtained by summing the contributions $\mathbf{U}_{ij}(i, j = 1,2,3)$ as in eq. 5.2.14.

5.2.12, 5.2.14, and 5.2.15 so that

$$\mathbf{U}_{12}(h) = -\frac{H_{12}}{12\pi}\left(\frac{1}{\delta^2} - \frac{1}{h^2}\right)A \qquad (5.2.17)$$

with an analogous expression for $\mathbf{U}_{23}(h)$. If we combine all these expressions, $\mathbf{U}_{123}(h)$ reduces to

$$\mathbf{U}_{123}(h) = n_1 u_1(\text{bulk}) + n_2 u_2(\text{bulk}) + n_3 u_3(\text{bulk}) + (\gamma_1 + \gamma_3 + 2\gamma_2)A$$

$$-\frac{A}{12\pi}(H_{13} + H_{22} - H_{12} - H_{23})\frac{1}{h^2} - \frac{A}{12\pi}(H_{12} + H_{23})\frac{1}{\delta^2}$$

$$= \mathbf{U}_{123}(\text{bulk}) + (\gamma_{12} + \gamma_{23})A - \frac{H_{123}}{12\pi}\frac{1}{h^2}A \qquad (5.2.18)$$

As in eq. 5.2.11, the interfacial tension γ_{ij} is

$$\gamma_{ij} = \frac{1}{12\pi}\frac{1}{\delta^2}\left\{\frac{1}{2}(H_{ii} + H_{jj}) - H_{ij}\right\} \qquad (5.2.19)$$

and the effective Hamaker constant H_{123} is

$$H_{123} = H_{13} + H_{22} - H_{12} - H_{23} \qquad (5.2.20)$$

which can be rearranged to reveal how the effective interaction emerges as a difference between the respective dispersion interactions

$$H_{123} = \left\{H_{13} - \frac{1}{2}(H_{11} + H_{33})\right\} - \left\{H_{12} - \frac{1}{2}(H_{11} + H_{22})\right\} - \left\{H_{23} - \frac{1}{2}(H_{22} + H_{33})\right\} (5.2.21)$$

We see in eq. 5.2.21 that H_{123} contains contributions from three terms. Each consists of $H_{ij} - \frac{1}{2}(H_{ii} + H_{jj})$—that is, the Hamaker constant for the interaction between two different media minus the average value of the interaction of the media with themselves. We encountered this combination of interaction constants in eq. 1.4.16 when we described the regular solution theory in Chapter 1. For a symmetric situation in which $1 = 3$, eq. 5.2.21 reduces to

$$H_{121} = H_{11} + H_{22} - 2H_{12} \tag{5.2.22}$$

showing even more clearly the similarity to eq. 1.4.16. The constant H_{121} can be shown to be a positive quantity using eq. 5.2.1, and an attraction always exists between similar particles. We also note in eq. 5.2.22 that $H_{212} = H_{121}$ so that the dispersion contribution to the Hamaker constant is the same, for example, for water across hydrocarbon as for hydrocarbon across water.

Instead of providing a general proof that H_{121} is positive, we introduce an assumption that leads us to some useful simplifications. Using eqs. 5.2.1 and 5.2.3 and the relation

$$H_{12} = \pi^2 C_{12} \rho_1 \rho_2 \tag{5.2.23}$$

for the Hamaker constant, we see that approximately

$$H_{12} \sim \rho_1 I_1 \alpha_1 \rho_2 I_2 \alpha_2 \tag{5.2.24}$$

where 1 and 2 are either similar or different media. When $1 \neq 2$ we see that H_{12} is the geometrical mean of H_{11} and H_{22} so that

$$H_{12}^2 \simeq H_{11} H_{22} \tag{5.2.25}$$

$$H_{121} \simeq (\sqrt{H_{11}} - \sqrt{H_{22}})^2 \tag{5.2.26}$$

and the positive sign of H_{121} becomes apparent.

When all three media illustrated in Figure 5.10 are different, H_{123} can be negative if H_{22} is intermediate between H_{11} and H_{33}. In this case, combining eq. 5.2.21 and the approximation in eq. 5.2.25 gives

$$H_{123} = \{\sqrt{H_{22}} - \sqrt{H_{11}}\}\{\sqrt{H_{22}} - \sqrt{H_{33}}\} \tag{5.2.27}$$

which is negative if $H_{11} > H_{22} > H_{33}$ or $H_{11} < H_{22} < H_{33}$.

This situation occurs most commonly when a liquid (2) wets a solid surface (1) in air (3). Then $H_{33} \simeq 0$ and often $H_{11} > H_{22}$. For liquid helium the Hamaker constant $H_{22} \simeq 6 \times 10^{-22}$ J is unusually small. For a solid with a typical value of $H_{11} \simeq 6 \times 10^{-20}$ J, we can estimate H_{123} as -6×10^{-21} J. This means that a helium film will be attracted to the wall of a vessel and will form a reasonably thick film. Because liquid helium has unusually low viscosity, it tends to flow out of containers through this film.

5.2.4 The Derjaguin Approximation Relates the Force Between Curved Surfaces to the Interaction Energy Between Flat Surfaces

Apart from the complications of effective dispersion interaction in a condensed medium, in practice we also must deal with geometries more complex than those shown in Figures 5.5 and 5.9. The integral of eq. 5.2.5 is more difficult to solve for more complex integration volumes. Although this problem is relatively trivial to solve with present-day computing facilities, we need good analytical approximations to understand the general principles.

Derjaguin realized that when surfaces are uniformly curved, their attractive force relates to the interaction energy between planar surfaces. For example, consider a sphere with radius R centered at a distance $R + h$ from a half-plane, as in Figure 5.10. Let us assume that the radius is much larger than the distance of closest proximity to the surface ($R \gg h$). Then we can calculate the force F acting on the sphere by considering the interaction energy U at two positions displaced by an infinitesimal amount dh, as follows:

$$F = \frac{U(h) - U(h + dh)}{dh} \tag{5.2.28}$$

Figure 5.10
Calculation of the force between a half-space and a sphere of radius R. The calculation considers the interaction between the half-space and the shell of thickness dh, obtained as the difference between a sphere at $R + h$ and $R + h + dh$, where h stands for the shortest distance between the planar surface and the surface of the sphere. The thin shell can be divided into rings of radius r, cross-sectional area $dr\,dh$, volume $2\pi r\,dr\,dh$, and distance $h_1 = h + r^2/2R$, to the planar surface. Integrating over all radii r makes it possible to obtain the force, assuming that it is nonnegligible only when $h_1 \ll R$. The figure displays a cut through the center of the sphere.

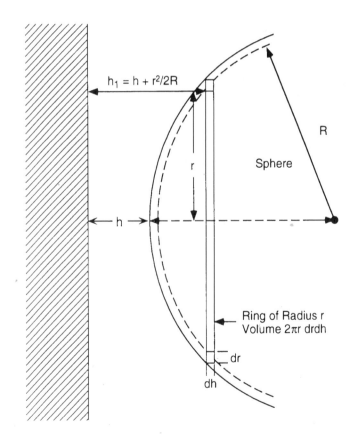

When h/R is small, we need consider only the shell of thickness dh (in the z direction) closest to the planar body; the shell on the back side of the sphere makes a negligible contribution to F because the large separation makes the interaction potential negligible.

As a lateral distance r from the point of closest approach, the shell is at a distance h_1 from the surface. A circular strip of volume $2\pi r\, dr\, dh$ exists in the range h_1 to $h_1 + dh$. By simple geometry $h_1 = h + r^2/2R$ as long as $R \gg r$. (See Figure 5.10.) At a distance h_1 from the surface, eq. 5.2.8 gives the interaction energy per unit volume $V_{12}(h)$. As in eq. 5.2.28, we obtain the total force by summing over all the circular strips — that is, by integrating over r so that

$$F_{12} = \int_0^R 2\pi r\rho V_{12}(h_1)dr$$

$$= 2\pi R\rho \int_h^\infty V_{12}(h_1)dh_1 \qquad (5.2.29)$$

using the geometrical relation between h_1 and r. We have extended the upper limit in the integration because we assume that $V_{12}(h)$ is negligible for h_1 of order R.

Let us compare eq. 5.2.29 with the expression for the interaction $\mathbf{U}_{12}$ between two planar surfaces, as given in eq. 5.2.7. The final integration step in eq. 5.2.5 amounts to an integration over planar sheaths of thickness dz_2 so that

$$\mathbf{U}_{12}(h) = \text{area} \int_h^\infty \rho V_{12}(z)dz \qquad (5.2.30)$$

with $V_{12} = H_{12}/(\pi\rho_2 z_2^{-3})^{-1}$ as in eq. 5.2.8. The integrals in eqs. 5.2.29 and 5.2.30 are the same, except for the integration variable, and

$$F_{12}(h) = 2\pi R \frac{\mathbf{U}_{12}(h)}{\text{area}} \qquad (5.2.31)$$

as long as radius R is large enough: that is, R should be much larger than the range of $V_{12}(h)$.

Equation 5.2.31 demonstrates how the force operating between a sphere and a plane relates to the interaction between two planar surfaces. The derivation used to form eq. 5.2.29 makes use of an additivity assumption, but we assume nothing particular about the potential $V_{12}(z)$ except that the range is short relative to R. In fact, the result has a far-reaching validity. In a general application of the Derjaguin approximation we should replace the energy $\mathbf{U}(h)$ in eq. 5.2.31, as well as in eqs. 5.2.31 and 5.2.32 below, with the free energy $\mathbf{G}(h)$ (or $\mathbf{A}(h)$). A relation equivalent to eq. 5.2.31 exists for the interaction between two crossed cylinders with radii R_1 and R_2

$$F(h) = 2\pi\sqrt{R_1 R_2}\frac{\mathbf{U}(h)}{\text{area}} \qquad (5.2.32)$$

As we will see in Chapter 6, this relation provides the basis for interpreting the results obtained by the surface forces apparatus.

For two spheres, a derivation along the same lines results in

$$F(h) = 2\pi \frac{R_1 R_2}{R_1 + R_2} \frac{\mathbf{U}(h)}{\text{area}} \qquad (5.2.33)$$

For short separations, $h \ll R_1, R_2$, we can integrate eq. 5.2.33 to obtain the dispersion interaction energy between the spheres using $\mathbf{U}(h)$ as in eq. 5.2.7 and

$$\mathbf{U}_{\text{sphere}}(h) = -\frac{R_1 R_2}{6(R_1 + R_2)} H_{12} \frac{1}{h} \qquad (h \text{ small}) \qquad (5.2.34)$$

Thus, a slow variation of the interaction with distance emerges in clear contrast to the interaction for spheres at long separations, $h \gg R_1, R_2$, where the R^{-6} molecular dependence of the force enters directly and

$$\mathbf{U}_{\text{sphere}}(h) = -\left(\frac{4}{3}\right)^2 R_1^3 R_2^3 H_{12} (h + R_1 + R_2)^{-6} \quad (h \text{ large}) \qquad (5.2.35)$$

5.2.5 The Lifshitz Theory Provides a Unified Description of van der Waals Forces Between Colloidal Particles

In his theory of nonideal gases, van der Waals realized that some attractive interactions operate between atoms or molecules. He also noted that these attractions establish the gas–liquid equilibrium. As we have seen, several mechanistic sources contribute to these attractive forces. Collectively, however, they are called van der Waals forces.

Sections 5.2.3 and 5.2.4 discussed the dispersion interaction (which usually provides the main contribution to the van der Waals force) in some detail. However, in molecular systems, polar (electrostatic) interactions also make substantial contributions to the van der Waals force. Even though we can derive the electrostatic interactions for isolated pairs of molecules, major difficulties arise if we want to calculate the net force in a condensed state.

As noted in Section 3.4, the dipole–induced dipole Debye interaction is nonadditive, so we cannot simply integrate an expression like eq. 5.2.3 to obtain the total force. Similarly, the dipole–dipole interaction depends on the relative orientation of the two molecules, and in a liquid we must average over the rotational motion. Because we cannot average separately for each pair of molecules, the dipole–dipole Keesom interaction is effectively nonadditive as well.

The pronounced many-body character of the electrostatic component to the van der Waals force between colloidal particles compels us to look for a different approach. For electrolyte solutions we solved this dilemma by describing the medium as a dielectric. Not surprisingly, the same solution applies in the present case, but going from a dielectric picture to actually calculating the force turns out to be much more difficult.

A group of Russian physicists led by Eugene Lifshitz solved this problem in the 1950s. To give a detailed account of the Lifshitz theory would be far beyond the scope of this text, but it represents a continuum electrodynamics point of view in which each medium is characterized solely by its frequency-dependent dielectric permittivity $\varepsilon_r(\omega)$ (cf. Section 3.5).

Although the Lifshitz theory and the Hamaker theory use very different approaches, the net results are similar. The same molecular interactions are operating, and the continuum description merely expresses them in a different way. The Lifshitz theory gives the same distance dependence of the interactions. For example, eq. 5.2.7 remains valid, although it requires a different interpretation of the Hamaker constant.

For two similar bodies (1) interacting across a medium (2), the approximate Lifshitz expression for the Hamaker constant is

$$H_{121}(\text{Lif}) = \frac{3}{4}\,kT\left\{\frac{\varepsilon_1(0) - \varepsilon_2(0)}{\varepsilon_1(0) + \varepsilon_2(0)}\right\}^2 + \frac{3\hbar}{4\pi}\int_{kT/\hbar}^{\infty}\left\{\frac{\varepsilon_1(i\omega) - \varepsilon_2(i\omega)}{\varepsilon_1(i\omega) + \varepsilon_2(i\omega)}\right\}^2 d\omega \qquad (5.2.36)$$

In this equation the first term on the right-hand side derives from the electrostatic interactions (dipole–dipole or dipole–induced dipole). The static (zero-frequency) dielectric constant determines its magnitude, and the factor kT indicates that the term is largely entropic. The fact that $(\varepsilon_1 - \varepsilon_2)^2/(\varepsilon_1 + \varepsilon_2)^2$ cannot exceed unity sets an upper limit of $\frac{3}{4}kT$ on the electrostatic contribution to the Hamaker constant. At $T = 300\,\text{K}$, for example, $\frac{3}{4}kT \simeq 3 \times 10^{-21}\,\text{J}$, which is an order of magnitude smaller than the dispersion contributions to H_{121}. Cancellations that occur in the dispersion part of H_{121} make the polar interaction more significant, but it seldom constitutes the dominant factor.

Dispersion interaction dominates the second term in eq. 5.2.36. Its dependence on the dielectric constant at imaginary frequencies may seem odd, but these imaginary frequencies are only the result of a mathematical trick. Since $\varepsilon(\omega)$ is an even function of ω, a series expansion contains only even powers of ω, which makes $\varepsilon(i\omega)$ real.

To see how the general formula of eq. 5.2.36 connects to the simpler Hamaker theory we presented, let us concentrate on the second term in the equation and consider an explicit example. We can relate the relative dielectric permittivity ε_r to the molecular polarizability α by the Clausius–Mossotti formula

$$\frac{\varepsilon_r - 1}{\varepsilon_r + 2} = \frac{\alpha\rho}{3\varepsilon_o} \qquad (5.2.37)$$

where ρ represents the number density of molecules. This formula is valid if α is not too large.

In the optical frequency range, where ε_r approximately equals the square of the refractive index, ε_r lies in the range of 2–3 for most substances. According to eq. 5.2.37, this means that $\alpha\rho(3\varepsilon_o)^{-1}$ will be in the range of 0.2–0.4. From this equation we also find that

$$\varepsilon_r = \frac{1 + 2\alpha\rho/(3\varepsilon_o)}{1 - \alpha\rho/(3\varepsilon_o)} \qquad (5.2.38)$$

We can use this result in the integral of the second term in eq. 5.2.36.

Let us consider the specific case when the intermediate medium 2 is a vacuum so that $\varepsilon_2 = 1$ at all frequencies. Then

$$H_{11}(\text{disp}) = \frac{3\hbar}{4\pi} \int_{kT/h}^{\infty} \left\{ \frac{\alpha(i\omega)\rho/\varepsilon_o}{2 + \alpha(i\omega)\rho/(3\varepsilon_o)} \right\}^2 d\omega$$

To a reasonable approximation we can neglect $\alpha(i\omega)\rho/(3\varepsilon_o) < 0.4$ in the denominator relative to 2 and

$$H_{11}(\text{disp}) \simeq \frac{3(\rho/\varepsilon_o)^2}{16\pi} \int_{kT}^{\infty} \alpha^2(i\omega)d(\hbar\omega) \qquad (5.2.39)$$

Using expression 5.2.2 for $\alpha(iw)$, we can obtain the ω dependence in the integrand of this equation. First we note that the difference between $\alpha(i\omega)$ at imaginary frequencies and $\alpha(\omega)$ is simply that the sign of the ω^2 term in the denominator is changed. Therefore, the imaginary frequencies in the Lifshitz expression are due to mathematical convenience, and we should not seek a physical significance for them.

With the explicit expression for $\alpha(i\omega)$, integrating eq. 5.2.39 becomes straightforward, although it remains cumbersome. The net result is that within the approximation, $\alpha\rho/(3\varepsilon_o) \ll 1$ is the dispersion contribution to the Hamaker constant

$$H_{11}(\text{disp}) \simeq \frac{1}{24} \left(\frac{\rho}{\varepsilon_o} \right)^2 \sum_{n,k} \frac{|\langle n|\bar{m}|0\rangle|^2 |\langle 0|\bar{m}|k\rangle|^2}{(E^n - E^o + E^k - E^o)} \qquad (5.2.40)$$

If we specify eq. 5.2.1 to two molecules of the same type and use the relation $H_{11} = \pi^2\rho^2 C_{11}$, we again find eq. 5.2.40 for the Hamaker constant.

This calculation demonstrates that the simple Hamaker theory can account for the dispersion contribution to the force if excitation energies are high relative to kT and molecular polarizabilities are moderate. For example, we must use the more elaborate Lifshitz theory for metals because of their high polarizability, but the Hamaker theory often suffices for insulators.

Figure 5.11 shows how $\varepsilon(iv)$ and the integrand in eq. 5.2.36 vary with frequency for water and hydrocarbon. Similarities between the values of n^2 for water and hydrocarbon lead to substantial cancellations in the dispersion interaction, although this is not the case for the zero-frequency term. Thus,

Figure 5.11
Frequency dependence of $\varepsilon_r(iv)$ for water and hydrocarbon. The dispersion contribution to the Hamaker constants water–hydrocarbon–water or hydrocarbon–water–hydrocarbon depends on the difference, $\varepsilon_w - \varepsilon_{hc}$, as in the integrand of eq. 5.2.36. The area under the dashed curve represents the integral of eq. 5.2.6. Note the logarithmic scale in the figure and that $2\pi v = \omega$ and furthermore, $\int f(\omega)d\omega = \int \omega f(\omega)d\ln\omega$. (J. Israelachvili, *Intermolecular and Surface Forces*, 2nd ed., Academic Press, London, 1992, p. 189.)

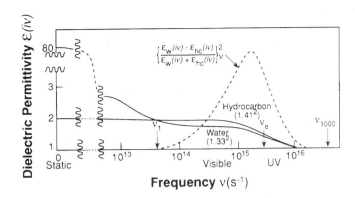

from the electrostatic interaction term. Hamaker constants H_{11} and H_{123} are shown in Table 5.1 for a range of insulators. Metals that interact strongly have larger values of H because of their special conductance and optical properties.

The Lifshitz theory also brings out the so-called retardation effect. The relativistic origin of this effect can be ascribed qualitatively to the finite velocity of propagation for electromagnetic interactions. When the distance between particles is of the same order as the wavelength of the excitations $E_n - E_0$ involved in the dispersion interaction, the distance dependence changes. Interactions become more long-ranged by one power in the inverse distance. Within this limit, the interaction potential

TABLE 5.1 Values of Hamaker Constants $(\times 10^{+20}/\text{J})$ for Some Common Combinations of Materials M

Material	M|Air|M	M|Water|M	M|Water|Air	M|Air|Water
Water	3.7	0.0	0.0	3.7
Alkanes				
$n = 5$	3.8	0.3	0.15	3.6
$n = 6$	4.1	0.4	0.0	3.8
$n = 10$	4.8	0.5	−0.3	4.1
$n = 14$	5.1	0.5	−0.5	4.2
$n = 16$	5.2	0.5	−0.5	4.3
Fused quartz	6.5	0.8	−1.0	4.8
Fused silica	6.6	0.8	−1.0	4.8
Sapphire	16.0	5.0	−3.8	7.4
Polymethyl methacrylate	7.1	1.1	−1.3	5.0
Polystyrene	6.6	1.0	−1.1	4.8
Polyisoprene	6.0	0.7	−0.8	4.6
Polytetrafluoroethylene	3.8	0.3	0.1	3.7
Mica (green)	10.0	2.1		

between two half-spaces is proportional to h^{-3} rather than h^{-2}. Retardation effects seldom concern us, however, because the interactions are very weak at the distances at which they become an important factor.

5.3 Electrostatic Interactions Generate Attractions by Correlations

In Section 5.1, we analyzed the electrostatic double-layer force on the basis of the mean field Poisson–Boltzmann equation and found that a repulsion always exists between similar surfaces. In Section 5.2, on the other hand, we concluded that correlations between electrons of separate molecules always give rise to an attraction between similar surfaces/particles. These two basic forces dominate in many colloidal systems, but even within the framework of electrostatic interactions, more subtle effects can be of major importance under some circumstances. To appreciate the significance of these effects, we must go beyond the description in which we treat electrostatic charge distributions in terms of their mean value.

Charges in the colloidal system are localized to ionic or dipolar groups, and the local charge varies strongly within each configuration of ions and molecules. An averaging due to the molecular motions leads to a more uniform, mean field, charge distribution. In general, motions between interacting molecules/ions are correlated, and if we consider only the mean charge distribution, we leave out an important physical effect due to these correlations.

The previous section showed how correlations between quantum electronic motions give rise to a strong attractive force. Through the Lifshitz formulation, we also briefly discussed the attractive force component that derives from correlation between molecular dipoles in the bulk medium. The Lifshitz theory presupposes homogeneous media and neglects special interfacial effects because they do not contribute to the leading asymptotic interaction term. However, force components operating at an intermediate range can be highly relevant in determining the physical properties of the colloidal system. We discuss three cases in which correlations between classical, thermally excited, degrees of freedom can contribute a significant attractive component to the force.

5.3.1 Ion Correlations Can Turn the Double-Layer Interaction Attractive

In the Poisson–Boltzmann description of the interaction between two similar planar charged surfaces, the mean ion distribution is laterally homogeneous. In reality, the charges are localized to individual ions, and at each instant (or for each configuration), both the lateral and the transverse distributions of the net charge are very inhomogeneous.

Consider one half of the system containing one charged wall, the neutralizing counterions, and possibly some additional electrolyte. Gauss's law says that a zero net electric field passes through the plane separating the two halves. However, lateral inhomogeneity causes each configuration to produce instantaneous fields that contain both negative and positive regions at the mid-plane as illustrated in Figure 5.12. Thus, the second half of the system experiences external electric fields and potentials that vary in space and it can respond by adjusting the current charge distribution to obtain an energy lower than for the mean field. This response implies an attractive contribution to the force. The mechanistic analogy with dispersion interaction is clear, but here we are dealing with a response in classical degrees of freedom.

The stronger the interaction between the ions, the easier it is to generate a correlation between them and the stronger the attractive ion correlation force. Aqueous systems have a high dielectric permittivity, but as illustrated in Figure 5.13, when the surface charge density is high and the counterions are divalent, the correlation effect can totally overpower the double-layer repulsion. For highly charged systems and monovalent ions, this effect can contribute a substantial correlation correction to the repulsive double-layer force.

In colloidal systems, this correlation effect has practical importance in a number of situations. Divalent ions like calcium often induce precipitation; the ion correlation effect is not the only possible cause in this case, it often plays a major role. For example, clays do not swell in the presence of calcium. Similarly, it is difficult to account for the properties of black films of charged surfactants without invoking ion correlation effects. The same molecular arrangement is found in concentrated oil-in-water emulsions. For liquids of lower polarity, the ion–ion interactions are stronger, and substantial ion correlation effects appear under more general conditions.

Figure 5.12
The neutralizing counterions are never uniformly distributed outside the negatively charged surface. For each instantaneous configuration of these counterions there will be a field $E(r)$ generated at r in the opposing half plane. On overage $\langle E(r) \rangle = 0$ but $\langle E^2(r) \rangle \neq 0$. The figure illustrates schematically for one particular configuration the electrical field lines, i.e. lines of constant magnitude $|E(\vec{r})|$ and with the arrow pointing in the direction of $E(\vec{r})$.

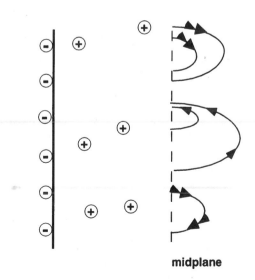

midplane

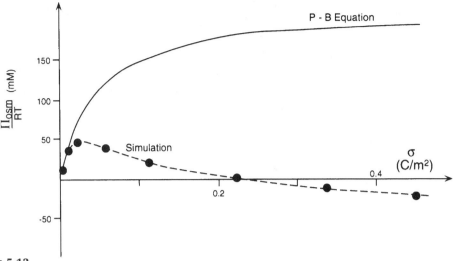

Figure 5.13
Force calculated between
two parallel identical
charged surfaces separated
by a distance $h = 21\text{Å}$ in
a solvent $\varepsilon_r = 80$ containing
only divalent counterions.
The solid line represents
the Poisson–Boltzmann
equation, circles the Monte
Carlo simulations. Force
is expressed in terms of the
osmotic pressure divided
by RT ($T = 298\,\text{K}$),
$1\,\text{mM} \leftrightarrow 2500\,\text{N/m}^2$. The
Monte Carlo simulation
explicitly considers the
instantaneous ion–ion
interactions to account for
the additional attractive
ion–ion correlation
component to the force. For
low surface charge
densities σ, the attractive
component also is
negligible for divalent
counterions, but becomes
substantial for $\sigma \gtrsim 0.03\,\text{C/}$
m^2. Under the given
conditions, the net force is
zero around $\sigma = 0.2\,\text{C/m}^2$,
and for even larger charge
densities, the attractive
component dominates and
the Poisson–Boltzmann
calculation is totally
misleading. (L. Guldbrand,
B. Jönsson, H.
Wennerström, and P. Linse,
J. Phys. Chem. **80**, 2221
(1984).)

5.3.2 Surface Dipoles Correlate to Yield an Attraction

In a colloidal system, the chemical composition of the surface
layer often differs markedly from the bulk of the particle. Two
approaching particles will see each other through these surface
layers. Interaction between surface layers becomes more important, on a relative scale, as the separation between the particles
decreases. The Lifshitz theory includes an asymptotically leading electrostatic bulk correlation term. However, strongly polar
groups carried by the surface layers can cause an additional
dipole–dipole correlation contribution that can dominate at
intermediate to short separations.

Zwitterionic and dipolar amphiphiles have highly polar
surface layers where electrostatic interactions play a dominant
role. This applies to solvent–headgroup as well as headgroup–
headgroup interactions within the aggregate. Consequently, we
should also expect headgroup–headgroup interactions between
aggregates to be important at close approach. As an example, we
consider an amphiphile with a zwitterionic headgroup like a
betain or a phospholipid. On a planar interface, we can resolve
the zwitterion dipolar vector into perpendicular and parallel
components which give rise to somewhat different types of
correlations.

Let us first consider the parallel component. A dipole
oriented along a surface is free to rotate because it has no
intrinsic directional preference. Averaged over all the in-plane
orientations, the net dipole moment vanishes, so the mean
electric field component is zero from the parallel dipole. A
second dipole on an opposing surface can interact favorably
with the source dipole by preferentially adopting an opposite
orientation. The result is an attractive correlation force similar
to the Keesom interaction we discussed in Section 3.3. We can
obtain an order-of-magnitude estimate of this force by assuming
that the dipoles are sufficiently separated on the interface so

that we can treat them as approximately independent. An integration using the Hamaker approach results in

$$\frac{F}{A} = - \frac{\pi \left(\dfrac{\rho m^2}{4\pi\varepsilon_r\varepsilon_o} \right)^2}{kTh^5} \tag{5.3.1}$$

where ρ represents the surface density of dipoles and m the in-plane dipolar component.

In the Litshitz theory, we also obtain the h^{-5} distance-dependence for two interacting thin layers. In fact, for more realistic cases of strong intrasurface correlations the magnitude of the force will change, but the h^{-5} distance-dependence remains valid. Actually, in this case, the force is very sensitive to molecular details and the magnitude can vary by an order of magnitude.

For the perpendicular component of the zwitterionic dipole, the only accessible degree of freedom is a translation along the surface. As a result, the thermal averaging is very different in this case relative to that of the parallel component. When two surfaces approach, their perpendicular dipolar components are oppositely oriented. At first glance, one might be tempted to conclude that this orientation produces a repulsive interaction between the surfaces. While this is true for two dipoles in juxtaposition, electrostatic interaction is long-ranged, and it changes sign when the dipolar orientation and the vector joining the dipoles form the so-called magic angle of 54° (3 cos^2 (54) − 1 = 0). For a uniform distribution of dipoles — as in a mean field description — the repulsive interaction between dipoles close to juxtaposition is canceled exactly by an attractive interaction between dipoles that are laterally farther apart. (See Figure 5.14).

The interaction between the uniform dipolar distribution is zero, so an attractive component appears when we allow for

Figure 5.14
A dipole at a distance h from a plane of oppositely oriented dipoles will sense a repulsion from dipoles located closest on the opposing surface. However, dipoles located laterally further, more than $h \tan 54°$, away give an attractive interaction. For a smeared out uniform dipolar distribution on the surface the net interaction averages to zero.

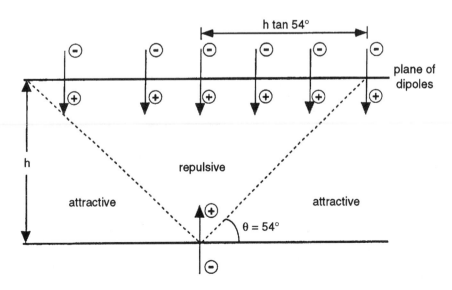

correlations in the lateral displacements of the dipoles on their respective surfaces. If we fix a dipole at some position on one surface, the correlations will fix the dipolar density at the corresponding position on the other surface. For a weakly correlated system in which we can treat the dipoles as independent, the net force is the same as for the parallel dipoles. Equation 5.3.1 applies, provided we let m denote the perpendicular dipolar component.

In the other extreme of very strong intrasurface correlations, as in a lattice of dipoles, we can obtain an attractive interaction by displacing the lattices by half a unit cell. This correlation in only one degree of freedom produces an energy gain that easily exceeds the thermal energy kT. The electric field outside an ordered surface decays exponentially on a length scale given by the lattice spacing a. Consequently, the force is short-ranged, but strong. An approximate expression for the force is

$$\frac{F}{A} \simeq -\frac{\text{const. } m^2}{\varepsilon_r \varepsilon_o a^6} \exp(-h/\lambda) \qquad (5.3.2)$$

where the decay length $\lambda \approx a/6.88$ and the constant is of order 10^2. This force, which often exceeds the dispersion contribution at short range, provides extra cohesion to polar solids. Ultimately, a strongly correlated system also has the asymptotic h^{-5} dependence of the force, but in this case, the asymptotic regime is attained only at long separations where the force is insignificant.

5.3.3 Domain Correlations Can Generate Long Range Forces

In the two previous subsections, we discussed attractive interactions that occur due to correlations between individual molecular degrees of freedom, that is, the positions of ions or the orientations/position of dipolar groups. The same basic electrostatic correlations also can operate with entities of colloidal dimensions rather than molecular ones. Some quantitative differences occur because the length scale changes, and with larger entities, the interaction per degree of freedom is more likely to exceed the thermal energy. For such a mechanism to operate, the surfaces must show inhomogeneities on the colloidal scale and these can be generated in several ways.

Surfactants can adsorb on a surface in the form of a micellar aggregate to an oppositely charged surface. The result might be a surface that is net neutral yet has regions of opposite charge. A similar situation can occur when a highly charged polyelectrolyte adsorbs to partial coverage on an oppositely charged surface. Figure 2.16 illustrates a third possibility: lateral phase separation with domains. In this case, a perpendicular dipole moment is associated with domains that interact laterally by electrostatic forces. Thus, two such surfaces may interact across a medium. A fourth alternative is for the surface

Figure 5.15

A laterally nonuniform charge distribution can arise when a positively charged particle or aggregate adsorbs on a negatively charged surface of a charge density of lower magnitude. Electrical fields are generated outside the surface and these extend out from the surface on the same length scale as the in-plane inhomogeneity as shown by the electrical field lines in the figure.

to form a solid structure of low symmetry. In this case, different domains of the same phase can carry an in-plane dipole moment.

In all these cases, a nonuniform charge or dipolar distribution along the surface gives rise to an electric field in the medium outside the surface. This field varies in sign laterally on the length scale of a domain size and decays in amplitude in the perpendicular direction on the same length scale, as illustrated in Figure 5.15. A second surface can respond to this field to generate correlation attraction just as in the case of the zwitterions. However, because larger entities produce stronger interactions, correlations also can develop in the distance regime of 10–100 nm. On the other hand, kinetic processes are slow, so that correlations may not have time to develop during the timescale of an experiment.

Asymptotically, a formula analogous to eq. 5.3.1 is valid with an h^{-5} distance-dependence. However, a more relevant case is the strongly correlated situation described by eq. 5.3.2. Up to this point, our quantitative description has neglected the influence of a screening electrolyte. This complication is not a major concern for molecular dipoles because the electric fields decay rapidly anyway. With the domain mechanism, however, we are considering a different length scale and screening should have a strong effect. From a model based on in-plane dipolar domains of radius R and a charge distribution $\sigma = \sigma_o \cos \phi$, we can solve the linearized Poisson–Boltzmann equation to obtain an explicit equation for the interaction between two similar surfaces. The calculated force is

$$\frac{F}{A} \simeq -\frac{\sigma^2}{2\varepsilon_r\varepsilon_o}\left\{\exp\left[\left(\kappa^2 + \frac{3.4}{R^2}\right)^{1/2}\frac{h}{2}\right] + \exp\left[-\left(\kappa^2 + \frac{3.4}{R^2}\right)^{1/2}\frac{h}{2}\right]\right\}^{-2} \qquad (5.3.3)$$

The Debye screening length κ^{-1} or the domain radius R—whichever is shorter—determine the range of the force. Figure 5.16 demonstrates this competition between R and κ in determining the range of the interaction. Finding a force like the one represented in eq. 5.3.3 requires rather special conditions, but when these are present, they produce striking effects because the force is very strong and easily dominates over the dispersion force, even at separations that are large on the colloidal scale.

5.4 Density Variations Can Generate Attractive and Oscillatory Forces

In analyzing the double-layer and van der Waals forces at long range, we have treated the medium as a continuum and con-

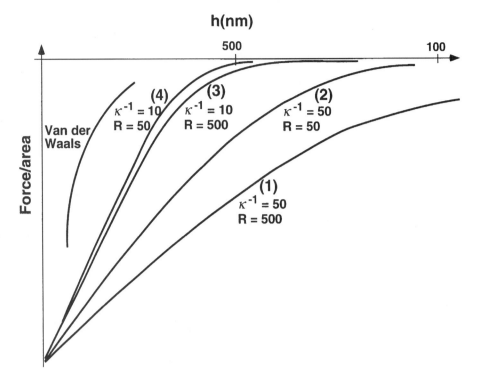

Figure 5.16
The calculated force per area according to eq. 5.3.3 for two planar surface with correlated dipolar domains of radius R. Curves are shown for the four possible combinations of screening lengths $\kappa^{-1} = 10$ nm (~ 1 mM NaCl in water), and $\kappa^{-1} = 50$ nm ($\sim 40 \,\mu$M NaCl) and domain radii $R = 50$ nm and $R = 500$ nm. The similarity between curves (3) and (4) show that the range of the force for $\kappa^{-1} = 10$ nm is determined by the electrolyte screening while for $\kappa^{-1} = 50$ nm the domain size of $R = 50$ nm is small enough to affect the range for the force. For domains of a colloidal size the magnitude of the force is much larger than the van der Waals force at the relevant separation distances.

sidered the surfaces as flat. In reality, the molecularity of the medium should have an effect, and we expect that this factor becomes more important the shorter the separation between the surfaces. Moreover, surfaces or interfaces are never perfectly flat. Here we distinguish between two major types of surfaces/interfaces: a solid surface that has a fixed geometry, whether it is flat, bumpy, or cracked, and a liquidlike surface that adopts whatever form will minimize the free energy, exhibits thermal fluctuations, and can deform in response to an external force. In this section, we will ignore the complications that arise from the flexibility of liquidlike surfaces and focus on the fixed solid surface. However, most of the effects we will discuss also can occur for liquidlike surfaces although they will be modified by the flexibility of the system.

One phenomenon peculiar to the solid system is what we will call the steric energetic effect. Virtually all solid surfaces are rough at the molecular scale. When two such surfaces approach, their roughness prevents them from establishing close molecular contact over an extended area. Even when only an attractive van der Waals force is operating, the adhesion energy is much smaller than we would estimate on the basis of the surface free energy. Clearly, the adhesive force depends on the roughness of the surface. In fact, pieces of steel will adhere strongly to one another if they are sufficiently polished, and the van der Waals force can cause a slightly elastic body like a thin foil of gold to adhere to most surfaces. Some substances, like mica and graphite, cleave naturally to form large planar surfaces, and this property has made them very useful for scientific studies where one wants to eliminate the steric energetic effect. In the Hamaker and Lifshitz theories the media are treated as homogeneous, lacking molecular character. In particular, one makes the assumption that the media have a homogeneous density right up to the surface/interface. This is generally not true and we should expect that this causes deviations from the Lifshitz theory. Below we will discuss four different types of density inhomogeneities and how they affect the force between surfaces.

5.4.1 Molecular Packing Forces Produce Oscillatory Force Curves with a Period Determined by Solvent Size

One assumption made explicitly in the Hamaker theory of dispersion forces and implicitly in the Lifshitz theory is that the density of a liquid remains constant up to the surface or interface. However, a liquid exhibits short range correlations, and the system is inhomogeneous on the molecular scale. At an interface, similar correlations exist, leading to variations in density on the scale of some molecular diameters. Such effects are most pronounced at smooth solid surfaces, where surface perturbation is constant in time and also along the surface. Because short range repulsion effects bear the main responsibility for the structure in a liquid, we can model them with reasonable accuracy by a hard-sphere system. Figure 5.17 shows the calculated density profile in a hard-sphere fluid outside a hard repulsive wall. The density approximately follows a damped oscillation $\sim \cos(2\pi z/\lambda_1) \exp(-z/\lambda_2)$. Here the wavelength λ_1 corresponds closely to a hard-sphere diameter and is relatively insensitive to density. On the other hand, decay length λ_2, which is on the order of a molecular size at liquid densities, varies with density and is not simply related to λ_1.

For a liquid between two solid surfaces, an interference will occur between the two density profiles at sufficiently short separations, as illustrated in Figure 5.18. Separations h that are integral (n) multiples of the length λ_1 — that is, $h = n\lambda_1$ — display an optimal arrangement in which the free energy is at a local minimum. We can see this as a partial layering of the liquid between the surface. Packing problems occur for separation values such as $h \simeq (n + \frac{1}{2})\lambda_1$, and the free energy displays a local maximum. These variations in free energy give rise to an oscillatory force of the approximate form

$$F_{\text{pack}}(h) = \text{const} \cdot \cos\left(\frac{2\pi h}{\lambda_1}\right) \exp\left(\frac{-h}{\lambda_3}\right) \qquad (5.4.1)$$

where the decay length λ_3 closely relates to the decay length λ_2 for the density oscillations and the constant is of the order of the external pressure. For short separations $h < \lambda_1$, solvent molecules cannot enter the gap, and in addition to direct adhesive forces, the packing force corresponds to overcoming the external pressure.

Figure 5.17
Schematic of a hard-sphere
liquid adjacent to a smooth
solid substrate and the
calculated density
distribution ρ as a function
of molecular diameter σ
away from the wall. The
density profile shows a
damped oscillation that
decays with a form $\cos(z/\lambda_1)\exp(-z/\lambda_2)$. (J.
Israelachvili,
*Intermolecular and Surface
Forces*, 2nd ed., Academic
Press, London, 1992, p.
262.)

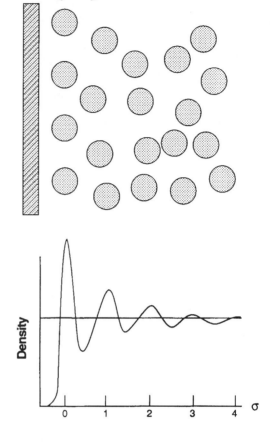

Figure 5.18
Schematic of a hard-sphere
fluid contained between
two smooth solid surfaces
separated by a distance
h. Separation by $h = n\lambda$
lead to optimal molecular
packing and a local
minimum in the free
energy; other separations
lead to an increase in free
energy with a maximum
occurring at $h = (n + 1/2) > 1$. (J. Israelachvili,
*Intermolecular and Surface
Forces*, 2nd ed., Academic
Press, London, 1992, p.
262.)

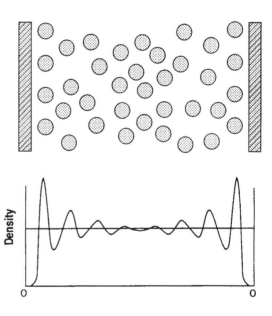

Figure 5.19
Measured force between the
mica surfaces in
octamethylcyclotetrasiloxane,
a nearly spherical molecule
with a diameter ~ 10 Å. Stable
regions in the surface forces
apparatus measurements (●)
are shown by continuous
curves, while unstable regions
are shown in dashed curves.
Inset: peak-to-peak
amplitudes of the oscillations,
$P_n - Q_n$, as a function of
D. The exponential decay
length is 1.0 nm. (R. G. Horn
and J. N. Israelachvili, *J.
Chem. Phys.* **75**, 1403 (1981).)

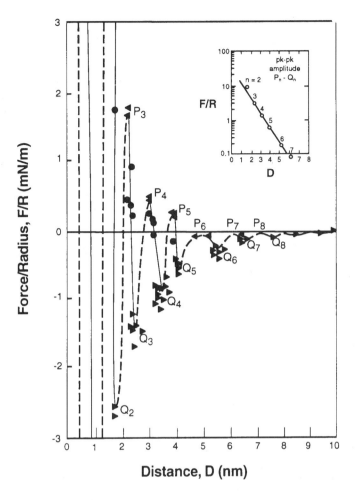

Figure 5.19 shows an experimental example of the packing force operating in a liquid consisting of approximately
spherical molecules. The oscillatory force found in this
example can be measured up to $n = 10$ periods. Nonspherical
molecules show a less pronounced but still measurable layering
effect. Straight-chain hydrocarbons can display an oscillatory
force where the length λ_1 is independent of chain length and
corresponds to the width of the molecule. Chain branching
tends to destroy these oscillations. So will surface roughness
that is due to an uneven solid surface or to capillary waves on
a liquid surface.

Even though the integrated effect of these packing forces
is small, we can encounter substantial kinetic barriers. Usually
the most sluggish step is to push away the last layer of solvent
on approach. Similarly, to pull surfaces away from adhesive
contact, we must overcome a barrier arising from the layering
effect as well as the attractive van der Waals force.

5.4.2 Capillary Phase Separation Yields an Attractive Force

In Section 2.5, we discussed how surface free energy could trigger the condensation of a liquid in the cracks and clefts on a particle's surface at a vapor pressure below saturation. Capillary condensation can be seen as a manifestation of a more general phenomenon that we call capillary phase separation. When two surfaces approach, they create an arrangement similar to that of a crack in a single particle. The contribution from the surface free energy can be substantial for the medium in the gap and may induce a transition to a phase with lower surface energy. Once the new phase is formed, it will give rise to a strong attractive capillary force. The most well-known manifestation of such a capillary force is the sand castle. This cannot be built underwater and collapses when it dries. Surface tension at the air–water interface keeps the castle standing when it is built from moist sand with a bridge of water between the sand grains.

Although gas-to-liquid is the most familiar transition in capillaries, we also can have the reverse, a liquid-to-gas transition, and also, for binary liquids, a liquid-to-liquid phase separation as illustrated in Figure 5.20. Even transitions to liquid

Figure 5.20
Schematic illustration of five different cases of capillary induced phase transitions leading to the formation of a lens of a second phase between a sphere and a plane in a bulk medium of another phase: (a) capillary condensation of water between hydrophilic surfaces (as in cracks in rock and other hydrophilic materials); (b) capillary vaporization of water between strongly hydrophobic surfaces (such as the material used in boiling chips); (c) separation of water from moist oil between hydrophilic surfaces (common in oil reservoirs); (d) oil separation from oil-contaminated-water (a challenge in cleaning processes); and (e) a lamellar liquid crystal separating from an isotropic sponge phase (see also Figure 5.23).

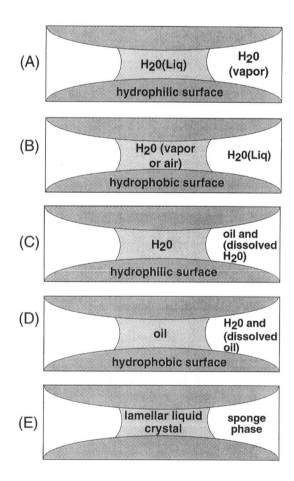

crystalline phases have been observed. Let us first analyze the force between two planar surfaces of area A that approach in a bulk medium α. A surface free energy $\gamma_{s\alpha}$ is associated with the surfaces. Now assume that molecules of medium α can form a phase β (vapor, liquid, or liquid crystalline). The surface free energy of this phase is $\gamma_{s\beta}$ per unit area. When $\gamma_{s\alpha} > \gamma_{s\beta}$, the surface free energy is decreased by

$$\Delta G_S = 2A(\gamma_{s\beta} - \gamma_{s\alpha}) = -2A\gamma_{\alpha\beta}\cos\theta \qquad (5.4.2)$$

if the β-phase is formed in the gap. The second equality follows from the Young equation (2.1.13). However, because α is the stable bulk phase, creating the new phase also creates an increase $G_{\beta\alpha}$ of free energy per unit volume. The net change in free energy is

$$\Delta G(h) = 2A(\gamma_{s\beta} - \gamma_{s\alpha}) + hAG_{\beta\alpha} \qquad (5.4.3)$$

and at the separation

$$h = -2(\gamma_{s\beta} - \gamma_{s\alpha})/G_{\beta\alpha} \qquad (5.4.4)$$

$\Delta G = 0$, and a transition to the β-phase can occur in the gap. At this separation, we find the onset of an attractive force

$$\frac{F}{A} = -\frac{\partial \Delta G(h)}{\partial h}\bigg/ A = -G_{\beta\alpha} \qquad (5.4.5)$$

This force is independent of the separation between the planar parallel surface for h less than the critical value given by eq. 5.4.4. This can be understood from the fact that once the β-phase is formed in the gap a change in separation dh causes a transformation β to α of a constant volume Adh irrespective of the value of h.

In the case of a liquid condensing from its vapor α at pressure p, we already calculated the free energy term $G_{\beta\alpha}$ in our discussion of nucleation in Section 2.3.3 as

$$G_{\beta\alpha} = -\frac{kT}{V_L}\ln(p/p_o) \qquad (5.4.6)$$

Here p_o is the saturation pressure and V_L the molecular volume.

Problem: Calculate the force curve between two planar surfaces where $(\gamma_{s\beta} - \gamma_{s\alpha}) = 10\,\text{mJ/m}^2$ in water vapor at 90% relative humidity and 25°C.
Solution: For water $V_L = 30 \times 10^{-30}\,\text{m}^3$ so the force is

$$\frac{F}{A} = \frac{kT}{V_L}\ln 0.9 = \frac{4.1 \times 10^{-21}}{30 \times 10^{-30}}\ln 0.9 \simeq -1.3 \times 10^7\,\text{N/m}^2$$

for

$$h < \frac{2 \times 10^{-2}}{1.3 \times 10^7} \simeq 1.8\,\text{nm}$$

Problem: Compare the strength of the capillary phase separation force of the previous problem with the van der Waals force at a separation of 1 nm. Assume that the Hamaker constant is $1 \times 10^{-20}\,\text{J}$.

Solution: The van der Waals force per unit area between planar surfaces is

$$\frac{F}{a} = -\frac{H}{6\pi h^3} = -\frac{1 \times 10^{-20}}{6\pi \times 10^{-27}} = -5 \times 10^5\,\text{N/m}^2$$

which is an order of magnitude smaller than the force $F/a \simeq -1.3 \times 10^7\,\text{N/m}^2$ found in the previous problem.

Applications usually involve particles that have curved surfaces. Previously, we dealt with this complication by using the Derjaguin approximation. It is far from evident from the derivation presented in Subsection 5.2.4 that we can use this approximation when we have two phases present in the gap. However, our detailed derivation shows its validity as long as equilibrium is maintained. The derivation also will illustrate the physical mechanisms involved and permit us to discuss nonequilibrium effects.

Consider a spherical particle of radius R_0 approaching a planar surface of the same material, as illustrated in Figure 5.21. Phase β now forms a lens with radius R_m around the point of closest approach, h_0. The β-phase radius will greatly exceed h_0, and the dominant interfacial energy is due to the contact between the lens and the solid surfaces. On forming the lens, we have a change in surface free energy

$$\Delta G_S \simeq 2\pi R_m^2 (\gamma_{s\beta} - \gamma_{s\alpha}) + 2\pi R_m h_0 \gamma_{\alpha\beta} \qquad (5.4.7)$$

ignoring the curvature of the lens when calculating areas. Typically, the radius of the lens is much larger than separation h_0, and we can neglect the second term, which is much smaller than the first. The excess free energy of the lens is given

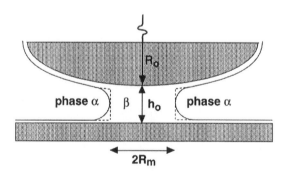

Figure 5.21
A capillary induced separation of phase β from a bulk phase α when a spherical particle of radius R_0 approaches a planar surface with a distance of closest approach of h_0. The phase β forms a concave lens of radius R_m. Note that the phase β typically also wets the surfaces even in bulk solution.

phase α   β h_0 phase α

$2R_m$

following the method of eq. 5.4.3

$$\Delta G_{tot}(h_0) \simeq 2\pi R_m^2(\gamma_{s\beta} - \gamma_{s\alpha}) + \pi R_m^2(h_0 + R_m^2/4R_0)G_{\beta\alpha} \qquad (5.4.8)$$

consisting of a negative surface free energy term and a positive contribution from transforming the volume $\pi R_m^2 \, (h_0 + R_m^2/4R_0)$ of the lens from the α to the β phase. An optimal value of the radius of the lens exists which varies with separation h_0. Taking the derivative and applying the condition $R_0 \gg R_m \gg h_0$, the optimum value is

$$R_m = \sqrt{R_0 R_K (1 - h_0/R_K)} \qquad (5.4.9)$$

where R_K is the Kelvin radius we encountered when discussing nucleation in Section 2.3.3.

$$R_K \equiv -2(\gamma_{s\beta} - \gamma_{s\alpha})/G_{\beta\alpha} = \frac{2\gamma_{\alpha\beta} \, \cos\theta}{G_{\beta\alpha}} \qquad (5.4.10)$$

Now we can evaluate the force in a straightforward manner by taking the derivative of eq. 5.4.8 with respect to h_0

$$F = -\frac{\partial \Delta G_{tot}}{\partial h_0} = 4\pi R_0(\gamma_{s\beta} - \gamma_{s\alpha})\left(1 - \frac{h_0}{R_K}\right)$$
$$= -2\pi R_0 R_K G_{\beta\alpha}(1 - h_0/R_K) \qquad (5.4.11)$$

In fact, we also can obtain this result from eq. 5.4.5 and the Derjaguin approximation, which states that the force between a sphere R_0 and a plane equals $2\pi R_0$ times the interaction free energy per unit area between two planar surfaces. Thus, from eq. 5.4.3

$$F(h_0) = 2\pi R_0\{2(\gamma_{s\beta} - \gamma_{s\alpha}) + h_0 G_{\beta\alpha}\}$$
$$= 4\pi R_0(\gamma_{s\beta} - \gamma_{s\alpha})(1 - h_0/R_K) \qquad (5.4.12)$$

exactly as in 5.4.11. Equation 5.4.11 applies when $h_o < R_K$ and it predicts a linearly varying attractive force. Figure 5.22 shows an example of such an attractive force due to a capillary phase transformation from a bicontinuous microemulsion water-oil-AOT to an aqueous phase. Since the mica surface is hydrophilic the mica–water surface free energy is lower than the mica–microemulsion surface free energy. However, the microemulsion contains large aqueous domains that most probably wet the mica surface. Consequently the difference in surface free energy is small so the force is relatively weak.

Figure 5.23 illustrates a more complex example of the interplay between the force and a capillary induced phase transition. In this case the bulk phase is an isotropic solution called the sponge phase, which will be further discussed in Chapters 6 and 11. This phase contains a disordered network of bilayers. Between the confining surfaces the bilayers transform into an ordered stack. Thus the primary effect is the observation

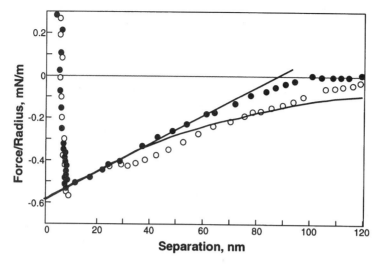

Figure 5.22
The measured force between two cylindrical mica surfaces in a system with a microemulsion bulk phase. The attractive force is due to a capillary-induced formation of a lens of an aqueous phase between the surfaces. Filled dots represent the force measured on approach while the unfilled dots represent the force measured on separation of the surfaces. The straight line is calculated using eq. 5.4.11 while the curved line follows eq. 5.4.13 using the same parameters. The repulsion observed at short separations is due to the double-layer force between the charged mica surfaces.
(From P. Petrov, U. Olsson and H. Wennerström *Langmuir 13*, 1000, 1997).

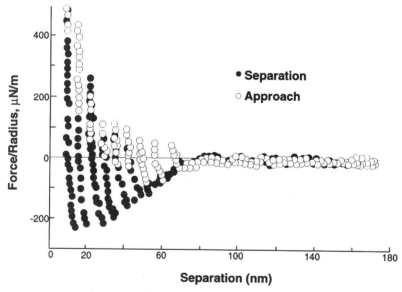

Figure 5.23
The measured force between two curved mica surfaces in an aqueous solution of a nonionic surfactant forming a sponge phase (see Chapter 6). In this case a lamellar phase is forming a lens between the surfaces. The successive repulsive force barriers are due to the elimination of lamellae one by one in the gap. On separation an attractive force appears that is consistent with eq. 5.4.11. (From P. Petrov, S. Miklavcic, U. Olsson and H. Wennerström *Langmuir 12*, 1000, 1996).

of a series of force barriers due to successive displacement of one bilayer after the other between the mica surfaces. However, on separating the surfaces an attractive force appears and the minima closely fall on a straight line as predicted by eq. 5.4.11. Also in this case the attractive force is weak indicating only a small difference between the surface energies for the mica-sponge phase and the mica-lamellar phase.

Our analysis of the force assumed that equilibrium was maintained at all separations. However, some processes can be slow to equilibrate. First, we know that nucleation of a separate phase can be sluggish indeed. Second, as the surfaces approach one another, material must be transported to and from the slit to establish the difference in concentrations. Thus, we can easily find time-dependent forces. A case of particular importance is when two surfaces that have been in contact for some time are quickly separated. As a first approximation, we can assume that no material is transported on a short timescale and that the volume of the lens stays constant. Taking the derivative of the free energy at constant lens volume yields a force

$$F = 4\pi R_0 (\gamma_{s\beta} - \gamma_{s\alpha}) \left\{ 1 - \left(1 + \frac{R_K^2}{h_0^2} \right)^{-1/2} \right\} \qquad (5.4.13)$$

For short separations, this force is the same as in eq. 5.4.11, but it is more long-ranged. The lower line in Figure 5.22 is a fit to eq. 5.4.13 to the force measured on separating the surfaces. This nonequilibrium force is clearly more long-ranged than the equilibrium one.

5.4.3 A Non-Adsorbing Solute Creates an Attractive Depletion Force

In a colloidal system, we often have a solute in the liquid medium. The solute can either adsorb on the surface—and consequently alter its surface properties significantly—or it can have less affinity for the surface than the solvent itself. In the latter case there is a zone close to the surface in which the solute is depleted. In Gibbsian terminology, we have a negative adsorption. We devote this subsection to an analysis of the consequence of a negative adsorption on the force between particles.

Even a negative adsorption affects the force between particles in the solution. Outside the surface, a nonuniform concentration profile ultimately reaches the bulk value. As two particles approach, the concentration profiles start to overlap, which will generate a force between the particles. We encountered a similar situation for charged surfaces where a force appears as the diffuse electrical double layers start to overlap.

When the surfaces have come closer than twice the thickness, h_{dep}, of the depletion zones, no solute can be accommodated between them. In such a situation, further decrease of the separation simply results in a transfer of solvent molecules from pure solvent to a solution of an osmotic pressure Π_{osm}.

This process implies a transfer of solvent from a region of high chemical potential to one of a lower value. The situation is quite analogous to the capillary condensation case discussed in the previous subsection. For planar surfaces, the force-per-unit-area equals the free energy difference per unit volume between the two environments, which in the present case is the osmotic pressure

$$\frac{F}{A} = -G_{\beta\alpha} = -\Pi_{osm} \qquad h < 2h_{dep} \qquad (5.4.14)$$

At $h \gg 2h_{dep}$ the force should go to zero, and the behavior of F/A for distances $h \gtrsim h_{dep}$ depends on the characteristics of the concentration profile outside a single surface. Figure 5.24 illustrates three qualitatively different cases.

To generate a strong depletion interaction, we need a combination of two properties of the solute. First, it should produce a high osmotic pressure so that the force is large, according to eq. 5.4.14. Second, the repulsive interaction with the surface should be long-ranged so that h_{dep} is large.

Figure 5.24
The distance dependence of the depletion force depends on the nature of the solute concentration profile outside a single surface. For a monotonically decaying solute concentration as in (A) the force also decays monotonically to the limiting value $-\Pi_{osm}$. Sometimes the depletion layer is followed by a peak in the concentration at $z \gtrsim h_{dep}$ as in (B). Then there is a force barrier prior to the attractive regime at $z < h_{dep}$. A third possibility (C) is that there are several oscillations in the solute concentration and then the force also shows oscillations similar to those shown in Figure 5.19.

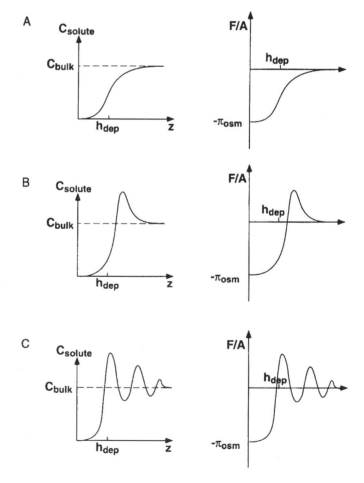

Most applications of depletion forces employ a polymer molecule as the active solute. In such cases, the size of the polymer coil determines the range of the force. Due to the polymer's conformational degrees of freedom, it also contributes appreciably to the osmotic pressure, as we will discuss in Chapter 7. For example, polyethylene glycol is used to induce fusion of biological cells. Finding such an effect could tempt one to conclude that the polymer molecules interact directly with the membrane during the fusion process. In reality, their function is to change conditions in the bulk solution, and they only affect the chemical potential of water at the scene of action.

Polymers often exhibit a peak in the density outside the depletion zone. This density peak results in the schematic force curve shown in Figure 5.24B in which a repulsive barrier occurs prior to the attractive depletion region.

Another possibility is to employ a charged micelle (or polyelectrolyte). In this case, a repulsive electric double-layer force can exist between the wall and the micelle while the counterions set up the osmotic pressure. This effect can be important in emulsions. In this case, there is also a strong micelle–micelle repulsion. At higher micelle concentrations, an oscillatory density profile can develop outside the single surface, producing an oscillatory force as in Figure 5.24C. A third alternative occurs for a simple electrolyte solution outside an apolar surface. In this case, the so-called image charge interactions (within the range of the Debye screening length) repel the ions from the surface. For reasonable electrolyte concentrations, the ion density goes smoothly toward the bulk value, and we have a force profile as shown in Figure 5.24A.

We can obtain an approximate expression for the depletion force between two spherical particles of radius R through the Derjaguin approximation (see eq. 5.2.33)

$$F(h_0) = \pi R \frac{G_{\text{plane}}(h_0)}{A} = \pi R \int_{h_0}^{\infty} \frac{F_{\text{plane}}(h)}{A} \, dh$$

$$\approx -2\pi R \Pi_{\text{osm}} h_{\text{dep}} (1 - h_0/2h_{\text{dep}}) \qquad (5.4.15)$$

where we have used eq. 5.4.14 as an approximate description of the depletion force. Equation 5.4.15 brings out the analogy between the depletion force and the force of eq. 5.4.12 due to capillary phase separation. The functional form is the same and the depletion thickness takes the role of the Kelvin radius R_K and Π_{osm} that of the difference in free energy density $G_{\beta\alpha}$.

5.4.4 Adsorption Introduces on Average an Increased Repulsion

Although some solutes are depleted from a surface, adsorption on a surface is far more common. Clearly, adsorption profoundly affects the properties of the surface and thus the forces between particles in a way that depends on the detailed chemistry. Adsorption of a charged surfactant normally leads to the formation of a charged surface that generates a double-layer

TABLE 5.2 Summary of Forces in Colloid Systems[a]

Type of Force	System	Expression	Equation	Figure	Comments
Electrostatic double layer	Force on a charged wall	$F/\text{area} = kT\left\{\sum_i c_i^*(\text{wall}) - \dfrac{\sigma^2}{2\varepsilon_r\varepsilon_o} - \Pi(\text{bulk})\right\}$	5.1.38		General results within the PB approximation. The osmotic bulk term Π(bulk) should be subtracted only when the system is in equilibrium with a bulk solution.
	Force on a neutral wall	$F/\text{area} = kT\sum_i c_i^*(\text{wall}) - \Pi(\text{bulk})$	5.1.12	5.1	See above.
	Two surfaces with identical charge	$F/\text{area} = kT\sum_i c_i^*(\text{midplane}) - \Pi(\text{bulk})$	5.1.26	5.2	See above.
	Two surfaces with identical charge	$F/\text{area} \approx 64\,kT\,c_o^*\Gamma_o^2\exp(-\kappa h)$	5.1.31		Valid only for $\kappa h \gg 1$.
	Two charged surfaces	$F/\text{area} \approx 64\,kT\,c^*(\text{bulk})\,\Gamma_{10}\Gamma_{20}\exp(-\kappa h)$	5.1.37	5.5	Also for oppositely charged surfaces. Valid only for $\kappa h \gg 1$.
	Two surfaces with surface charges	$F/\text{area} \approx \dfrac{1}{2\varepsilon_r\varepsilon_o}\dfrac{\sigma_1^2 + \sigma_2^2 + 2\sigma_1\sigma_2\cosh(\kappa h)}{\sinh^2(\kappa h)}$			Constant charge. PB equation linearized: $ze(\Phi/kT) < 1$ Valid irrespective of sign of surface charges
	Two surfaces with small surface charges	$F/\text{area} \approx \dfrac{\kappa^2\varepsilon_r\varepsilon_o}{2\sinh^2(\kappa h)}[2\Phi_{10}\Phi_{20}\cosh(\kappa h) - \Phi_{10}^2 - \Phi_{20}^2]$			Constant potential. PB equation linearized. Valid irrespective of sign of surface charges.
	Two charged spheres	$F \approx 128\pi\dfrac{R_1 R_2}{R_1 + R_2}\dfrac{kT}{\kappa}c^*(\text{bulk})\,\Gamma_{10}\Gamma_{20}\exp(-\kappa h)$			Vaid only for $\kappa h \gg 1$. R_1 and $R_2 \gg h$. h is the smallest separation between the surfaces of the spheres.
	Two charged spheres, small surface potentials	$F \approx (\sigma_1^2 + \sigma_2^2)\dfrac{\exp(-\kappa h) + 2\sigma_1\sigma_2}{\sinh(\kappa h)}$			Constant surface charge valid for Φ_{10}. $\Phi_{20} \leqslant kT/ze, h \ll R_1$ and R_2.
	Two charged spheres, small surface potentials	$F \approx \dfrac{R_1 R_2\pi}{(R_1 + R_2)}\dfrac{\kappa\varepsilon_r\varepsilon_o}{\sinh(\kappa h)}[2\Phi_{10}\Phi_{20} - \Phi_{10}^2 - \Phi_{20}^2]\exp(-\kappa h)$			Constant surface potential $\Phi_{10}, \Phi_{20} \leqslant kT/ze, h \ll R_1$ and R_2.
	Two identical charged spheres at constant charge	$F \approx 2\pi R_1\dfrac{c_0^* kT}{\kappa}\left\{2\Phi_0\ln\left[B + \Phi_0\cosh\left(\dfrac{\kappa h}{2}\right)\right] - \ln[\Phi_0^2 + \cosh(\kappa h) + B\sinh(\kappa h)] + \kappa h\right\}$			$B = \{1 + \Phi_0^2\sinh^{-2}(\kappa h/2)\}^{1/2}$. Approximation valid also for small κh.
London dispersion forces	Two atoms 1 and 2	$F = -6C_{12}\dfrac{1}{R_{12}^7}$	5.2.1		F obtained by differentiating eq. 5.2.1. C_{12} is the coefficient for atom−atom interaction defined in eq. 5.2.1.

TABLE 5.2 / Continued

Type of Force	System	Expression	Equation	Figure	Comments
	Molecule (2) and surface (1) in vacuum or vapor	$F = -\dfrac{\pi C_{12}\rho_1}{2}\dfrac{1}{h^4}$			Obtained by differentiating eq. 5.2.8.
	Two surfaces 1 and 2 in vacuum or vapor	$\dfrac{F_{12}}{\text{area}} = -\dfrac{H_{12}}{6\pi}\dfrac{1}{h^3}$	5.2.9	5.6	Attractive force, where $H_{12} = \pi^2 C_{12}\rho_1\rho_2$, eq. 5.2.5.
	Two spheres (R_1, R_2) small separation	$F = -\dfrac{H_{12}}{6h^2}\dfrac{R_1 R_2}{R_1 + R_2}$			R_1 and $R_2 \gg h$ $h = R - R_1 - R_2$
	Sphere (1) and surface (2), small separation	$F = -\dfrac{H_{12}}{6h^2}R_1$			$R_1 \gg h$
London dispersion forces	Two spheres, large separation	$F = -\dfrac{32\,H_{12}}{3}\dfrac{R_1^3 R_2^3}{R^7}$	5.2.35		Separation $R \gg R_1$ and R_2
	Two aligned cylinders of radius R_1 and R_2 and length L	$F = -\dfrac{H_{12}L}{8\sqrt{2}\,h^{5/2}}\left(\dfrac{R_1 R_2}{R_1 + R_2}\right)^{1/2}$			R_1 and $R_2 \gg h$
	Two crossed cylinders	$F = -\dfrac{H_{12}}{6}\dfrac{(R_1 R_2)^{1/2}}{h^2}$			R_1 and $R_2 \gg h$
	Condensed phase (2) between identical materials 1 and 3	$H_{12} = (H_{11}^{1/2} - H_{22}^{1/2})^2$	5.2.26	5.8	
	Condensed phase (2) between different materials 1, 3	$H_{123} = \{\sqrt{H_{22}} - \sqrt{H_{11}}\}\{\sqrt{H_{22}} - \sqrt{H_{33}}\}$	5.2.27	5.10	
Packing force	Liquids between smooth solid surfaces	$F/\text{area} \approx A\cos\left(2\pi\dfrac{h}{\lambda_1}\right)\exp\left(-\dfrac{h}{\lambda_3}\right)$	5.4.1	5.19	$A \leqslant kT\rho n$ (liquid), $\lambda_1 \approx$ molecular diameter, $\lambda_3 \approx$ molecular diameter at liquid densities
Capillary phase separation force	Phase β between surfaces in medium of phase α	$F/\text{area} = -G_{\alpha\beta}$	5.4.5	5.25	$G_{\beta\alpha} =$ difference in free energy density between phases β and α
Short range repulsive force	Liquids and other fluids interfaces in a polar solvent	$\dfrac{F}{\text{area}} \approx A\exp\left(-\dfrac{h}{\lambda_1}\right)$		5.25	$\lambda \approx 0.15 - 0.3$ nm, $A \approx 10^6 - 10^8$ N/m^2
Undulation force	Bilayers in liquid state	$\dfrac{F}{\text{area}} = \dfrac{3\pi^2}{64}\dfrac{(kT)^2}{K_b}\dfrac{1}{h^3}$	5.5.1		$\kappa =$ bending rigidity
Depletion force	Force in solvent containing a nonabsorbing polymer or aggregate	$\dfrac{F}{\text{area}} = -\Pi(\text{solute})$			For $2h <$ depletion layer. For larger h the force goes to zero.

[a]In this table, relationships are given in terms of the force; some of the referenced equations in the text are given in terms of energy.

repulsion. However, if a surfactant adsorbs on an oppositely charged surface, it may make the surface electrically neutral and hydrophobic. Conversely, adsorption of a nonionic surfactant makes a hydrophobic surface more polar, and the force between two such surfaces typically shows a short range repulsion.

We find a similar scenario with polymers. Adsorption can result in repulsion or attraction, depending on the balance between the polymer–polymer and polymer–solvent interactions. In the next section—when we discuss liquid–liquid interfaces with internal degrees of freedom—we will present a more detailed discussion of the mechanisms for the forces involved in these cases. Here we concentrate on describing a general relation between force and surface free energy that is useful for the conceptual understanding of forces in systems with adsorbing species.

In Section 2.4, we discussed how the surface tension of a liquid–vapor or a liquid–liquid interface changed when a solute was added to the system. We found a fundamental relation in the Gibbs adsorption equation stating that changes in surface tension (or surface free energy) directly relate to the surface excess of the solute. A solute that shows negative adsorption (surface depletion) increases the surface free energy, γ, while adsorption leads to a decrease in γ.

The same basic relation applies at a solid–liquid interface. Let us now consider the situation where we bring two planar solid surfaces from a large separation up to molecular contact in a solvent containing the solute A. For the separated surfaces, we have an excess surface free energy $2A\gamma_{SL}(c_A)$ where we have explicitly indicated that the interfacial energy depends on the solute concentration c_A. The free energy change in bringing the surfaces into contact is the work of cohesion, as expressed in eq. 2.1.11

$$\Delta G/A = -2\gamma_{SL}(c_A) \qquad (5.4.15)$$

Alternatively, we can write the free energy changes as an integral of the force as defined in eq. 5.1.1 ($\Delta G \simeq \Delta A$)

$$\frac{\Delta G}{A} = \int_{contact}^{\infty} \frac{F(h)}{A} dh = -2\gamma_{SL}(c_A) \qquad (5.4.16)$$

For the depletion case, $\gamma_{SL}(c_A) > \gamma_{SL}(c_A = 0)$ according to the Gibbs equation so that eq. 5.4.16 shows that, on average, the force should be more attractive for $c_A > 0$ than in the pure solvent. For adsorbing solutes, on the other hand, $\gamma(c_A) < \gamma(c_A = 0)$, and, on average, the force is more repulsive (less attractive) in the presence of the solute. Note that to obtain direct molecular contact between the solid surfaces, the adsorbed solute must be squeezed off the surface. We expect the major extra repulsive contribution to the force to come at separations where there is a forced desorption.

A practical concern in systems with adsorbents is whether there is enough time when two surfaces approach to equilibrate the adsorption. Equation 5.4.16 applies at equilibrium, but another possibility is that the surfaces approach so rapidly that the amount adsorbed remains constant. How does this force relate to the equilibrium one? We discussed an analogous problem for the electrical double-layer force when we distinguished surfaces approaching at constant charge from those at constant potential. The condition of constant charge is similar to that of constant adsorbed amount. In the case of the electrical double-layer force, we concluded that the forces were the same at large separations where the double-layer overlap is small. A difference appeared only when the constant potential condition implied a substantial change in the surface charge, in which case the repulsion became weaker. We can generalize this case to the adsorption situation where we conclude that provided the separated surfaces are in equilibrium with the solution, the force is the same whether one considers the constant adsorbed amount or equilibrium as long as the interaction is weak. If the force at constant amount is repulsive, equilibrating shows a weaker repulsion closer in. We cannot support the often-stated view that for adsorbing polymers, a fundamental difference exists between the force at constant adsorbed amount (rapid approach or covalently bound polymers) compared to the state when adsorption equilibrium is maintained. On the contrary, we find that the force is asymptotically the same under both conditions.

5.5 Entropy Effects Are Important for Understanding the Forces Between Liquidlike Surfaces

Emulsion stability depends on the presence of a repulsive force between the droplets. These surfaces are liquid in character, and the interfaces themselves possess thermally excited degrees of freedom that profoundly influence the forces. The same argument applies to forces in foams. Forces between self-assembled aggregates are similarly controlled to a considerable extent by the liquidlike character of the aggregates. We also find such conditions on solid surfaces covered by a layer of adsorbed surfactant or polymer which has a liquidlike disorder. Irrespective of molecular details, the net force has an important repulsive component because two interacting surfaces always have reduced configurational freedom relative to free surfaces. This fact constitutes a conceptual difference in comparison to the solid surfaces discussed in the previous subsection. Another complication that we do not address explicitly here is that liquidlike surfaces easily deform under the action of interparticle forces. In Chapter 11, we will see that this property has dramatic consequences for the structure of concentrated emulsions and foams. In fact due to the deformations it poses fundamental complications to uniquely define a force between two liquid drops or gaseous bubbles.

5.5.1 Reducing Polymer Configurational Freedom Generates a Repulsive Force

A polymer layer can cover a surface either by adsorption from solution or through covalent attachment. If the polymer is soluble in the liquid medium, the parts of the polymer molecule that are not in direct contact with the surface tend to extend out from the surface into the solution in order to maximize the number of polymer configurations. In Chapter 7, we will discuss the issue of polymers on surfaces in more detail. Here we concentrate on the basic mechanism behind the force between surfaces covered by polymers. When two surfaces approach one another, the polymer tails reaching out from one surface start to impede the configurational freedom of the polymers of the other surface, reducing the configurational entropy. Unless direct interaction between the monomer units compensates for this effect, the results is a repulsive force. Because polymer molecules can have large sizes, the range of this force easily can be larger than the van der Waals and double-layer forces.

In the literature, this repulsive force usually is called the steric force. This terminology is somewhat misleading because it obscures the insight that the force is due to a rather subtle entropy effect resulting from the greater difficulty of packing connected objects relative to disconnected ones. Although the terminology is established, we must distinguish this effect from the steric force operating between solid objects that we discussed in the introduction to Section 5.4. Thus, we suggest that a repulsive steric entropic force operates for polymer systems in contrast to the steric energetic force for solid objects.

5.5.2 Short Range Forces That Encompass a Variety of Interactions Play Key Roles in Stabilizing Colloidal Systems

When we measure forces between surfaces, we first attempt to account for the experimental results quantitatively in terms of a combination of van der Waals and electrostatic double-layer forces. If deviations between theory and experiment fall outside the estimated experimental uncertainty, we invoke the presence of additional forces.

In practice, at short range it is much easier to verify the presence of an extra repulsive component than an attractive one. An additional short range attractive force serves only to augment the van der Waals attractive interaction, which brings the colloidal surface into contact anyway. However, the presence of a short range repulsion has a much larger impact on the system's properties because if it is sufficiently large, it can overpower the attractive force and prevent the surfaces from moving into contact. Consequently, much more attention has been paid to short-range repulsive forces than to short range attractive ones.

One of the most dramatic and also best-studied examples of a short-range stabilizing force occurs with zwitterionic liquid bilayers (see Chapter 6 for a discussion of their general properties). Figure 5.25 shows the measured force between bilayers in the liquid crystalline state. For these locally neutral surfaces, a short-range repulsive force dominates the van der Waals attraction at bilayer separations less than ≈ 25 Å. The force increases by several orders of magnitude as the water content is decreased and the surfaces are driven together. The force curve decays exponentially with distance; the decay constant lies in the range of 1.3–3.0 Å. In the absence of this repulsive force, stable liquid crystals and zwitterionic bilayer membranes would not exist.

There is no consensus on the molecular origin of this extra stabilization. Historically, these forces have been called hydration or solvation forces and discussed in terms of solvent orientation at surfaces and/or water structural effects. However, other molecular mechanisms more likely contribute. In liquid-like bilayers, protrusions of the surfactant or lipid molecules in a direction normal to the bilayer readily occur. Confinement of the conformational freedom in large head groups, such as those found in phospholipids, also contribute. The freedom associated with these molecular motions is lost when two surfaces move together, and this loss of freedom implies a decrease in entropy, which in turn generates a repulsive force. As with any phenomenologically based observation, several different molecular mechanisms (which depend on the detailed molecular properties of the surface) may contribute to the measured force.

Short range stabilizing forces are also observed between solid surfaces, such as silica and mica, across water. In this case the support is molecularly rigid and the fluctuations discussed

Figure 5.25
Bilayer separation and molar ratio (water/lipid) for egg phosphocholine plotted as a function of osmotic pressure illustrating short range stabilizing force as determined by osmotic stress measurements. (R. P. Rand and V. A. Parsegian, *Biochim. Biophys. Acta* **988**, 351 (1989).)

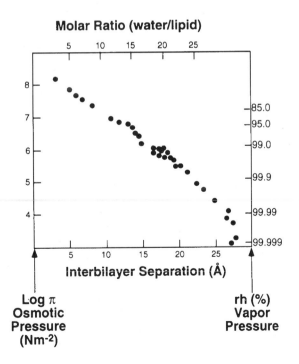

in the preceding paragraph should be of minor importance. However, in dealing with surfaces at molecular separation, surface morphology becomes an important issue. Since surfaces are rarely perfectly smooth and also show some chemical heterogeneity a molecular interpretation of the force at short range becomes a very delicate matter under such circumstances. Mica, on the other hand, is smooth and has a long range order parallel to the surface. The most likely cause of the deviation from the DLVO potential observed at short separations is the quantitative breakdown of the Poisson–Boltzmann description of the repulsive force.

5.5.3 Undulation Forces Can Play an Important Role in the Interaction of Fluid Bilayers

When aggregates or macromolecules are present, we can expect deviation from Lifshitz theory at large separations as well as small ones. In a lamellar liquid crystal, the bilayers are under zero tension; that is, the derivative of the free energy with respect to bilayer area is zero, while the second derivative is positive. This makes the (free) energy costs low for bending the bilayers, which undergo undulatory motions that contribute to lowering the free energy of the system. The larger the separation, the more important the undulation motion becomes. This is equivalent to saying that the undulations contribute an effective repulsive force of entropic origin. A detailed derivation by Helfrich gives the decaying force as the inverse cube of the separation h

$$\frac{F}{\text{area}} = \frac{3\pi^2}{64} \frac{(kT)^2}{\kappa_b} h^{-3} \tag{5.5.1}$$

Here κ_b represents the so-called bending rigidity and measures the stiffness of the bilayer which will be further discussed in Chapter 11. It has the dimension of energy and typically lies in the range of $1-20\,kT$ for amphiphile bilayers.

Asymptotically, this force is more long range than the attractive van der Waals force between bilayers, which varies as h^{-5} when the bilayer thickness is small relative to the separation (see Exercise 5.8). Depending on the temperature, κ_b, and the Hamaker constant, the attractive or the repulsive force can dominate in the bulk liquid crystal.

In this section we have described the steric entropic force found in polymer systems, the undulation force for lipid bilayers, and the short range repulsive force for liquidlike surfaces. In every case, the force is repulsive and entropic in origin (although this issue is debated for the short range repulsive force). The view that we have presented reveals a fundamental similarity in the mechanism of these forces. In fact, double-layer repulsion has an analogous mechanistic origin. Although difficult to prove in a general sense, we conjecture that repulsive forces between similar surfaces always have an entropic origin.

5.6 The Thermodynamic Interpretation of the Hydrophobic Interaction is Problematic Due to Entropy-Enthalpy Compensation

Whether the applications of colloidal science are technological or biological, most processes occur in aqueous medium. So far in this chapter, we have characterized the medium in terms of simple constants like dielectric permittivity, refractive index, and molecular size. We call the strong cohesion that exists in liquid water due to electrostatic interaction hydrogen bonds, and they make water an associated liquid. How the structural effects of water influence interactions has become a classical subject of scientific discussion. We encountered an example in our analysis of the short range repulsion force.

The best-documented case in which water's structural effects are influential is the hydrophobic interaction we introduced in Chapter 1. Because hydrophobic interaction plays a central role in colloid science, we return to a more detailed discussion of this phenomenon. Now that we have established the basic intermolecular interactions, we are in a better position to understand some of the subtleties that have generated considerable confusion in the past. We identify two levels of hydrophobic interaction. The first is the molecular level where the interaction between individual molecules leads to such effects as micellar self-assembly. The second level relates to the force operating between colloidal entities that have hydrophobic surfaces.

5.6.1 Understanding the Mysteries of Water

In Chapter 1, we established the fact that in a variety of solvents, the solubility of nonpolar molecules decreases as the solvent's cohesive energy increases. In polar solvents like water, hydrazine, and ethylammonium nitrate, we described this interaction as solvophobic. In Chapter 4, we showed how solvophobic interactions act as the driving force behind the formation of micelles by amphiphilic molecules. Within the framework of the regular solution theory introduced in Section 1.4, the solvophobic effect is described primarily by an interaction parameter w (water–solute) larger than $2RT$. In eq. 1.4.16, w is expressed in terms of a difference $W_{AB} - \frac{1}{2} W_{AA} - \frac{1}{2} W_{BB}$. Denoting water with A and B as the apolar solute, the large positive value of w arises from $W_{AA} \gg W_{AB} \approx W_{BB}$. In other words, the cohesive interaction W_{AA} is large in water and also other polar solvents. In Chapter 3, we analyzed in detail the electrostatic interactions that provide the strong cohesive energy to polar media and we emphasized the electrostatic origin of the hydrogen bond.

Regular solution theory predicts that a solute's solubility should increase with temperature. This pattern is observed in virtually all solvents, but for water one often finds the reverse trend below room temperature. Figure 5.26, which was published by Shinoda in 1977, clearly demonstrates water's unusual behavior in a plot of the logarithm of the solubility of the alkylbenzenes in water as a function of the reciprocal temperature. It shows that at elevated temperatures, where water behaves like a normal hydrogen-bonded solvent, the solubility follows the predictions of regular solution theory with a linear dependence of $\log x_s$ on $1/T$. In an intermediate temperature range, we see that the solubility deviates from the linear behavior while it still decreases with decreasing temperature; then it goes through a minimum near 25°C before increasing again. Shinoda pointed out that a linear extrapolation of the high temperature behavior to 25°C allows us to identify the results we can attribute to the special structural effects of water. Below approximately 90°C, the standard free energy of solution of nonpolar solutes in water actually decreases. We observe this same pattern when we identify the hydrophobic contribution to micellization (Figure 4.13).

Every cook who has ever watched a pot of water come to to a boil has observed bubbles forming on the side of the pan as the temperature approaches 100°C. Nearly every scientist who makes measurements on aqueous solutions as a function of temperature implicitly accommodates water's unusual behavior. In aqueous solutions, if we begin (with liquid) at 25°C and

Figure 5.26
Water solubility of alkylbenzenes (I, C_6H_6; II, $CH_3C_6H_5$; and III, $C_2H_5C_6H_5$) versus $1/T$. As temperature decreases, the solubility of alkylbenzenes in water initially decreases but then becomes almost constant. For many substances, like the rare gases, the solubility displays a minimum around 20°C and then actually increases with decreasing temperature. The differences between extrapolated and measured solubility in Figure 1.9 provide a measure of water structure effect that becomes more pronounced with decreasing temperature. (C. K. Shinoda, *Principles of Solution and Solubility*, Dekker, New York, 1978, p. 130.)

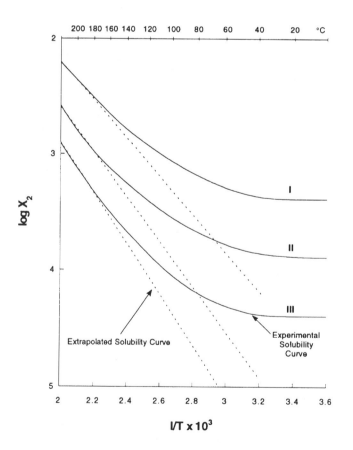

then increase the temperature, gas bubbles form as a consequence of the decreasing gas solubility. To avoid this effect, we must degas aqueous solutions by heating and then cooling them before starting our measurements. On the other hand, degassing is unnecessary in most nonaqueous solvents because they obey the predictions of the regular solution theory.

What molecular interactions between nonpolar solutes and water molecules account for their features in aqueous solution? A traditional approach to answering this question is to analyze the solubilization process in terms of changes in free energy, enthalpy, and entropy and then draw inferences from our observations. We can proceed in two ways:

1. Examine directly how the free energy changes with temperature. Since $\Delta G = -kT \ln X_2$, where X_2 is the mole fraction solubility, Shinoda's plot displays qualitatively how the free energy changes with temperature.

    The main conclusion derived from this first analysis is that in an aqueous solution, the structural effects of water, which are larger with decreasing temperature, promote an increase in the solubility of nonpolar groups.

2. Analyze the change in the free energy with temperature, using the Gibbs–Helmholtz equation to obtain values of ΔH and ΔS. This approach, which was first published by Frank and Evans in 1945, has dominated the thinking of most scientists for the past 50 years. We can most conveniently accomplish this analysis by observing the transfer of nonpolar gases from a reference solvent like cyclohexane into water. Table 5.3 shows that for such a transfer, the positive free energies, ΔG^o, simply reflect the truism that oil and water don't mix. However, the surprisingly small magnitude of the enthalpies and the large negative entropies suggest that in aqueous solutions, entropic changes primarily govern solubility, rather than the enthalpy changes usually observed in most other solvents and which follow from the regular solution theory.

Frank and Evans suggested that nonpolar groups induced an increase in water structure in their own immediate vicinity by increasing either the strength or the number of hydrogen bonds.

TABLE 5.3 Thermodynamic Function for Transfer of 1 mol of Argon from Cyclohexane to Water and to Hydrazine

	ΔG (kJ/mol)	ΔH (kJ/mol)	ΔS (J/K·mol)
1. $Ar(X_2 = 1$ in $C_6H_{12}) \rightarrow Ar(X_2 = 1$ in $H_4N_2)$	11.9	9.5	−7.9
2. $Ar(X_2 = 1$ in $H_4N_2) \rightarrow Ar(X_2 = 1$ in $H_2O)$	−1.5	−20.7	−64.0
3. $Ar(X_2 = 1$ in $C_6H_{12}) \rightarrow Ar(X_2 = 1$ in $H_2O)$	10.4	−11.2	−72.0

From R. Lumry, E. Battistel, and C. Jolicoeur, *Faraday Symp. Chem. Soc.* **17**, 93 (1984).

They proposed the formation of "icebergs" around the argon molecules. Solid stoichiometric clathrates in which each nonpolar group is completely caged inside a water cavity provide a well-defined analog for Frank and Evans' "ice-bergs." Figure 5.27 shows the structure of two typical clathrates. Several hundred of these structures are known, involving rare and diatomic gases, short-chain hydrocarbons, and salts such as $(C_4H_9)_3SF$ as solutes. In all these structures, each water molecule hydrogen-bonds to four adjacent water molecules by varying the lengths and angles of the hydrogen bonds. The sheer variety of clathrate structures with cavities of different sizes and shapes illustrates water's versatility in forming accommodating structures. Even upon melting, remnants of clathratelike cages continuously form and disperse around nonpolar groups. Since the lifetime of such structures is $\approx 10^{-11}$ sec, the term "icebergs" is somewhat misleading. How can we reconcile the Shinoda free energy argument with the Frank and Evans hypothesis of local structure formation? The following set of arguments provides a way to resolve this issue. First, we note that the peculiar thermodynamic pattern we observe in water is not unique. We can find the same pattern in other hydrogen-bonding solvents such as EAN (cf. Table 5.4) and in ethylene glycol.

Second, we can devise a thought experiment in which we divide the dissolution process into two steps, as illustrated in

Figure 5.27
Clathrate structures. Clathrates are stoichiometric crystalline solids in which water forms cages around solutes. Each line represents an O–H–O bond. (a) When this clathrate contains Cl_2, the larger tetrakaideca-hedral cavities are occupied, but not the dodecahedral ones. With the smaller CH_4 solute, both cavities are filled. (L. Pauling, *The Nature of the Chemical Bond*, Cornell University Press, Ithaca, NY, 1960 p. 471.) (b) clathrate structure of $(i–C_5H_{11})_4NF$ (left) and position of the $(i–C_5H_{11})_4N^+$ ion (right). The F^- ion replaces one of the water molecules. (D. W. Davidson, in *Water, A Comprehensive Treatise*, Vol. 2, F. Franks, ed., Plenum Press, New York, 1973, p. 115.)

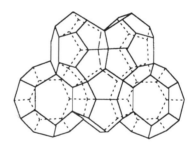

A

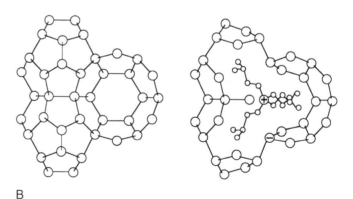

B

TABLE 5.4 Thermodynamics of Transfer of Nonpolar Gases from Cyclohexane to Ethylammonium Nitrate (EAN) and Water

Gas	Solvent	$\Delta G°$ (kJ/mol)	$\Delta H°$ (kJ/mol)	$\Delta S°$ (J/K·mol)
Kr	EAN	6.7	−3.8	−38
	H_2O	11.7	−12.0	−79
CH_4	EAN	6.7	−2.1	−17
	H_2O	12.1	−11.0	−75
C_2H_6	EAN	8.4	−4.0	−42
	H_2O	16.3	−9.0	−84
$C_4H_{10}{}^a$	EAN	(15.1)	(−24)	(−131)
	H_2O	(26.6)	(−26)	(−176)

From D. F. Evans, *Langmuir* **4**, 3 (1988).
[a]The values for butane refer to the transfer from the gas phase to EAN and to water.

Figure 5.28. In step 1, we create a hole in the solvent and then insert the nonpolar molecule. In step 2, we allow the solvent molecules in the vicinity of the solute to respond to its presence. We can draw two significant conclusions from this analysis: (a) solubility in water is determined mainly by a

Figure 5.28
Schematic of a two-step process for transferring an apolar solute from oil into water on a polar solvent. In the first step, a hole is created in the solvent and the apolar solute is inserted. Because of the similarity of the ΔH_{vap} and γ for H_4N_2 and H_2O, we assume that the cost of making the hole is similar in these two solvents (Table 5.3, line 1). Next, we allow the solvent molecules to respond to the presence of the solute. We assume the difference between the two solvents (Table 5.3, line 2) measures water's hydrophobic hydration. Steps 1 and 2 combine to give the total change in ΔG, ΔH, and ΔS for transferring polar solutes into aqueous solutions (Table 5.3, line 3).

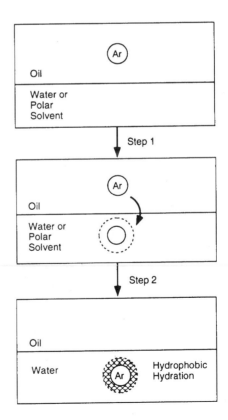

solvophobic interaction arising from high polarity, a property that water shares with other polar solvents and which is in accord with the predictions of regular solution theory; and (b) the structure of water subtly affects the dissolution process with profound consequences for the thermodynamic parameters.

For the first step, we examine the thermodynamics associated with transferring argon molecules from cycohexane into hydrazine (Table 5.3, line 1). Hydrazine and water share astonishingly similar physical properties, including their melting points (1.7 vs. 0°C), viscosities (0.889 vs. 0.90 cP, 25°C), and other properties listed in Table 5.3. On average, both can form four hydrogen bonds per molecule. In fact, they differ only in the properties we associate with the unique structural properties of water, such as heat capacity, density maximum, and low heat of fusion. The process of creating a hole (in our solvent) requires energy because we must overcome the attractive interaction between hydrogen-bonded (solvent) molecules. Therefore, the value of ΔG of 11.9 kJ/mol arises mainly from the enthalpic contribution of 9.5 kJ/mol. Due to the physicochemical similarities between hydrazine and water (particularly the enthalpy of vaporization and the surface tension), the costs of making a hole in either one are similar.

We can evaluate the second step, which measures the hydrophobic hydration of a nonpolar solute in water, by transferring argon from hydrazine to water (Table 5.3, line 2). The free energy charge associated with this step is much smaller than for step one and it carries the opposite sign. In contrast to the free energy there are large changes in enthalpy and entropy so that $\Delta H_2 \approx -\Delta H_1$ and $|\Delta S_2| \gg |\Delta S_1|$. For the total process $\Delta H_3 = (\Delta H_1 + \Delta H_2)$ is small and $\Delta S_3 = (\Delta S_1 + \Delta S_2)$ are dominated by the changes associated with step two. The net effect being that the free energy change ΔG_3 of the total process appears to be dominated by the entropy change in the second structural rearrangement step since $\Delta G_3 \approx -T\Delta S_3$. This occurs even though the combined values of ΔH_2 and ΔS_2 make only a very small contribution to the total ΔG. Thus, we find ourselves in a peculiar and counterintuitive situation: literal interpretations of the signs and magnitudes of ΔH and ΔS associated with process in water can lead to an incorrect interpretation of ΔG. This demonstrates the difficulty with obtaining a molecular interpretation of ΔH and ΔS values of a complex process.

In retrospect, we can see that if researchers had directed more attention toward the interpretation of changes in ΔH and ΔS over a broad temperature range (like that displayed in Figure 5.26) instead of focusing on changes below 90°C or on comparisons with other polar liquids (like those illustrated in Figure 1.11), they could have avoided much of the confusion regarding water's role in amphiphilic self-assembly. Furthermore, the implication of this analysis is not limited to water's interaction with nonpolar groups; it is much more general. Virtually all processes involving water near room temperature exhibit some structural response that is accompanied by large changes in ΔH

and ΔS, but only small changes in ΔG. Processes involving large cancellation between ΔH and $T\Delta S$ are called compensated processes.

We can make the specific example given above more general by considering the Gibbs free energy change

$$\Delta G = \Delta H - T\Delta S \qquad (1.4.1a)$$

associated with an isothermal, constant pressure process. We evaluate ΔH at temperature T from

$$\Delta H = \Delta H_m + \int_0^T \Delta C_p(T')dT' = \Delta H_m + \Delta H_c \qquad (5.6.1)$$

where ΔH_m is the enthalpic change at absolute zero and the integral gives the heat necessary to raise products and reactants to temperature, T.

Similarly, we can write the entropic contribution to ΔG as a sum of two terms

$$T\Delta S = \int_0^T \frac{\partial}{\partial T'} [T'\Delta S(T')]dT' =$$

$$\int_0^T \Delta S(T')dT' + \int_0^T \Delta C_p(T')dT' \equiv T\Delta S_m + T\Delta S_c \qquad (5.6.2)$$

where we have used the thermodynamic relation

$$\left(\frac{\partial \Delta S}{\partial T}\right)_p = \frac{\Delta C_p}{T} \qquad (5.6.3)$$

The equation shows that we can write $T\Delta S$ as the sum of two terms, the second of which we can identify as a contribution to the entropy upon heating the products and reactants to temperature, T. Combining eqs. 1.4.1a, 5.6.1 and 5.6.2 gives

$$\Delta G = \Delta H_m + \int_0^T \Delta C_p dT - \left[\int_0^T \Delta S \, dT + \int_0^T \Delta C_p dT\right]$$

$$= \Delta H_m + \Delta H_c - T\Delta S_m - T\Delta S_c \qquad (5.6.4)$$

Inspection of eqs. 5.6.2 and 5.6.4 shows that $\Delta H_c - T\Delta S_c = 0$, that is, the parts of ΔH and ΔS associated with adding heat to the system do not contribute to the free energy. Thus, we obtain

$$\Delta G = \Delta H_m - T\Delta S_m \qquad (5.6.5)$$

where the subscript m indicates the parts of ΔH and ΔS that do contribute to ΔG. This analysis is general and places a fundamental limit on our ability to directly interpret ΔG in terms of its first derivatives ΔH and ΔS in all thermodynamic processes.

Do you find this confusing? If you do you probably are not alone. After all, you have been taught that you can gain physical insights into processes (such as changes in $\Delta \mathbf{G}$) by analyzing the corresponding changes in $\Delta \mathbf{H}$ and $\Delta \mathbf{S}$. Now you are presented with arguments suggesting that the complexity of changes in $\Delta \mathbf{H}$ and $\Delta \mathbf{S}$ precludes such literal interpretation.

The issues involved in analyzing compensated behavior are subtle. Equation 5.6.5 illustrates general thermodynamic compensation, while the process shown in Figure 5.2.8 exemplifies another type of compensation in which one process, that is, the transfer of a nonpolar group into water, engenders a nonstoichiometric response, that is, the structuring of water around the nonpolar group.

A detailed explanation of this area would take us far beyond the scope of this book. However, the implications are clear:

1. In any thermodynamic process, we must be very cautious when interpreting $\Delta \mathbf{G}$ in terms of $\Delta \mathbf{H}$ and $\Delta \mathbf{S}$.

2. In aqueous solutions with large heat capacities (see Table 1.6) the compensation enthalpy and entropy, that is, $\Delta \mathbf{H}_c$ and $\Delta \mathbf{S}_c$, may be of opposite sign and larger in magnitude than $\Delta \mathbf{H}_m$ and $\Delta \mathbf{S}_m$. As illustrated by the analysis in Figure 5.28, molecular interpretation of $\Delta \mathbf{G}$ in terms of $\Delta \mathbf{H}$ and $\Delta \mathbf{S}$ can be very misleading.

3. In relating molecular processes to colloidal interactions, it is important to develop theoretical expressions in which the separate entropic and enthalpic contributions to the free energy are clearly delineated. The insights that derive from this approach, as opposed to obtaining an expression for $\Delta \mathbf{G}$ directly, are invaluable in understanding complex processes.

5.6.2 Strong Attraction Exists Between Hydrophobic Surfaces Although Experiments Have Failed to Establish the Distance Dependence of this Force

Hydrophobic interactions also operate between colloidal particles. They are of prominent importance for organizing biological systems on the colloidal scale. As discussed in the previous subsection, the molecular source of the hydrophobic interaction is found in the strong electrostatic cohesion in water (including hydrogen bonding). How much of this effect is taken into account in the Lifshitz theory? In Section 5.2.2 we estimated the surface tension of a hydrocarbon to be $17 \, \mathrm{mJ/m^2}$. A similar calculation for water would yield $\gamma_{\mathrm{H_2O}} \simeq 15 \, \mathrm{mJ/m^2}$, which differs widely from the experimental value of $72 \, \mathrm{mJ/m^2}$.

Polar interactions, such as hydrogen bonds, are generally recognized as the main source for the high surface tension of water. Clearly, substantial deviations from the Lifshitz theory must be taken into account if we attempt to analyze two water droplets approaching one another at short range across a slit of vapor.

We arrive at a similar conclusion by considering γ_{H_2O-HC}, the interfacial tension between water and a hydrocarbon. If we combine eqs. 5.2.11, 5.2.12, and 5.2.21, we can write

$$\gamma_{H_2O-HC} = \gamma_{H_2O} + \gamma_{HC} - \frac{H_{H_2O-HC}}{12\pi} \frac{1}{\delta^2}$$

$$= \frac{1}{24\pi} \frac{1}{\delta^2} \{H_{H_2O} + H_{HC} - H_{H_2O-HC}\}$$

$$\simeq \frac{1}{24\pi} \frac{1}{\delta^2} \{\sqrt{H_{H_2O}} - \sqrt{H_{HC}}\}^2 \qquad (5.6.6)$$

Using the last equality and the values $H_{H_2O} = 3.7 \times 10^{-20}$ J and $H_{HC} \approx 5 \times 10^{-20}$ J from Table 5.1, we obtain $\gamma_{H_2O-HC} \simeq 0.3$ mJ/m^2. This large deviation from the experimental value of 50 mJ/m^2 again demonstrates the limitation of the Lifshitz theory at short range for polar systems.

On the other hand, if we use the first equality in eq. 5.6.6 and take surface tensions γ_{H_2O} and γ_{HC} and H_{H_2O-HC} as the experimental values, the estimated value becomes $\gamma_{H_2O-HC} \simeq 70$ mJ/m^2. This result is of the same magnitude as the experimental value of 50 mJ/m^2. Estimating both γ_{H_2O} and γ_{H_2O-HC} using the Lifshitz expression neglects a dominant component. This component also appears in the interaction between two apolar surfaces across water, for an aqueous film in air and for aqueous drops in a hydrocarbon medium.

The interaction between two hydrocarbon surfaces in an aqueous medium relates to the hydrophobic interaction between hydrocarbon molecules in water discussed in Chapter 1. Bringing two such surfaces together from infinity to contact should lower the free energy by $2\gamma_{H_2O-HC} = 0.1$ J/m^2, which is a very large effect. There is thus a strong attractive force, but the argument says nothing about its range. The Lifshitz theory implies that these strong polar interactions have a small effect at long range and from this perspective the strong attraction should have a short range character.

However, there are several reports in the literature of observations of very long range attractive surfaces in water. The best-studied case is that of mica surfaces made hydrophobic by depositing a monolayer of a dialkyldimethylammonium cation on the surface. Figure 2.21 shows atomic force microscopy (AFM, see Section 2.7) pictures of such monolayers, demonstrating that a well-ordered layer actually exists in water. The force between two such surfaces can indeed be very long-ranged, as shown in Figure 5.29 for a $C_{20}H_{41}$ saturated chain of the quaternary ammonium. However, as shown in Figure 5.30, the force depends on temperature and alkyl chain length in a way that strongly suggests that for these systems the long range force is present only when the alkyl chains are in an ordered (solid) state. Since the interfacial free energy γ_{H_2O-HC} depends

Figure 5.29
Interaction forces between hydrophobic monolayers in aqueous solution at 25°C as determined by surface forces measurements. Data are plotted as the derivative of the force $d(F/R)/dD$ with respect to the distance of separation of the two surfaces. Comparison with the predictions of van der Waals forces (dotted line) establishes that the measured force is orders of magnitude stronger than predicted and extends over a much longer range. DHDA, DODA, and DEDA denote monolayers of dialkyldimethylammonium with C_{16}, C_{18}, and C_{20} saturated chains respectively. (Y.-H. Tsao, S. X. Yang, D. F. Evans, and H. Wennerström, *Langmuir* **7**, 3154 (1991).)

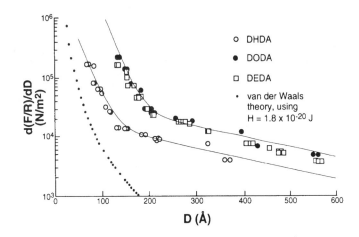

only slightly on the solid versus melted nature of the hydrocarbon, the temperature dependence shown in figure 5.30 argues that the long-range attractive force is not directly related to the hydrophobic interaction.

In Sections 5.3 and 5.4, we have discussed two types of interaction either of which can produce an attraction that potentially exceeds that of the van der Waals force and also operates at long range. They are the electrostatic domain correlation force described in Section 5.3.3 and the capillary phase separation force described in Section 5.4.2. Both possibilities have experimental support. With a cationic surfactant adsorbed on mica one can envisage a lateral inhomogeneity on the surface as in Figure 5.15 which would generate a force with the range of either the size of the domains or the inverse Debye length, whichever is smaller. (See Figure 5.16.) Gas bubbles have a strong tendency to adhere to hydrophobic surfaces; and when a second surface approaches, a lens of gas is formed that connects the two surfaces as illustrated in Figure 5.20. This yields an

Figure 5.30
The long range attractive force as a function of surfactant chain length at 50°C. The measured force decreases with temperature in a manner that correlates with the melting of the hydrocarbon chains. Above the chain melting temperature, the attractive force becomes comparable to that predicted by the van der Waals force. (Y.-H. Tsao, S. X. Yang, D. F. Evans, and H. Wennerström, *Langmuir* **7**, 3154 (1991).)

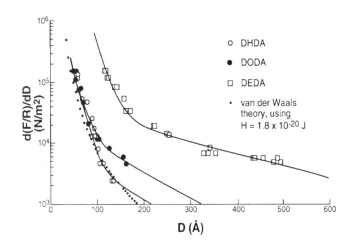

attractive force by the mechanism described in Section 5.4.3. Although further experimental work is needed to clarify the situation, from the theoretical arguments presented, we would expect a relatively short range attractive "true" hydrophobic force. In this case, the experimental failure to establish a distance dependence reflects the continued difficulty of measuring strong attractive forces.

5.7 Hydrodynamic Interactions Can Modulate Interaction Forces

So far we have discussed forces between particles or surfaces at rest. One additional effect occurs when the particles move relative to one another. We can distinguish between two kinds of dynamical effect. The most obvious is that if particles move rapidly enough, there is not enough time for equilibrating the internal degrees of freedom. For example, electrostatic potentials might adjust only slowly to the equilibrium values at charged surfaces because the adsorption/desorption of ions can be a slow process. Similarly, the ions of the electrical double layer can be left slightly behind when a charged particle moves. For particles covered with adsorbed polymers, the rearrangements at the surface are often slow, causing the force between two such surfaces to depend on the rate of the relative motion.

Dynamic effects of this kind tend to be system dependent and determined by the specific chemistry at or near the surfaces. Thus, it is difficult to state general rules concerning their importance. However, if relaxation processes at the surface of the single particle are of the same magnitude as the time for encounters between particles, the effect of the internal relaxation process on the force between particles should be sought.

The second, more general, dynamical effect arises when particles, macromolecules, or aggregates move in a liquid medium. As a big unit moves, it creates a flow in the surrounding incompressible liquid. The presence of a second particle will influence the flow patterns in the liquid and will result in the two particles seeing each other through the flow of the intervening liquid. This effect is called hydrodynamic interaction.

Let us illustrate the effect by considering two similar spherical particles of radius R in a medium. Assume, for simplicity, that no force exists between the particles when they are at rest unless they are in direct contact. If one of the spheres moves toward the other with a velocity v, a velocity field will be created in the liquid. Liquid must move away from the front end of the sphere and flow back into the rear end. This flow field extends far into the liquid. When a second particle is present in the liquid, it will interfere with the liquid flow and a greater force must be applied on the moving particle to maintain the constant velocity $\vec{v}$. If we now define a friction coefficient

$$\mu = \frac{F}{v} \tag{5.7.1}$$

we see that the friction coefficient increases when the liquid flow is perturbed by a second particle.

For a single sphere, Sir George Stokes derived

$$\mu = 6\pi R\eta \tag{5.7.2}$$

where η stands for the viscosity of the liquid. Combined with the Einstein relation

$$D = \frac{kT}{\mu} \tag{5.7.3}$$

relating the diffusion coefficient and the friction coefficient, we find the expression

$$D = \frac{kT}{6\pi\eta R} \tag{5.7.4}$$

for the self-diffusion of a sphere in a pure liquid.

In the presence of a second particle, we concluded that the friction coefficient increased. By eq. 5.7.3, this increase implies that the diffusion coefficient depends on the distance between particles. We can calculate the distance dependence of the diffusion coefficient by solving the hydrodynamic equations with boundary conditions, taking into account the second stationary particle. The problem does not have a simple analytical solution, but for radial motion we can write

$$D(r) = D_0 G(r) \tag{5.7.5}$$

where the unperturbed diffusion coefficient D_0 is given in eq. 5.7.4. The correction factor $G(r)$ can be written in terms of a numerical approximation valid for $x > 0.01$.

$$G(r) = \frac{C_1 x + C_2 x^2 + x^3}{C_3 + C_4 x + C_5 x^2 + x^3} \tag{5.7.6}$$

where the coefficients C_i are summarized in Table 5.5. The variable $x = h/R = (r - 2R)/R$ represents the ratio between the surface-to-surface distance and the radius. The most notable feature of eq. 5.7.6 is that $G(r)$ becomes very small for small x values. This fact implies that as two particles come close, their

TABLE 5.5 Coefficients in Rational Fraction Approximation (eq. 5.7.6) to the Hydrodynamic Function $G(r)$

	C_1	C_2	C_3	C_4	C_5
Sphere–Sphere	0.154030	1.29993	0.0782416	1.10529	2.81955
Sphere–Plane	2.04185	5.60414	2.06393	8.59190	6.72180

From D. Y. Chan and B. Halle, *J. Colloid Interface Sci.* **102**, 400 (1984).

relative motion becomes very slow, a property that can have considerable importance for coagulation kinetics.

Although the liquid flow transmits a mechanical force from one particle to another, it affects only the dynamics of the system. These hydrodynamic forces possess a different character from the forces discussed in Sections 5.1 –5.6 in that they influence not the equilibrium distribution of the particles, but only the dynamics of the processes leading to equilibrium.

As a second example of the hydrodynamic effects, consider a sphere of radius R moving toward (or away from) a planar wall. Clearly, the planar wall perturbs the flow of liquid much more than a second sphere. However, this effect is mainly quantitative, and we can still use eqs. 5.7.5 and 5.7.6, but with different (larger) values for the coefficients C_i, in the latter equation. (See Table 5.5.) Thus, the diffusional motion of a colloidal sphere slows down considerably for surface separations that are of the same order as the radius of the particle. For large particles this effect sets in much further from the surface than any direct interaction forces of the double-layer or van der Waals type.

Similarly, if the colloidal particle is moved by some external force toward the planar surface, it will sense an increasing friction coefficient

$$\mu(r) = \mu_0 G^{-1}(r) \tag{5.7.7}$$

At constant velocity $\vec{v}$, this friction can be expressed as an interaction force $\vec{F}(r)$

$$\vec{F}(r) = (\mu(r) - \mu_0)\vec{v} = \mu_0(G^{-1}(r) - 1)\vec{v} \tag{5.7.8}$$

where we have subtracted the frictional force at infinite separation. The hydrodynamic force is repulsive when the sphere approaches the surface, but attractive when it moves away. The force in eq. 5.7.8, which is of utmost importance in lubrication, is often referred to as the lubrication force. Because $\mu_0 \sim R$, smooth surfaces that possess the largest possible R values are desirable to obtain the good lubrication achieved when the two surfaces do not come into direct contact.

Exercises

5.1. A suspension of clay particles is dewatered by applying an external pressure of $10^5\,\mathrm{N/m^2}$ ($\sim 1\,\mathrm{atm}$) across a membrane impermeable to the clay particles. At what relative humidity should the piece of clay be stored at room temperature to ensure that it neither dries nor takes up moisture?

5.2. Two identical disklike clay particles of radius $2 \times 10^{-5}\,\mathrm{m}$, thickness $5 \times 10^{-6}\,\mathrm{m}$, and density $3 \times 10^3\,\mathrm{kg/m^3}$ move in a stream. One of them settles at the bottom at a calm stretch while the other goes all the way to the sea. Calculate the thickness of the water layers separating the (highly charged) clay particles from the sediment on the bottom (also carrying a high negative charge) at equilibrium. Assume that the stream is free from salt while the seawater has a NaCl content of 3.5% w/w.

5.3. A film of water can condense from humid air on a quartz surface with a high density of ionized groups. Calculate the equilibrium thickness of the film as a function of the relative humidity. Assume that both the electrical double layer and the van der Waals forces are important. The Hamaker constant quartz–water–air is -1.0×10^{-20} J. ($T = 298$ K)

5.4. Sodium chloride has been deposited on a glass surface by evaporation of a thin electrolyte solution film. The resulting surface density of NaCl is one per 1000 Å^2. This surface is placed in a closed container thermostated at 25°C containing an open beaker with a 1×10^{-5} M NaCl solution. What happens?

5.5. Israelachvili and Adams measured the interaction energy per unit area between two charged mica surfaces across an aqueous solution of 1.0×10^{-4} M KNO$_3$. The table summarizes measured values at long range.

Separation (nm)	Interaction Free Energy ($\mu J/m^2$)
29	141.0
41	89.0
50	67.0
60	45.0
84	28.0
91	15.8
112	8.4
135	3.3

Are these measurements consistent with the theory for weakly overlapping double layers? In such a case, what is the value of Γ_0? ($T = 298$ K)

5.6. At the air–water interface of an SDS solution above the CMC (8 mM) there is a highly charged monomolecular film. Micellar aggregates are electrostatically repelled from the surface. Estimate the distance of separation between the monolayer and a micelle at which interaction equals kT ($T = 300$ K) by assuming either (a) that the Derjaguin approximation applies and both surfaces are highly charged or (b) that the micelle of an effective charge of $-20e$ approaches a surface with a surface charge density corresponding to -0.2 C/m^2. Calculate the electrostatic potential from the Gouy–Chapman theory.

5.7. Solve the linear Poisson–Boltzmann equation for two identical parallel homogeneously charged surfaces with charge density σ, separated by a distance h, and in equilibrium with a bulk aqueous solution with a concentration c_b of a 1:1 electrolyte. Evaluate the force between the surfaces.

5.8. Using the Hamaker theory, calculate the dispersion

interaction between two parallel sheets of thickness D separated by a distance h.

5.9. Estimate the equilibrium concentration profile of water molecules above a water surface at 25°C. Assume that the liquid density is uniform up to the liquid surface and calculate the profile starting from 10 Å above the surface. The Hamaker constant is 3.7×10^{-20} J.

5.10. Calculate the dispersion interaction between a half-plane and a sphere of radius R, using the Hamaker method. Comment on the validity of the Derjaguin approximation at different separations.

5.11. Calculate the translational diffusion coefficient of a spherical particle with radius 10 nm in water at 25°C. How is the diffusion coefficient changed when the sphere-approaches a planar wall 20 nm away?

5.12. A spherical colloidal particle of radius 10^{-6} m in an aqueous solvent is moved away from a planar surface at a constant velocity of 10^{-3} m/s. Calculate the required force in the range of distance of closest separation of 10^{-8}–10^{-6} m, considering only hydrodynamic interactions. Calculate also the ratio between the hydrodynamic force and the van der Waals force, using $H = 6 \times 10^{-21}$ J.

5.13. Derive an approximate expression for the adhesion energy between two rigid spherical particles of radius R, interacting only by van der Waals forces. (Assume a distance of closest approach of δ.)

5.14. Two flat surfaces are in adhesive contact in a liquid with molecules of radius R. The surfaces are pulled apart under atmospheric pressure to a separation $h \leqslant 2R$ so that no liquid can penetrate the gap. Estimate the force in the range $0 \leqslant h \leqslant 2R$, assuming that the surfaces only interact by van der Waals forces.

Literature

References	Section 5.1 Electrostatic Double Layers	Section 5.2 van der Waals Forces	Section 5.5 Forces Between Liquidlike Surfaces	Section 5.7 Hydrodynamic Interactions
Adamson & Gast	CHAPTER VI Long-range forces	CHAPTER VI Long-range forces		
Davis	CHAPTER 12.3 DLVO Theory	CHAPTER 12.3 DLVO Theory		
Hunter (I)	CHAPTER 9.1 Particle interactions	CHAPTER 9.2 Long-range attractive forces		
Hunter (II)	CHAPTER 7.1–7.5 Plates and Spheres CHAPTER 7.7.1 Adsorbed liquid films	CHAPTER 4 London and Liftshitz theories, Derjaguin approximation		CHAPTER 9.1–9.10 Transport and fluid flow
Israelachvilli	CHAPTER 12.7– 12.17 Double-layer forces	CHAPTERS 6, 7 & 11 Van der Waals forces CHAPTER 10.5 Derjaguin approximation	CHAPTER 14.7–14.9 Fluctuation forces	
Kruyt	CHAPTER IV Gouy–Chapman Theory	CHAPTER VI 9–11 London dispersion forces		
Meyers	CHAPTER 5 Double layer forces	CHAPTER 7 Long-range attractive forces		

[a]For complete reference citations, see alphabetic listing, page xxxi.

6

BILAYER SYSTEMS

CONCEPT MAP

Bilayers

Bilayer structures constitute a midpoint between normal and reversed amphiphilic structures and play crucial roles in biological and industrial processes.

This chapter focuses on the lipids of biological membranes (Figure 6.2), which constitute the building blocks for many biological structures. Surfactants relevant for industrial applications are also discussed.

Manifestations of Bilayer Structures

Bilayers are the basic structural element of many different global structures including:

- **Lamellar liquid crystals,** which are stacks of bilayers separated by solvent and containing hydrocarbon chains in the melted state. Defects in the bilayers cause them to close up on themselves to form onionlike shells, called liposomes, and multilayer tubes (Figure 6.16).

- **Gel phases** containing bilayers with crystallized hydrocarbon chains separated by liquidlike solvent. Mismatches between the cross-sectional areas of chains and head groups lead to tilted or interdigitated chains or rippled surfaces (Figure 6.3).

- **Bicontinuous phases** in which the bilayer is draped on a surface of zero mean curvature (Figure 6.4) or forms a sponge or L_3 structure, which is a disordered version of the cubic phase.

- **Vesicles** (Figure 1.6F), which are usually metastable and whose size depends on the preparation method.

- Various bilayer structures often differ in their free energies by less than kT and transform from one structure to another in response to small changes in temperature or composition.

Key Features of Bilayer Systems

The solubility of bilayer-forming amphiphiles is usually low (10^{-5}–10^{-10} M), and processes that proceed via monomer exchange occur very slowly; attaining equilibrium requires days or even years (Table 6.3).

When the hydrocarbon chains are melted, amphiphile diffusion in the in-plane, lateral direction occurs quickly, but rotation of the polar head groups through the center of the oil bilayers is slow.

Characterization of Bilayer Systems

X-ray diffraction provides information on bilayer structure and thickness, head group area, solvent thickness, whether the chains are melted or frozen, and electron density profiles across the bilayer.

Polarized light microscopy yields characteristic patterns identifying various types of phases.

Video-enhanced microscopy (VEM) provides real-time images (Figure 6.9) displaying bilayer structures and dynamics such as undulation and phase transformations.

Transmission electron microscopy (TEM) gives high resolution direct images (Figure 6.10) approaching molecular resolution but requires that samples be dehydrated or frozen.

Nuclear magnetic resonance (NMR) yields a structural parameter (Figure 6.12) relating hydrocarbon chain orientation in the bilayer and how it is changed by membrane additives (Figures 6.13 and 6.14).

Differential scanning calorimetry (DSC) detects changes in phase when an amphiphilic sample is heated and gives values of ΔH_m associated with these transformations (Figure 6.15).

Osmotic stress (OS) and **surface forces apparatus (SFA)** [Figure 6.16] techniques provide direct measurements of bilayer interaction forces (Figures 6.20–6.23), demonstrating short range stabilizing and undulation forces.

Role of Bilayer Structures in Biological Processes

A biological membrane is a complex structure containing lipids, proteins and polysaccharides in which lipids provide the basic organizing structural unit. Lipid bilayers provide

- A barrier to passive diffusion of small polar solutes (ions and sugars) and for macromolecules (proteins, nucleic acids, etc.).

- A unique two-dimensional solution environment for membrane proteins.

- An internal organizing structure which supports a multitude of chemical processes.

Role of Bilayers in Diffusion Processes

Even though a bilayer hydrocarbon core is only 30–40 Å thick, it provides an effective barrier to diffusion of ions and polar compounds across it.

Black lipid membranes (BLM, Figure 6.25) are useful for measuring a solute's permeability P_s, which is determined by its diffusion coefficient D_s, as well as the partition coefficient K_s between bulk solvent and membrane and the membrane thickness d_2, $P_s = D_2 K_s / d_2$ (eq. 6.3.6).

Because D_s is similar for all simple solutes, P_s is determined mainly by membrane solubility K_s and accounts for the range of P values: 10^{-6} m/s for water, 10^{-10} for sugars, and 10^{-14} for ions.

Diffusion of ions across membranes involves electrostatic interactions, and we can relate fluxes to differences in potential (eq. 6.3.13), which in turn can be related to permeabilities and concentrations.

Finally, we consider diffusion of solutes in lamellar liquid crystals in directions parallel (eq. 6.3.16) and perpendicular (eq. 6.3.22) to the bilayer structure to obtain weighted diffusion coefficients.

Roles of Bilayers as Solvent-to-Membrane Proteins

Proteins associate with membranes by (Figure 6.28):

- Spanning the width of the bilayer such as polar ends extending into the adjacent aqueous solution.

- Anchoring via hydrophobic interactions to one side of the membrane.

- Anchoring via a hydrocarbon tail.

- Binding electrostatically.

Interaction between identical proteins can be described in terms of a two-dimensional colloidal system involving

- Attractive van der Waals forces.

- Repulsive electrostatic forces.

- Short-ranged entropic repulsive forces arising from the flexibility of the apolar side chains.

- Attractive forces between the polar groups residing in the apolar lipid matrix.

Transmembrane Transport

The selective transport of solutes across membranes is a fundamental physiological process.

Transport can occur by:

- A membrane soluble carrier (Figure 6.32).

- A membrane spanning channel (Figure 6.33).

- A pump consuming ATP (eq. 6.4.1).

- Exocytosis/endocytosis.

In the **chemiosmotic mechanism** free energy is transformed from chemical bonds to electrostatic and entropic forms and then back into the chemical bond in the ATP molecule. This is accomplished by a series of selective transmembrane transport events.

The conduction of a **nerve signal** occurs through:

- A propagating membrane potential controlled by cross membrane ion transport.

- An endocytosis of transmitter substances at synapses (Figure 6.35).

6

Bilayers formed by amphiphilic molecules occupy a central position in the study of self-assembly for two reasons. First, they represent the midpoint between normal and reversed structures, and under suitable conditions, the majority of amphiphiles aggregate into bilayers. Second, the bilayer is the basic building unit of the biological membrane. A multitude of processes in the living cell rely on the unique and versatile properties of the membrane lipid bilayer.

This chapter focuses on the description of bilayer membranes formed by biological lipids. We also will discuss studies of bilayers formed by synthetic surfactants, particularly when illustrating experimental techniques as well as the general principles behind the formation of bilayers.

In the living cell the lipid membrane (see Figure 6.1) performs three basic functions. First, it serves as a barrier for molecular diffusion processes. Ions and small polar metabolites like sugars penetrate a bilayer very slowly to pass from the inside to the outside or vice versa. Macromolecules, like proteins, polynucleotides, or polysaccharides, cannot penetrate an intact bilayer by diffusion. The second role of the lipid bilayer is to act as a two-dimensional solvent for hydrophobic membrane proteins. A substantial fraction of the proteins in a cell depends on the lipid membrane for their function. The membrane proteins have adapted to the unique amphiphilic milieu of a 4 nm thick hydrophobic layer with an aqueous environment on both sides. The third function is to generate an intracellular structure that serves to organize the multitude of metabolic processes that occur in the living cell. It is becoming increasingly evident that this is a very important role of the membrane in all but the most primitive organisms.

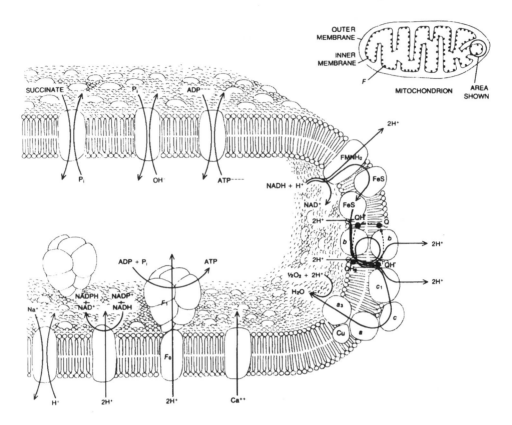

Figure 6.1
Schematic picture of the membrane of a mitochondrion. A number of proteins are dissolved in the lipid bilayer or adsorbed to the surface. Most of the specific functions involve the transport of an ionic molecule or atom across the membrane. (P.C. Hinkle and R.E. McCarty, *Sci. Am.* March 1978, p. 109.)

6.1 Bilayers Show a Rich Variation with Respect to Local Chemical Structure and Global Folding

6.1.1 Many Amphiphiles Form a Bilayer Structure

In Chapter 1, we discussed the self-assembly of amphiphiles and found that for a surfactant number $N_s = v/la_o$ around unity, the preferred aggregate structure is a bilayer. Thus, in relation to a micelle-forming amphiphile, one that produces a bilayer structure should have a bulkier apolar region and/or a smaller head group. Chemically, this structure can be obtained in several ways. One approach for obtaining a bilayer-forming amphiphile involves adding a second apolar chain of approximately the same size to a surfactant molecule. This strategy works irrespective of the nature of the polar head group. Common examples are the cationic dialkyl dimethyl quaternary ammonium surfactants that we discussed briefly in the previous chapter. These surfactants have widespread use in conditioners for hair and textiles. Aerosol OT (di-2-ethylhexylsulphosuccinate) is an anionic double-chain surfactant that has been much used in scientific studies.

Another approach involves reducing head group size. For nonionic surfactants of the alkyl oligooxyethylene type, a shorter ethylene oxide chain promotes bilayer formtion. For example, $C_{12}E_4$ forms a lamellar phase at ambient temperatures. Upon adding another ethylene oxide unit, the lamellar state forms first at an increased temperature, around 50°C. This behavior correlates with the strong temperature effects discussed in Section 4.3.5. Monoglycerides, which have small head groups, aggregate into bilayers or even inverted structures, suggesting a surfactant number slightly above unity. These amphiphiles have a broad use as food additives and in pharmaceutical formulations.

A micelle-forming surfactant can be induced to self-assemble into a bilayer by adding a long-chain cosurfactant such as an alcohol. It has a small polar head and makes the average size of the head group smaller, and ultimately, a bilayer emerges as the most stable structure. The so-called catanionic surfactants which contain a mixture of cationic and anionic amphiphiles, constitute a more intricate variation on this theme. Since these amphiphiles attract one another electrostatically, the effective head group size becomes much smaller than the sum of the head group areas for the separate systems, and the single-chain surfactants will assemble into a bilayer. In a perfectly matched system, the resulting properties resemble those of a double-chain zwitterionic system.

6.1.2 Membrane Lipids Exhibit Chemical Variations on a Common Theme

For the major fraction of membrane lipids, chemical structure can be seen as variations on a common theme. The apolar group typically consists of two hydrocarbon chains originating from various fatty acids. These chains are attached via ester or ether covalent bonds to a linker group, usually a glycerol, which also is bonded to the polar head group. Figure 6.2 summarizes the chemical structure of some common lipids, including conventional acronyms.

From Figure 6.2 we see that the chemical diversity of polar groups is similar to that found for synthetic surfactants. We find ionic, zwitterionic, and sugar head groups. All ionic lipids carry a negative charge, a fact that has considerable significance for the electrostatic interactions. Typically, 10—20% of the lipids of a membrane are charged. The nature of the head group depends on the organism. Glycolipids with mono- or disugar polar groups dominate in plants. In animals, phospholipids are most abundant, while in bacteria both types of lipids are common.

The apolar parts of the membrane lipids originate from the fat metabolism, and the chemical diversity in the fatty acid chains is the same in the membrane lipids as in the fat. The hydrocarbon chain length typically ranges from 16 to 22, with the degree of unsaturation increasing with increasing chain length. Table 6.1 shows the chemical formulae and names of some common fatty acids, and Table 6.2 shows the distribution of fatty acids in phospholipids of different organs in some animals. This distribution varies between species, between

Figure 6.2
(A) Chemical structure of some common classes of lipids. The phospholipids dominate in animals, while the glycolipids dominate in plants and bacteria.

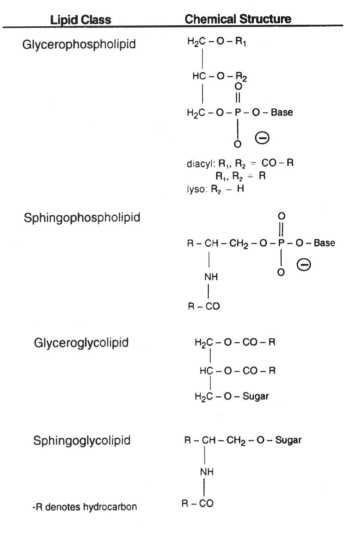

Lipid Class	Chemical Structure
Glycerophospholipid	
Sphingophospholipid	
Glyceroglycolipid	
Sphingoglycolipid	

-R denotes hydrocarbon

A

organs of the same species, and even between different locations in the same organ.

In addition to the types of lipids shown in Figure 6.2, higher organisms also possess lipid of a quite different chemical nature: the infamous cholesterol. For example, the red blood cell has a very high content of cholesterol ($\approx 50\,\text{mol}\%$ lipid). Cholesterol consists of rigid, relatively planar steroid skeleton with a single OH as the polar group. By itself, it is not a bilayer-forming lipid at ambient temperatures, and it plays a role similar to that of an alkanol cosurfactant in surfactant systems.

Given the variations in head groups, linker and fatty acid chains, and also including substitution patterns, a single cell contains on the order of 100 different lipid species. One of the major enigmas of lipid biochemistry is to understand the func-

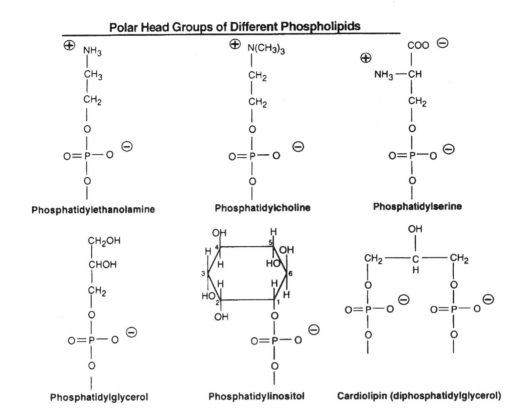

Polar Head Groups of Different Phospholipids

Phosphatidylethanolamine

Phosphatidylcholine

Phosphatidylserine

Phosphatidylglycerol

Phosphatidylinositol

Cardiolipin (diphosphatidylglycerol)

B

Figure 6.2
(B) Chemical structure of the polar head groups of some common phospholipids.

tional significance of this diversity. Lipid composition varies between organisms, between cells, between different organelles within the cell, and even between the two sides of the bilayer. We can readily understand some trends. The degree of chain unsaturation increases with decreasing growth temperature. The values for tuna muscle given in Table 6.2 exemplify this effect. Saturated alkane chains tend to crystallize, and unsaturation is introduced to ensure a liquidlike character of the membrane. As a contrast, Table 6.2 also shows a high fraction of saturated C_{16} chains in the membrane of pig lungs. Their functional significance may be that unsaturated chains are susceptible to oxidation by O_2. Having saturated chains predominate in the cells that have the highest oxygen exposure minimizes the chemical degradation.

6.1.3 Comparing the Properties of Spherical Micelles and Bilayers Provides Useful Insight into the Many Distinctive Molecular Properties of Bilayers

Table 6.3 summarizes many of the properties of spherical micelles and bilayer systems. Monomer solubilities of typical zwitterionic and polar surfactants (as measured by their CMCs)

TABLE 6.1 Common Name of Acyl Chains and Melting Point of the Corresponding Fatty Acid

Acyl Chain	Common Name	Symbol	Melting Point of Acid (°C)
$CH_3(CH_2)_6-\overset{\overset{\textstyle O}{\|\|}}{C}-O-$	capryl	8:0	16
$CH_3(CH_2)_8-\overset{\overset{\textstyle O}{\|\|}}{C}-O-$	capric	10:0	32
$CH_3(CH_2)_{10}-\overset{\overset{\textstyle O}{\|\|}}{C}-O-$	lauroyl	12:0	44
$CH_3(CH_2)_{12}-\overset{\overset{\textstyle O}{\|\|}}{C}-O-$	myristoyl	14:0	54
$CH_3(CH_2)_{14}-\overset{\overset{\textstyle O}{\|\|}}{C}-O-$	palmitoyl	16:0	63
$CH_3(CH_2)_{16}-\overset{\overset{\textstyle O}{\|\|}}{C}-O-$	stearoyl	18:0	70
$CH_3(CH_2)_{18}-\overset{\overset{\textstyle O}{\|\|}}{C}-O-$	arachidoyl	20:0	77
$CH_3(CH_2)_7\overset{H\quad\ H}{\underset{}{\diagdown C=C \diagup}}(CH_2)_7-\overset{\overset{\textstyle O}{\|\|}}{C}-O-$	oleoyl	$18:1^{\Delta 9}$	13
$CH_3(CH_2)_4(\overset{H\quad\ H}{\underset{}{\diagdown C=C \diagup}}CH_2)_2-(CH_2)_6-\overset{\overset{\textstyle O}{\|\|}}{C}-O-$	linoleoyl	$18:2^{\Delta 9,12}$	−5
$CH_3CH_2(\overset{H\quad\ H}{\underset{}{\diagdown C=C \diagup}}CH_2)_3-(CH_2)_6-\overset{\overset{\textstyle O}{\|\|}}{C}-O-$	linolenoyl	$18:3^{\Delta 9,12,15}$	−11
$CH_3(CH_2)_4(\overset{H\quad\ H}{\underset{}{\diagdown C=C \diagup}}CH_2)_4-(CH_2)_2-\overset{\overset{\textstyle O}{\|\|}}{C}-O-$	arachidonoyl	$20:4^{\Delta 5,8,11,14}$	−50

are $\sim 10^{-4}$ M, but those of the phospholipids that form bilayers are much lower, $\sim 10^{-10}$ M. This difference of a factor of 10^6 in monomer solubilities reflects the difference between having one or two hydrocarbon chains and leads to profound differences in many of the molecular properties of micellar and bilayer systems.

As described in Section 4.2, the formation of micelles has been characterized by a number of experimental techniques.

TABLE 6.2 Distribution of the Most Common Fatty Acids of Phospholipids in Different Membranes

Source	$m:n(\%)^{a,b}$				
	16:0	18:0	18:1	20:4	22:6
Phosphatidylcholine					
Brain (human)	34	13	45	1	0
Lung (pig)	47	9	26	5	0
Muscle (tuna)	19	5	8	7	55
Liver (rat)	30	17	10	18	3
Phosphatidylethanolamine					
Brain (human)	5	30	9	13	30
Liver (rat)	25	14	6	10	21

Data from D. Marsh, *Handbook of Lipid Bilayers*, CRC Press, Boca Raton, FL, 1990.
[a] m denotes the number of carbons and n the number of cis double bonds.
[b] These fatty acids account for $\approx 90\%$ of the total content in the cases listed.

Micelle formation clearly illustrates the characteristics of the aggregation process described in Section 1.1. The CMC defines the "start" of the process, the point at which micelles begin to form spontaneously. Adding more surfactant leads to the formation of more micelles with well-defined aggregation numbers rather than leading to larger micelles. This "stop" process limits aggregate size. The micelle's radius, which is determined by the length of the fully extended hydrocarbon chain, leads to aggregates with well-defined properties.

Our direct information on the bilayer aggregation process is more limited because most experimental techniques are not sensitive enough to yield useful data at such extremely low monomer concentrations. Monomer solubility still defines the start process, and the bilayer's thickness is a well-defined

TABLE 6.3 Comparison Between Properties of Micellar and Bilayer Aggregates

Property	Micelles	Bilayers
Monomer solubility	$\sim 10^{-2}$ M	$10^{-5}-10^{-10}$ M
τ for monomer exchange	$10^{-3}-10^{-6}$ s	$10^{2}-10^{-3}$ s
τ for aggregate lifetime	$10^{-1}-10^{-3}$ s	days to years
Characteristic temperature	Krafft temperature	Chain-melting temperature
Structural directionality	All directions equivalent	Lateral diffusion rapid, flip-flop slow
Aggregation pattern	Forms well-defined aggregate at well-defined CMC	Basic structural unit appears in a variety of global structures

quantity determined by the length of the hydrocarbon chains while there is no molecular limit to bilayer growth in the lateral direction. In an ideal situation, bilayers attempt to minimize the degree to which the hydrocarbons at their edges are exposed to the aqueous environment by extending indefinitely in the lateral direction. In reality, finite samples show defects, and bilayers tend to close up on themseves, leading to a variety of macroscopic structures described in the next section.

In micellar solutions, the relaxation times for monomer exchange range from 10^{-3} to $10^{-6} s^{-1}$, and they involve a diffusion-limited process that depends directly on the monomer concentration. In bilayers, monomer exchange is considerably slower, as predicted by their much lower monomer solubility.

Lifetimes for typical micelles vary from 10^{-3} to $10^{-1} s$. With bilayers, such transformations involve rearranging the entire bilayer structure, and a particular specific macroscopic state often reflects the methods used to prepare the sample. Thus, micelles respond rapidly to changes in their environment, while bilayers are much more sluggish. It can require hours, days, or even years for a bilayer to achieve an equilibrium state.

The transport properties of micelles and bilayers also display major differences. With spherical micelles, all directions are equivalent, and a surfactant molecule can diffuse very rapidly within a micelle. A bilayer, on the other hand, has two distinct diffusion modes. The first involves lateral diffusion within the bilayer plane. It occurs rapidly since it mainly involves the amphiphile chains moving through the oillike region of the bilayer. The second diffusional mode is slow since it involves moving the head group from one side of the bilayer to the other. This "flip-flop" requires the polar head group to rotate through the oillike interior of the bilayer. In pure bilayers, the time required to execute this flip-flop process is typically hours to days.

6.1.4 Pure Amphiphiles Form a Range of Bulk Bilayer Phases

The lamellar liquid crystal is the generic bulk phase of a bilayer-forming amphiphile. As illustrated in Figure 1.6c, a lamellar liquid crystal consists of a stack of bilayers separated by aqueous films. In the liquid crystalline state, the molecules diffuse freely in the lateral direction, while they move in the perpendicular direction only on a very long time scale. The apolar chains behave as in a liquid with a sizable fraction of gauche conformations. The forces acting between bilayers determine the thickness of the aqueous layer. As discussed in the previous chapter, a range of forces is operating, and bilayer systems are very useful in the study of these forces.

When a lamellar phase is cooled, it undergoes a phase change that can lead to the formation of a crystal or sometimes to a stable or metastable gel phase. Figure 6.3 shows some typical gel phase structures. Their characteristic feature is that the alkyl chains are crystallized while liquidlike solvent still

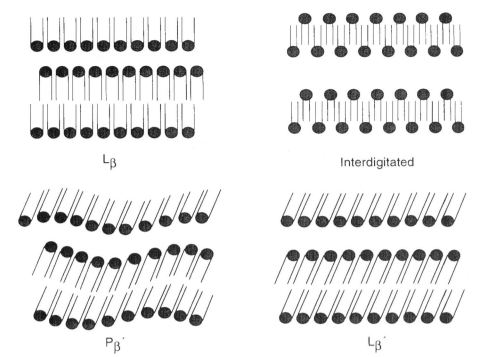

L_β

Interdigitated

$P_{\beta'}$

$L_{\beta'}$

Figure 6.3
When alkyl chains
crystallize, they have a
cross-sectional area of
$\approx 20\,\text{Å}^2$ per chain in a
plane perpendicular to the
direction of the chain. If
the area of the polar groups
in the bilayer matches that
of the chain(s), a structure
L_β is adopted. In the more
likely case of a mismatch
between polar head and
chain areas, a tilted
structure, as in $L_{\beta'}$, or a
rippled structure, as in $P_{\beta'}$,
can appear. For single-
chain amphiphiles, chains
may even interdigitate,
making the thickness of the
apolar layer the same as
the length of a single chain.

exists between the bilayers. The crystalline state of the chains
implies that more stringent conditions are imposed on the
packing. In this case, a mismatch can easily arise between the
cross-sectional area per chain and the head group area. To
accommodate such a mismatch, the chains may tilt relative to
the bilayer plane, the bilayer may develop a ripple, or the
chains may even interdigitate, depending on the degree of
mismatch.

Bilayers form the basic structural element in other ordered
phases. Figure 6.4 shows a representative bicontinuous phase of
cubic symmetry in which the bilayer is draped on a surface of
zero mean curvature. Because the structure is bicontinuous, the
amphiphile and the solvent each diffuse in three dimensions
within its own medium, displaying an effective diffusion coefficient only slightly smaller ($\sim 2/3$) than its value in a homogeneous solution. Under these circumstances, the bilayer
separates the solvent into two distinct regions. Molecular transport from one region to the other involves passage through the
bilayer, a fact that has particular relevance for biological membrane systems.

In isotropic solutions, bilayers can appear in the form of
vesicles shaped like spherical shells (see Figure 1f). Depending
on the nature of these vesicles, we usually distinguish between
small unilamellar vesicles (SUVs; radius $\lesssim 100\,\text{nm}$), large unilamellar vesicles (LUVs), and multilamellar liposomes. There is
no sharp distinction between liposomes and a lamellar phase
with excess solvent. Vesicles tend to occur most readily for
systems with surfactant numbers N_s slightly below one. They
are seldom true equilibrium systems, but they are often meta-

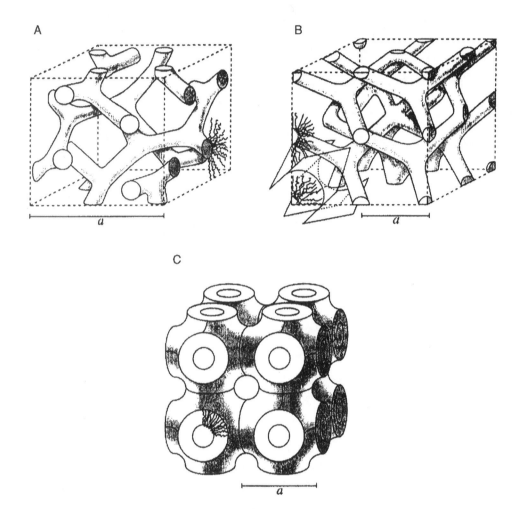

Figure 6.4
Illustration of three different arrangements of bilayers in cubic symmetry: (a) a gyroid structure; (b) a double diamond (as in two interwoven diamond lattices); and (c) plumbers nightmare with the same symmetry as in the monolayer version of Figure 1.6e. (Seddon, *Biochem. Biophys. Acta* **1031**, 1–69 (1990).)

stable enough to allow detailed physical and chemical studies that have proved very valuable. In vitro studies of membrane proteins are one example.

An intriguing bilayer arrangement commonly found in surfactant systems is the so-called sponge or L_3 phase. In this phase, the bilayer extends for an unlimited distance in three dimensions, forming a single macroscopic aggregate with isotropic properties. This phase corresponds to a disordered version of a cubic phase shown in Figure 1.6e and it occurs for surfactant numbers N_s slightly larger than one.

The difference in free energy between the various bilayer structures often is merely on the order of a small fraction of kT per molecule. Thus, small changes in temperature or composition can cause the transformation from one structure to another.

6.1.5 Vesicles Can be Formed by Several Methods

Vesicles commonly occur in the living cell and many physiological processes involve steps where vesicles are formed or consumed by fusion with another bilayer system. The importance of vesicles is not restricted to living systems and they are also very useful for in vitro studies of membrane physical and chemical properties. Because most vesicles are thermodynamically unstable, the properties of a vesicle solution depend on how it is prepared.

A simple but nonselective way of producing vesicles is to subject a lamellar dispersion to ultrasound, such as a lamellar phase containing excess water. The sound waves tear the lamellae, which then can reseal into vesicles. The smaller the vesicle, the less the perturbation affects it; so the result of the ultrasound treatment is a broad but skewed size distribution of vesicles that peaks at rather small aggregates, typically $R \simeq 150\text{---}200\,\text{Å}$. A disadvantage of the sonication process is that the large energy input may chemically degrade the lipid.

Cholate dialysis represents a more controlled method for preparing vesicles. In this technique we prepare a dilute solution of lipid–cholate mixed micelles and dialyze it. High monomer solubility enables the cholate to readily pass the dialysis membrane. As a result, the cholate content of the lipid solution will decrease slowly but steadily, and the lipid content of the micelles will increase until the large aggregates finally close to form vesicles. These vesicles possess a more uniform size, which we can control by varying parameters such as salt content, temperature, and pH.

Another way of producing vesicles is to inject a solution of the lipid in an organic solvent or solvent mixture into an excess of water. Large unilamellar vesicles form as the organic solvent evaporates or is dissolved in the water. Because the experimental conditions can vary greatly, obtaining vesicles of specified properties with this method is to some extent an art rather than a science.

In many instances, vesicles appear "spontaneously" as we prepare the lipid–water system. We can obtain large unilamellar vesicles by simply taking a dry lipid powder and adding the aqueous solvent. With charged lipids, continuous dilution of the system can lead to the formation of vesicles under the action of double-layer repulsions.

The kinetic stability of a vesicle system depends on the rate of the process by which two vesicles fuse to form a larger one. Vesicle stability is really an example of the general problem of colloidal stability, discussed further in Chapter 8.

The fusion process is easier to understand if we think of it as occurring in two steps. In the first step, the respective bilayers come into proximity. The occurrence of this step depends strongly on interbilayer forces. For charged systems with low electrolyte concentrations, double-layer repulsion largely prevents two vesicles from meeting. For zwitterionic phospholipid vesicles, a sizable shortrange stabilizing force also tends to prevent close molecular contact.

The second step in the process actually fuses the two bilayers. Any substantial barrier to this process may cause the two vesicles to aggregate and stay together for long periods without ever fusing. Factors that influence the fusion process are the internal strength of the bilayer, the ease with which nonbilayer structures form, and mechanical tensions in the bilayer.

Some chemicals, such as peptides, as well as lipids like lysolecithin and ions like calcium, act as so-called fusogens to promote fusion. The chemical diversity of the fusogens indicates that no single mechanism causes fusion; rather we can induce it through different molecular pathways.

6.2 Complete Characterization of Bilayers Requires a Variety of Techniques

6.2.1 X-Ray Diffraction Uniquely Identifies a Liquid Crystalline Structure and Its Dimensions

Liquid crystalline phases are amenable to detailed structural studies by x-ray or neutron diffraction. In a lamellar phase, symmetry planes are parallel so that higher order reflections are simple multiples of the first-order one. Thus, we observe sharp peaks when the Bragg condition ($2d \sin \theta = n\lambda$) is satisfied, where d is the lamellar repeat distance, $n = 1, 2, 3$, and θ is the diffraction angle and λ the wavelength (Figure 6.5). In phases of other symmetries, all diffraction planes are no longer parallel, which results in different characteristic diffraction patterns. For a typical lamellar phase, the repeat distance d lies in the range 4 nm and up. With $\lambda = 0.14$ nm for the typical Cu(K_α) radiation source the scattering angle $\theta \simeq \sin \theta \leqslant (0.14/8)(180/\pi) \leqslant 1.0$ is small, and special devices are needed to record the scattering so close to the primary beam.

The structural difference between the liquid crystal and the gel phases lies in the packing of the hydrocarbon chains; the difference in scattering properties appears at high scattering

Figure 6.5
Bragg reflection in a cubic NaCl crystal. Constructive interference occurs when the distance
$ABC = 2d \sin \theta$ equals the wavelength of the radiation.

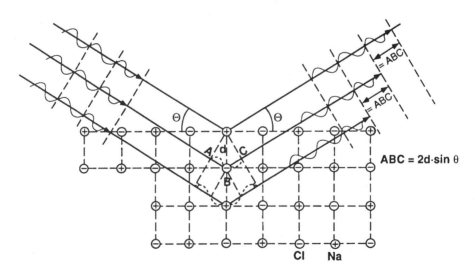

ABC = 2d·sin θ

Cl   Na

Figure 6.6
Representation of a lamellar phase with a repeat distance d in terms of pairs of layers of thickness d_1 and d_2 containing solvent and amphiphile, respectively. With a total lamellar area of A, the total volume $V = dA$, and the volumes of the respective regions are $d_iA = V_i = n_iv_i$, where n_i equals the amount and v_i the partial molar volume. Eliminating the area A, we find $d_2 = d(1 + n_1v_1/n_2v_2)^{-1}$.

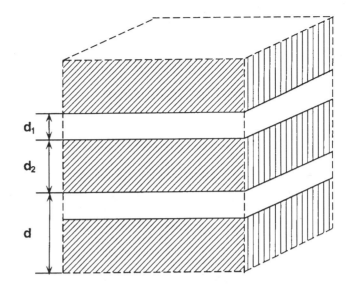

angles. For frozen chains, a sharp peak is observed at an angle that corresponds to ordered chains separated by 0.42 nm. For liquid crystalline systems, this peak shifts to slightly higher diffraction angles, corresponding to an increased average chain separation of 0.45 nm. An increase in molecular disorder also increases the width of the diffraction peak.

If we know the partial molal volumes v_1 and v_2 of the constituents, we can use the measured repeat distance d and the composition (n_1, n_2) to calculate the thicknesses of the individual polar (d_1) and apolar (d_2) layers (see Figure 6.6). We have $d = d_1 + d_2$ and $d_1/d_2 = n_1v_1/n_2v_2$, giving $d_2 = d/(1 + n_1v_1/n_2v_2)$, provided the two components are completely separated into the two domains. This analysis also gives us the area per amphiphile $(a_0 = 2v_2/d_2)$. Figure 6.7 shows such an analysis

Figure 6.7
Double logarithmic plot of repeat distance (Å) versus inverse amphiphile volume fraction $\Phi_a = n_2v_2/(n_1v_1 + n_2v_2)$ for CTAB–hexanol–water (molar ratio 2.38:1). The straight lines have a slope of 1.0, showing that $d \sim \Phi_a^{-1}$, which is characteristic for one-dimensional swelling. (K. Fontell, A. Khan, D. Maciejewska, and S. Puang, *Colloid Polym. Sci.* **269**, 727 (1991).)

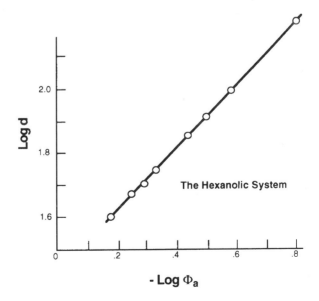

Figure 6.8
Density profiles in a fluid
bilayer. The glycerol
groups precisely mark the
water–methylene
boundary for DOPC at
66% *rh*. The double arrow
indicates the minimum
instantaneous thickness of
the bilayer, taken as the
transbilayer separation of
the extreme edges of the
water distributions defined
by their intersections with
the double-bond
distributions. We define
this as the dynamic
thickness. Here it is 28.6 Å,
which is several angstroms
smaller than the equivalent
hydrocarbon slab
thickness. However, as the
hydration of the bilayer
increases, the dynamic
thickness is expected to
decrease. Notice that
thermal motion causes a
small but significant
overlap of the water and
double-bond distributions.
(M.C. Wiener and S.H.
White, *Biophys. J.* **61**, 434
(1991).)

performed for varying amphiphile concentrations. The data
illustrates types of behavior known as one-dimensional swelling. A repeat distance inversely proportional to the volume
fraction of amphiphile demonstrates that the amphiphile bilayers remain unchanged in the swelling process. With two-
dimensional swelling, the forces between the bilayer surfaces
become strong enough to affect the intrabilayer packing as well,
resulting in a varying bilayer thickness and area per amphiphile.

By carefully measuring the intensity of the scattered
x-rays or neutrons for an oriented sample, we can determine the
density profile of various components across a repeat unit.
Figure 6.8 shows such profiles for dioleoylphosphatidylcholine
(DOPC) in a slightly swollen liquid crystalline state. The water
density profile shows two Gaussians separated by approximately 48 Å, which corresponds to the repeat distance. At the
relative humidity of 66%, water with a bulk density is not
present in any region. The shaded glycerol region overlaps with
the water and so does the tail of the CH_2 distribution. A double
bond located between the ninth and tenth carbon atoms in the
alkyl chain can be discerned. The figure shows a clear dip in
the alkane density profile at the center, which indicates that the
bilayer is really composed of two distinct monolayers with
some overlap. Indeed, in freeze-fracture electron microscopy
studies lamellar structures preferentially break at this bilayer
midplane. In the same way some overlap exists across the water
between the head groups of one bilayer with those of the
adjacent one.

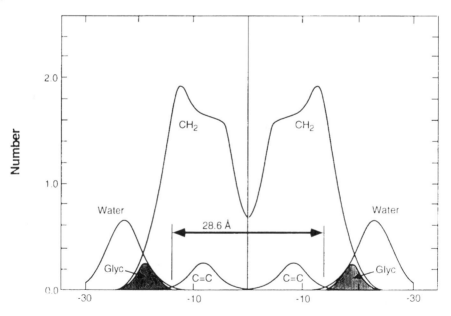

Distance from HC Center (Å)

6.2.2 Microscopy Yields Images of Aggregate Structures

A simple method that often enable us to obtain qualitative information about phase structure uses a microscope equipped with crossed polarizers to study the sample. Liquid crystalline phases are inherently anisotropic, making the index of refraction depend on orientation. In general, such birefringent systems change the plane of polarization of polarized light. An isotropic solution appears black between crossed polarizers, but a liquid crystalline system is more transparent.

A lamellar phase partly orients between two microscope coverslips. Polarized light propagating along the optical axis perpendicular to the lamellae retains its direction of polarization. However, defects in the packing have a dramatic effect on the polarized light. Depending on the nature of the defects, we can observe characteristic patterns, which can be used provisionally to identify different types of phases. When the anisotropy in the refractive index is strongly wavelength dependent, these patterns can display beautiful colors.

Several microscopy techniques provide direct visualization of bilayer structures. One advantage of light microscopy is that it permits us to examine objects in their natural aqueous environment. However, two major limitations reduce the effectiveness of the technique for characterizing amphiphilic microstructures. First, contrast limits our ability to distinguish delicate structures from the surrounding fluid and thereby places a lower limit on the size of the object we can study. Second, resolution limitations restrict the amount of structural information we can extract from microscopic observations; we cannot distinguish objects separated by a distance of less than 0.5λ. Video-enhanced microscopy (VEM), which employs a light microscope equipped with differential interference contrast (DIC) optics, a video camera for image detection, and a computer for image analysis, overcomes some of these limitations. Replacing the human eye with a video camera for image detection and using a computer for real-time image processing considerably increases our ability to detect small structures of inherently low contrast and at the same time relaxes the resolution restriction to 0.3λ. VEM lets us see isolated latex spheres as small as 600 Å and detect extended individual bilayers only 60 Å thick. We cannot obtain direct structural dimensions with such small objects because the images are enlarged by diffraction. However, we can see real-time behavior and obtain some feeling for the dynamics of such structures.

Figure 6.9 reproduces a photograph of a VEM image of a 1.7 wt % double-chain ionic surfactant sample that is 6 months old. It displays the diversity of structures present in the sample and confirms the tendency of most bilayers to be curved back on themselves, eliminating any exposed bilayer hydrocarbon edges. The large structures at points A and B show the characteristic birefringence of liquid crystalline structures when the VEM optics are switched to polarizing optics without moving the microscope slide. We can follow the undulations of individ-

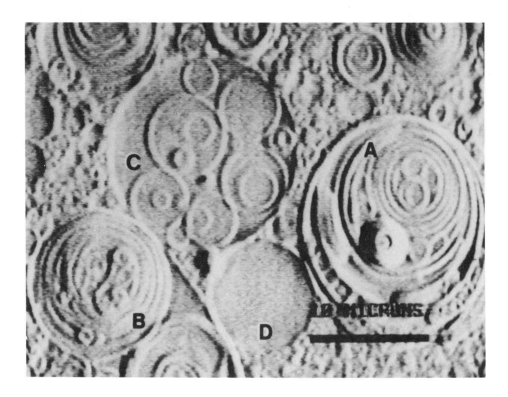

Figure 6.9
VEM micrograph of 1.7%
SHBS: bar = 10 μm; A, B,
liposomes (birefringent
under crossed polars); C,
large vesicle containing
entrapped smaller vesicles;
D, vesicle with a "dust-
storm" appearance in real
time, indicating that it
is filled with small,
unresolvable particles.
(D.D. Miller, J.R. Bellare,
D.F. Evans, Y. Talmon, and
B.W. Ninham, *J. Phys.
Chem.* **91**, 674 (1987).)

ual layers within the larger birefringent structure in real time at
point B and the caged movement of smaller vesicles entrapped
within larger vesicles at point C. The "unstructured" spherical
region of diameter $\approx 9\ \mu$m (point D) is of particular interest. In
real time, it shows the "sandstorm" appearance associated with
structures beyond the resolution limit of the microscopy. We
will return to this point later.

In principle, high resolution direct images provided by
transmission electron microscopy (TEM) can yield information
on individual molecules. TEM experiments require compatibil-
ity with the high vacuum in the microscope ($<10^{-6}$ torr), and
samples must be thin enough ($<0.25\ \mu$m) to permit electrons to
pass through them. Also, to prevent image blurring, all motion
on the supramolecular scale must be arrested. Until recently
these factors severely limited our ability to apply TEM to the
study of amphiphilic microstructures.

TEM samples containing amphiphilic molecules usually
are prepared in one of two ways. The first method is borrowed
from the biological sciences. It involves dehydration of the
specimen, followed by staining or metal coating to obtain
image contrast. We obtained our original pictures of relative
nonlabile amphiphilic structures, such as liposomes and bilayers prepared from long-chain phospholipids, using this technique. However, surfactant systems containing labile amphiphiles rearrange when dehydrated or stained, which produces
artifacts.

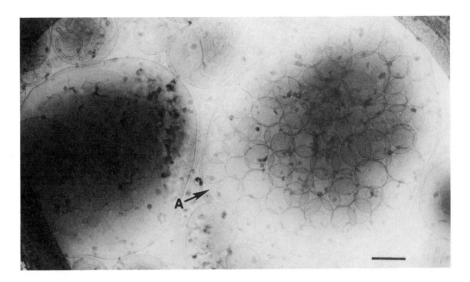

Figure 6.10
Vitrified hydrated unstained (VHU) cryo-TEM picture of a 1.7 wt% SHBS sample. Note large (1.75 μm) single-walled vesicle (A) encapsulating many smaller vesicles; bar = 0.25 μm. (D.D. Miller, J.R. Bellare, D.F. Evans, Y. Talmon, and B.W. Ninham, *J. Phys. Chem.* **91**, 674 (1987).)

The second approach involves rapid freezing of a sample. This technique has two variations. In the freeze-fracture method, small droplets of the sample are projected onto a cryogenically cooled metal surface and mechanically fractured after freezing. The microscope images a template of the fractured surface produced by vapor deposition of a thin metal layer over the sample's surface. Alternatively, we can suspend a thin layer of the liquid sample on a holey carbon grid and plunge it into liquid ethane. The sample freezes so rapidly that amorphous ice is formed. Ice crystals give diffraction patterns that would dominate the TEM image, but with amorphous ice the background is as clear as a glass window. We transfer the frozen specimen to the TEM in a cold stage cooled by liquid nitrogen. Both these techniques considerably increase our ability to image surfactant microstructures. Both depend on the ability to freeze a structure so rapidly that it does not have time to reorganize. We use cryo-TEM images to illustrate this chapter; those produced by the freeze-fracture method appear in Chapter 11.

Figure 6.10 shows several images of a sample with the same composition as used for the optical image in Figure 6.9. Of particular interest is the enclosed vesicle structure of diameter 1.75 μm (point A, Figure 6.10), which contains a large number of small unilamellar vesicles. We believe this structure is similar to the structure at point D in Figure 6.9. The cryo-TEM picture also shows that freezing has been rapid enough for spherical micelles with aggregation number of 65 to survive the freezing process.

6.2.3 Nuclear Magnetic Resonance Techniques Provide a Picture of Bilayer Structure on the Molecular Level

Several nuclear magnetic resonance (NMR) techniques allow us to study the molecular arrangement in a bilayer. We can obtain the most detailed information by using deuterium NMR spectroscopy in the lamellar liquid crystal.

Because of the anisotropic nature of the bilayer, a so-called quadrupolar splitting arises in a ^{2}H NMR spectrum. For deuterions on an alkyl chain, we can relate the magnitude of this splitting to the order parameter

$$S_{CD} = \tfrac{1}{2}\langle 3\cos^2\theta_{CD} - 1\rangle \qquad (6.2.1)$$

where θ_{CD} represents the angle between the C—D bond and the normal to the bilayer and the brackets denote a thermal average. The order parameter S_{CD}, is specific for each C—D bond of a deuterated hydrocarbon chain. By measuring S_{CD} at all the different positions in a chain, we can obtain a picture of how it is arranged in the bilayer. Figure 6.11 shows a deuterium NMR spectrum of a perdeuterated palmitoyl chain.

A typical order parameter profile for a saturated chain in a phospholipid bilayer is shown in Figure 6.12. The parameter is nearly constant for the first eight positions, reflecting the preferential orientation of the chain perpendicular to the surface. Farther away from the polar head, a drop in the order parameter indicates a near isotropic environment in the center of the bilayer. Figures 6.13 and 6.14 illustrate the effect of two different additives on the order parameter. Cholesterol, which induces a marked increase in S_{CD} all along the lecithin hydrocarbon chain, is anchored at the polar–apolar interface by its hydroxy polar group. The stiff steroid skeleton of cholesterol limits the conformational freedom of neighboring chains and thus favors a trans relative to a gauche conformation, with a concomitant increase in the order parameter.

Cholate influences the order in the bilayer in a quite different way. In addition to its charged group, its steroid part contains three OH groups, all located on the same side of the molecule. Unlike the cholesterol molecule, cholate cannot pen-

Figure 6.11
Deuterium NMR spectrum of a lamellar liquid crystalline phase in the system d_{31} palmitoyl–oleoylphosphatidyl-choline–sodium cholate–water. Signals from all the 15 different deuteriums along the palmitoyl chain are resolved through the difference in the quadrupolar splitting. The numbering refers to the carbon atom to which deuteriums are bound and starts at the carboxylate end of the palmitate (G. Lindblom, H. Wennerström, K. Fontell, O. Soderman, and G. Arvidson, *Biochemistry* **21**, 1553 (1982).)

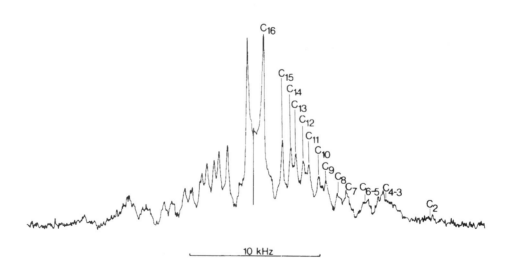

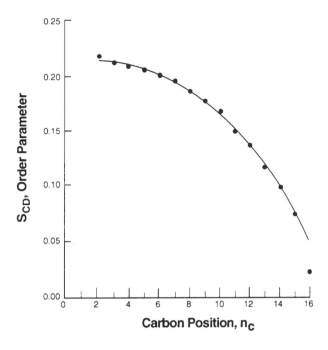

Figure 6.12
Order parameter profile for palmitoyl (d_{31}) chain palmitoyl–oleoyl-phosphatidylcholine (POPC) in a lamellar liquid crystal at 30°C. (Data from G. Lafleur, P. Cullis, and M. Bloom, *Eur. Biophys. J.* **19**, 55 (1990).)

etrate the bilayer readily, but it can adsorb flat on the surface. When a cholate molecule rests flat on the surface, a major perturbation is induced in the alkyl chains as they fill the space below it. Therefore, cholate drastically decreases the order parameters, and as the amount of cholate increases, the plateau region in the order parameter profile disappears. In fact, intro-

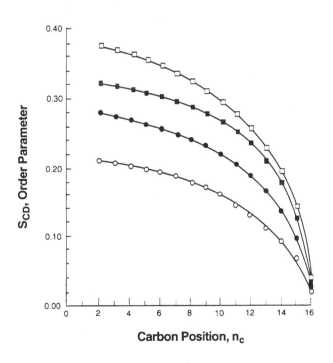

Figure 6.13
Order parameter in a palmitoyl chain increases as cholesterol is added to the lamellar phase: pure POPC (○); 20 mol % cholesterol (●); 30 mol % (■); 45 mol % (□) at 30°C. (G. Lafleur, P. Cullis, and M. Bloom, *Eur. Biophys. J.* **19**, 5 (1990).)

Figure 6.14
The order parameter decreases drastically for the palmitoyl chain of DOPC when sodium cholate is added, cholate/DOPC ratio: ▲ 0.084; ■ 0.78. (G. Lindblom, H. Wennerström, K. Fontell, O. Soderman, and G. Arvidson, *Biochemistry* **21**, 1553 (1982).)

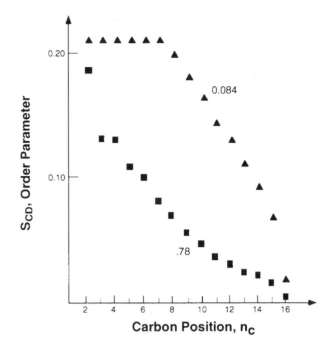

ducing a sufficient amount of cholate will disintegrate the bilayer into finite micelles, a process that plays an important role in fat digestion (as discussed in Section 4.5.2).

6.2.4 Calorimetry Monitors Phase Transitions and Measures Transition Enthalpies

Differential scanning calorimetry (DSC) can be used to study phase changes in bilayer systems. The technique is particularly important for analyzing the gel-to-liquid crystal transition. We can observe a strong peak at this transition due to the high enthalpy associated with the melting of alkyl chains.

The DSC experiment monitors the sample temperature as a function of time for a constant heat transfer to the system. This process measures the heat capacity C_p of the system, which peaks sharply at phase changes.

Ideally C_p should be infinite at a first-order phase transition, but in practice mechanisms always exist that lead to a finite but strongly peaked heat capacity. Figure 6.15 shows this effect for a phospholipid–water system: the peak at 41°C is due to the gel to liquid crystal transition while the bump at 34°C signals a transition between two different gel phases.

By integrating the peak in Figure 6.15 (curve B) we can determine the enthalpy change ΔH, associated with the phase transition. Table 6.4 shows measured ΔH, values of the gel-to-liquid crystal transition for a series of phospholipids. For the saturated chains, ΔH_t increases by ~ 2.5 kJ/mol per added carbon atom. These ΔH_t values are somewhat lower than the heats of fusion for pure alkanes of similar size, which indicates that although the gel differs from a crystalline phase and the liquid

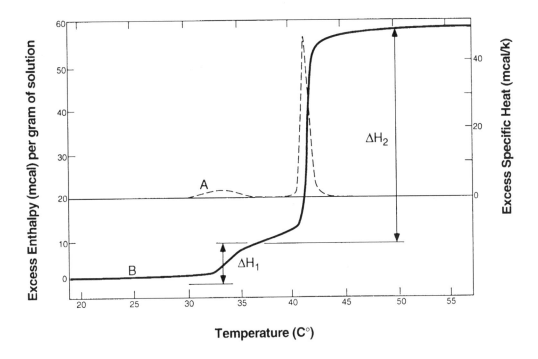

Figure 6.15
The variation with temperature of the excess specific heat (curve A, right-hand ordinates) and the excess enthalpy (curve B, left-hand ordinates) during the gel-to-liquid crystal transition of dipalmitoyl–lecithin in aqueous suspension at a lipid concentration of 1.88 mg/mL. ΔH_1 is the enthalpy for the transition from $L_{\beta'}$ to P_{β} at $\approx 34°C$. ΔH_2 is the transition enthalpy for P_{β} to L_α chain melting at 41°C. (H.J. Hinz and J.M. Sturtevant, *J. Biol. Chem.* **247**, 6071 (1971).)

TABLE 6.4 Thermodynamic Parameters, Transition Temperature T_t, Enthalpy ΔH_t, and Entropy ΔS_t for the Transition from Gel (L_β, P_B) to Liquid Crystal (L_α) for a Number of Phospholipids

Head Group	One Chain	Two Chains	$T_t(°C)$	ΔH_t (kJ/mol)	ΔS_t (kJ/mol)
Choline	Dodecyl	Dodecyl	−1	12	45
	Tetradecyl	Tetradecyl	23.5	25	83
	Octadecyl	Octadecyl	55	42	129
	Tetradecyl	Hexadecyl	35	33	107
	Tetradecyl	Icosano	40	34	108
	Hexadecyl	cis-9-Octadeceno	3	30	100
Ethanolamine	Dodecyl	Dodecyl	30	16	53
	Hexadecyl	Hexadecyl	64	35	105
Serine (Na$^+$ salt)	Tetradecyl	Tetradecyl	36	31	94
	Hexadecyl	Hexadecyl	54	38	115

Data from D. Marsh, *Handbook of Lipid Bilayers*, CRC Press, Boca Raton, FL, 1990.

crystal is not quite an isotropic liquid, the process is mainly one of chain melting.

In the experiment recorded in Figure 6.15, the amphiphile is in equilibrium with excess water. If only a limited amount of water is present or if more than one amphiphile is a component, the transformation from one phase to another must involve a region with two coexisting phases. Under these circumstances, heat capacity no longer shows a distinct peak at a given temperature; instead, a melting process takes place over a temperature range. We will analyze this effect further in Chapter 10.

6.2.5 We Can Accurately Measure Interbilayer Forces

Measurement of colloidal forces between surfaces raises an enigmatic problem of control over their properties on a molecular level. To get reproducible results, the surfaces must be clean, homogeneous, and smooth. These conditions are difficult to achieve on the surface of a bulk material.

We can use self-assembled lamellar systems to solve this practical problem. The very nature of the assembly process generates a planar homogeneous interface that is rough only on a molecular scale. Because the system constitutes part of the bulk phase, impurities are not selectively localized at the interface under study. The price we pay for allowing intermolecular interactions to generate the surface is the loss of some degree of external control over its properties. In contrast to our ability to manipulate monolayers, we can vary the surface density of a bilayer only within narrow limits.

Bilayer systems have provided many of the most accurate measurements of forces between surfaces. Two quite different experimental methods are used to take these measurements.

The osmotic stress technique makes full use of the possibility of controlling a bulk bilayer system through thermodynamic parameters. Varying the osmotic pressure—that is, the solvent chemical potential—changes the swelling of the lamellar liquid crystal. We can monitor structural changes by x-ray diffraction, and the measurements yield a relationship between osmotic pressure and the lamellar repeat distance. Using the procedure described in Section 6.2.1, this can be converted in turn into a thickness of the solvent layer.

In a typical measurement, the lamellar liquid crystal is in contact with a polymer solution with known osmotic pressure, Π_{osm}. The polymer cannot penetrate the lamellar structure because of the presence of a semipermeable membrane, but the solvent (usually water) exchanges freely. Equilibrium is reached when the force per unit area between the surfaces in the liquid crystal is the same as the osmotic pressure in the polymer solution. Thus, by measuring the separation between the surfaces as a function of the osmotic pressure, we obtain a force versus distance relation. An example of force curves obtained by osmotic stress measurements was given in Figure 5.25.

We can control the osmotic pressure of the sample in

several other ways. For low values of Π_{osm}, as an alternative to using the depletion effect, we can simply apply an external pressure on the bilayer to squeeze out the solvent. This approach ceases to be feasible, however, when the force per area corresponds to tens or hundreds of atmospheres. In this situation, we can achieve high values of Π_{osm} by permitting the amphiphile to equilibrate with solvent (water) vapor at a controlled pressure. An amusing way to obtain very low values of Π_{osm} makes use of gravity. The sample is mounted in a closed container, where the gas phase is saturated with solvent vapor and with excess solvent at the bottom of the container. Swelling of the lamellar liquid crystal involves a change, mgh, of the gravitational energy of solvent, where h is the height of the sample above the liquid surface.

An alternative to the osmotic stress technique uses the surface forces apparatus (SFA). We can use the SFA to measure forces interacting over distances from 0 to 10^3 nm, with a resolution of 0.2 nm. To obtain this degree of refinement, the surfaces must be molecularly smooth. We can achieve this by covering them with mica, which cleaves into large areas of thin molecularly smooth planar sheets. The sheets are silvered on the back before being glued onto glass lenses ground as cylindrical sections with radii $R \approx 1$ cm. Then the lenses are mounted at right angles in the SFA, attaching the lower one to a leaf spring and the upper one to a piezoelectric crystal. (See Figure 6.16).

Figure 6.16
Surface forces apparatus (SFA), which permits a direct measurement of forces between molecularly smooth surfaces. (J.N. Israelachvili, *Intermolecular and Surface Forces*, Academic Press, London, 1992, p. 170.)

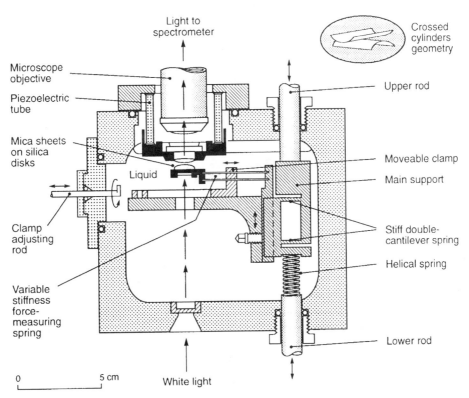

To understand how the SFA works, we need to know how to control and monitor the separation between the two mica surfaces. We can measure separation distance by passing white light through the thin mica surfaces. Partial reflection at the silvered surfaces generates interference patterns of equal chromatic order known as fringes. We can analyze these fringes in a spectrometer and measure the change in separation by the change in the wavelength of a particular fringe as two surfaces approach one another.

A three-stage mechanism controls the distance between the surfaces. The upper rod permits us to move the lower surface over large distances of 1 μm to 1 cm, but it is not used during the actual force measurement. A double-cantilever spring (which is about 1000 times stiffer than the helical spring attached to the lower lens) operates the lower rod. This differential spring mechanism makes it possible to reduce a 1 μm movement of the lower rod to a 1 nm movement between the two mica surfaces. The piezoelectrical crystal attached to the upper mica surface controls small movements of 1–10 Å.

If we advance the lower micrometer rod when the two surfaces are so far apart that no interaction force exists between them, the main support of the double-cantilever spring and the mica surface move at the same rate. We use this knowledge to calibrate the lower micrometer. However, when we push the mica surfaces so close together that a repulsive force arises, the weak leaf spring bends downward. As a consequence, if we optically measure the distance actually moved, it is less than we would expect (e.g., 40 Å rather than 50 Å). The leaf spring, therefore, must have bent 10 Å. If we push the surfaces still closer together, we can record a degree of deformation in the leaf spring corresponding to each separation D. Because we can measure the spring constant independently, we can obtain the total force between the mica surfaces as a function of separation distance.

Usually we divide the measured force F by the mean radius of the curved surface R and plot it versus D. If we use the Derjaguin approximation (eq. 5.2.32), F/R equals $2\pi\mathbf{U}$, where $\mathbf{U}$ represents the corresponding interaction potential per unit area between flat surfaces. The physical interpretation of this equivalence is that in an SFA experiment, we measure an average of interaction forces determined by all distances larger than D between the curved mica surfaces. As a result of this averaging, the total force relates to the integral of forces between surface segments from infinity to D, that is, the energy at D.

Native mica crystallizes in layer structures. On cleavage it has a claylike surface with one potassium ion site per 0.50 nm^2. In low dielectric constant liquids, mica behaves as a neutral surface and permits us to make measurements like those described in Figure 5.29. However, in ionizing solvents like water, potassium counterions dissociate from the surface, generating an electrical double layer, and we can measure the interaction forces between the charged surfaces as a function of salt concentration. The theoretical description of double-layer forces given in Chapter 5 has been tested extensively in this way. We obtain a quite different surface in the presence of cationic surfactants, which exhibit a strong Coulomb interaction with the negatively charged mica surface. This interaction leads to a strong surfactant adsorption on the surface. Depending on the nature of the surfactant and on its concentration, adsorption can lead to the formation of a hydrophobic monolayer, that is, a monolayer with the hydrocarbon chains exposed to the (aqueous) solution, or to the formation of a bilayer. With single-chain surfactants, a nonuniform self-assembly on the surface with micellelike aggregates is also probable. Bilayers of insoluble amphiphiles also can be deposited onto the mica through the use of the Langmuir-Blodgett technique (Section 2.5.2).

Although the osmotic stress and the SFA techniques measure essentially the same properties, some fundamental differences exist between them. Unlike the SFA method, the osmotic stress technique yields the force between planar surfaces directly and thus more accurately. To compare data obtained by the two techniques, we must either integrate the osmotic stress data with respect to distance (if we want to find the interaction potential) or differentiate the SFA data (if we want the interaction force). However, the osmotic stress technique is feasible only when the force is repulsive and only this part of the interaction curve can be measured. Thus, it is not possible to integrate and obtain absolute values for the interaction potential.

When two surfaces interact strongly, stresses also occur in the surfaces. These surfaces respond differently because of the difference in boundary conditions. In the SFA setup, a repulsive force pushes molecules away from the point of close contact. In contrast, a repulsive force in the osmotic stress measurement causes the area per molecule to decrease with increasing repulsion; that is, when the surfaces are forced toward each other by an increase in the osmotic pressure, the bulk lamellar phase responds by making the distance between the surfaces as long as possible using as little solvent as possible, a manifestation of Le Chatelier's principle.

The two techniques are further differentiated because osmotic stress only determines the average separation between surfaces, while the bilayers are free to fluctuate around the mean separation. In the SFA, surfaces are fixed by the solid mica support; there is no undulation force.

The pipette aspiration technique permits us to determine the mechanical properties of isolated vesicles and the contact energies of adhering vesicles. When anhydrous lipids are hydrated, they spontaneously form a small number of vesicular capsules of varying size. Large isolated unilamellar vesicles ($10-20 \times 10^{-6}$ m) can be visualized easily in the video-enhanced microscope if they are captured by manipulating a micropipette into the vicinity of the vesicle and drawing part of it into the pipette (Figure 6.17) by applying a small suction pressure ΔP_{asp}. If the vesicle bilayer is liquid, $\Delta P_{asp} \approx 10^{-6}$ atm, while if it is solid, $\Delta P_{asp} \approx 10^{-4}$ atm.

These values of ΔP_{asp} are several orders of magnitude smaller than those required to filter water from the vesicle

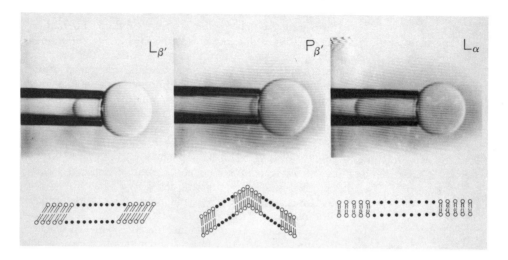

Figure 6.17
Video micrographs of a
DMPC bilayer vesicle in
three structural phases:
left, $L_{\beta'}$ state at 8°C; middle,
$P_{\beta'}$ state at 16°C; right,
L_α state at 25°C. Length in
pipette represents
projected area of bilayer
structure. (E. Evans and D.
Needham, *J. Phys. Chem.*
91, 4219 (1987).)

against the osmotic activity of trapped solutes. As a consequence, the vesicle's volume remains constant over a pressure
range. Increasing the suction draws the vesicle further into the
pipette and increases the area by

$$\Delta A \simeq 2\pi R_p \left(\frac{1 - R_p}{R_0} \right) \Delta L \qquad (6.2.2)$$

where R_p and R_0 are the dimensions of the pipette and vesicle
and ΔL represents the change in projected length inside the
capillary. This change is proportional to the change in the
projected area of the bilayer structure. Thus, we can generate
force versus area data to characterize the mechanical behavior
acompanying bilayer area dilation, surface extension, and bending. Inducing a rapid change in pressure and following the
resulting changes in vesicle area as a function of time provides
information on bilayer viscosity and surface dynamics.

Figure 6.18 illustrates one application of this technique in
which an individual dimyristoylphosphatidylcholine (DMPC)
vesicle is carried through temperature cycles resulting in the
formation of L_β, $P_{\beta'}$, and L_α bilayer structures. As shown in
Figure 6.8, changes in the area as a function of temperature
provide information on both the elastic properties of these
structures and the tilt associated with the $P_{\beta'}$, rippled phase.

Vesicle adhesion is accomplished by maneuvering two
vesicles (each attached to a micropipette) into the same vicinity
(Figure 6.19). One vesicle is held at high pressure so that it
forms a rigid sphere, while the other is held at a pressure
sufficiently low that it remains flexible. The extent of adhesion
can be controlled by changing the pressure on the flexible
vesicle, and reversibility can be tested by cycling the pressure. Analyzing the extent of coverage of the rigid vesicle
versus pipette suction permits us to estimate the free energy of
adhesion.

Figure 6.18
Effect of bilayer stress on solid DMPC bilayer structure and phase formation. Relative vesicle area is plotted versus temperature for fixed bilayer tensions. Values of the acyl chain tilt to the bilayer normal (angles in parentheses) were derived from the ratio of projected areas for the rippled–solid and planar–solid surfaces at the same temperature. (E. Evans and D. Needham, *J. Phys. Chem.* **91**, 4219 (1987).)

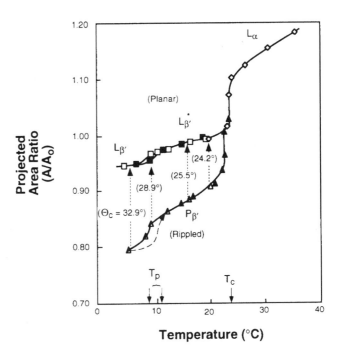

Figure 6.19
Video micrographs of vesicle–vesicle adhesion test. (left) Vesicles maneuvered into close proximity but not forced to contact. (center and right) Spontaneous adhesion allowed to progress in discrete steps controlled by suction applied to lower vesicle. (E. Evans nd D. Needham, *J. Phys. Chem.* **91**, 4219 (1987).)

6.2.6 Measurements of Interbilayer Forces Play a Key Role in Testing Theories of Surface Interactions

For charged surfactants, the electrical double-layer force dominates the interbilayer interaction. Figures 6.20 and 6.21 illustrate this property by showing the measured (using the SFA technique) interaction potential for the DHDAA $((C_{16})_2N(Me)_2OHC)$ and DHDAB $((C_{16})_2N(Me)_2Br)$ systems (see Figure 6.5 for structure). As expected, the repulsive energy varies exponentially with a decay length corresponding to the Debye length. However, the magnitude of the force depends on the counterion.

Figure 6.20
Forces measured between adsorbed DHDAA, $(C_{16})_2N(Me)_2OAC$, bilayers in 5×10^{-4} M DHDAA and 3×10^{-4} M sodium acetate solution. The observed Debye length now corresponds reasonably well to that expected for the sodium acetate solution. However, at separations less than a Debye length the observed forces are still significantly below the calculated values for fully ionized bilayers. It may be that the presence of aggregates still affects the forces even in this concentration of added salt. (R.M. Pashley, P.M. McGuiggan, B.W. Ninham, J. Brady, and D.F. Evans, *J. Phys. Chem.* **90**, 1637 (1986).)

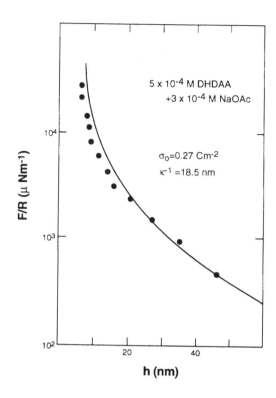

Figure 6.21
Measured forces between DHDAA bilayers adsorbed from a 5×10^{-4} M solution on addition of 10^{-5} M potassium bromide. These forces are very similar to those observed when no KBr was added (inset). The solid line is the calculated curve for the case of interaction at constant charge. The inset shows the force curve for distances less than 9 nm. (R.M. Pashley, P.M. McGuiggan, B.W. Ninham, J. Brady, and D.F. Evans, *J. Phys. Chem.* **90**, 1637 (1986).)

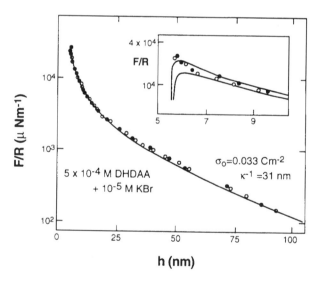

With Br^- counterions, the measured force implies that 90% of the counterions associate to the surface, leaving only approximately 10% of the bilayer surface charge to generate the diffuse double layer. Acetate, a more solvated ion, has a smaller affinity for the surface. Consequently, the measured force is more repulsive, consistent with 50% of the surface charge contributing to the diffuse part of the double layer.

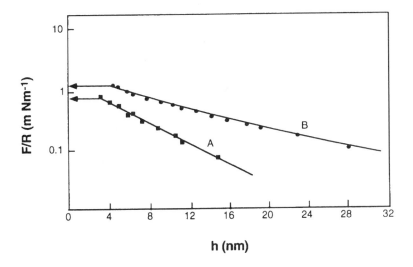

Figure 6.22
Measured forces between distearoylphosphatidyl-glycerol bilayers (22°C, pH 55) using the SFA technique: curve A, 1.2 mM aqueous CaCl$_2$ solution; curve B, 0.12 mM CaCl$_2$ solution. Note the jump into contact that occurs at a separation of approximately 4 nm. (J. Marra and J. Israelachvili, *Biochemistry* **24**, 4608 (1985).)

A biological membrane typically includes between 15 and 20% of charged lipids, which clearly play a role in modifying long range interactions with other membranes and charged species. Figure 6.22 shows an SFA measurement of a system with charged distearoylphosphatidylglycerol bilayers adsorbed on the mica. In this case, Ca^{2+} are the counterions and the use of a divalent counterion produces two consequences. First, the screening length decreases because of the z^2 term in eq. 3.8.13. A second more subtle effect calls into question the mean field approximation underlying the Poisson–Boltzmann description of double-layer forces. Because of correlations between ions, an attractive term in the force appears in a more exact description (see Figure 5.13). At short separations this attractive term can dominate and cause instability to appear in the measurement. Figure 6.22 shows its appearance at a distance of 3–4 nm.

In SFA measurements bilayers are rigidly attached to mica and undulations are suppressed. One way to investigate the role of undulations is to compare SFA and osmotic stress measurements on the same bilayer system. Figure 6.23 gives an example for the bilayer interactions in dihexadecyldimethylammonium acetate (DHDAA) at 40°C above the chain melting temperature. The dashed line represeents the differential SFA data, which coincide with the osmotic stress data. It appears that undulation forces do not play a significant role in this particular system. In Chapter 10, we cite examples in which undulation forces significantly influence bilayer interactions.

6.3 The Lipid Bilayer Membrane Has Three Basic Functions

The biological membrane is a complex structure containing lipids, proteins and polysaccharides. We will adopt the point of view that the lipids provide the basic structural unit. Proteins

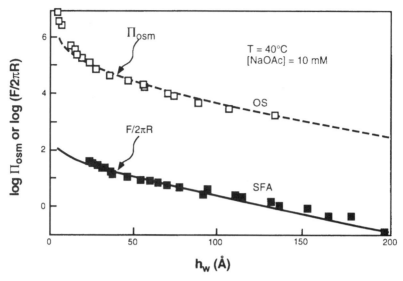

Figure 6.23
Force data from osmotic
stress (OS) and surface
forces apparatus (SFA)
measurements of DHDAA,
$(C_{16})_2N(Me)_2OAC$, in
NaOAc (10 mM), at 40°C.
The SFA data points are
fitted to the DLVO theory
using a surface potential of
210 mV with a Debye
length of 34 Å. The dashed
line, which is equivalent
to the pressure curve as
measured in the OS
measurements (in units of
N/m^2), is obtained by
differentiating the SFA
fitting curve (solid line) (in
units of 20^{-5} N/m). (Y.H.
Tsao, D.F. Evans, R.P.
Rand, and V.A. Parsegian,
Langmuir **9**, 233 (1993).)

and polysaccharides attached to this unit are responsible for
most of the chemical function, although they carry out the
chemical processes in the specific environment created by the
lipids. Lipids play three major roles. First, they provide a
barrier for passive diffusional motion of small polar solutes like
ions, sugar, and low molecular weight matabolites as well as all
macromolecules; proteins, nucleic acids, and other polysaccharides. Second, the bilayer provides a unique solvation environment for membrane proteins. The third function of the
membrane, which has attracted less attention, is its use in the
internal organization in the cell. In all organisms except the
most primitive ones, the plasma membrane of the cell envelope
only represents a small fraction of the total membrane content.
The major fraction is engaged in the internal organization of the
multitude of chemical processes that occur in a cell.

6.3.1 Diffusional Processes Are Always Operating in the Living System

In a liquid, small molecules diffuse rapidly on short length
scales. A typical value for the translational diffusion coefficient
D in water is 10^{-9} to $10^{-10}\,m^2/s$. According to equation 1.6.6,
diffusion is effective in eliminating concentration differences
over distances $(tD)^{-1/2}$. Taking 1 s as a typical diffusion time,
the corresponding distance is 10 μm. This distance corresponds
to the size of a typical cell, which shows that diffusional
barriers are necessary to sustain concentration differences. The
lipid bilayer provides such a diffusional barrier. Even though
the hydrocarbon core of a bilayer may be only 30–40 Å thick, it
provides an effective barrier to ions and other polar compounds.
Solutes dissolved in the aqueous region will quickly establish
an equilibrium with their bilayer surface facing the solution.
Concentration of polar substances in the bilayer interior is low.

Although the mobility of a particular molecule is relatively high in the membrane, this low average concentration results in low net transport.

We can characterize the ability of a molecule A to cross a membrane by the permeability P (ms^{-1}) defined by

$$J_A = -P_A \Delta c_A \qquad (6.3.1)$$

Here J_A(mol/m^2/s) represents the flow per unit area of molecule A when a concentration difference Δc_A(mol/m^3), exists across the membrane. (For the solvent water Δc_A should be replaced by $\Delta \Pi_{osm}/RT$).

One technique for measuring permeability involves forming a black lipid membrane (BLM) in a small orifice separating two aqueous reservoirs. (See Figure 6.24.) We can prepare such a bilayer film by gently brushing an organic solution containing the amphiphile across the orifice. As the organic solvent diffuses into the adjacent aqueous phases, the film thins (Figure 6.25) and exhibits interference colors in reflected light. Ultimately, it turns black when the thickness is much smaller than the wavelength of the light, yet it possesses surprising stability. If BLMs are prepared with meticulous care and scrupulous cleanliness, they will last for hours, provided that the solubility of the amphiphile is sufficiently low.

The permeability coefficient for water in BLMs is 5–100 × 10^{-6} m/s. For polar solutes such as glucose, glycerol, and urea it is much smaller (in the range 5–300 × 10^{-10} m/s). For small ions such as Na$^+$, K$^+$, and Cl$^-$ the values lie in the range 1–100 × 10^{-14} m/s: that is, several orders of magnitude smaller.

Figure 6.24
Experimental setup for measurements of permeability properties of black lipid membranes (BLM). Two aqueous compartments are separated by a partition with a hole (∼1 cm diameter). Lipid dissolved in an organic solvent is applied to the hole with a brush. A film is formed, which thins gradually with time as shown in Figure 6.28. (J.H. Fendler, *Membrane Mimetic Chemistry*, New York, Wiley, 1987, p. 101.)

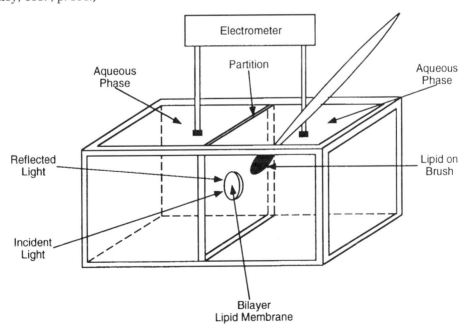

Figure 6.25
Thinning of the film in the orifice occurs because the organic solvent has a finite solubility in the aqueous phase. The bilayer is much thinner than the wavelength of light and the film is totally transparent. If we shine light on the orifice, it appears black as a result of the absence of reflections. (J.H. Fendler, *Membrane Mimetic Chemistry*, Wiley, New York, 1987, p. 101.)

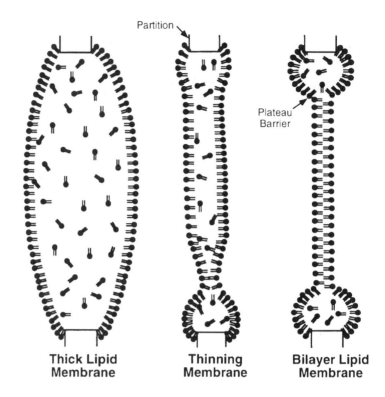

Thick Lipid Membrane   **Thinning Membrane**   **Bilayer Lipid Membrane**

Figure 6.26 presents an idealized model of the molecular cause of the small permeability of polar molecules in a membrane. Here a (uniform) membrane of thickness d_2 separates two bulk phases with a concentration of solute S of c_I and c_{II}. If the bulk phases are properly stirred, their concentration of the solute is uniform. According to Fick's first law, the flow of molecules per unit area is

$$J_S(z) = -\frac{D_S(z)}{RT} c_S(z) \frac{d\mu_S}{dz} \tag{6.3.2}$$

At steady state—where there is no buildup of the S inside the membrane—the flow, J_S, must be independent of the coordinate z. Similarly, if the membrane has uniform properties, D_S is independent of z and thus

$$c_S(z) \frac{d\mu_S}{dz}$$

is constant at steady state. If the solute is present in low concentrations and external fields are absent

$$\mu_S = \mu_S^\theta + RT \ln c_S$$

and

$$J_S = -\frac{D_S}{RT} c_S(z) \frac{d\mu_S}{dz} = -D_S \frac{dc_S(z)}{dz} = -\frac{D_S(c_S(z_{II} - \delta) - c_S(z_I + \delta))}{d_2} \tag{6.3.3}$$

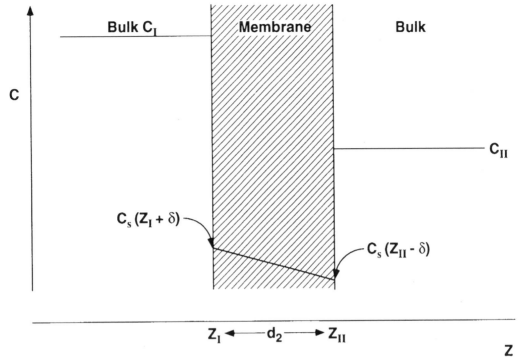

Figure 6.26
The steady state
concentration profile for
the diffusion of a solution S
across a membrane of
thickness d_2, separating
two bulk phases with
solute concentrations c_I
and c_{II}.

where $c_S(z_{II} - \delta)$ and $c_S(z_I + \delta)$ are the solute concentrations just a small distance δ into the membrane on either side. At the membrane surface, the solute is in equilibrium with the bulk at the respective sides so that

$$c(z_{II} - \delta) = c_{II}(\text{bulk}) \cdot K_S$$
$$c(z_I + \delta) = c_I(\text{bulk}) \cdot K_S$$

(6.3.4)

where K_S represents the partition coefficient between bulk solvent and membrane. Thus,

$$c(z_{II} - \delta) - c(z_I + \delta) = K_S\{c_{II}(\text{bulk}) - c_I(\text{bulk})\} \quad (6.3.5)$$

and by comparing with the definition in eq. 6.3.1, the permeability coefficient is

$$P_S = \frac{D_S K_S}{d_2}$$

(6.3.6)

In a fluid membrane, diffusion coefficients of small molecules possess the same magnitude as in bulk solvents, and the important factors determining the permeability factor are the bilayer thickness and, in particular, the partition coefficient K_S. This fact explains why small ions, whose low solubility in an apolar environment makes the partition coefficient K_S very small, exhibit very low values of P.

When ions are transported, an additional effect arises from the electrostatic interactions. We can see the bilayer surrounded by an electrolyte solution on either side as a capacitor (cf. Figure 3.6b). The capacitance C is (cf. eq. 3.5.5)

$$C = \frac{\varepsilon_r \varepsilon_0}{d} \text{ area} \qquad (6.3.7)$$

With an estimate of $\varepsilon_r \simeq 3$ and $d_2 \simeq 40$ Å, $C = 7 \times 10^{-3}$ A·s/V·area. Thus, to obtain a potential difference of 100 mV, we need to transport

$$\frac{Q}{\text{area}} = 7 \times 10^{-3} \times 10^{-1} = 7 \times 10^{-4} \text{ As/m}^2 \leftrightarrow 4.4 \times 10^{15} \text{ e/m}^2 \leftrightarrow 1e/2 \times 10^4 \text{ Å}^2$$

For a colloidal system, this transport corresponds to a low charge density. A potential difference $\Delta\Phi$ across the membrane also implies that an electric field, $-\Delta\Phi/d_2$, exists in the membrane.

Ions moving across the membrane will sense this field, which can either facilitate or hinder the motion of the ion. When applying Fick's first law, eq. 6.3.2, we have to incorporate the electrostatic interaction in the (electro)chemical potential and for an ion i of charge q_i, the flux is

$$J_i = -D_i \left(\frac{dc_i(z)}{dz} + \frac{q_i}{kT} c_i(z) \frac{d\Phi}{dz} \right) \qquad (6.3.8)$$

When we can neglect the charge inside the membrane, the electrical field is constant and

$$\frac{d\Phi}{dz} = \frac{(\Phi_{II} - \Phi_I)}{d_2} = \frac{\Delta\Phi}{d_2} \qquad (6.3.9)$$

At steady state the flux is constant; as a result of the electrostatic term, however, the concentration gradient is no longer constant as in eq. 6.3.3. By solving the differential equation 6.3.8, we find

$$c_i(z) = K_i c_I \exp\left(\frac{-qi\Delta\Phi}{kT} \frac{z}{d_2} \right) + \frac{J_i kT d_2}{q_i D_i \Delta\Phi} \left\{ \exp\left(\frac{-q_i\Delta\Phi}{kT} \frac{z}{d_2} \right) - 1 \right\} \qquad (6.3.10)$$

where we have fixed the boundary condition at $z = z_I = 0$

$$c_i(0) = K_i c_I \qquad (6.3.11)$$

At the other side, $z = z_{II} = d_2$, the boundary condition is

$$c_i(d) = K_i c_{II} \qquad (6.3.12)$$

as in eq. 6.3.4. When the conditions are applied to eq. 6.3.10, we obtain a relation between the flux J_i and the potential difference $\Delta\Phi$

$$J_i = \frac{q_i \Delta\Phi}{kT} P_i \left\{ \frac{c_{II} - c_I \exp(-q_i \Delta\Phi/kT)}{\exp(-q_i \Delta\Phi/kT) - 1} \right\} \quad (6.3.13)$$

If $c_{II} = c_I \exp(-q_i \Delta\Phi/kT)$, the flux is zero and the ion is in equilibrium under the constraint that the potential difference be fixed. This condition is called a *Donnan equilibrium*. However, consider a more general case in which the individual ion flows are nonzero. At the steady state (when the potential difference is constant), the net electric current must be zero and the individual ion currents must cancel so that

$$\sum_i q_i J_i = 0 \quad (6.3.14)$$

This condition can be satisfied only at a unique potential difference. If we consider only the monovalent ions Na^+, K^+, and Cl^-, we find by combining eqs. 6.3.13 and 6.3.14

$$\Delta\Phi = \Phi_{II} - \Phi_I = \frac{kT}{e} \ln \left\{ \frac{P_{Na} c_{Na}^{II} + P_K c_K^{II} + P_{Cl} c_{Cl}^{I}}{P_{Na} c_{Na}^{I} + P_K c_K^{I} + P_{Cl} c_{Cl}^{II}} \right\} \quad (6.3.15)$$

This so-called Goldman equation shows how the electrostatic transmembrane potential depends on ion permeability and concentration. Typically, the Donnan equilibrium of the most permeable ion determines the potential at given ion concentrations.

In this nonequilibrium situation the potential difference is controlled by the permeabilities, so we can regulate $\Delta\Phi$ by making the permeabilities time dependent. A nerve signal is propagated along the nerve by opening and closing sodium and potassium channels, that is, by changing the ion permeabilities.

So far we have discussed the transport of polar molecules across a single membrane. Some new features appear when we consider the transport in a bulk liquid crystal although the basic principles remain the same. A lamellar liquid crystal is anisotropic, and we should consider the diffusion parallel and perpendicular to the lamellae separately.

In the parallel direction (see Figure 6.27), the effective diffusion coefficient $D_{\parallel}$ is simply the weighted average of the diffusion in two domains, which we assume to be isotropic. The average diffusion represents the diffusion in the respective domains weighted by the fraction of molecules there

$$D_{\parallel} = (d_1 D_1 + K_S d_2 D_2)/(d_1 + K_S d_2) \quad (6.3.16)$$

where K_S is the distribution coefficient between domains.

To arrive at the expression for the average perpendicular diffusion, we must develop a somewhat longer derivation. At steady state the flow J in the two regions must be the same and

$$-J_\perp = \frac{D_1}{kT} c_1 \left(\frac{d\mu}{dz} \right)_1 = \frac{D_2}{kT} c_2 \left(\frac{d\mu}{dz} \right)_2 \equiv \frac{D_\perp}{kT} c_{av} \left(\frac{d\mu}{dz} \right)_{av} \quad (6.3.17)$$

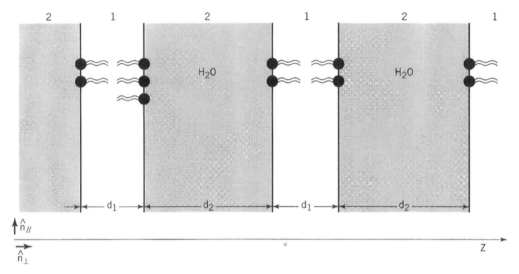

Figure 6.27
Diffusion in a lamellar liquid crystal. In the parallel direction, solutes can diffuse without passing the interface, while in the perpendicular direction, a diffusional process involves repeated transition between regions 1 and 2.

where we have introduced c_{av} and $(d\mu/dz)_{av}$ as the respective properties seen on an average macroscopic scale and used in eq. 6.3.2 to identify $D_\perp$. Thus

$$c_{av} = \frac{d_1 c_1 + d_2 c_2}{d_1 + d_2} \tag{6.3.18a}$$

$$\left(\frac{d\mu}{dz}\right)_{av} = \frac{d_1 \left(\frac{d\mu}{dz}\right)_1 + d_2 \left(\frac{d\mu}{dz}\right)_2}{d_1 + d_2} \tag{6.3.18b}$$

We solve for $D_\perp$ by assuming that we can neglect the electrostatic complications. Then $d\mu/dz$ is constant within each layer as in eq. 6.3.3 and

$$\left(\frac{d\mu}{dz}\right)_2 = \left(\frac{d\mu}{dz}\right)_1 \frac{D_1 c_1}{D_2 c_2} \tag{6.3.19}$$

using eq. 6.3.17. If we combine the second equality of this equation with eq. 6.3.18, we find

$$\boldsymbol{R}TJ_\perp = \left(\frac{d\mu}{dz}\right)_1 D_1 c_1 = D_\perp \left\{ d_1 \left(\frac{d\mu}{dz}\right)_1 + \frac{d_2 D_1 c_1}{D_2 c_2} \left(\frac{d\mu}{dz}\right)_1 \right\} (d_1 + d_2)^{-2} (d_1 c_1 + d_2 c_2) \tag{6.3.20}$$

and the desired average diffusion coefficient is

$$D_\perp = D_1 \frac{(d_1 + d_2)^2}{(d_1 + K d_2)(d_1 + (D_1/KD_2) d_2)} \tag{6.3.21}$$

If we assume for the sake of simplicity that the local dynamics in the two environments are similar so that $D_1 \simeq D_2$, then

$$D_\perp \simeq D_1 \frac{(d_1 + d_2)^2}{(d_1 + K d_2)(d_1 + K^{-1} d_2)} \tag{6.3.22}$$

which has its maximum value $D_\perp = D_1$ when $K = 1$. For both $K \gg 1$ and $K \ll 1$, that is, when the solute strongly prefers one region relative to another, the perpendicular diffusion is very low. Thus, an oriented lamellar phase can present a substantial diffusional barrier to most solutes, even though the local molecular motion is as fast as in a liquid.

The outer part of the human skin, the *stratum corneum* or horny layer, has penetration properties relative to small molecules that resemble the properties of $D_\perp$ in eq. 6.3.22. In particular, the skin protects the organism against penetration from both strongly polar molecules, such as water, and strongly apolar molecules, such as hydrocarbons. However, molecules that are soluble in both water and hydrocarbon—such as dimethyl sulfoxide (DMSO)—readily penetrate the skin. Organic chemists know that a drop of DMSO spilled on the skin partly ends up in the lungs, producing a garliclike smell and a foul taste in the mouth.

Although the *stratum corneum* has a complex structure, one of its essential elements seems to be bilayers stacked as they are in a lamellar liquid crystal. One way to administer drugs is to apply them on the skin either in salves or on a bandage. The rate of the diffusive transport of the active component through the skin clearly influences the efficiency of the treatment.

6.3.2 The Lipid Membrane is a Solvent for Membrane Proteins

A biological membrane consists of anywhere between 25 and 75% w/w lipid. The remaining 75 to 25% is protein, glycoprotein, or lipoprotein. Thus, the protein species appears to be as dominant in the membrane as the lipids. However, proteins penetrate into the aqueous medium outside the bilayer plane. This fact implies that if one specifies the concentrations by occupied area, the lipid content increases substantially and lipids emerge as the main component. Indeed, previous discussions in this chapter have been based on this assertion.

Proteins can associate with the membrane in several different ways, as illustrated in Figure 6.28. The archetypical membrane protein is an integral part of the bilayer. It spans the width of the bilayer and is exposed to the aqueous medium on both sides. A protein also can be anchored on one side of the bilayer by a hydrophobic interaction. A third possibility is that

Figure 6.28
Illustrations of different modes of protein–lipid bilayer interactions: (a) integral membrane spanning protein; (b) protein binding to one side of the bilayer through a hydrophobic patch; (c) mainly hydrophilic protein attaching hydrophobically through a covalently bound alkyl chain; (d) mainly hydrophilic protein with a hydrophobic pocket that can attract a lipid chain; and (e) hydrophilic protein binding electrostatically.

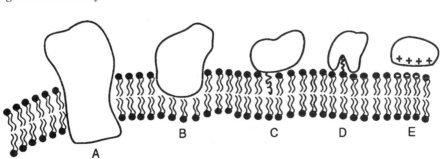

the protein contains one, or a few, covalently attached fatty acid chains, which anchor an otherwise polar protein to the membrane. A protein also can bind electrostatically to the membrane.

Proteins are inserted into the membrane with a prescribed orientation, and the reorientation is sufficiently slow that once the proteins have been inserted, they keep their orientation during the biologically relevant time. However, they can undergo chemical transformations. The plasma membrane contains proteins that become heavily glycosylated, with a substantial fraction of charged sugar units. By this mechanism, a protective shield of entangled sugar units covers the cell. This cover plays an important role in the contact communication between different cells. It is also a target for the adhesion of infectious bacteria.

The bilayer interior provides a very apolar environment for a membrane protein. To fit in, these proteins have a minimum of polar amino acid side-chains for the part of the molecule that does not protrude from the bilayer surface. Charged amino acids are very rare in this region; even OH and NH_2 groups are uncommon. In the backbone, the amide linkage necessarily has a polar character. As a result, the membrane-spanning parts usually form α-helices (and to a lesser extent β-sheets) so that the amide and carboxy groups can be engaged in intramolecular hydrogen bonds. The medium provides neither a hydrogen bond donor nor an acceptor. An α-helix elongates 0.15 nm per residue so that a membrane-spanning protein consists of a series of sequences containing 20 to 25 hydrophobic amino acid residues spaced by shorter or longer stretches of more hydrophilic ones. Each such sequence of twenty-odd apolar residues forms a membrane-spanning helix, and typically several of those fold to form the tertiary structure of a membrane protein.

The biological membrane with its lipids and integral membranes can be seen as a two-dimensional concentrated colloidal solution. Which factors contribute to the lateral protein–protein interaction? Consider two identical integral protein molecules. They will interact with

1. An attractive van der Waals force

$$F = -\frac{\mathbf{H}_{12}L}{16}\left(\frac{R_p}{(r - R_p)^4}\right)^{1/2} \tag{6.3.23}$$

where R_p represents the radius, L the length, and r the separation between the proteins. For eq. 6.3.23 to apply, $r - R_p \ll R_p$. In the other limit, $r \gg R_p$, we have the normal r^{-7} dependence of the force. In eq. 6.3.23 we have also neglected the role of the protein part, which is in the aqueous medium.

2. A repulsive electrostatic interaction. The proteins typically carry some charge on the parts that are exposed to the aqueous medium on either side. It is important to realize that the bilayer also has a charge due to the

lipids with average charge density σ_L. Thus, the long range electrostatic interaction vanishes when the average protein charge density $\sigma_P = q_P/(\pi R^2)$ matches that of the lipids so that

$$F \sim (\sigma_P - \sigma_L)^2 \qquad (6.3.24)$$

The electrolyte in the aqueous solution will screen this interaction.

3. At short range in the lipid matrix, a repulsive component resulting from the flexibility of apolar side chains resembles the steric entropic force discussed in Chapter 5.

4. If the proteins also contain polar groups exposed in the apolar lipid matrix, these proteins will generate an attractive interaction of the general solvophobic type. This type of force also operates between α-helices of the same protein, and it is important for arranging the α-helices and thus for establishing the tertiary structure of the membrane protein.

6.3.3 Cell Membranes Fold into a Range of Global Structures

The simplest procaryotic organisms (those organisms which lack a cell nucleus) have as their sole membrane system a plasma membrane that separates the organism from the exterior. From careful studies of such simple bacteria, we can obtain yet another perspective on the role of the chemical diversity of lipids. Growth conditions such as temperature, salt content in the medium, or availability of essential fatty acids affect the lipid composition. By studying lipid phase behavior, we can conclude that lipid composition is regulated to preserve the effective surfactant number N of the mixture at a value slightly above 1. At present, we understand neither the physiological reason for this regulation of the composition nor the detailed mechanism by which it operates.

Higher organisms with eucaryotic cells (those with a cell nucleus) exhibit a rich variation of membrane functions. In addition to the plasma membrane, we find a number of intracellular membrane systems that support a range of key cellular functions, as schematically illustrated in Figure 6.29. Let us begin by discussing some of the functions present in virtually all cells. In the mitochondria nutrients react with O_2 to produce the basic free energy source ATP required by the cell. As we will discuss in more detail in Section 6.4, in this case, the membrane has the dual role of solubilizing the key enzymes and generating two compartments. The membrane is folded to increase the area/volume ratio. In the endoplasmatic reticulum (reticulum means network), the bilayers are folded into a network-like system reminiscent of the cubic phase shown in Figure 1.6e, but with a larger spacing between the connection points. Usually, the network appears disordered, as in the sponge phase, but numerous observers have found ordered

Figure 6.29
Overview of mammalian cell ultrastructure with sketches of some of the organelles. The size of typical mammalian cell is on the order of 20 μm in diameter. RER, rough endoplasmic reticulum; SER, smooth endoplasmic reticulum. (T. Landh, Thesis, University of Lund, Sweden, 1996.)

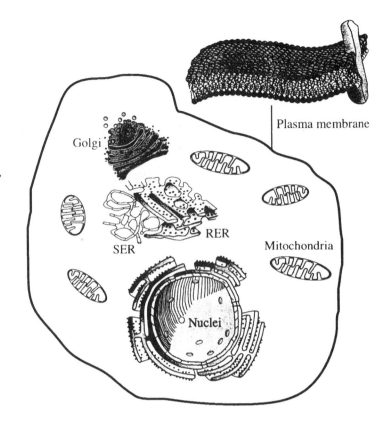

cubic-like arrangements in the endoplasmatic reticulum. Figure 6.30 shows one very clear example. The functional role of this intricate membrane arrangement is unclear, but two of its aspects appear to be: (1) the area-to-volume ratio is proportional to the inverse repeat distance in the network and thus independent of system size facilitating a scale-up from a small to a large cell; (2) within the network, solutes on the outside, on the inside, and in the membrane all have free transport through space.

Ribosomes, in which protein synthesis occurs, attach to the endoplasmic reticulum, and the nascent membrane protein chain can be inserted into the bilayer as it emerges. Then the newly synthesized protein can be transported in the bilayer network to its intended destination. Similarly, proteins intended for extracellular compartments can be exported on site and transported through the channels of the network.

Often the endoplasmic reticulum connects to a looser membrane network called the nuclear envelope, which surrounds the cell nucleus. One function of the nuclear envelope appears to be simply to keep the chromosomes of the nucleus in place while at the same time allowing for free transport of macromolecules like m-RNA.

Lysosomes perform the cleaning function in a cell. These vesicular compartments contain proteolytic enzymes which would be harmful to the native components of the cell. Therefore, the vesicular arrangement is ideal for this purpose.

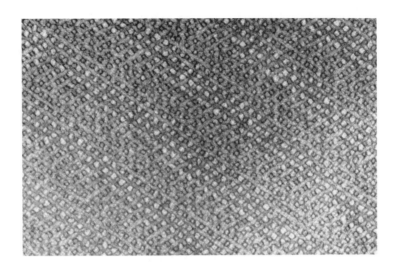

Figure 6.30
Electron micrograph of a patch of the smooth endoplasmic reticulum from a cell of a lemur, showing the membranes forming a structure of cubic symmetry. (Sisson and Fahrenbach, *Am. J. Anat.* **121**, 337 (1967).)

Figure 6.31
Illustration of the membrane arrangement in a chloroplast with interconnected stacks of coin-like double bilayer patches. The photosystem II is concentrated in the stacks and photosystem I in the bare bilayer regions.

In addition to these general functions, each cell plays a specific role in the organism. For these purposes membrane folding is used to obtain a structural organization relevant for specific processes. Here we use photosynthesis to illustrate the role that membranes play in a fundamental physiological process. In the chloroplasts in the leaves of green plants, the energy of visible light is converted into chemical free energy. The crucial enzymatic systems, called photosystem I and photosystem II, are located in the thylakoid membrane. Figure 6.31 shows the structure of this folded membrane. The thylakoid contains sections of stacked membranes, but on addition of salt, it unfolds into a single giant vesicle, showing that it is a single compartment in the folded state. The membrane's sensitivity to electrolyte demonstrates that electrostatic attractive interbilayer

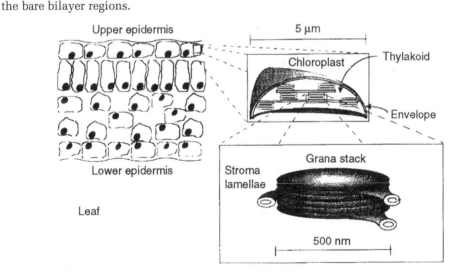

forces promote the folding. Figure 6.31 also shows that large fractions of the bilayer remain exposed to the cytoplasm. Careful studies reveal that photosystem II prefers to locate in the folded regions, while photosystem I prefers the freely exposed membrane patches. This preference demonstrates not only how membrane folding results in a high area to volume ratio, but also how it can be used to trigger lateral organization of the membrane protein systems.

Through these examples we have seen how the living cell exhibits all of the bilayer arrangements that have been observed for bulk phases of amphiphiles. Even more intriguing than the delicate control of these different modes of folding are the dynamic processes leading from one type of folding to another.

6.4 Transmembrane Transport of Small Solutes is a Central Physiological Process

6.4.1 Solutes Can Be Transported Across the Membrane by Carriers, in Channels, by Pumps, or by Endocytosis/Exocytosis

In Section 6.3.1, we saw that the permeability P of a solute across a lipid bilayer is determined primarily by the partition coefficient between the bilayer and the aqueous phase. One way to increase P is to increase the partition coefficient. We can accomplish this by introducing a carrier that has a large solubility in the membrane and also associates specifically with a solute. A well-known example is valinomycin, a cyclic peptide that forms a complex with potassium ions, as illustrated in Figure 6.32. With valinomycin present, the effective solubility of potassium ions in the bilayer increases dramatically, while the effective diffusion coefficient is reduced slightly because the valinomycin–potassium complex is now the diffusing en-

Figure 6.32
Valinomycin is a cyclic decapeptide that gives a strong hydrophobic complex with potassium ions. (a) Molecular structure: A = L-lactate; B = L-valine; C = D-hydroxyisovalerate, and D = D-valine. (b) Potassium transport across a bilayer. Both loaded and unloaded valinomycin diffuse across the bilayer. In the presence of a concentration difference of K+, $(c_\mathrm{I}(\mathrm{K}^+) > c_\mathrm{II}(\mathrm{K}^+)$, there is a net flux of K^+.

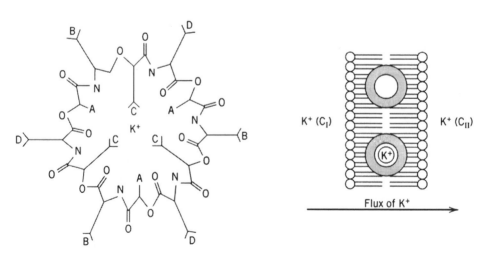

tity. The potassium-loaded valinomycin partitions into the bilayer due to a combination of two effects. First, when the ion is bound, the polar side-chains of the cyclic peptide coordinate the charge in the center, leaving an apolar periphery. Second, according to the Born model of ion solvation discussed in Section 3.6, the electrostatic solvation energy of an ion is inversely proportional to its radius. In the present case, the radius should be rather large because the peptide is included, and as a result, it greatly reduces the penalty in electrostatic energy for an ion entering the bilayer. Many ion carriers are based on a cyclic structure like that of valinomycin. Some of them have found a use as antibiotics. The basis for such a physiological effect will become apparent later in this section.

Protons have unique properties as a cationic species in that they always associate covalently with some other molecular entity. Thus, carriers for proton translocation across the bilayer are based on a different chemical principle from the principle at work in the coordination in a cyclic ring. A typical example is 2,4-dinitrophenol, in which the phenolic proton is acidic and both the protonated and acidic forms are present a neutral pH. In the ionized form, the charge is delocalized onto the aromatic ring and the nitro groups. As the Born model predicts, this delocalization gives the charged species a large radius that enables the charged dinitrophenolate to dissolve into the bilayer. Thus, the molecule can diffuse across the bilayer in both its neutral and in its charged form. Even though the physical transport involves a negative charge, the net result of a 2,4-dinitrophenol molecule diffusing in one direction in the charge form and back again in the neutral form is the transport of a proton in the direction opposite to that of the negative charge. Compounds that have the ability to transport proton through this mechanism are highly toxic to cells.

A more elaborate strategy for increasing the permeability of small ions across the bilayer is the formation of a channel. In this case, a polypeptide or protein forms a rigid structure consisting of a polar channel that spans the bilayer. This channel prevents the ion from making contact with the apolar interior as it moves across the bilayer. The efficiency per structural unit is much higher for channels than for carriers because the ion is the only moving species, possibly with some hydration water. Gramacidin is a channel-forming polypeptide that is often studied, see Figure 6.33. It adopts a helical structure in the bilayer, but because gramacidin is only 15 amino acids long, two oppositely oriented helices must associate in a dimer in order to form a bilayer-spanning channel. Once a channel is open, however, ions diffuse through with such ease that it is possible to measure the conductivity of a single channel in a black lipid membrane. The presence of a non-specific channel is highly damaging to a cell. Venoms make use of this circumstance. For example, bee venom contains the channel-forming protein mellitin. Channel-forming proteins also occur as integral parts of the living cell. However, usually they show high chemical selectivity, and some regulatory mechanism opens and closes the channels.

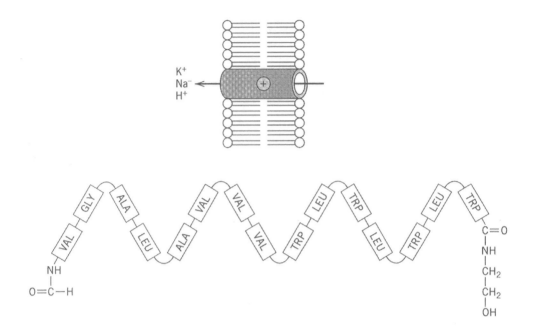

Figure 6.33
Gramidicin A is a
polypeptide that forms a
helical structure in a
bilayer. When two helices
anchored in the respective
monolayers of the bilayer
meet, a polar pore
spanning the bilayer is
formed. In that case, ions
that are small enough can
pass freely across the
bilayer.

An even more elaborate way of transporting a small polar
molecule/ion C across bilayer membrane is through a pump. In
a pump mechanism, translocation requires a concerted chemical transformation of A to B

$$A + C(I) \rightleftharpoons B + C(II) \qquad (6.4.1)$$

where I and II denote the two sides of the membrane. Note that
the reaction is reversible and can go in either direction, depending on the free energy conditions. Due to the coupling of a
reaction to the transport, the pump is a membrane-bound
enzyme. In fact, reaction 6.4.1 represents one of the most
fundamental biochemical reaction schemes and one of the
simplest of a whole class of reactions coupled to transport.
Clearly, the stoichiometric ratio can be different so that more
than one C is transported per transformation $A \rightarrow B$. Several
species also can be involved in the transport so that when C
goes from I to II, another molecular species (possibly with a
different stoichiometric coefficient) goes from II to I.

Our discussion so far has focused on the transport of
small solutes. Transporting a larger or even a macromolecular
solute across a membrane requires another strategy. The most
common is a process called endocytosis or exocytosis, depending on whether the solute goes from the outside to the inside of
the cell or vice versa. The total process consists of two distinct
steps. Let us take endocytosis as an example. In the first step, a
vesicle containing the external macromolecule is formed by
budding off from the plasma membrane. The vesicle can diffuse
into the cell, but topologically the macromolecule remains in an
outside compartment. In the second step, the vesicle is emptied
and its bilayer fragment is recycled. In exocytosis, loaded
vesicles formed in the cell interior fuse with the plasma mem-

brane. Alternatively, material can be exported in the form of vesicles budded off from the plasma membrane which then empty their content at the destination.

6.4.2 The Chemiosmotic Mechanism Involves Transformations Between Chemical, Electrical, and Entropic Forms of Free Energy through Transmembrane Transport Processes

Replication as well as all other activities we associate with life requires that the living organism has the ability to convert an external free energy source, like chemical nutrients or sunlight, to a useful form. Such free energy conversion processes are, in the living cell, far more sophisticated than the corresponding technical solutions devised in the form of car engines or solar cells. In the cell, both photosynthesis, producing oxygen, and respiration, consuming it, use the same molecular principle. Termed the chemiosmotic mechanism, this principle nicely illustrates several of the main themes of this book. Let us take the specific example of the respiration of an aerobic organism. The central chemical processes, called oxidative phosphorylation, that lead to the conversion of oxygen to water and the formation of ATP from ADP occur in the mitochondrion, see Figure 6.1. We can write the overall chemical reaction as

$$2NADH + O_2 + 2H^+ + x(ADP + HPO_4^{2-} + H^+) \rightarrow 2NAD + 2H_2O + x(ATP + H_2O) \quad (6.4.2)$$

where x denotes the unknown stoichiometric ratio between the NADH oxidation and the ATP synthesis. In this scheme, NADH denotes a metabolic intermediate that is produced from NAD, for example, through degradation of fat in the citric acid cycle, and it represents the chemical fuel. The product ATP acts as a general source of free energy for a multitude of processes throughout the cell. This free energy is released when the acid anhydride ATP is hydrolyzed to ADP and phosphate.

The design challenge in the scheme represented in 6.4.2 is to complete the reaction so that most of the free energy is converted from NADH to ATP rather than being lost as thermal heat. In contrast to the conditions in a car engine, the process must occur at constant temperature. Therefore, the cell cannot use two compartments with a temperature difference to store free energy.

Chemical free energy comes in quantized units determined by the free energy change of transforming a single molecule. Oxidation of NADH is the equivalent of a combustion process

$$2NADH + O_2 + 2H^+ \rightarrow 2NAD + 2H_2O \qquad (6.4.3)$$

whose free energy change is $-430\,kJ/mol$ or $174kT$ released per converted oxygen molecule, at pH7, an oxygen pressure of 0.1 atm., and equal NADH and NAD concentrations. Thus, the reaction appears irreversible with the equilibrium completely in favor of the products. For the ATP synthesis, the free energy changes are less dramatic, a fact that accords with the knowledge we derive from chemistry that acid anhydrides can form under some circumstances. For the reaction

$$ADP + HPO_4^{2-} + H^+ \rightarrow ATP + H_2O \qquad (6.4.4)$$

the free energy change is approximately $28 \, \text{kJ/mol}$ or $10kT$ per ATP, at pH7, $1 \, \text{mM} \, HPO_4^{2-}$, and equal concentrations of ATP and ADP. To use the available free energy, the stoichiometric coefficient x in scheme 6.4.2 should be of the order 10. Clearly, many individual chemical steps are necessary to arrive at a molecular mechanism that can give such a high value of x.

In the mitochondrion, the lipid membrane has been used as the basic device to solve this problem. This membrane simultaneously divides space into two compartments and solubilizes the enzymatic machinery used to catalyze the crucial reactions. Figure 6.34 provides a simplified illustration of the organization of the system. Through a series of reactions, NADH is oxidized to NAD, while the electrons are transferred to the oxygen molecule, which is turned into water. Concomitant with these chemical transformations is a crucial net transport of several protons across the membrane from the inside to the outside. The reaction of scheme 6.4.4 is organized in a similar way. It takes place on an enzyme that protrudes into the inner part of the mitochondrion. The net reaction also involves the crucial transport of protons, but in this case, from the outside to the inside. Synthesized ATP is exported to the cytoplasm through an exchange reaction with ADP, and phosphate is imported into the mitochondrion in an analogous manner through an exchange with hydroxide ions. This series

Figure 6.34
Simplified model of the reaction sequence in the mitochondrion. Here m protons are transported in NADH oxidation reactions, producing a pH and electrical potential difference across the bilayer which in turn, are utilized when n protons are translocated in the ATP synthesis reaction.

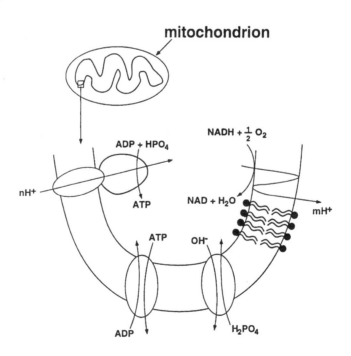

of reactions represent a closed loop that can operate at steady state, producing the net reaction of scheme 6.4.2.

The most crucial ingredient in the chemiosmotic mechanism is the transport of protons across the membrane. This process has two consequences. First, transport of a charged species generates an electric potential difference $\Delta\Phi$ across the membrane so that the outside becomes positive and the inside negative. As the potential builds up, the NADH oxidation reaction senses increasing resistance from the electrical forces. Second, proton transport also generates a difference ΔpH with the inside becoming more basic than the cytoplasm. For each proton transported from the inside to the outside, a free energy change occurs, often denoted the proton motive force

$$\Delta\mathbf{G}_{\text{trans}} = N_{Av}(\Phi_{\text{in}} - \Phi_{\text{out}}) - \mathbf{R}T\ln 10(\text{pH}_{\text{in}} - \text{pH}_{\text{out}}) = N_{Av}e\Delta\Phi - 2.3\mathbf{R}T\Delta\text{pH} \qquad (6.4.5)$$

where the first term on the right-hand side is an electrical free energy, while the second comes from the difference in entropy of mixing in the two compartments. In this way, the chemical free energy from the oxidation of NADH can be converted and stored as electrical and entropic free energy. This free energy can be used in the synthesis of ATP where it is converted back into chemical free energy.

The effective electrical capacitance and buffer capacity of the mitochondrion determine the distribution of free energy between $\Delta\Phi$ and ΔpH. Typical magnitudes are $\Delta\Phi = 100\,\text{mV}$ and $\Delta\text{pH} = -2$, corresponding to the free energies of $4\mathbf{R}T$ and $4.6\mathbf{R}T$ for transported protons, respectively. Measuring the number of protons transported in a certain elementary reaction in the sequence leading to the overall reaction 6.4.2 is experimentally difficult. Controversies still exist concerning the precise numbers, and to arrive at quantitative estimates we have to make a particular choice. If no NADH is produced and no ATP is consumed, the system will reach a conditional equilibrium with $\Delta\mathbf{G}_{\text{trans}} \approx 10\mathbf{R}T$ and the ratios NAD/NADH $\approx$ ATP/ADP $\approx$ 10^3 if we assume that 16 protons transported in reaction 6.4.2 and $1 + 1$ in the ATP synthesis/translocation. In the natural state, ATP is consumed constantly in the cytoplasm and NADH is produced. If this occurs at a constant rate, the system finds a steady state with a stoichiometry of $x = 16/2 = 8$ and a lower value of $\Delta\mathbf{G}_{\text{trans}}$ and smaller ratios NAD/NADH and ATP/ADP than we find at equilibrium. The higher the external ATP consumption, the farther away the steady state is from equilibrium and the larger are the losses in free energy. The inevitable passive proton leakage across the membrane also causes a loss that generates heat. In warm-blooded animals, this effect can contribute to the control of the body temperature in so-called brown fat tissue.

6.4.3 Propagation of a Nerve Signal Involves a Series of Transmembrane Transport Processes

The transmission of a nerve signal represents a specialized physiological process in higher animals. During evolution, such specialized functions often emerge through a synthesis of fundamental mechanisms that also are present in primitive organisms. Basically, propagation of a nerve signal occurs through a fascinating combination of different transmembrane transport events. As an example, let us take the propagation of a signal to the motoric system. The nerve cells have protrusions, called axons, that can be long (1 cm–1 m), and the propagation of a signal has two separate aspects: intracellular propagation along axons and transmission of the signal from the axon of one cell to the next. From a molecular perspective, these aspects of propagation steps occur in very different ways.

Within a nerve cell, the major free energy consumption is due to a sodium potassium ion pump. In a concerted process, ATP is hydrolyzed to ADP while three sodium ions are transported from the cytoplasm to the extracellular fluid and two potassium ions from the outside to the inside. The ion pump maintains an ion concentration difference across the nerve membrane with low sodium and high potassium concentrations on the inside. Due to a net transport of charge, a membrane potential develops with the inside negative at around -70 mV relative to the outside. Except for a slow passive leakage of ions across the membrane, an equilibrium develops in the pump reaction in a resting state. Potassium ions are close to a Donnan equilibrium, while the sodium ions are out of equilibrium by approximately $5kT$ per ion. The role of the ion pump is to make the free energy available in a form suitable for signal transmission, which inevitably involves a free energy loss.

The axon membrane contains channel proteins through which sodium and potassium ions can pass. In the resting state, these channels are closed, but this regulatory gate is sensitive to the transmembrane potential. If the potential increases from -70 mV to a critical value of about -40 mV, the gates start to open. Opening the gates allows for sodium transport, which in turn will make the transmembrane potential less negative and ultimately even positive. When this happens, more gates open, and the situation becomes unstable. This mechnism creates a rapid but local change in the membrane potential. Ions diffusing along the axon propagate this change in the potential, opening ion channels further down the axon, and so on. In a functional system, a regeneration mechanism must complement the firing system. When the gates have opened and the local membrane potential has increased to a positive value, the gates close again, and the action of the ion pump regenerates the initial state.

Within the axon, the nerve signal is basically of electrical origin. An intricate interplay exists between controlled ion currents perpendicular to the membrane and passive ion currents along it. Clearly, having a rapid signal transmission for the motoric system provides a great evolutionary advantage. To this end, vertebrates have developed an additional membrane system, called the myelin sheath, wound around the axon. In this system, gated ion channels are located at small bare areas between successive myelin sheaths. Myelin increases the capacitance so that a transport of fewer charges are needed to change the membrane potential. Because ion current determines the rate of signal propagation, the myelin ensures propagation at a lower current and thus speeds signal propagation.

Figure 6.35
Diagram showing chemical synapse, that is, the junction between two nerve cells. When an electrical signal reaches the presynaptic membrane in the cleft, vesicles fuse with the plasma membrane and deliver the chemical transmitter—for example, acetylcholine—into the synaptic cleft. The transmitter diffuses across the cleft and reaches receptors on the postsynaptic membrane.

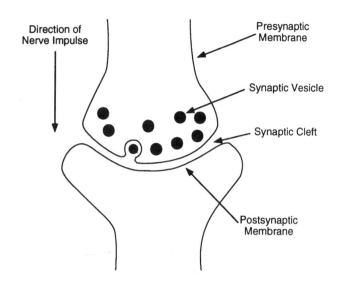

Ultimately, the nerve signal reaches the end of the axon, and a different transmembrane transport mechanism takes over to transmit the signal to the next axons (or muscle cell). This link is called a synapse, and Figure 6.35 shows its structure. At the end of the axon, vesicles store so-called transmitter substances. The motoric system's transmitter substance is acetylcholine. When the spike in the membrane potential reaches the end of one axon, it triggers a fusion process in which the vesicles merge with the plasma membrane and release the transmitter into the synaptic cleft that separates the two axons. The transmittance diffuses to the other side of the left and binds to protein receptors located in the membrane of the receiving (or postsynaptic) axon. The receptors in turn trigger a new spike in the membrane potential to propagate the signal. Enzymatic decomposition of the transmitter substance and the budding off of unloaded vesicles into the presynaptic axon readies the system to receive a new signal. From a membrane point of view, this signal transduction involves a delicately controlled fusion event as well as recycling to form new vesicles ready to be loaded by fresh transmitter substance. Chemically inert substances such as ethanol, diethylether, and even xenon have a general anesthetic effect. Their presence inhibits signal transduction across the synaptic cleft. Because these compounds are chemically inert, the most likely explanation of their effect is that they change the surfactant number N_s of the lipid bilayer, affecting the delicate vesicle fusion and fission events.

Exercises

6.1. A small unilamellar vesicle is formed at pH 6 in an aqueous solution. How many free protons are there inside the vesicle if the inner radius is 15 nm? Interpret the result.

6.2. A liquid crystalline sample in an open tube is placed in a closed container having a bulk pure water phase at the bottom. What is the osmotic pressure in the liquid crystal if the tube is placed 10 cm above the water surface?

6.3. A characteristic signature of a gel phase is the appearance of a sharp x-ray diffraction peak corresponding to a distance of 4.2 Å between symmetry planes. The chains are packed in a two-dimensional hexagonal array. What is the area per chain?

6.4. For DPPC in the gel $L_{\beta'}$, state, the repeat distance measured at 19°C by x-ray diffraction is 63.4 Å when the crystal is in equilibrium with excess water. Under these conditions there are 12 water molecules per lipid. Calculate the angle of tilt of the hydrocarbon chains, given that the chains pack at an area of 20 Å^2 measured perpendicularly to the chain direction. The lipid has an effective specific volume of 9.36×10^{-4} m^3/kg.

6.5. The table shows the measured repeat spacing in a lamellar phase of water–sodium octonate (NaC$_8$) decanol (Dec) for a series of compositions (% w/w).

NaC$_8$	Dec	H$_2$O	d(Å)
14.7	42.1	43.2	41.0
13.9	39.9	46.2	43.1
13.0	36.9	50.1	45.8
12.0	32.9	55.1	52.0
10.4	29.4	60.2	58.2
9.0	25.8	64.2	66.9
8.0	22.0	70.0	77.0
7.0	20.0	73.0	84.9

Calculate the thickness of the water and amphiphile layers, assuming that their respective specific volumes are 0.996×10^{-3} and 1.11×10^{-3} m^3/kg. Is the area per head group constant?

6.6. In Exercise 5.8 we found that the van der Waals force between two bilayers of thickness L separated by a solvent layer of thickness d is

$$F_{vdW}(d) = -\frac{H}{6\pi}\left\{\frac{1}{d^3} - \frac{2}{(d+L)^3} + \frac{1}{(d+2L)^3}\right\}$$

At what separation d is this van der Waals force balanced by the repulsive undulation force when $T = 300$ K, $H = 6 \times 10^{-21}$ J, $L = 4$ nm, and $\kappa_b = 8 \times 10^{-20}$ J? (Assume that the forces are independent.)

6.7. The short range repulsive force for DPPC in the liquid

phase has been measured to be

$$\frac{F}{\text{area}} = 1 \times 10^9 \exp\left(\frac{-h}{\lambda}\right) (\text{J/m}^3)$$

where $\lambda \simeq 2.2\,\text{Å}$. Assuming that the Hamaker constant is $5 \times 10^{-21}\,\text{J}$, what is the separation h for a phase in equilibrium with pure water? How many water molecules per lipid are there at equilibrium if we assume that the area per lipid is $68\,\text{Å}^2$?

6.8. Using the expression for the force between two DPPC bilayers in water, calculate the adhesion energy per unit area between two flat bilayers at equilibrium with pure water.

6.9. Using the results of the preceding exercises, estimate the interfacial free energy of a lamellar DPPC liquid crystal and pure water for an interface parallel to the bilayers.

6.10. Calculate the adhesion energy between two equal-sized DPPC vesicles of radius R. Assume that they do not undulate or deform upon contact and use constants of preceding exercises.

6.11. When an ionic surfactant like $C_{14}N(Me)_3Br$ is added to a DPPC lamellar liquid crystal, the surfactant ion is solubilized in the bilayer. Solubilization introduces an electrostatic double-layer force between bilayers. What ratio of CTAB to lipid is needed to induce an extensive swelling of the lamellar liquid crystal (a) when the medium is pure water and Br^- counterions and (b) in $1 \times 10^{-2}\,\text{M}$ NaBr?

6.12. A phospholipid vesicle of inner radius 15 nm is formed in a 1 mM KCl solution. It is then placed in a pure water solution. Estimate the time required for the ions to leak out into the solution if the permeability coefficient is $10^{-13}\,\text{m/s}$ for both ions and assuming that the vesicle remains intact despite the change in osmotic conditions.

6.13. Consider the vesicle system of Exercise 6.12. Assume that the pure aqueous medium now contains some valinomycin, leading to a drastic increase in the permeability coefficient for K^+. What is the change in membrane potential $\Delta\Phi$ for each transported K^+ ion? What is the maximum value of $\Delta\Phi$ that can be generated?

6.14. Estimate the parallel $D_\parallel$ and perpendicular $D_\perp$ diffusion coefficients for water and KCl in a phospholipid lamellar phase with $d_{H_2O} = 30\,\text{Å}$ and $d_{lipid} = 40\,\text{Å}$ Use the bilayer permeability coefficients $P_{H_2O} = 1 \times 10^{-5}\,\text{m/s}$ and $P_K \approx P_{Cl^-} \approx 1 \times 10^{-13}$. The diffusion coefficients in water are around $2 \times 10^{-9}\,\text{m/s}$ for all three species.

6.15. To a solution of vesicles, radius 25 nm, prepared in pure water, one quickly adds KCl to yield a $10^{-3}\,\text{M}$ solution. Calculate the time evolution of the vesicle's internal volume using $P_{H_2O} = 1 \times 10^{-5}\,\text{m/s}$ and $P_{K^+} = P_{Cl^-} = 10^{-13}\,\text{m/s}$. Consider only osmotic effects.

6.16. The squid has unusually large nerve axons, which are suitable for experimental studies. Across the axon membrane there is an electrostatic potential drop. Inside the axon $[K^+] = 0.41\,M$ and $[Na^+] = 0.49\,mM$, while on the outside $[K^+] = 22\,mM$ and $[Na^+] = 0.59\,M$. What is the steady state value of the potential when the ion permeabilities are such that

(a) $P_{K^+} \gg P_{Na^+}$
(b) $P_{K^+} \ll P_{Na^+}$
(c) $P_{K^+} = P_{Na^+}$

Neglect the role of other ions (implying that their permeabilities are smaller).

6.17. Assuming that in Exercise 6.16 $P_{K^+} = 5 \times 10^{-10}\,m/s$ and $P_{Na^+} = 1 \times 10^{-12}\,m/s$, what is the flow of ions at steady state? Calculate the free energy loss per time and area due to ion flux. (The steady state is supported by an Na^+/K^+ ion pump protein that consumes ATP).

6.18. Calculate the capacitance per unit length of a myelin structure in which a central cylindrical (radius R_1) part of a nerve cell is covered by n concentric bilayers as in a liposome. Assume that the bilayers are 40 Å thick and separated by water layers of thickness 400 Å. Note that $n(40 + 400)$ can be larger than R_1.

6.19. Using Exercise 6.18, calculate how many monovalent ions per unit length are transferred from the outside of the myelin structure to inside the nerve cell to change the potential from $-70\,mV$ to $+30\,mV$ using $R_1 = 1 \times 10^{-6}$ and $n = 25$.

Literature

References[a]	Section 6.1: Bilayers— Formation and Structure	Section 6.2: Experimental Techniques	Section 6.3: Lipid Bilayer Membranes	Section 6.4: Transmembrane Transport
Hiernenz & Rajagopalan				CHAPTER 8.10 Biological membranes
Hunter		CHAPTER 7.7.4 SFA measurements		
Israelachvili	CHAPTER 17 Bilayer structures	CHAPTER 12.19 SFA measurements	CHAPTER 18 Bilayers and biological membranes	
Kruyt		CHAPTER I.4 Optics and scattering		
Shaw		CHAPTER 3 Microscopy		
Tanford	CHAPTERS 11–12 Biological lipids— motility & order		CHAPTERS 13–19 Proteins, membranes	

[a]For complete reference citations, see alphabetic listing, page xxxi.

7

POLYMERS IN COLLOIDAL SYSTEMS

CONCEPT MAP

Polymers

- Polymers contain N_p covalently bonded monomers whose collective effect leads to unique and highly useful properties.

- Polymers are synthesized by condensation or by free radical reactions that join monomers together to form a macromolecule. Usually, such reactions produce a range of N_p values, which makes the polymer polydisperse. Polydispersity is measured by the ratio of M_n/M_w — the number- and weight-average molecular weights. Many bipolymers are monodisperse.

Homopolymers	Block Copolymers	Heteropolymers
N_p identical monomers with uniform interactions along the chains; they often form random coils in which $M_{\text{polymer}} = N_p M_{\text{monomer}}$.	Two or more covalently linked homopolymer chains	Different monomers joined in a nonregular pattern. Proteins contain 20 different monomers.

Polymer Configurations and the Radius of Gyration

The size of a polymer coil is characterized by its radius of gyration R_g (eq. 7.1.6).

- In dilute solution, polymers can fold (Figure 7.1) in many ways because they possess many internal degrees of freedom. In the compact state $R_g \propto M^{0.33}$, with a linear configuration $R_g \propto M$, while as a random coil, $R_g \propto M^\alpha$, where $1 > \alpha > 0.33$.

- The radius of gyration R_g is smaller than the average end-to-end distance R_{1N} (eq. 7.1.7).

- R_g is determined by: (1) local interactions like trans/gauche configurations; (2) excluded volume effects resulting from overlap repulsion between monomers; and (3) balance between solvent–monomer and monomer–monomer interactions as measured by the χ-parameter (eq. 7.1.10).

- Based on χ, three polymer solution regimes can be defined when: $\chi < 0.5$, good solvent where configurational entropy dominates leading to an open polymer chain; $\chi > 0.5$, bad solvent were monomer–monomer interactions are strong, leading to a dense polymer coil; and $\chi = 0.5$ a theta solvent where polymer steric interactions and solvent repulsion balance to form a random coil with $R_g \propto M^{0.5}$.

- The persistence length, l_p, describes the stiffness of a polymer chain (eq. 7.1.15).

- When polymers dissolve in a solvent many more coil configurations become accessible which adds a significant entropic contribution to the free energy of solution (eq. 7.1.21).

- Polyelectrolytes can form rods with large values of l_p (eq. 7.1.23), while proteins often form compact globular structures in their native configurations.

Polymer Solutions

Three concentration regimes exist (Figure 7.2): dilute — intermolecular interactions unimportant; semidilute — chain overlap becomes important; and concentrated — chain entanglements dominate.

   Transition between dilute and semidilute occurs at the overlap concentration $C^* \approx 0.1$ to $5\,\text{wt}\%$ (eq. 7.2.2).

- The Flory–Huggins theory describes the semidilute regime in terms of regular solution theory and provides a basis for calculating ΔG_{mix} (eq. 7.2.5). The interaction parameter χ provides a basis for classifying polymer–solvent interactions.

Characterization of Polymer Solutions

- Light, neutron and x-ray scattering techniques provide R_g and B_2, the second virial coefficient (Figure 7.4).

- In dilute solution, for self-diffusion coefficient $D \propto M^{-0.5-0.6}$ while for intrinsic viscosity (eq. 7.1.33), $[\eta] \propto M^{0.5-0.8}$ (eq. 7.1.38 and Figure 7.5). The intrinsic viscosity provides a convenient way to determine M.

- In the semi-dilute and concentrated solutions both D (eq. 7.2.11) and η (eq. 7.2.13 and Figure 7.8) show a strong dependence on polymer size as predicted by the reptation model.

Many polymer solutions display rheology properties in which the response to mechanical perturbation involves both elastic deformation and viscous flow.

Newtonian fluids display a linear relation between shear stress ($\sigma = F/A$, see Figure 7.9) and shear rate ($\gamma = d/dt(dx/dy)$, which defines the viscosity. Measurements of change in viscosity with polymer concentration yield molecular weights and particle shape.

Non-Newtonian fluids display a time-dependent response to shear stress that involves both elastic and viscous contributions and can be related to characteristic relaxation times (Figure 7.10).

Polymer Association

Polymers can associate to form a number of useful structures.

- When both units of a diblock copolymer $B_m C_n$ are in a good or bad solvent, the copolymer behaves like a homopolymer. When the solvent is good for one unit but bad for the other, its amphiphilic character leads to association and to formation of micelles, bilayers, etc. (Figure 7.11).

- Helices form when individual monomers possess amphiphilic properties. Transformation between helices and random coils is a cooperative process (Figure 7.12).

- Gels result from self-association or chemical crosslinking of polymers. On a macroscopic level, gels resemble solids, but on a microscopic level they retain fluidlike properties (Figure 7.13).

- Polymers facilitate the self-assembly of amphiphiles (Figure 7.17) and polymer/surfactant interactions can lead to complex viscosity behavior (Figure 7.18).

Polymers on Surfaces
Adsorption Processes

Thermodynamics	Kinetics
• Physical adsorption leads to many polymer–surface contacts (Figure 7.16). If $E_{abs} > kT$, the process is usually irreversible. • With amphiphilic block copolymers, one block selectively adsorbs onto the surface, while the other interacts with the solvent. • Polyelectrolytes adsorb strongly onto oppositely charged surfaces, but adding electrolyte can desorb them.	• Rate of adsorption is determined by the slow rate of polymer diffusion to a surface. • Polydispersity can affect the adsorption process. Small polymers adsorb first because they diffuse faster than large ones, but the final equilibrium favors adsorption of large polymers. • Equilibrium can take a very long time to achieve and may never be reached.

Forces Between Surfaces with Adsorbed Polymers

- Surfaces covered with polymer will begin to interact when the surface separation $\geqslant 2R_g$. At this distance (≈ 10–100 nm), van der Waals and double-layer forces are negligible. (Figure 7.17).

- In a good solvent ($w < 0.5\,RT$), a repulsive force (called steric stabilization, Figure 7.18) arises because the configurational entropy of the interacting chains is reduced, while in a bad solvent ($w > 1/2\,RT$), polymers concentrate at the surface to minimize contact with the solvent, leading to an attraction.

- Colloidal particles covered with a polymer will coagulate when the solvent changes from good to bad by variation in temperature, etc. This aggregation process is usually reversible.

7

7.1 Polymer Molecules in Solution

Polymers are molecules whose size falls within the colloidal range. Historically polymer science developed as a branch of colloid chemistry, but with the advent of modern polymer chemistry, studies of polymers have become a subject in their own right, with entire textbooks and university courses devoted to them. In this chapter we focus only on aspects of polymer science that are particularly relevant to the general properties of colloidal systems.

In colloid science we encounter polymers of several different kinds. **Homopolymers** consist of a small monomer unit repeated N_p times, where N_p is called the degree of polymerization. In a **block copolymer** the polymer chain consists of blocks of one repeating monomer unit followed by one or more blocks of other repeating unit(s). **Heteropolymers**, in which several different monomer units are joined in a nonregular pattern, are often found in biopolymers: proteins and nucleic acids. If the monomer units carry a charge, we can refer to a homo-, hetero-, or block copolymer as a **polyelectrolyte**. In Table 7.1 we list polymers of particular importance for colloidal systems and indicate their use. The difference in chemical complexity of these different polymers also is reflected in their self-assembly properties in solution. The homopolymer, in which the chemical interactions are uniform along the chain, typically forms a random coil conformation. On the other hand, block copolymers resemble amphiphiles, and as we shall see, they show all the different self-assembly structures found with surfactants. Finally, heteropolymers, like proteins, can organize into a very specific conformation determined by the detailed sequence of amino acids. Clearly, polymer systems show a rich behavior with respect to the conformation of the single molecule and also in the assembly of the different molecules.

Polymers are made by linking monomer units to the growing chain in a stepwise fashion

$$M_n + M \rightarrow M_{n+1}$$

TABLE 7.1 Some Polymers of Special Importance in Colloidal Systems

Common Name	Repeating Unit	Example of Use
Synthetic Polymers		
Polyethylene oxide (PEO), polyethyleneglycol (PEG)	(CH_2-CH_2-O)	Steric stabilization in water; inducing depletion forces
Polyacrylamide	$(CH_2-\overset{\displaystyle H}{\underset{\displaystyle \underset{\displaystyle NH_2}{\overset{\displaystyle C=O}{C}}}{C}}-O)$	Gels, (chromatography, diapers), rheology modifiers
Polystyrene	$(CH-CH_2)$ $\mid$ ϕ	Latex particles, rigid plastic containers, rigid component in polymer alloys
Polyimines	R $\mid$ $(C=N)$	Flocculant in papermaking
Pluronic polyethylene oxide and polypropylene oxide block copolymer	(CH_2-CH_2-O) $(CH-CH_2-O)$ $\mid$ CH_3	Drug delivery, detergency
Polystyrene sulfonate	$(CH-CH_2)$ $\mid$ SO_3^-	Reverse osmosis membranes

Common Name	Repeating Unit	Source	Example of Use
Biopolymers			
DNA	Adenosine, guanosine, cysidine, thymidine nucleotides	Any cell	
Cellulose	Glucose	Plants	Thickeners, conditioners (modified)
Starch	Glucose	Plants	Paper making
Gelatin	Glycerin, valine, alanine	Collagen, bones	Photographic lining, food
Agarose	Glucose	Seaweed	Chromotography gel
Hyaluronic acid	Glucose, glucoronic acid	Connective tissue	Eye surgery, veterinary medicine
Carageenans	Glucose, glucuronic acid, glucose sulfate	Algae	Food additives

In the biological system, the link is formed by enzymatic catalysis. For example, in protein synthesis precise control of the length of the chain gives the protein a specific molecular weight.

For all synthetic polymers and some biological ones, some partly random event terminates the chain reaction, resulting in a distribution of molecular weights. The size polydispersity of polymer molecules constitutes a serious complication in all fundamental studies of their properties and also in many applications. By affecting the termination step in the chain reaction, we can control the average size of polymer, but the synthesis must be followed by some chromatographic procedure or fractionated precipitation to achieve a relatively narrow size distribution.

Determining the full size distribution $F(N_p)$ of a given polymer sample is so difficult that we must be content with specific averages. The most accessible values are the degree of polymerization, which is also called the number average

$$\langle N_p \rangle_{na} = \int N_p F(N_p) dN_p \tag{7.1.1}$$

together with the so-called weight-averaged polymerization number

$$\langle N_p \rangle_{wa} = \int \frac{N_p^2 F(N_p) dN_p}{\langle N_p \rangle_{na}} \tag{7.1.2}$$

When the size distribution function simply represents a peak at a given N_p value, such as N_p' for a protein,

$$\langle N_p \rangle_{wa} = \frac{N_p'^2}{N_p'} = N_p' = \langle N_p \rangle_{na} \tag{7.1.3}$$

the number average and the weight average are the same. As soon as a true distribution exists, the weight average is always larger than the number average. Longer polymers contribute more to the weight than to the number average value.

Different methods for determining molecular weights yield different averages when a distribution in size exists. In Chapter 4 we discussed how to determine micelle sizes by light scattering. The same method can be applied to polymers, as discussed later in this chapter, and this measurement yields a weight-averaged molecular weight M

$$\langle M \rangle_{wa} = M_1 \langle N_p \rangle_{wa} \tag{7.1.4}$$

where M_1 equals the molecular weight of the monomer. In contrast, osmometry counts individual molecules and

$$\langle M \rangle_n = M_1 \langle N_p \rangle_{na} \tag{7.1.5}$$

Determining such different average molecular weights provides a useful measure of the width of the size distribution function. The ratio $\langle N_p \rangle_{wa}/\langle N_p \rangle_{na}$ can vary in the range of 1–3, where 1.1 represents a value obtained when great care has been taken to obtain a narrow size distribution, while 2 is obtained for a purely random termination reaction. Values larger than 2 have been observed for some biopolymers.

7.1.1 Chain Configurational Entropy and Monomer–Monomer Interactions Determine the Configuration of a Single Polymer Chain

When a polymer is dissolved in a high excess of solvent, we can neglect interactions between polymer molecules. However, many internal degrees of freedom are present, and the polymer can fold in different ways. How can we characterize the chain conformation? We can distinguish three different types of chain folding (see Figure 7.1). In the compact state, the chain folds back on itself to minimize polymer–solvent contact in a way that resembles a micellar configuration. The linear size of the polymer globule (the radius) then increases as $N_p^{0.33}$, where N_p represents the degree of polymerization. Soluble proteins in their native state are often found in this globular form.

Some polymers adopt a stiff linear configuration, often dictated by the formation of a helical structure. In this case the longest linear dimension increases linearly with N_p. A short DNA double helix typifies this behavior.

Most synthetic polymers, as well as proteins and DNA in denatured states, show a much less well-defined structure, which is referred to as a (random) coil. In this case the characteristic linear dimension R increases with the degree of polymerization as some power, $R \sim N_p^\alpha$, where α is larger than 0.33 but smaller than unity.

In the coil state the polymer chain adopts many different conformations, and we can characterize the state only by average properties. The most common measures are the radius of gyration R_g, which for a homopolymer is defined by

$$R_g^2 = \frac{\left\langle \sum_{i=1}^{p} |\vec{r}_i - \vec{r}_{cM}|^2 \right\rangle}{N_p} \tag{7.1.6}$$

Figure 7.1
The triangle of Haug, illustrating three extreme types of polymer configuration: the stiff rod ($L \sim N_p$), the compact globule ($R \sim N_p^{0.3}$), and the coil ($R_g \sim N_p^\alpha$ with $0.5 < \alpha < 1$).

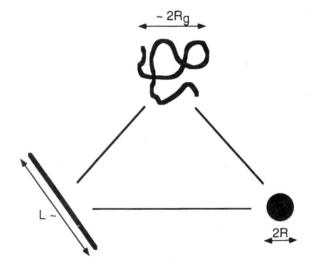

and the average end-to-end distance R_{1N}

$$R^2_{1N} = \langle |\vec{r}_{N_p} - \vec{r}_1|^2 \rangle \tag{7.1.7}$$

where $\vec{r}_i$ stands for the position of the ith segment and $\vec{r}_{cM}$ is the location of the center of mass. The angular brackets denote an average over all allowed coil conformations. The end-to-end distance is larger than the radius of gyration, but they scale similarly with changes in molecular weight.

Even though the chain conformation is characterized by one length in eqs. 7.1.6 and 7.1.7, we should not think of the typical chain configuration as a loose homogeneous spherical ball. The density of monomer units decreases out from the center. Furthermore, in a particular configuration, the decrease is more rapid in some directions than in others, so that although the coil is spherical on average, each configuration has a nonspherical distribution of monomers and typically exhibits a distinct asymmetry.

A polymer can adopt a large number Ω of configurations. As an example we can consider a polymer where the monomer–monomer bond can adopt three basic orientations, as in a carbon–carbon single bond with two gauche and one trans conformation. With a thousand monomers we have

$$\Omega \simeq 3^{1000} \simeq 10^{477} \tag{7.1.9}$$

configurations of a single molecule! Not all of these configurations have the same energy. The size of the polymer coil, as measured by R_g or R_{1N}, results from an average over all these 10^{477} weighted by their respective Boltzmann factors. We can distinguish between three types of interactions affecting these weight factors.

First are short range intrachain interactions. For example, in polyethylene, the trans conformation is lower than the gauche by approximately 2.4 kJ/mol, which favors the formation of locally straight chains. A second, more subtle, effect is that a chain cannot fold back on itself due to the overlap repulsion between monomeric units. Even though this interaction has a very short range in space, it takes on a long range character when counting along the chain since monomer units that are well separated along the chain can come close in space as the molecule folds. The net result is to favor more extended conformations of the chain. In the polymer literature, this is referred to as an excluded volume effect.

The third, and most significant, contribution to the energy in the Boltzmann factor comes from the monomer-solvent interactions. Even though the polymer molecule might be at a high dilution, the local concentration of monomer units may be high within the polymer coil. From a molecular point of view, we have a local situation that is much the same as in an ordinary binary liquid mixture, and we can use the same basic model to analyze it that we used for such mixtures in Chapter 1. In the regular solution model the basic parameter is the interaction constant, w, introduced in eq. 1.4.16. When we apply it to a apply it to a polymer system we only need to reinterpret the bulk coordination number, z_b, since each monomer unit is linked covalently to two other monomers.

For the typical case of $w > 0$, monomer–monomer contacts are energetically favorable, and the more compact the coil, the larger the number of such contacts. Balancing this organizing effect is the entropy promoting randomness, that is, equal weights given to all possible configurations. The crucial quantity determining the equilibrium properties of the coil is, in analogy with regular solutions

$$\chi = w/RT \qquad (7.1.10)$$

commonly referred to as the χ parameter. When analyzing polymer coil properties at different χ values, we can identify three important regimes:

1. For $\chi < 0.5$ the intermolecular interactions are weak compared to the thermal energy, and chain configuration entropy dominates. This regime is called good solvent conditions.
2. For $\chi > 0.5$ monomer–monomer interactions are strong. To obtain many such contacts, the polymer coil becomes dense and this regime is called bad solvent conditions.
3. At $\chi = 0.5$, a transition occurs from good solvent to bad solvent behavior. The temperature when $\chi = 0.5$ is called the theta (θ) temperature. It represents the transition from a good to a bad solvent and we denote this a theta solvent in which theta (θ) conditions apply. These have important applications in fundamental studies of polymer solutions.

7.1.2 Persistence Length Describes the Stiffness of a Polymer Chain

A polymer chain is very flexible on a length scale that is large compared to the size of a monomer unit, and this flexibility leads to a coil conformation. The overall shape of the coil does not depend on the local stiffness of the chain, while the size does. In the same way as a coil of a thin flexible thread is much smaller than a coil of a stiff rope of equal length, the size of a polymer coil not only depends on the length of the chain but also on its stiffness. Moving along a polymer chain in a particular coil conformation is like following the path of a diffusing molecule. The random walk is a mathematical model that describes some of the basic features of both polymer configurations and diffusive motion. In the simplest form the random walk is a sequence of N_r steps of length l_r each in a random direction. The root mean square length $\langle L_r^2 \rangle^{1/2}$ of such a walk is

$$\langle L_r^2 \rangle^{1/2} = N_r^{1/2} l_r \qquad (7.1.11)$$

If we impose the additional restriction that the path may not cross itself within a distance of order l_r one has the self-avoiding random walk which is less compact on average and

$$\langle L_r^2 \rangle^{1/2} \sim N_r^{0.6} l_r \qquad (7.1.12)$$

When applying this equations to a polymer, it is tempting to identify the number of steps N_r with the degree of polymerization N_p, and the step length with a monomer size. However, this identification does not properly account for the local stiffness of the polymer chain. The stiffness of the chain is characterized by the persistence length l_p, which is the length over which two parts of the chain keep their orientational correlation. In a simplified picture we can see the chain as a sequence of freely linked stiff segments, so-called Kuhn segments, whose length corresponds to twice the persistence length. With a length L of a fully stretched chain, the contour length, there are $L/2l_p$ Kuhn segments.

We can make use of the random walk properties by identifying the step size in the walk with the persistence length of a particular chain. An important characteristic of the θ-condition is that the polymer coil has random walk properties and the end to end distance

$$R_{1N}^2 = \frac{L}{l_p} l_p^2 = 2Ll_p \qquad (7.1.13)$$

using eq. 7.1.11. We will see later in this chapter that in scattering experiments the radius of gyration is experimentally more accessible than the end-to-end distance. It is generally true that for long chains R_g and R_{1N} are proportional and for the random walk model

$$R_{1N}^2 = 6R_g^2 \qquad (7.1.14)$$

When we measure the radius of gyration at θ-conditions for a polymer of known contour length we obtain the persistence length

$$l_p = 6R_g^2/L^{1/2} \qquad (7.1.15)$$

using eqs. 7.1.13 and 7.1.14. For two different polymers with the same length L, the size of the coil will increase with increasing persistence length, as we can see from eq. 7.1.13. Persistence length can vary from values in the range 0.5–1.0 nm for very flexible polymers, such as $(-CH_2-)_n$, $(-O-Si(CH_3)_2-)_n$, and $(-O-C_2H_4-)_n$, to more than 100 nm for some polyelectrolytes in low salt conditions.

We can now express how the size of the polymer coil depends on the relevant physical parameters. The monomer solvent interaction enters implicitly through the solvent quality, while the others enter explicitly. At θ-conditions

$$R_g = (6l_p l_M)^{0.5} N_p^{0.5} \sim M^{0.5} \qquad (7.1.16)$$

Here we have used that the contour length $L = l_M N_p$.

Under good solvent conditions, $\chi < 0.5$, the polymer coil configurations behave as a self-avoiding random walk and in analogy with eq. 7.1.12

$$R_g \sim N_p^{0.6} \sim M^{0.6} \tag{7.1.17}$$

with a proportionality constant that increases with an increasing persistence length. In a bad solvent, $\chi > 0.5$,

$$R_g \sim N_p^{0.33} \sim M^{0.33} \tag{7.1.18}$$

One should note that the proportionality $R_g \sim N_p^x$ in eqs. 7.1.16–7.1.18 only applies for large N_p. For shorter molecules nonnegligible correction terms appear.

7.1.3 When Polymers Dissolve into a Solvent Many More Coil Configurations Become Accessible

The solvent quality not only affects the size of the polymer coil, but it clearly has a strong effect on the polymer's solubility. Our first question is: Should we expect a polymer to be soluble to any appreciable extent? Today we know the answer can be yes, but the question was one of major concern in the early days of polymer science. When a polymer is dissolved from a pure state, for example, a crystal, into a dilute solution, there are three contributions to the free energy of the process. One is the entropy of mixing of the polymer molecule with the molecules of the solvent. For dilute systems in which we can neglect interactions between polymer molecules, the ideal entropy of mixing expression, eq. 1.4.8, applies for polymers as it does with any other solute. This contribution is determined solely by the number of dissolved molecules, but in the polymer case, it is independent of the molecular weight and N_p. A second contribution derives from bringing the polymer from the crystalline state, where it is surrounded by like polymers, into solution, where it is surrounded by solvent molecules to a large extent. For $w > 0$ this contribution involves an increase in energy on dissolution and thus limits the solubility. Since these interactions involve the monomer units, the energy increase per molecule increases linearly with degree of polymerization N_p ($\Delta U_{\text{solution}} \simeq z_b w N_p$). Since typically $w > 0$ the energy will win over the entropy of mixing if the molecules are sufficiently large. For this reason, large polymers seemingly cannot be soluble to any appreciable extent. The resolution of this dilemma that faced the polymer scientists around 1930 lies in a third free energy contribution. In the crystal, the polymer ideally adopts a unique conformation, so that $\Omega = 1$ and

$$S_{\text{crystal}} = k \ln \Omega = 0 \tag{7.1.19}$$

We can associate an entropy with the polymer molecule's many internal degrees of freedom. On dissolution, many more configurational states become available. Earlier we estimated that for three available states per monomer–monomer bond $\Omega \simeq 3^{N_p}$ and for this case

$$S_{\text{solution}} = k \ln 3^{N_p} = kN_p \ln 3 \qquad (7.1.20)$$

Even though we have overestimated Ω somewhat, the resulting proportionality between S_{solution} and N_p remains in more accurate descriptions.

Combining the three free energy contributions, the final expression for the free energy change of dissolution is

$$\Delta G(\text{dissolution}) = RT\{n_s \ln(1 - X_p) + n_p \ln X_p\} + z_b w N_p n_p - RT \ln 3 N_p n_p \qquad (7.1.21)$$

The fact that the two last terms dominate over the first for large N_p values except at extreme dilutions implies that if

$$w/RT < \ln 3 / z_b \qquad (7.1.22)$$

a polymer is likely to be soluble whatever its molecular weight, while if $w > RT \ln 3 / z_b$, it is hardly soluble.

This calculation is not numerically accurate, but it brings home two essential qualitative points. First, in contrast to colloidal entities that lack conformational freedom, a polymer molecule can be dissolved even in the limit of high molecular weight. This qualitative property is largely independent of the degree of polymerization. Second, we can relate the criterion for solubility to the same χ parameter, eq. (7.1.10), that determines the nature of the polymer coil. It also follows that understanding the changes in available polymer configurations provides a key to the conceptual understanding of polymer solutions.

Above we have for simplicity discussed the solubility of a crystalline polymer. In a melt we have polymer coils with superficially the same distribution of coil conformations as in a θ-solvent with $R_g \sim N_p^{1/2}$. However, there are stringent constraints on packing coils of long chains so that they fill space completely and the configurational entropy of the chain is much smaller in the melt than for separated coils.

Another manifestation of this effect is that different polymers as a rule phase separate in the melt. In a mixing process there is no increase in available coil configurations since the packing problem remains. The small ideal entropy term in mixing polymer molecules cannot compensate for the energy increase due to the positive w interaction.

7.1.4 Charged Polymer Chains Display a More Extended Conformation

Charged polymers play an important role in colloid chemistry. In addition to the practical aspect that charges on the chain greatly increase the aqueous solubility of polymers, polyelectrolytes strongly influence the electrostatic forces between colloidal particles.

These polyelectrolytes can be synthetic, like polyacrylic acid, polystyrene sulfonate, or polyimides, or they can have a biological source, like DNA or hyaluronic acid, one of many charged polysaccharides. In general, they are soluble in water and possibly other highly polar solvents but not in apolar solvents.

The presence of the charged groups influences both intra- and intermolecular interactions. Intramolecular repulsions between the charged groups lead to a more extended conformation of the chain compared to the corresponding neutral polymer. This means that the persistence length is mostly determined by the electrostatic effects. Odijk derived an approximate relation for the persistence length

$$l_p = l_p^0 + l_p^{el} = l_p^0 + \frac{e^2}{16\pi\varepsilon_r\varepsilon_o kT\kappa^2\alpha^2} \qquad (7.1.23)$$

Here l_p^0 represents the persistence length of the chain in the absense of the long range electrostatic interactions, l_p^{el} is the electrostatic contribution, and α is the distance between charges along the chain. However, the persistence length concept is problematic for polyelectrolytes and equation 7.1.23 should be used as a qualitative guideline rather than as a quantitative relation.

Clearly, the density of charges along the chain also will influence the role of the electrostatics. A common model used to describe a polyelectrolyte starts from a charged cylinder and solves the Poisson–Boltzmann equation. This cylindrical geometry represents an intermediate case between the planar charged surfaces and spherical aggregates discussed in earlier chapters.

In the cylindrical geometry, a phenomenon called ion condensation appears as a consequence of the long range interaction. To a good approximation, the counterion population can be divided into two categories. One category, called condensed ions, is electrostatically bound close to the charged cylinder, while the remaining ions are diffusely bound. The point of this division is that we can easily calculate the fraction of condensed ions. They have the effect of reducing the effective linear charge density, $\xi = \pm e/\alpha$, to a critical value

$$\xi_{crit} = \pm \frac{4\pi\varepsilon_r\varepsilon_o kT}{z^2 e} \qquad (7.1.24)$$

Thus, the fraction of condensed counterions equals $1 - \xi_{crit}/\xi$ ($\xi \leqslant \xi_{crit}$). For monovalent ions ($z = 1$) in water ($\varepsilon_r = 80$) at room temperature, $\xi_{crit} = 1e/(0.71\,\text{nm})$. The distribution of the non-condensed ion can then be calculated using the linearized Poisson–Boltzmann equation. It should be stressed that the ion condensation model provides a conceptually simple way of dealing with the electrostatic consequences of ion distribution. In the form presented here it simply represents an approximate way to describe the solution of the nonlinear Poisson–Boltzmann equation in a cylindrical geometry.

Owing to their long persistence length, polyelectrolyte coils are very extended and tend to overlap at very low concentrations. When coils entangle, the viscosity increases. For example, the use of hyaluronic acid in eye surgery is based primarily on its ability to increase the viscosity of the aqueous medium even at low concentrations.

7.1.5 Protein Folding Is the Result of a Delicate Balance Between Hydrophobic and Hydrophilic Interactions and Configurational Entropy

Proteins are heteropolymers in which the monomer building blocks are 20 different amino acids attached via amide linkages. Figure 7.2 shows some representative structures formed by adding alkyl, polar or ionic groups to glycine, the simplest amino acid. The specific amino acid sequence of a protein which defines its primary structure is coded by the DNA. The secondary structure refers to the regular arrangement of the chain as in α-helices or β-sheets, and also in sidechain–sidechain disulfide bonds. The tertiary structure refers to the protein's unique three-dimensional structure is determined by the interactions between individual amino acids.

Using the 20 different amino acids in specific sequences of a polymer chain, nature has managed to make protein molecules with a remarkable range of functions. Enzymes, that occur either as globular water soluble molecules or bound in the lipid membrane, catalyze a multitude of specific biochemical

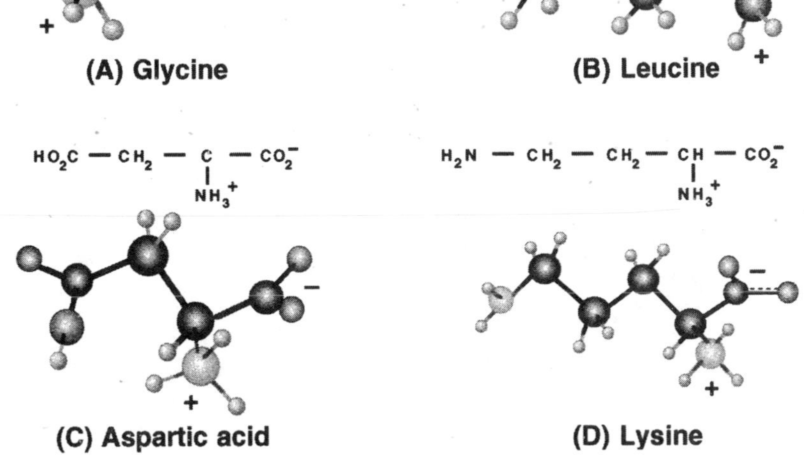

Figure 7.2
Illustration of modifications to the simple zwitterionic aminoacid glycine (A), which lead to incorporation of a hydrocarbon chain (B), a carboxylic acid (C), and an amine group (D). Upon a change in pH, C and D can impart either negative or positive charges to a protein molecule.

(A) Glycine

(B) Leucine

(C) Aspartic acid

(D) Lysine

reactions. Proteins like hemoglobin transport and store metabolites. Others like insulin and calmodulin regulate physiological processes. Collagen or silk are used as materials. Gluten in seeds and egg white in eggs store amino acids. Histones control DNA packing in chromosomes. Antibodies act as specific complexing agents. Actin and myosin form a molecular motor.

It is one of the future challenges of polymer technology and colloid science to, however slightly, approach this degree of sophistication in material design. The protein example demonstrates the great variety of properties that can be obtained by using monomers of different physicochemical characteristics in a controlled sequence. Using the block-copolymer concept one has recently made some advances in tailor making the properties of synthetic polymers. This should only be the beginning.

In the language of polymer chemistry, the compact globular conformation typifies a polymer molecule in a bad solvent, yet the protein remains soluble. The reason is clearly that the different side chains represent a spectrum of χ parameters with respect to water. Thus, the protein chain strives to adopt a conformation that hides the apolar side chains from the aqueous solvent while retaining the polar side chains at the surface of the folded molecule. However, some compromises have to be made in the folded state. Polar groups, such as the amides linkages of the protein backbone, must enter the apolar interior. To do as well as possible in an intrinsically uncomfortable environment, they tend to link up by hydrogen bonds either in α-helices or β-sheets.

Thus, the driving force for the formation of a globular protein conformation in water is the hydrophobic interaction and is the same as for the association of surfactants into a micellar aggregate. The packing possibilities are much more constrained in the protein, however, because the units are covalently linked and because polar or semipolar groups are distributed along the chain and in the backbone. Although the globular state does not form because of the hydrogen bonds in the interior of the protein globule, these bonds are necessary to prevent the folded state from becoming too unfavorable.

Heating a globular protein denatures it. Typically denaturation is completed over a relatively narrow ($\sim 10\,K$) temperature range. Thus, protein folding is a cooperative process, and the cooperativity has the same basic source as in a micellization. The alternative to the native folded state is a more or less random coil with a large number of accessible conformations. Denaturation occurs when the temperature is raised because of the increased importance of the TS term of the chain conformation entropy in the random coil. Thus, we see how the same principles govern the assembly of the amino acids in a globular protein that operate in a surfactant aggregate.

7.1.6 Scattering Techniques Provide Information about Molecular Weight and Chain Conformation

Scattering of light, neutrons, or x-rays is the most generally applicable method for studying the structure of polymers as well as colloidal systems in general. Polymer solutions are well suited to illustrate the general principles of scattering studies.

In a scattering experiment, a monochromatic beam of light, neutrons, or x-rays is focused on the liquid sample and the intensity of the scattered beam is measured by a detector as illustrated in Figure 7.3. Collisions between photons, neutrons, or x-rays and the molecules of the sample are elastic (with no energy exchange), but a transfer of momentum takes place. The larger the momentum transfer, the higher the scattering angle θ. We can measure the momentum transfer by the scattering vector $\vec{Q}$ (see Figure 7.3), such that

$$|Q| = \frac{4\pi \sin(\theta/2)}{\lambda} \tag{7.1.25}$$

For neutrons, the wavelength is determined by the velocity v, through the de Broglie relation

$$\lambda = \frac{h}{mv} \tag{7.1.26}$$

where m equals the mass and h is Planck's constant. The straightforward scattering experiment consists of measuring the intensity of the scattered beam as a function of the scattering vector.

Even though the scattering, as such, has different molecular causes, to a large extent we can describe the scattering of light, x-rays, and neutrons within one and the same framework. Light scattering is caused by local variations in the refractive index, while the x-ray scattering comes from variations in the electron density; and neutrons scatter from a nonuniform distibution of nuclei, particularly protons and deuterons.

Figure 7.3
Radiation scattered from two centers separated by a vector $\vec{r}$ interferes to produce scattering at an angle θ. The incoming beam has a wave vector, $k_0 = (2\pi/h)\vec{p}_0$. For elastic scattering the magnitude of the momentum $\vec{p}$ is conserved and $|k_0| = |k_p| = 2\pi/\lambda$. The magnitude of the scattering vector $Q = k_p - k_0$ is $4\pi \sin(\theta/2)/\lambda$ (eq. 7.1.21).

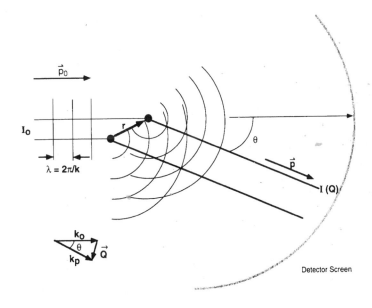

In addition to differences in the molecular scattering mechanisms, the scattering techniques differ with respect to the accessible Q values. The long wavelengths used in light scattering imply small Q values by eq. 7.1.25. In contrast, both x-rays and neutrons operate with wavelengths of $1-10\,\text{Å}$, which are small relative to the characteristic sizes of colloidal systems. Therefore, we can usually obtain the relevant scattering information for small scattering angles. Consequently, these techniques are commonly referred to with the acronyms SAXS (small-angle x-ray scattering) and SANS (small-angle neutron scattering).

Chapter 4 discussed how light scattering can be used to determine micelle size. We pointed out that the equations applied only when micelles had dimensions very much smaller ($<\lambda/20$) than the wavelength of the light so that they acted as a point scattering center. In this case the scattering intensity is angle dependent only through the trivial factor $(1 + \cos^2\theta)$.

However, interference effects of light scattered from different monomer units along the chain can appear in the scattering from a polymer molecule. These interference effects depend on how the monomer units are distributed in space and thus potentially provide structural information in the scattering data. Interference effects can also occur between monomer units on different molecules, so information about the relative distribution of different polymer molecules is also available. How can we extract these different pieces of information?

We can write the total scattering intensity as the product of four factors:

total scattering = instrumental constant × contrast factor × concentration × sample structure dependent factor

The first three factors can be measured independently, and the relevant information is found in the sample scattering function, which we denote $I(Q)$.

If we can consider the intra- and interparticle interferences as independent, the scattering function becomes a product of a form factor $P(Q)$ and a structure factor $S(Q)$

$$I(Q) = P(Q)S(Q) \qquad (7.1.27)$$

The form factor depends only on intraparticle interferences and is independent of concentration as long as the molecules or particles remain unchanged. Given the distribution of groups in the polymer or particle we can determine $P(Q)$ through a Fourier transform.

The structure factor accounts for interparticle interferences. It relates to the pair distribution function $g(r)$ of particles through

$$S(Q) = 1 + \frac{4\pi c_p^*}{Q} \int_0^\infty (g(r) - 1)r\sin Qr\,dr \qquad (7.1.28)$$

where c_p^* stands for the number of particles per unit volume.

Thus, if we know the particle structure, we can calculate $P(Q)$, measure $I(Q)$, and then determine the particle distribution $g(r)$. On the other hand, if we want to determine the particle/molecule structure, we should try to dilute the system so that $S(Q)$ is practically independent of Q and then measure $P(Q)$ through $I(Q)$. From the measured $P(Q)$, we can determine the particle structure. For example, for a uniform sphere of radius R

$$P(Q) = \left\{ \frac{3(\sin QR - QR \cos QR)}{(QR)^3} \right\}^2 \qquad (7.1.29)$$

Because $P(Q) = 0$ for Q values satisfying $\sin QR = QR \cos QR$, the form factor oscillates, and we can readily determine the radius R from the location of the zeros of $P(Q)$. (In practice, $P(Q)$ exhibits pronounced minima at only one or two Q values.) These idealized schemes do not work in many cases because it is difficult to perform measurements over a sufficient range of Q or to disentangle $P(Q)$ and $S(Q)$ in some regions.

In the region of low Q values, both $P(Q)$ and $S(Q)$ behave in a characteristic way. $P(Q)$ is defined so that $P(0) = 1$. In a series expansion, the term that is linear in Q is zero and

$$P(Q) \simeq 1 - \frac{R_g^2 Q^2}{3} \qquad (7.1.30)$$

for small Q. We see that the coefficient of the Q^2 term yields the radius of gyration as defined in eq. 7.1.6. Clearly, the rationale for defining R_g in this way is precisely so that it can be measured in a scattering experiment.

The structure factor $S(Q)$ approaches the inverse osmotic compressibility as Q goes to zero (c.f. eq. 4.2.6)

$$S(0) = kT \left(\frac{\partial \Pi_{osm}}{\partial c^*} \right)^{-1} \qquad (7.1.31)$$

For dilute systems

$$\Pi_{osm} = kTc^*$$

according to eq. 1.5.13 and $S(0) = 1$ according to eq. 7.1.31. Furthermore $S(Q)$ becomes independent of Q. In more concentrated systems with interacting macromolecules/particles Π_{osm} deviates from the value of an ideal solution and furthermore, the interaction also causes a structure to develop in the solution. The $S(Q)$ starts to deviate from the value of unity at $Q = 0$ and it becomes Q dependent for Q values of order $2\pi \xi^{-1}$ where ξ is the range of the correlations in the solution.

In a typical light scattering experiment on a dilute polymer solution, we can determine $I(Q)$ at a range of Q values. Extrapolation to $Q = 0$ yields $S(0)$ at the given concentration. It is advisable to repeat the experiment at a new concentration and extrapolate to $Q = 0$ again. Repeating this procedure a few times gives a set of data that allows for the simultaneous

Figure 7.4
Zimm plot for measuring polymer molecular weight. Measurements are performed at a number of scattering angles θ for a series of concentrations. According to eq. 7.1.25, the form factor depends on $Q^2 \sim \sin^2 \theta/2$, which gives a theta dependence to the scattering intensity from intramolecular interferences. Conversely, the structure factor at $Q \sim \sin \theta/2 = 0$ is concentration dependent as a result of interparticle interferences as in eq. 7.1.31. By plotting the concentration c over the Rayleigh ratio, ΔR_θ (eq. 4.2.8), multiplied by the optical constant K_0, versus $\sin^2 \theta/2 + \text{constant} \times c$, the results can be extrapolated simultaneously to $\theta = 0$ and $c = 0$ to yield the weight-averaged molecular weight. The respective slopes provide the radius of gyration and the second virial coefficient. Data are for a cellulose nitrite sample. (H. Benoit, R. M. Holtzer, and P. Doty, *J. Phys. Chem.* **58**, 635 (1954).

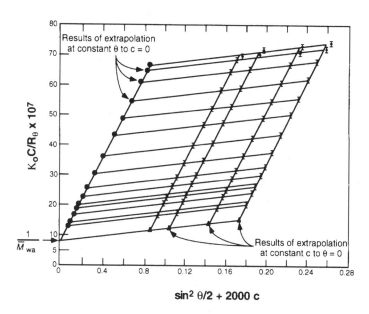

determination of the molecular weight, the second virial coefficient B_2, and the radius of gyration R_g. Usually evaluation of the data is performed in a so-called Zimm plot as illustrated in Figure 7.4.

For polymer systems light scattering is unsuitable for getting structural information beyond the measurement of R_g. The method of choice is neutron scattering, which is sensitive to scattering at higher Q values corresponding to structure on a shorter length scale. An example in Chapter 9 features the use of neutron scattering for studying structure in a solution of colloidal latex particles.

7.1.7 Polymer Self-Diffusion and Solution Viscosity Reflect the Dynamic and Structural Properties of a Polymer Coil

In solution, a polymer coil has a range of dynamic modes from translation and rotation of the whole molecule to local conformation changes. The monomer units interact locally with the solvent, but when the molecule moves—in part or in its entirety—this motion involves solvent molecules. They in turn affect other solvent molecules, which can influence a more distant segment of the polymer. We discussed this mechanistic description of hydrodynamic interactions briefly in Section 5.7. Hydrodynamic interactions are important for polymer coil dynamics, and they tend to correlate motions of different segments of the coil.

Let us first consider the translational self-diffusion of a polymer coil. For compact spherical particles of radius R in a solvent of viscosity η, the self-diffusion coefficient is given by the Stokes–Einstein relation

$$D = \frac{kT}{6\pi\eta R} \tag{5.7.5}$$

which is derived by considering a liquid flow past a spherical object with the stick boundary condition that the flow is zero on the surface of the sphere. A polymer coil is on average spherical and we have expressions for the mean radius of the coil, but to use eq. 5.7.5, we must also justify the stick boundary condition that the liquid velocity is zero at a distance R from the center of the stationary coil. This condition implies that the liquid flow inside the coil is zero. Although individual liquid molecules clearly diffuse relatively freely inside the coil, hydrodynamic interactions could cause the internal solvent to follow the coil. Partial drainage of the solvent in the coil is occurring, but experiments indicate that this effect is relatively small and in a theta solvent, we find

$$D \sim M^{-0.5} \tag{7.1.32}$$

which is the relation we expected by combining eq. 5.7.5 and the relation $R_g \sim M^{0.5}$ (eq. 7.1.16). In a good solvent, the same reasoning suggests an exponent -0.6, and experimentally one finds values in the range -0.55 to -0.6. Self-diffusion coefficients of polymers are rather demanding to measure experimentally, so while we could use eq. 7.1.32 to measure molecular weights, this is not done in practice.

A much more easily accessible parameter is the viscosity of a polymer solution. It serves as a simple route for determining molecular weight and particle shape. As the solution is diluted, the viscosity, η, approaches that of the solvent, η_s, linearly with solute volume fraction, Φ, so that

$$\eta = \eta_s(1 + [\eta]\Phi) \tag{7.1.33}$$

where $[\eta]$ is called the intrinsic viscosity. For spheres Einstein derived that $[\eta] = 2.5$. On the other hand, for stiff rods with length L and radius R ($L \gg R$)

$$[\eta] = \frac{L^2}{\pi R^2} \tag{7.1.34}$$

This relationship shows that stiff rods with $L \gg R$ give a much more viscous solution than spherical particles, even in the limit where no direct interaction takes place between the rods. We can understand the molecular origin of the intrinsic viscosity based on the fact that no velocity gradients can exist inside the particles. With stick boundary conditions, a particle in a velocity gradient will sense a torque and rotate. A rotating rod generates more dissipation than a rotating sphere explaining the larger value of $[\eta]$ for the rods.

According to the Einstein result, for compact spheres the intrinsic viscosity is independent of the size and only determined by the volume fraction of particles. Polymer coils are spherical on average, but at a given volume fraction, their intrinsic viscosity does depend on molecular size. To see this, we first adopt the assumption established in our analysis of diffusional motion that we can neglect the drainage of the coil. The radius of the coil varies as

$$R \sim M^{\alpha} \qquad (7.1.35)$$

where $\alpha = 0.5$–0.6, depending on solvent quality. The volume, V_c, of a coil then varies as $V_{coil} \sim M^{3\alpha}$. The effective volume fraction of coils is

$$\Phi(coil) = c^* V_{coil} \qquad (7.1.36)$$

where c^* equals the concentration of polymer molecules.

Now a coil contains both polymer and solvent so that $\Phi(coil)$ is much larger than Φ_p, the volume fraction of polymer. In a measurement we prepare samples of known concentration of polymer on a weight basis equal to c^*M. Knowing the density of the polymer this can be converted to a volume fraction of polymer molecules $\Phi_p \sim c^*M$. By measuring the viscosity for a series of concentrations we determine $[\eta]\Phi_p$ experimentally. By Einstein's argument this should equal $2.5\Phi(coil)$ so that

$$[\eta] = \frac{2.5\Phi(coil)}{\Phi_p} \sim \frac{c^*M^{3\alpha}}{c^*M} \sim M^{(3\alpha - 1)} \qquad (7.1.37)$$

with the exponent $3\alpha - 1$ varying between 0.5 and 0.8 as α varies between 0.5 and 0.6. We can use this relation to determine molecular weights, provided that we can calibrate against some known samples. Experimentally the exponent 0.5 is confirmed for theta conditions and Figure 7.5 provides an illustration. In good solvents the exponent is distinctly higher; we often find a value around 0.75, but rarely 0.8. This discrepancy could indicate that drainage of the coil is not totally negligible for these more extended coils.

7.2 Thermodynamic and Transport Properties of Polymer Solution Change Dramatically with Concentration

7.2.1 Different Concentration Regimes Must Be Distinguished to Describe a Polymer Solution

A polymer molecule with a characteristic radius R_g contains mostly solvent in its random coil. The larger the exponent α, the longer the persistence length l_p, and the higher the degree of polymerization N_p, the less dense the coil. The expanded nature of the coils implies that they start to overlap and interact at low polymer concentrations.

Figure 7.5
The specific intrinsic viscosity [η] for polystyrene as a function of molecular weight in two solvents, toluene and tetrahydrofuran (THF), close to theta conditions. The slope of the straight lines is 0.50. (G. Meyerhoff and B. Appelt, *Macromolecules* **12**, 2103 (1979).)

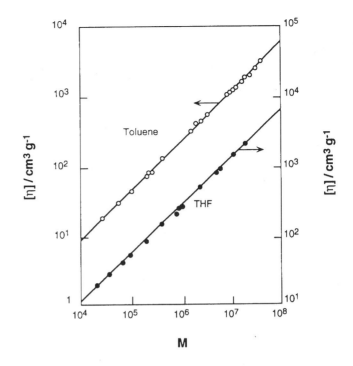

Figure 7.6
The three concentration regimes of a polymer solution. When the concentration is less than the overlap concentration C^*, we have a dilute solution. In the semidilute system $c > c^*$, but the volume fraction Φ is still small ($\Phi \ll 1$). In the concentrated system $c \gg C^*$ and Φ approaches unity.

Polymer–polymer interactions play somewhat different roles in distinct concentration regimes (see Figure 7.6). When discussing the polymer chain conformation, we explicitly considered the single chain, that is, the dilute solution. In practice, most polymer solutions are in the so-called semidilute regime, where solvent exceeds polymer on a weight-on-weight basis,

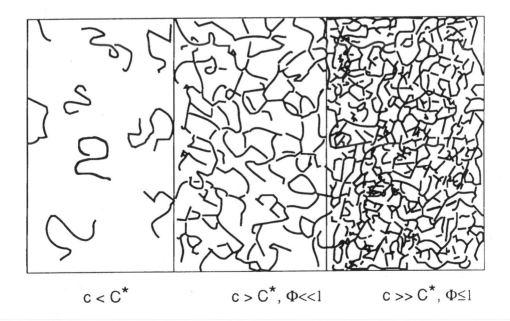

$$c < C^* \qquad c > C^*, \Phi \ll 1 \qquad c \gg C^*, \Phi \leq 1$$

but the coils of individual molecules overlap so that a monomer on one chain is also likely to make contact with monomers of other chains. In semidilute solutions, the chains form a nonconnected network with a mesh size smaller than the radius of gyration. Furthermore, the mesh size decreases with increasing concentration because the coils are pushed even more into one another.

The crossover from a dilute solution to a semidilute regime is described by the characteristic concentration C^*. Clearly, C^* does not have a precise value, but we can estimate it by assigning a volume V_{coil} to each polymer and calculating when the total number of coil fills the volume of the solution.

If we assume

$$V_{coil} = \frac{4}{3}\pi R_g^3 \tag{7.2.1}$$

the overlap concentration C^* (molar monomers) is

$$C^* = \frac{3N_p}{4\pi N_{Av}}\frac{10^{-3}}{R_g^3} \tag{7.2.2}$$

A more experimentally accessible criterion for the overlap concentration is obtained from eq. 7.1.38 for the intrinsic viscosity. If we require $\Phi(coil) \approx 0.5$ the first equality shows that the overlap concentration in volume fraction is

$$\Phi_p^* \approx \frac{1.25}{[\eta]} \approx [\eta]^{-1}$$

simply the inverse of the intrinsic viscosity. The factor of 1.25 is often dropped as in the second equality, since we are dealing with an order of magnitude estimate.

Because R_g varies with N_p with a power 0.5 in a theta solvent and a power 0.6 in a good solvent, $C^* \sim N_p^{-0.5}$ for a theta solvent and $C^* \sim N_p^{-0.8}$ in a good solvent. Thus, the higher the molecular weight of the polymer, the lower the overlap concentration. Estimates of C^* based on measured $[\eta]$ values show that the overlap concentration typically lies in the range of 0.1–10% (w/w). Low values occur for long polymers with stiff chains in a good solvent.

A higher concentrations of polymer, where solvent and polymer are present in approximately the same amount on a weight basis, we encounter yet another concentration regime. In these concentrated solutions, the picture of a loose polymer network is no longer relevant. Instead the system consists of highly entangled chains, with properties similar to those of a polymer melt. This concentration regime is relevant, for example, in describing the latex particles so important for modern paints.

The polymer-polymer interaction also affects the coil conformation, which tends to become more compact at higher concentrations in good solvents so that one finds $R_g \sim N_p^{0.5}$ irrespective of solvent quality in this concentration regime.

7.2.2 The Semidilute Regime Is Well Described by the Flory–Huggins Theory

Mixing polymer and solvent produces two types of entropy effects. As with any binary mixture, entropy increases from mixing two different molecular species. However, few polymer molecules are present on a molar basis and this entropy contribution is small. The more important effect is that the entropy of the polymer configuration increases the more space is available in the solution. Having small solvent molecules as neighbors allows for much more chain flexibility than being surrounded by equals.

By accounting for the connectedness between molecules, Paul Flory and Maurice Huggins (Flory 1953) independently succeeded in generalizing the regular solution theory to semidilute polymer solutions. Using a lattice model, as in Figure 1.9 and assuming a random distribution of the monomer units on the lattice sites, with the constraint that the polymer chains remain intact, the entropy of mixing adopts the deceptively simple form

$$S_{\mathrm{mix}} = R(n_s \ln \Phi_s + n_p \ln \Phi_p) \qquad (7.2.3)$$

where s and p denote solvent and polymer, respectively. The number of moles of species i is denoted by n_i, and Φ_i represents the volume fraction they occupy. A solvent molecule and a monomer unit are assumed to occupy the same volume so that, for example,

$$\Phi_s = \frac{n_s}{n_s + n_p N_p} \qquad (7.2.4)$$

It is important to realize that although eq. 7.2.3 is formally closely analogous to the ideal entropy of mixing eq. 1.4.8, the main contribution to S_{mix} comes from the change in configurational entropy of the polymer chain, rather than from the mixing of molecular units.

We can derive the interaction contribution to the free energy of mixing exactly as we did for the regular solution theory in Section 1.4. The only difference is that the number of nearest neighbors is reduced by two because the monomer units in the chain are covalently linked to two other monomers. The resulting expression for the free energy of mixing is

$$\Delta \mathbf{G}_{\mathrm{mix}} = \mathbf{R}T(n_s \ln \Phi_s + n_p \ln \Phi_p) + (n_s + N_p n_p)w\Phi_p\Phi_s \qquad (7.2.5)$$

in close analogy with eq. 1.4.18 for a regular solution. A notable feature of ΔG_{mix} is that for a given value of the polymer volume fraction Φ_p, it varies only slightly with the degree of polymerization N_p. In fact, only the small term, $n_p \ln \Phi_p$, changes with N_p at constant Φ_p.

Physically, therefore, the relevant quantity in the solution is the mesh size in the network of chains, and this size is determined solely by the polymer volume fraction. Thus in practice it is very difficult to determine polymer molecular weights in the semidilute regime by measuring a thermodynamic quantity, while a number of transport properties are sensitive to the size of the molecules.

We can obtain the chemical potential of the solvent by differentiating eq. 7.2.5.

$$\mu_s = \mu_s^\theta + RT \ln \Phi_p + RT \left(\Phi_p - \frac{\Phi_p}{N_p} \right) + w\Phi_p^2 \qquad (7.2.6)$$

When the volume fraction of polymer is small, we can write

$$\ln \Phi_s = \ln(1 - \Phi_p) \simeq -\Phi_p - \frac{1}{2}\Phi_p^2 \qquad (7.2.7)$$

so that the chemical potential μ_s equals

$$\mu_s = \mu_s^\theta - RT \left[\frac{\Phi_p}{N_p} - \left(\chi - \frac{1}{2} \right)\Phi_p^2 \right] \qquad (7.2.8)$$

introducing $\chi \equiv w/RT$. The term that is linear in Φ_p vanishes for long polymers (large N_p), while the term quadratic in Φ_p becomes positive when $w > 0.5\,RT$ or $\chi > 0.5$. In the limit $N_p \to \infty$ the polymer is insoluble, and we have a bad solvent. (The chemical potential μ_s increases with increasing polymer concentration.) For $\chi < 0.5$, the polymer is soluble over the whole semidilute regime, and we have a good solvent. In the regular solution theory of a binary mixture, demixing occurs when $\chi > 2$, so that polymers are intrinsically less soluble than monomers. Because of the internal entropy of the chain, however, a sizable positive value of the interaction parameter w is required to achieve phase separation.

When using the Flory-Huggins theory and the free energy expression eq. 7.2.5, we should remember that from a molecular point of view, the theory is based on the same simplified picture as the regular solution theory. The theory is conceptually appealing, but problems can arise when it comes to quantitative predictions.

7.2.3 In a Semidilute or Concentrated Solution, Polymer Diffusion Can Occur Through Reptation

Entanglements between different coils strongly inhibit the translational diffusion of the center of mass of a polymer molecule. In Section 7.1.7, we analyzed how an intact single coil moved through a bare solvent. This mode of motion is totally quenched in a more concentrated system in which extensive entanglements exist between coils. Set on a local scale, polymer solutions show extensive Brownian motion as in any liquid system, and this should lead to translational diffusion. Which motional mode makes the dominant contributions to the translation motion under these circumstances?

In one mode that is operative even at very high concentrations, the polymer slides along the position of its own chain as illustrated in Figure 7.7. Steric interference to this motion from other polymer molecules can occur only at the head end. The polymer motion must slide perfectly along the chain only at the highest polymer concentrations. In a semidilute system, the chain can make lateral excursions on the scale of the mesh size, and we can envisage the chain moving along in a snake-like motion. This movement forms the conceptual basis of the reptation model of diffusion and viscosity of entangled polymer solutions.

To arrive at the molecular weight dependence of the diffusion coefficient, we first note that the friction factor μ for sliding the polymer in the tube increases linearly with the length of the chain, so we can write

$$\mu = \mu_{\text{mon}} \cdot N_p \sim M \qquad (7.2.9)$$

Furthermore, a unit displacement, l_t, along the tube does not cause a similar displacement of the center of mass, r_M, of the polymer since the polymer is in a coil. If the two ends of the coil are located close together in space, moving material from one end to the other hardly affects the location of the center of mass, while if they are located on opposite sides of the coil there is a substantial translational component to the motion. The average mean square displacement of r_M is related to the average end-to-end distance through

$$\langle \Delta r_M^2 \rangle = l_t^2 \frac{\langle R_{1N}^2 \rangle}{L^2} \sim \frac{M^{2\alpha}}{M^2} \sim M^{-2(1-\alpha)} \qquad (7.2.10)$$

which can be derived using the random walk model. Combining eqs. 1.5.31 and 7.2.10, we find that for the self-diffusion coefficient

$$D \sim M^{-2} \qquad (7.2.11)$$

since α tends toward 0.5 for concentrated polymer systems. Comparing eq. 7.2.11 with eq. 7.1.29, we see that the diffusion coefficient depends much more strongly on molecular size in concentrated systems than in dilute ones.

Polymer self-diffusion is strongly concentration dependent. In eq. 7.1.32, we gave a value relevant for infinite dilution.

Figure 7.7
Illustration of reptation motion in a concentrated polymer solution (or polymer melt). Due to the constraints from other chains, shown as thin lines, the illustrated chain, thick line, can only slide along its own axis. A movement of length l_t causes a change in the center of gravity $\Delta r_m \sim l_t \cdot (R_{1N}/L)$.

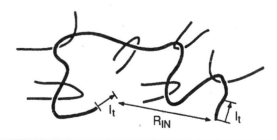

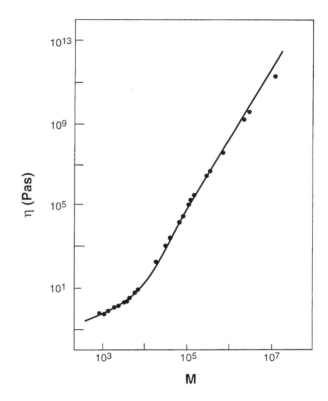

Figure 7.8
The molecular weight dependence of the viscosity of a polybutadiene melt at 25°. Above $M = 10^4$ the viscosity depends on the molecular weight to the power 3.4 as shown by the straight line. (From R. H. Colby, L. J. Felters and W. W. Graessley, *Macromolecules* **20**, 2226 (1987)).

As concentration increases, other polymer coils initially exclude volume, resulting in a decrease of D linear with concentration in analogy with eq. 7.1.33 for the viscosity. As the solution approaches the overlap concentration the concentration dependence increases, and in the semidilute regime, we find a relation

$$D \sim c^{-\beta} \tag{7.2.12}$$

where β has been found to be 1.75 for a good solvent and 3 for a theta solvent. At high concentrations, all systems approach theta conditions.

Concentrated polymer solutions typically show non-Newtonian behavior under high shear rates, but it is possible to study the Newtonian range at low shear rates to obtain the (Newtonian) viscosity. In a velocity gradient a polymer coil must constantly disentangle (and entangle) to sustain a different velocity from its neighboring coils. In a concentrated system, this process should become more sluggish when the molecular weight is higher. Figure 7.8, which shows an increase in η from 1 to 10^{12} poise between $M = 10^3$ and $M = 10^7$, illustrates the strong dependence on molecular weight. The slope of the line in the double logarithmic plot is 3.4, showing that

$$\eta \sim M^{3.4} \tag{7.2.13}$$

So far, this generally observed relation has defied a detailed theoretical interpretation. A straightforward application of the reptation theory yields an exponent 3.0, which is in reasonable, but not perfect, agreement with theory. In this theoretical description, the crucial molecular factor is the time it takes for a polymer coil to reptate completely out of its initial tube.

7.2.4 Polymer Solutions Show a Wide Range of Rheological Properties

One widespread use of polymers is to control the rheological properties of a solution or colloidal dispersion. Flow properties are crucial in food processing (jam, ice cream, sauces, etc.) as well as in paints. Rheological properties are also essential from a fundamental point of view because they can be used to characterize a system. As we will see, the difference between a polymer solution and a polymer gel becomes most apparent in light of the difference in the rheology of the systems.

In general, rheology deals with how a system responds to a mechanical perturbation in terms of elastic deformations and of viscous flow. Normally, we associate the elastic response with solids and the viscous response with liquids, although both systems show both types of behavior, albeit on very different time scales. The typical colloidal system shows a behavior intermediate between the solid and the simple liquid, and not surprisingly we find in these systems characteristic elastic and viscous behaviors. In a simple rheological experiment, the sample is put between two parallel walls at separation y. (See Figure 7.9). One of the walls is stationary, while the other can be displaced. Displacing this wall a distance x will generate a shear force F on the stationary wall. This force is

Figure 7.9
Schematic representation of a shear cell. Two parallel walls are separated by a sample of thickness y. When the upper surface is displaced a distance x from x_0 to x_d, a force F is generated on the lower surface.

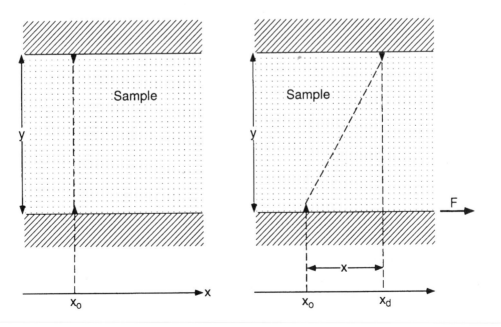

proportional to the area A of the walls, and we can define a shear stress σ as

$$\sigma = \frac{F}{A} \qquad (7.2.14)$$

For a perfect elastic solid, Hooke's law applies, and the shear stress is proportional to the displacement x divided by the thickness y, that is, the shear strain γ

$$\gamma = \frac{x}{y} = \frac{dx}{dy} \qquad (7.2.15)$$

where in the last equality we have considered both x and y to be infinitesimal quantities. For an ideal (Hookeian) solid, we can calculate the shear strain from macroscopic distances x and y because it is constant through the sample, but in general we will have to use the differential form. For a Hookeian solid

$$\sigma = G_0\gamma \qquad (7.2.16)$$

where G_0 is called the static shear modulus.

On the other hand, if the sample is a simple liquid, a displacement x will induce only a transient force on the stationary plate. A shear stress constant in time is observed only if the second plate is moved at a constant rate, dx/dt, which implies a constant shear rate

$$\dot\gamma = \frac{d}{dt}\left(\frac{dx}{dy}\right) \qquad (7.2.17)$$

in the sample. For simple fluids a linear relation exists between the shear stress σ (i.e., the force on the stationary surface) and the shear rate $\dot\gamma$ (i.e., the rate at which the other surface is moved) so that

$$\sigma = \eta\dot\gamma \qquad (7.2.18)$$

where η stands for the viscosity of the system. Liquids that obey this linear relation over a large range of shear rate $\dot\gamma$ are called Newtonian.

In the Newtonian regime, the viscosity alone determines the rheological behavior. However, the most relevant rheological properties in applications of polymer solutions and colloidal dispersions relate to their non-Newtonian behavior. To describe these properties we need a conceptual framework that brings us continuously from the ideal Hookeian solid to the Newtonian liquid.

With the setup of Figure 7.9, we can consider an experiment in which the upper wall is moved periodically with a frequency ω and an amplitude x_0. Under these conditions the shear strain γ and the shear rate $\dot\gamma$ will vary as

$$\gamma = \frac{x_0}{y}\sin\omega t = \gamma_0\sin(\omega t) \qquad (7.2.19a)$$

$$\dot\gamma = \frac{x_0}{y}\omega\cos\omega t = \gamma_0\omega\cos(\omega t) \qquad (7.2.20b)$$

The oscillatory motion will give rise to a time-varying shear stress at the stationary surface the same frequency. In general, however, a phase shift δ takes place so that

$$\sigma = \sigma_0 \sin(\omega t - \delta) \qquad (7.2.21)$$

For sufficiently small amplitudes γ_0, the response in shear stress will depend linearly on γ_0, and we can write

$$\sigma = \gamma_0(G'(\omega)\sin(\omega t) + G''(\omega)\cos\omega t) \qquad (7.2.22)$$

where $G' = (\sigma_0/\gamma_0)\cos\delta$ and $G'' = (\sigma_0/\gamma_0)\sin\delta$ are called the storage modulus and the loss modulus, respectively. These values generally depend on the frequency ω. For an ideal solid the storage modulus is $G'(\omega) = G_0$ and the loss modulus is $G''(\omega) = 0$, while for a Newtonian liquid $G'(\omega) = 0$ and $G''(\omega) = \eta\omega$.

By measuring the frequency dependence of the loss and the storage moduli, we obtain information on the time scales for mechanical relaxation processes in the sample. Figure 7.10 compares the rheological properties of a polymer solution and a polymer gel. At low frequencies the storage moduli remain substantial for the gel but not for the solution. Furthermore, the polymer solution has a Newtonian viscosity, while gel viscosity decreases dramatically with increasing frequency. This example indicates how we can use rheological measurement as a sensitive indicator for association phenomena in polymer systems.

Another example of non-Newtonian behavior is found for systems where the polymer chains crosslink reversibly, for example, through the formation of a chemical complex. On time scales short relative to the average lifetime of the crosslink, the system behaves as a connected rubber and responds elastically,

Figure 7.10
Rheological parameters: □, dynamic viscosity $\eta\Sigma = (G'^2 + G''^2)/\omega$ (Nsm^{-2}); ●, storage modulus $G'(\omega)$; and ○, loss modulus $G''(\omega)$ (Nm^{-2}) for (a) a polymer dextrane aqueous solution and (b) an aqueous agarose gel.
Note the nearly perfect Newtonian behavior of the solution and the strongly elastic properties of the gel. (E. R. Morris and S. B. Ross-Murphy, in *Techniques in the Life Sciences*, Vol. B3, Elsevier/North-Holland, Amsterdam, 1981, Chap. B310.)

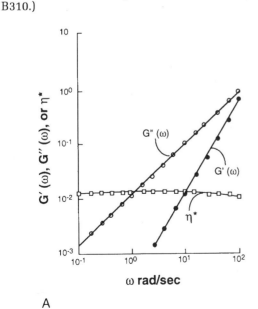

A

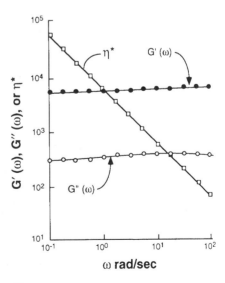

B

while for longer times, or low frequencies, the system flows as a liquid. This is the principle behind the properties of silly putty, which can be made from polyvinyalcohol and water by mixing in some borate to make labile crosslinks.

For polymers in the semidilute solution shear will create constant disentanglements/entanglements to allow for a velocity gradient. The perturbation from the shear will cause a change in the local structure of the solution, which in turn will cause deviations from the linear Newtonian behavior. To somewhat decrease the disentanglement the polymer coils will on average orient and deform to extend along the lines of flow. This facilitates the flow and leads to a shear thinning.

7.3 Polymers May Associate to Form a Variety of Structures

So far, we have focused on polymer properties that can be associated with the intrinsic structures of random coils, rods and compact globules. Interaction between polymer molecules promotes cooperative processes that lead to self-assembly of polymeric systems.

7.3.1 Block Copolymers Show the Same Self-Assembly Properties as Surfactants

A diblock copolymer composed of monomer units B and C has the general formula

$$\underbrace{B \cdots B}_{m} - \underbrace{C \cdots C}_{n}$$

that is, $B_m C_n$. In a solvent, the conformation of a single isolated chain depends strongly on the solvent quality with respect to types of monomer units. If it is a good solvent for both units, the configuration will resemble that of a homopolymer in a good solvent. If it is a bad solvent with respect to both units, it will not dissolve. However, when the solvent is good relative to only one of the monomers, say B, the molecule displays amphiphilicity. In the single molecule the C part of the chain is in a bad solvent and the C chain adopts a compact structure to which a random coil B chain is attached (see Figure 7.11). A strong attraction exists between the compact C parts. Indeed, when the concentration is increased sufficiently, an association results that is closely analogous to the micelle self-assembly process. Well-defined CMCs can be found, as well, although they are typically very small and thus difficult to detect.

The self-assembly process starts with the attraction between C units, while the stop occurs because B coils of different molecules overlap more and more, the higher the aggregation number. At some point the effective repulsion between the B

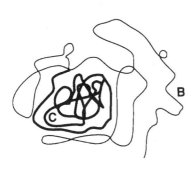

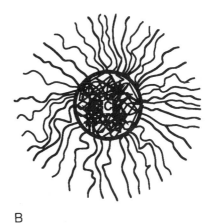

A   B

Figure 7.11
(A) Single block copolymer
BC in a solvent that is
bad for *C* but good for *B*.
The *C* part (thicker line)
adopts a compact state,
while the *B* part (thinner
line) is in a coil
configuration. (B) Micelle
formed by self-assembly of
BC block-copolymer
molecules. Crowding
caused by the aggregation
results in the stretching
of the *B* chains.

units balances the attraction between the C parts and the
aggregation process stops. The geometry of the resulting aggregate depends on the relative length of the two blocks as well as
on the persistence length of the B chain in the solvent.

As the ratio n/m between C and B units increases, the
structures progress from spherical micelles to ellipsoidal
micelles to hexagonal, cubic, and lamellar phases. If the C part
is largest, inverted aggregates—hexagonal and spherical—are
found.

7.3.2 Polymers with Amphiphilic Monomer Units Often Form Ordered Helix Structures

When discussing solvent quality, we have considered the
monomer as a single unit with a given interaction parameter
relative to the solvent. For molecularly complex monomers this
is clearly a simplification, and one group of the monomer may
interact favorably with the solvent while the rest of the molecule dislikes the solvent, as in an amphiphilic monomer unit.
How can we build a structure that solves the dilemma of having
solvent exposure for the solvophilic parts while protecting the
solvophobic parts of the monomer from solvent contact?

We discussed one possibility for the globular proteins, but
this solution of the problem really requires a distribution of
different monomer units. In the case of homopolymers, but also
for some heteropolymers, an alternative is the formation of a
helix, which can be a single strand as for the α-helix of proteins,
a double helix as in DNA, or a triple helix as in gelatin.

The protein α-helix is formed to shield the polar amide
backbone from an apolar environment. It is the dominant protein structure in the part of an integral membrane protein that
resides within the lipid bilayer.

The force that drives the formation of the double helix of
DNA is the hydrophobic interaction between base pairs. The
primary force acting against the dimerization is the electrostatic
free energy due to the charged phosphate groups. Association
also involves a loss of solvation energy of the polar hydroxy and

nitrogens of the purine and pyrimidine rings. However, to a large extent a hydrogen bond matching in the helix reduces this free energy cost.

Helix formation is a cooperative process, which is signaled, for example, by the occurrence of the transition from a helix to a random coil over a narrow temperature region. Cooperativity of the helix coil transition can be explained using a simple model with an initiation step and a propagation step. Let us illustrate the principle with the formation of a double helix. In the initiation step, two strands, S and S', with N_p units, are brought together to form a short helix with i monomers, where i might equal one

$$S + S' \rightleftharpoons (SS')_i \tag{7.3.1}$$

with an equilibrium constant K_i, so that

$$\frac{[(SS')_i]}{[S]^2} = K_i \tag{7.3.2}$$

This process is typically improbable, so for the fraction we write

$$\frac{[(SS')_i]}{[S]} = K_i[S] \ll 1 \tag{7.3.3}$$

However, once the helix has been initiated, a propagation step can follow in which the helix part of the chain increases

$$[(SS')_n] \rightleftharpoons [(SS')_{n+1}] \tag{7.3.4}$$

We assume the equilibrium constant K_p to be independent of the number of helix units n as long as $n \geqslant i$. If $K_p > 1$, the formation of helical parts will stabilize the dimer and when $K_p \gg 1$, the only species occurring in substantial amounts are the free chains and/or the double helix involving the whole chain. By repeatedly using eq. 7.3.4 we find

$$[(SS')_{N_p}] = K_p^{(N_p - i)}[(SS')_i] \tag{7.3.5}$$

and from eq. 7.3.3

$$\frac{[(SS')_{N_p}]}{[S]^2} = K_i K_p^{(N_p - i)} \tag{7.3.6}$$

So although $K_i[S] \ll 1$, we can still have $K_i[S]K_p^{(N_p - i)} \gg 1$. The transition point marked in Figure 7.12 occurs for $K_i[S]K_p^{(N_p - i)} = 1$.

What affects the different constants K_i and K_p? Configurational entropy is always higher in the coil than in the helix, so we expect both K_i and K_p to decrease with increasing temperature. For DNA another contribution affects the process. Because DNA is a polyelectrolyte, forming a double helix involves the

Figure 7.12
Examples of helix–coil
transitions of bacterial
DNA in a KCl solution.
Schematic representation
of the transition for a
mixture of two
polynucleotides with
complementary monomers
(solid line) and natural
DNA (dashed line). The
transition can be monitored
by measuring light
absorbance at 260 nm. It is
complete within 2°C for
the pure system, while in
the native state it occurs
over 5°C. (b) Two-step
melting for DNA with
uracil (U) and adenine (A)
in a 2:1 ratio. A complete
double helix exists below
45°C, but above 54°C we
find only single strands.
Betwen the two transitions
is a A·U double helix
with U single strands. (C.
Stevens and G. Felsenfeld,
Biopolymers **2**, 293 (1964).)
(c) The salt dependence
of the helix–coil transition.
Varying the KCl
concentration from 0.7 M to
0.01 M changes the
transition point by 20°C.
(C. Schildkraut and S.
Lifson, *Biopolymers* **3**, 195
(1965).)

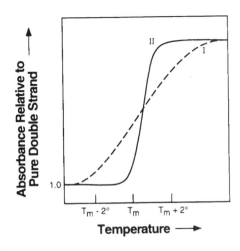

A

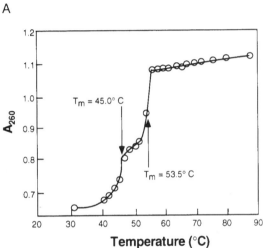

B

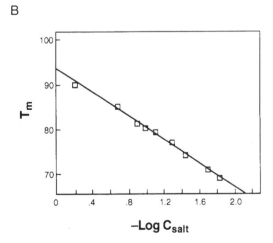

C

creation of an even more highly charged species. Therefore, the product $K_i K_p^{(N_p - i)}$ contains a contribution due to the free energy of the electrical double layer of the double helix. As a result, the higher the electrolyte concentration, the higher the helix–random coil transition temperature (Figure 7.12c).

When a polymer changes from a random coil to a helix conformation, it loses virtually all the chain configurational entropy. In Section 7.1.3 we concluded that the main mechanism that makes polymers soluble in a good solvent is the increase in configurational entropy. Consequently, we should expect helix formation to be followed by precipitation from solution. For DNA, precipitation does not occur because the charges make different double helices repel one another. However, in other systems, such as gelatin or polysaccharides like agarose or carrageenans, the charge stabilization mechanism does not operate to the same extent. In the following section we shall see that in these polymers precipitation can lead to the formation of a network gel rather than a compact solid.

7.3.3 Polymers Form Gels Through Chemical Crosslinking and by Self-Association

Above the overlap concentration C^*, the polymer chains are entangled and the mesh size of the entanglements changes with concentration. However, entire chains can move relative to one another. For example, even if the dynamics of chain diffusion are slow, the system behaves as a liquid. In many applications it is desirable to create a system in which the local dynamics are rapid, yet on a larger scale the system is stationary. This goal can be achieved by creating a polymer gel.

In a chemically crosslinked gel, overlapping polymer coils are covalently connected at a few positions. As the crosslinking progresses, larger and larger polymers are created until macroscopic molecular weights are reached, in which single molecules extend through the whole macroscopic sample. Once the crosslinked polymer has been formed, the solvent can be dried and the polymer will retain a memory of its macroscopic shape that reappears when it is redissolved. The polymer and the solvent form a gel (see Chapter 10 for a general discussion of the term) in which both elastic and viscous properties remain at low frequencies, that is, $G'(\omega) \neq 0$ for small ω.

Crosslinked polymer gels have many practical uses. Gel electrophoresis has become a routine procedure in analyzing protein solutions or for fractionating DNA. One main role of the gel is to prevent convection, which would destroy the resolution. In DNA the network in the gel also strongly influences the separation process.

Another use is in the formation of filter membranes for purification processes. The active part of a reverse osmosis system for water purification is a crosslinked polyelectrolyte gel membrane. By controlling the pore size, only the solvent and low molecular weight compounds can be made to pass the membrane.

Disposable diapers contain small granules of slightly ionized crosslinked polyacrylamide or polyacrylic acid. These granules can swell extensively (up to 1000 times) to incorporate an aqueous solution by electrostatic double-layer forces, so the degree of swelling depends on the electrolyte content in the water.

A number of polymers can form gels without chemical crosslinking. They are called physical gels. In these cases the gelling of a polymer solution can be induced by temperature change, change in salt concentration, or by adding some specific chemical that can adsorb on the polymer. The most well-known gel former is gelatin, a protein obtained from collagen, but a number of polysaccharides also form gels. These include agarose (widely used in chromatography), pectin (the basic gelling component of natural fruit jelly), and carrageenan (a common food additive obtained from seaweed).

All these polymers can form helices, and the molecular event that clearly seems to trigger gel formations is a coil–helix transition. As we remarked in Section 7.3.2, helical polymers have intrinsically low solubility because they possess little configurational entropy. Thus, we can expect a precipitation at the formation of helices. However, size polydispersity and possibly also chemical heterogeneity create a large probability that the ordered packing of helices will be disrupted and one strand can go from one helix to the next, leaving a coil segment between. This packing would lead to the network structures in Figure 7.13, which shows two different ways of generating a network.

A useful aspect of the physical gels is that the gel formation is practically reversible. If the gel is formed by cooling, it can be dissolved by heating. The process is not thermodynamically reversible, however, and the heating and cooling curves are slightly displaced relative to one another. The physical gel is typically not an equilibrium state, and in some cases we can observe aging processes in which the gel contracts and leaves solvent behind. This phenomenon is called syneresis.

7.3.4 Polymers Facilitate the Self-Assembly of Amphiphiles

When an amphiphile is added to a dilute polymer solution, we often observe a cooperative self-assembly process below the CMC of the pure amphiphile solution. A micellelike aggregate is formed which interacts with the polymer coil. Formation of these polymer-bound micelles occurs at a critical aggregation concentration (CAC), which varies with the nature of the amphiphile and the polymer.

The micelles associated with the polymer can form a coiled string of beads. However, micelle–micelle repulsion will occur for charged amphiphiles in particular, so each polymer coil can accommodate only a limited number of micelles. When the coils are saturated with micelles, new amphiphiles will go into the solution again until the CMC is reached. Figure 7.14

Figure 7.13
The formation of a gel network by double helices. (a) Helices are joined by single-strand sections, so that each molecule participates in two or more helices. (b) Bundles of helices join in an irregular pattern. Each molecule is involved in only one helix. (Courtesy of Lemmar & Piculell).

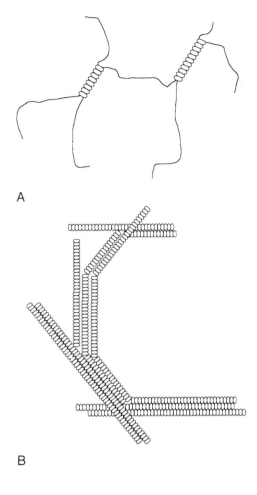

A

B

illustrates this point by a schematic plot of the amphiphile chemical potential versus concentration.

What is the nature of the combined amphiphile polymer aggregate? Table 7.2 lists CACs for a series of amphiphile–polymer pairs. Typically the CAC is smaller than the CMC by a factor of 10 to 1000, and in a homologous series it seems to decrease somewhat faster than the CMC with increasing chain length, particularly when the polymer is a polyelectrolyte of opposite charge.

Combined observations show that the amphiphile aggregate is a micelle, and we can understand that the CAC is lower than the CMC due to a stabilization of the aggregate by interaction with the polymer chain. The nature of this interaction varies from system to system. With SDS and polyethylene oxide, the rather polar polymer adsorbs on the micelle surface primarily to reduce hydrocarbon–water contact. Polyelectrolytes will bind micelles of opposite charge electrostatically. For a polyelectrolyte like polystyrene sulfonate, an additional hydrophobic interaction takes place between the styrene part and the interior of the micelle.

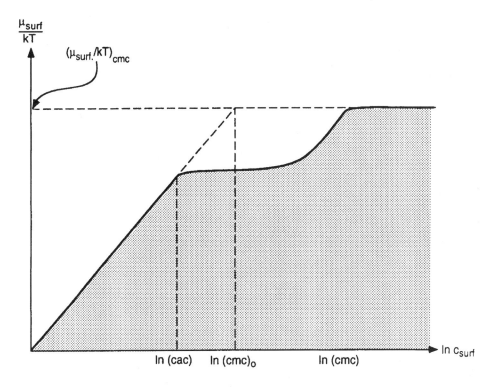

Figure 7.14
Schematic picture of the variation of the surfactant chemical potential with increasing surfactant concentration in a solution containing a constant amount of polymer. At the critical association concentration (CAC), surfactants self-assemble on the polymer. At the critical micelle concentration (CMC)$_0$, which can be measured in the absence of polymer, aggregates continue to form on the polymer. The surfactant chemical potential first reaches the value that allows for the formation of free micelles at higher concentrations. This CMC depends on the polymer concentration.

TABLE 7.2 Critical Association Concentration (CAC) for Some Surfactant–Polymer Systems

Surfactant	Polymer	CAC (mM)	CMC (mM)
$C_{12}H_{25}$—OSO_3Na	Polyethylene oxide	5.7	8.3
	Polyvinylpyrrolidone	2.5	8.3
$C_{10}H_{21}OSO_3Na$	Polyvinylpyrrolidone	10.0	32.0
$C_{12}H_{25}N(CH_3)_3Br$	Sodium hyaluronate	7.0	16.0
	Sodium alginate	0.4	16.0
	Sodium polyacrylate	0.03	16.0
$C_{14}H_{29}N(CH_3)_3Br$	Sodium hyaluronate	0.4	3.8
	Sodium alginate	0.03	3.8
	Sodium polyacrylate	0.0025	3.8

Figure 7.15
The viscosity of a 1% (w/w) solution of modified polyacrylate (PAA) as a function of added sodium dodecyl sulfate. The polymer, which has 500 or 150 monomer units, has been substituted with C_{18} alkyl chains to 3 and 1%, respectively. Note that for the $N_p = 500$ and 3% substitution, the viscosity increases by four orders of magnitude. (I. Iliopoulos, T. K. Wang, and R. Audebert, *Langmuir* **7**, 617 (1991).)

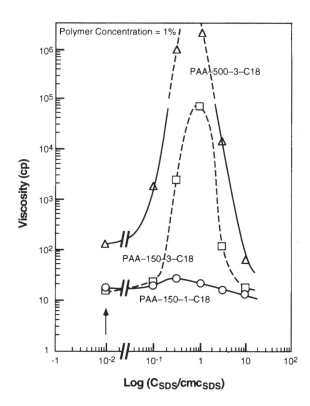

A more specific micelle–polymer interaction can be obtained by using a chemical modification process to add a few long alkyl side chains to the polymer main chain. These side chains will strengthen the polymer–polymer interaction and also serve as a strong binding site for micelles. Figure 7.15 shows the viscosity for solutions of such hydrophobically modified polymers. Dramatic changes in the viscosity signal that the micelles are able to link two polymer chains by incorporating side chains from the different main chains. As a result, a transient polymer network is formed which accounts for the dramatic changes in the viscosity. Such polymers are used as thickeners in paint and in other colloid dispersions for which a high viscosity is desirable.

7.4 Polymers at Surface Play an Important Role in Colloidal Systems

7.4.1 Polymers Can Be Attached to a Surface by Spontaneous Adsorption or by Grafting

A molecule in solution will adsorb to a surface provided the adsorption energy is substantially larger than kT. An adsorbing

Figure 7.16
Polymer configurations at a surface. (a) Grafted chain or block copolymer with one highly adsorbing segment. The main chain is assumed to be in a good solvent with no affinity for the surface. Thus, the chains try to avoid the surface. (b) Adsorbing homopolymer, but in a good solvent, indicating a loop, two trains, and the two tails that extend farthest into the solution. (J. Israelachvili, *Intermolecular and Surface Forces*, 2nd ed., Academic Press, London, 1992, p. 291.)

polymer molecule typically makes many contacts with the surface (see Figure 7.16). Even if the energy per monomer contact is relatively small, the number of contacts result in a high adsorption energy. To some extent this high adsorption energy is compensated for by a loss of configurational entropy but if the monomer has some affinity for a surface, the polymer usually adsorbs strongly and in practice often irreversibly.

We can also promote polymer adsorption through more specific effects. A block copolymer with one block experiencing bad solvent conditions will be very surface active and will have a strong tendency for adsorption. A polyelectrolyte will adsorb strongly on an opposite charged surface. Finally, the polymer chain can be grafted to a surface by a covalent bond.

Chain configuration at the surface depends on the forces that bring the polymer down to the surface. For grafted polymers, in which the monomer has no affinity for the surface, the chain will adopt a coil configuration, stretching out toward the solution (see Figure 7.16). Neighboring chains will repel one another, and this repulsion will influence the coil conformation when the grafting density is higher than R_g^{-2}, the inverse second power of the radius of gyration. For this dense so-called polymer brush, the monomer concentration c_M decays parabolically out from the surface

$$c_M(z) = \text{constant}(z - z_0)^2 \qquad (7.4.1)$$

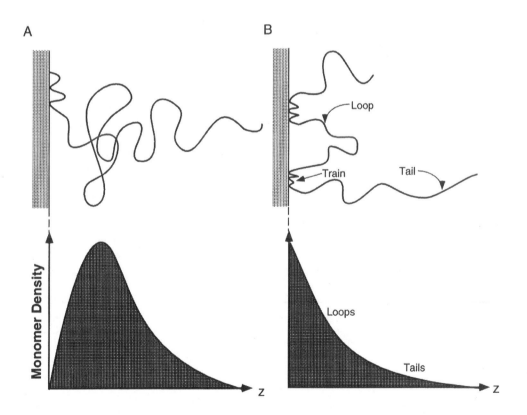

Similarly, for a block copolymer B_mC_n, where the C part adsorbs strongly on the surface, the conformation of the B coil depends on the surface density of polymer chains. In a series of polymers with constant n and increasing m, the coil–coil repulsion becomes more and more significant as m increases. For small m values the polymer density is determined solely by the packing of the C chains on the surface and the B coil extends on a scale of R_g in both the parallel and perpendicular directions adopting a so-called mushroom configuration. When the C coils become so small that the surface density of polymer chains exceeds R_g^{-2}, the packing density at the surface will decrease and the C coil will stretch away from the surface. When $n \ll m$, the chain conformation will be the same as in the dense polymer brush and the distribution function of eq. (7.4.1) will apply.

Electrostatic forces cause a polyelectrolyte to adsorb on an oppositely charged surface. The polymer acts like a giant counterion to the surface, and entropy increases greatly when the single polyelectrolyte replaces the small counterions at the surface and simultaneously releases its own counterions. Rinsing with water will never remove the polyelectrolyte from the surface, but the entropy effects have been greatly reduced in a concentrated electrolyte solution, and the polyelectrolyte can readily come off the surface.

Because the adsorption and the conformation of the polyelectrolyte in solution are governed mainly by electrostatic effects, surface conformation depends strongly on salt concentration in an aqueous medium. In pure water adsorption is very strong, and the polyelectrolyte adopts a conformation that provides an optimal match between polymer and surface charges. This process can lead to a chain that essentially lies flat on the surface. The surface density of chains will be close to $-\sigma/Z_p e$, where σ stands for the surface charge density of the bare surface and $Z_p e$ the charge of the polyelectrolyte. When the intrinsic charge density is higher on the polyelectrolyte than on the surface, the adsorption leads to a reversal of the effective surface charge.

As the salt concentration is increased, the effective strength of the electrostatic interactions is decreased. The polymer can gain conformational entropy by extending somewhat into the solution, and the ions of the electrolyte act as counterions. Increasing the salt concentration gradually loosens the electrostatic binding, and ultimately the polymer reaches a state in which it no longer binds to the surface or in which more chemical forces determine the adsorption behavior.

A noncharged homopolymer that adsorbs from a good solvent onto a surface will not change its coil conformation dramatically. Typically, the thickness of the adsorbed layer, as well as its lateral extension, is of the same magnitude as the radius of gyration R_g. The chain conformation can be described in terms of two tails (i.e., the two ends of the chain sticking out from the surface), a number of trains (sequences of the chain that have contacts with the surface), and a number of loops in which the chain makes long excursions into the solution between two trains. (See Figure 7.16).

7.4.2 Kinetics Often Determines the Outcome of a Polymer Adsorption Process

Exposing a clean surface to a solution containing an adsorbing polymer creates a nonequilibrium situation. The kinetics of the adsorption is typically transport controlled. In the unstirred boundary layer, outside the surface, diffusion determines the transport to the surface. Owing to the size of the polymer molecules, the diffusion coefficients are so small that the adsorbed layer builds up on a time-scale which can be on the order of seconds and minutes even for this diffusion-controlled process.

In practice, the distribution of polymer molecular weights usually plays a role. Thermodynamically, the largest polymers have a higher affinity for the surface because they make more surface contacts per molecule. However, the smaller polymers have a higher diffusion coefficient, and they will be the first to reach the surface. Thus, in the initial state, the shorter polymer molecules are preferentially concentrated at the surfaces, while at equilibrium the longer polymers are the ones that reside at the surface.

To reach equilibrium, longer polymers must replace the shorter ones. This process is slow because it requires desorption of the shorter chains. Depending on the circumstances, changes can be observed over hours and days, or the exchange may be so slow that for practical purposes the initial state remains stable.

Because kinetics are so slow, the performance of the adsorption process becomes an issue of great practical importance. Not only the transport to the surface, but also surface rearrangements can constitute very slow processes, so surface polymers may take a long time to respond to changes made in the solution. For this reason, polymer adsorption processes are delicate to study in the laboratory and in practical applications kinetic effects must be carefully watched.

7.4.3 Forces Between Surfaces Change Drastically When Polymers Adsorb

Surfaces covered with polymers will "see" one another through the outer part of the polymer coil. These coils will start to overlap at surface–surface separations on the order of $2R_g$. In practice, distances in the range 10–100 nm are so large that the effect of van der Waals and double-layer forces between the bare surfaces is negligible and polymer–polymer overlap provides the dominant contribution to the force.

In a good solvent the polymer coil expands away from the surface to gain configurational entropy. When it encounters a coil emanating from the second surface, its allowed conformations are restricted and the free energy increases. This mechanism generates a repulsive force between the surfaces as discussed in Section 5.5 (see Figure 7.17). The dimensions of the polymer coil determine the range of the force, and configurations that extend far from the surface are particularly important in determining the onset of the repulsive force.

Figure 7.17
Schematic illustration of configurations of polymers on two opposing surfaces restricting each other, resulting in a decrease in entropy and normally in a repulsion between the surfaces.

The repulsive force is in the literature commonly referred to as the steric interaction, although we have seen that its origin is basically entropic. For this reason we introduced the name steric entropic force in Chapter 5.

Figure 7.18 shows an experimentally measured force curve for mica surfaces covered with polyethylene oxide (PEO) in water at 20°C, where water is a good solvent for PEO. The force is indeed repulsive, increasing with increasing molecular weight as R_g does. The force is strong enough to be measured out to several multiples of R_g.

In a bad solvent, that is, $\chi \gtrsim 0.5$, the polymer wants to contract at the surface to avoid contact with the solvent. As a second surface approaches, a chain anchored at one surface can make an excursion to the vicinity of the other surface without too much solvent exposure, which leads to an increased configurational entropy and an attractive force. The attraction in a bad solvent is sensed at shorter separations than the repulsion in a good solvent for polymers of the same molecular weight (see Figure 7.19). Furthermore, the attraction rapidly turns into a repulsion because the polymer chains start to squeeze each other.

Table 7.3 demonstrates the crucial role played by solvent quality. It shows that suspensions of colloidal particles covered by adsorbed polymers became kinetically unstable at a temperature very close to the theta temperature for the polymer–solvent pair.

Figure 7.18
Force between two curved mica surfaces with adsorbed polyethylene oxide in water, which is a good solvent at room temperature. A monotonic repulsion is observed, and the range of the force scales with the radius of gyration. Furthermore, measurable forces are obtained at distances clearly larger than $2R_g$, showing the importance of the odd tails sticking far into the solution. (J. Klein and P. F. Luckham, *Nature* **300**, 429 (1982).)

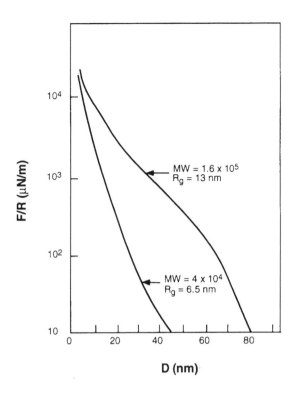

Figure 7.19
Force between curved surfaces with adsorbed polystyrene in cyclohexane at 24°C, which is below the theta temperature. In the bad solvent a strong attraction, measurable up to $\sim 2R_g$ with a maximum around R_g, can be observed. (J. Klein, *Nature* **288**, 248 (1980).)

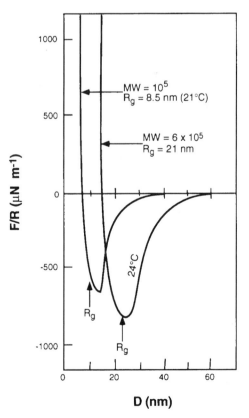

TABLE 7.3 Comparison of Theta Temperature with Critical Flocculation Temperature (CFT)

Stabilizer	Molecular Weight	Dispersion Medium	Temperature (K)	
			CFT	θ
Polyethelene oxide	10,000	0.39 M MgSO$_4$	318 ± 2	315 ± 3
Polyethelene oxide	96,000	0.39 M MgSO$_4$	316 ± 2	315 ± 3
Polyethelene oxide	1,000,000	0.39 M MgSO$_4$	317 ± 2	315 ± 3
Polyacrylic acid	9,800	0.2 M HCl	287 ± 2	287 ± 5
Polyacrylic acid	51,900	0.2 M HCl	283 ± 2	287 ± 5
Polyacrylic acid	89,700	0.2 M HCl	281 ± 1	287 ± 5
Polyvinyl alcohol	26,000	2 M NaCl	302 ± 3	300 ± 3
Polyvinyl alcohol	57,000	2 M NaCl	301 ± 3	300 ± 3
Polyvinyl alcohol	270,000	2 M NaCl	312 ± 3	300 ± 3
Polyacrylamide	18,000	2.1 M (NH$_4$)$_2$SO$_4$	292 ± 3	—
Polyacrylamide	60,000	2.1 M (NH$_4$)$_2$SO$_4$	295 ± 5	—
Polyacrylamide	180,000	2.1 M (NH$_4$)$_2$SO$_4$	280 ± 7	—
Polyisobutylene	23,000	2-Methylbutane	325 ± 1	325 ± 2
Polyisobutylene	150,000	2-Methylbutane	325 ± 1	325 ± 2

A slightly different version of the polymer overlap interaction occurs when the surfaces are covered with a dense polymer brush. Figure 7.20 shows a series of experiments with surfaces covered by a short block copolymer of the general formula $CH_3(C_2H_4)_{m-1}(OC_2H_4)_nOH$, where m and n range from block copolymer values to values typical of nonionic surfactants. Shlomo Alexander and Pierre-Gilles de Gennes (1987) derived an expression for the force per unit area

$$\frac{F}{A} = \text{const}\ \frac{kT}{S^3}\left[\left(\frac{2L}{h}\right)^{9/4} - \left(\frac{h}{2L}\right)^{3/4}\right] \qquad h < 2L \quad (7.4.2)$$

which fits the data if the thickness L of the brush is treated as a parameter, as illustrated in Figure 7.20.

If we create a system lacking enough polymers to saturate the surfaces, an attraction can occur even in a good solvent. In this case a polymer coil can extend from one surface to the other and make molecular contact on both sides. One surface then provides the missing polymer to the other, resulting in an attractive interaction usually called *bridging*.

A similar attractive mechanism can operate in a nonequilibrium state in which two surfaces have been pushed toward each other and left to equilibrate for some time. In the compressed state, the polymer coil from one side can adsorb on the opposing surface or become entangled in loops of the polymer adsorbed on the opposing surface. At this stage any attempt to separate the surfaces on a timescale short relative to the time required for disentangling the chains causes an attractive nonequilibrium force to appear. This force is one of many sources of hysteresis effects with polymer-covered surfaces.

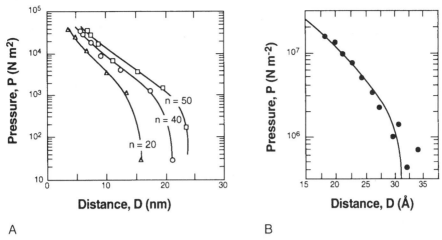

Figure 7.20
(A) Measured forces between adsorbed monolayers of the polyethylene oxide surfactant $C_{18}EO_n$ in water. The solid lines are fits to the theory of Alexander de Gennes for overlapping head groups modeled as polymer "brush" layers, using $L = 7.8$ m for $n = 20$, $L = 10.5$ nm for $n = 40$, and $L = 12$ nm for $n = 50$ (all fitted for $s = 13$ nm). (B) Measured forces between bilayers of $C_{12}EO_4$ in water and corresponding brush theory curve, using $L = 1.6$ nm (0.4 nm/EO group), $s = 0.93$ nm, and $T = 25°C$. (J. Israelachvili and H. Wennerström, *J. Phys. Chem.* **96**, 520 (1992).)

7.4.4 Polyelectrolytes Can Be Used Both to Flocculate and to Stabilize Colloidal Dispersions

Polyelectrolytes have a large practical use in controlling the stability of colloidal dispersions. They combine the effects of electrostatic and polymer mediated interactions. By a careful design one can generate stabilizing/destabilizing effects that are difficult to accomplish by other means.

When added to a suspension of charged colloidal particles, the polyelectrolytes act as giant counterions with potentially large spatial extension. If charged surfaces are first equilibrated with an excess of polyelectrolyte, they might be neutralized and the long-ranged double layer force disappears. It is replaced with an attractive force mainly caused by the additional configurational freedom available for the polyelectrolyte when it can form a bridge between the two charged surfaces.

For a sufficiently highly charged polyelectrolyte there is charge reversal and the stability of the system is seemingly unchanged with a repulsive double-layer force still operating. However, the sensitivity to addition of electrolyte might change. If the polymer adsorbs on the surface solely by electrostatic forces, then a high salt concentration will make it desorb. If, on the other hand, there is also some chemical interaction between surface and the monomer unit, the polyelectrolyte remains adsorbed and one has a situation similar to a noncharged polymer in a good solvent. Thus the steric entropic force will be operating and there is a repulsive force also under high salt conditions.

We observe dramatic effects when only small amounts of the polyelectrolyte are added to the system. Then one molecule easily ends up bridging two surfaces and inducing an attractive interaction. In practice, this behavior is very useful because adding only small amounts of the polyelectrolyte can have a major effect on the stability of colloidal solutions. For example, positively charged polyelectrolytes are used extensively in papermaking to flocculate cellulose fibers together with clay fillers and other negatively charged additives. Practical experience affirms the importance of the order in which the different components are mixed, again illustrating the important role of the nonequilibrium states in polymer adsorption.

Exercises

7.1. Consider a polymer sample in which the size distribution function $F(N_p)$ is a Gaussian so that $F(N_p) \sim \exp(-(N_p - 10,000)^2/10^7)$. Calculate the ratio between the weight-averaged and number-averaged aggregation numbers.

7.2. Using the random walk model, calculate the radius of gyration for a polyethylene coil at theta conditions. Assume that the Kuhn length is approximately 1 nm, corresponding to seven CH_2 units. Do the calculation for $(CH_2)_{1000}$ and $(CH_2)_{10,000}$.

7.3. Estimate the overlap concentration C^* for the polymers in Exercise 7.2.

7.4. Estimate the overlap concentration of an aqueous DNA solution with a persistence length of 15 nm and a contour length of the double helix of 200 nm.

7.5. Calculate the persistence length of a highly charged polyelectrolyte in 1×10^{-3} M NaCl.

7.6. Using the Flory–Huggins theory, calculate the structure factor $S(Q)$ at $Q = 0$ for a semidilute solution, $\Phi = 0.05$, of a polymer in a solvent with $\chi = 0.3$. What is the difference between $N_p = 1000$ and $N_p = 10,000$?

7.7. A protein with a molecular weight of 40,000 has 250 amino acid residues. It adopts a globular conformation in water. Estimate the number of residues in contact with the solution.

7.8. Two polymers of equal size are mixed in a melt. What is the free energy of mixing in the framework of the regular solution theory?

7.9. What is the maximum depletion force obtained between two phospholipid bilayers in a 10% (w/w) polyoxyethylene aqueous solution? Use the Flory–Huggins theory with $\chi = 0.4$.

7.10. Calculate how the depletion force contribution to the adhesion varies with the molecular weight of the polymer. Assume that full depletion occurs for separations smaller than $2R_g$ and neglect it for larger separations.

7.11. Colloidal particles of radius 10^{-5} m are dispersed in water and stabilized by the adsorption of $C_{18}EO_{20}$. Estimate the adhesion energy when the Hamaker constant is 1×10^{-20} J. (*Hint:* Use the data of Figure 7.20.)

7.12. Calculate the electrostatic contribution to the double-helix, single-coil transition of DNA at 1×10^{-2} and 1×10^{-1} M KCl. The double helix has a radius of approximately 12 Å and one negative charge per 1.7 Å projected on the axis. For simplicity, assume that the denatured state is a single helix with the same radius but half the charge density. (*Hint:* Use eq. 4.3.14.)

7.13. Use the result of Exercise 7.12 and the data of Figure 7.12 to estimate the enthalpy of denaturation of DNA.

Literature

References	Section 7.1 Polymer Molecules in solution	Section 7.2 Semi-dilute Polymer solutions	Section 7.3 Polymer Association	Section 7.4 Polymers at Surfaces
Hiemenz & Rajagopalan	CHAPTER 2.1 Radius Gyration	CHAPTER 3.4 Flory–Huggins Theory CHAPTER 4 Viscosity	CHAPTER 9 Viscosity	CHAPTER 13.5–13.8 Steric stabilization
Hunter (II)		CHAPTER 2.4 Mechanical stress CHAPTER 9.1–9.10 Transport flow in suspensions		
Israelachvili				CHAPTER 14 steric stabilization
Kruyt		CHAPTER II Rheology		
Meyers				CHAPTER 14 Polymer adsorption
Shaw		CHAPTER 9 Rheology		
Shihoda	CHAPTER 8 Polymer solutions			
Vold & Vold	CHAPTER 13 Macromolecules in solution CHAPTER 14 Neutral polymers CHAPTER 15 Polyelectrolytes	CHAPTER 10 Rheology CHAPTER 16 Gels CHAPTER 17 Membranes		CHAPTER 8.3 Steric stabilization

[a]For complete reference citations, see alphabetic listing, page xxxi.

8

COLLOIDAL STABILITY

CONCEPT MAP

Colloid Stability

When two colloidal particles undergoing Brownian motion approach each other, they experience two types of interaction: the static forces described in Chapter 5, which we consider first, and hydrodynamic forces mediated by solvent molecules, as described in Section 5.7.

Figure 8.1 depicts two important types of static interaction. When no repulsive energy barrier exists, particles come together into a primary minimum. A barrier creates a potential energy curve with a maximum separating a deep primary minimum at short separations and a shallow minimum further out.

Most colloids are thermodynamically unstable and remain in solution only if a potential barrier $\gg kT$ prevents them from moving into the primary minimum.

Flocculation involves reversible particle association. Coagulation is irreversible.

With emulsions and foams coagulation can lead to coalescence (i.e., fusion of liquid droplets).

DLVO Theory

In the DLVO theory, colloidal stability (eq. 8.2.1a) is determined by a balance between double-layer repulsion $V_R(h)$, which increases exponentially with decreasing distance h, and van der Waals attraction $V_A(h)$.

DLVO expressions for parallel plates (eq. 8.2.4a) and identical spheres (eq. 8.2.9) show that $V_R(h)$ depends on κ and Φ_0 thorough $\Gamma_0 = \tanh(ze\Phi_0/4kT)$ (eq. 3.8.14).

The critical coagulation concentration, CCC (eq. 8.2.11 for spheres), is a useful measure of colloid stability. It approximately corresponds to the electrolyte concentration at which the maximum value of $V(h) = 0$.

The CCC becomes independent of potential when $ze\Phi_0/kT \gg 1$, but it is very sensitive to counterion valency because CCC $\propto z^{-6}$, verifying the century-old empirical Schulze–Hardy rule.

When $ze\Phi_0/kT < 1$, the CCC $\propto \Phi_0^4/z^2$ and thus becomes extremely sensitive to potential, but less sensitive to counterion valency.

Kinetics of Aggregation: Rapid Coagulation

The initial rate of coagulation involves dimer formation between identical particles P and is expressed as $-d[P]/dt = k_r[P]^2$, where k_r is a second-order rate constant.

We obtain an expression for k_r by calculating the flux resulting from particles diffusing toward a central particle using Fick's first law, yielding $k_r = 8\pi RD_i$ (eq. 8.3.8) for identical spheres.

By using Stokes's law, $D = kT/6\pi\eta R$, we can rewrite $k_r = 4kT/3\eta$. The coagulation rate is independent of R because the increased probability that large particles will collide as a result of their size (as measured by R) is compensated by the increased probability that small particles will collide because of their greater mobility (as measured by D).

We can generalize the expression for k_r to include trimer, quatamer, etc. formations and obtain an expression (eq. 8.3.15) giving the change in the total number of particles $\Sigma_i[P_j]$ with time (Figure 8.5).

Slow Coagulation

When $V(h)$ has a barrier much larger than kT, coagulation rates become an order of magnitude slower. By including a potential-dependent term in Fick's first law, we obtain $k_s = k_r/W$ (eq. 8.3.27); W represents the stability ratio and is proportional to $\exp(V_{max}/kT)$ (eq. 8.3.31).

Figure 8.6 shows how changing the electrolyte concentration by a factor of 10 can change k_s by a factor of 10^7.

When coagulating particles become locked into a configuration set by their initial contact, a loosely packed, less dense precipitate forms which displays fractal properties (Figure 8.7).

Electrokinetics: Measurement of the Zeta Potential

Application of the DLVO theory requires knowledge of Φ_0. In many practical situations this is difficult to obtain. Instead, we can correlate the zeta potential, ζ, obtained from electrokinetic measurements, to the stability of colloidal systems (Table 8.4).

In a zeta potential experiment, we induce flow between a colloidal surface and the adjacent fluid and measure the potential at the position in the double layer where fluid begins to flow.

There are four electrokinetic effects: **electrophoresis** and **sedimentation potential** measurements involve motion of charged particles in a stationary liquid; **streaming potential** and **electro-osmosis** measurements involve flow of fluid past a stationary charged surface.

Electrophoresis

In an electrophoresis experiment (Figure 8.10), we impose an electrical field E and measure the colloidal particle's velocity v, which is more conveniently expressed in terms of the mobility $u = v/E$. The relation between u and ζ depends on the ratio $R/(1/\kappa)$, that is, particle radius to Debye length.

For the extreme limits $\kappa R \ll 1$, we obtain the Hückel equation 8.4.6, while $\kappa R \gg 1$ results in the Helmholtz–Smoluchowski equation (eq. 8.4.12). The relation between these two limits is given by the Henry equation, which is displayed in Figure 8.9.

Streaming Potentials and Electro-Osmosis

In a streaming potential measurement (Figure 8.11), we cause a solution to flow with a specific conductance of κ_b through a capillary by applying a pressure gradient ΔP across it. The zeta potential (eq. 8.4.18) arises because the flow of counterions induces a potential Φ_{st} which we measure with two electrodes located at the ends of the capillary tube.

In an electro-osmosis experiment (Figure 8.12), we apply a potential Φ_{eo} across two reversible electrodes connected by a capillary of length L, or a porous plug. The field exerts a force on the counterions in the double layer and they in turn carry solution in the tube along with them. The relation between volumetric flow rate and the zeta potential is given by eqs. 8.4.19–8.4.21.

8

One of the key issues underlying our ability to apply colloid science to technical problems, biological questions, or scientific experimental methods is the stability of the colloid state. This question, which fascinated Faraday 140 years ago, remains relevant today although with an awareness of more sophisticated implications.

The observant eye discovers many commonplace manifestations of the dichotomy between stable and unstable colloidal systems. Ordinary milk, for example, displays an amazingly rich range of colloidal behavior. It owes its white color to multiple light scattering from colloidal fat globules and protein aggregates. Milk fresh from the cow shows colloidal stability except for gravitational effects. If we use a centrifuge to separate the cream from the milk, we create two stable systems, both containing fat and protein. Whip the cream and it becomes a metastable foam. Whip it a bit more and the foam collapses, some water separates out, and it forms butter, which is a metastable colloidal system of water droplets in fat. We can keep butter for a very long time in the refrigerator, but if we heat it the fat and water separate. Starting as milkfat in water, the fat finally has separated completely from the water.

Milk's protein component shows a similar richness. The majority of milk proteins are present in protein aggregates called casein micelles. In ordinary milk they form an apparently stable solution. Even at higher temperatures, as in a cup of milky tea, stability prevails, but try to add a few drops of lemon juice! The skin we find on milk that has been heated to boiling is surface-precipitated protein, and protein adsorption makes the saucepan hard to wash. The latter effect causes great problems in the dairy industry, where surfaces foul during pasteurization. However, colloidal instability can be turned to an advantage. Delicate creamy yogurt results from microbial activity, producing lactic acid, lower pH, and colloidal coagulation. Alternatively, proteolytic enzymes transform the surface properties of the protein aggregates so that the fat globules and the casein micelles coprecipitate to form cheese curd, leaving the whey behind.

Because of their considerable economic importance, the properties of milk systems have been studied in great detail, and we will not pursue any of these special phenomena. However, they represent a rich spectrum of colloidal stabilization–destabilization processes, such as steric stabilization of fat globules, foam formation in whipped cream, emulsion inversion in the formation of butter, and emulsion breaking by heating.

The protein component shows electrostatic stabilization, coagulation by charge neutralization using lemon juice, adorption/heterocoagulation by heating, coagulation by partial neutralization in yogurt, and coagulation by surface modification throuh the action of rennet. We will encounter more detailed discussions of all these process types in this and subsequent chapters.

8.1 Colloidal Stability Involves Both Kinetic and Thermodynamic Considerations

8.1.1 The Interaction Potential Between Particles Determines Kinetic Behavior

A colloidal particle has a Brownian motion in solution. On approaching another particle, it senses not only the surrounding solvent, but also the effective interaction with this other particle.

Particle–particle interaction has two components. One is a result of the static forces discussed in detail in Chapter 5. The other is a dynamic effect that leads to a correlation of the motions of the two particles mediated by the motions of the solvent molecules. This hydrodynamic interaction influences the dynamic, but not the static, equilibrium properties of colloidal systems as discussed in Section 5.7.

Before we enter into a detailed description of this dynamic effect, let us qualitatively discuss the consequences of the static interaction potential. For a colloidal system, we can identify interaction curves of the two general types shown in Figure 8.1. In Figure 8.1A, a monotonically decreasing free energy persists up to a primary minimum. No energy barrier prevents the system from reaching this minimum, and it easily achieves this bound state. The extent to which the association between the particles is reversible depends on the depth of the potential well and on possible further processes in the aggregated state.

A reversible association is often called flocculation, while an irreversible association is called coagulation. The distinction between the two terms is a qualitative rather than a quantitative one. Nonionic micellar solutions are examples of a system whose potential energy minimum is particularly shallow. With some exceptions, micellar systems show little tendency toward aggregation, and they are thermodynamically stable in the dispersed state.

Figure 8.1
(A) Monotonically decreasing potential energy leads to reversible association or coagulation, depending on the depth of the potential energy well. Dashed line shows the corresponding force curve. Note that the force is zero at the equilibrium separation and attractive for larger h. (B) Potential energy curve showing two minima, a shallow secondary minimum at large distances, and a primary minimum at shorter distances, separated by a maximum. If the maximum is sufficiently high, it serves as a barrier to particle coagulation. Generally the primary minimum is so deep that once particles have passed over it, association becomes irreversible. Dashed line shows the force, which is zero at the secondary minimum on the top of the barrier and in the primary minimum (and at large separations).

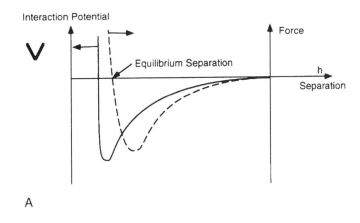

A

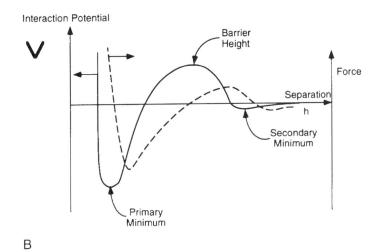

B

On the other hand, oil droplets in water manifest a strong adhesion force. Furthermore, once the two droplets come into contact with one another, they merge irreversibly into one bigger drop. In this case coalescence follows coagulation.

For vesicles of neutral phospholipids, the energy minimum can be so deep that a relatively stable dimer is formed. The dimer can grow to form a trimer, and so on, by addition of more vesicles. Mild mechanical agitation can redisperse such a floc, but if left unperturbed it will sediment, and ultimately irreversible fusion will occur.

The nonmonotonic interaction potential of Figure 8.1B allows for more intricate behavior. In this example, the potential curve displays two minima, a deep primary minimum at short separations and a shallow secondary minimum farther out. A potential energy barrier separates the two.

If the system reaches the primary minimum, the coagulation process typically is irreversible. In addition, the secondary minimum may be deep enough to support the type of flocculation we have already discussed. If the energy barrier far exceeds kT, the primary minimum becomes inaccessible and the system is kinetically stable. With a barrier on the order of only kT relatively rapid approach is possible. Stable aggregates are formed for sufficiently deep or wide interaction wells at sufficiently high particle concentrations.

To obtain a semiquantitative understanding of the conditions necessary for aggregation, let us analyze the first step, the formation of dimer P_2 from two particles P

$$P + P \rightleftharpoons P_2 \qquad (8.1.1)$$

The degree of association at equilibrium is determined by the equilibrium constant

$$K = \frac{[P_2]}{[P]^2} \qquad (8.1.2)$$

K closely relates to the second virial coefficient B_2 (cf. Section 1.5). For spherical particles of radius R, in a case in which the distance of closest approach is $2R$, the expression for K is

$$K \simeq \int_{2R}^{\infty} \left\{ \exp\left(\frac{-V(r)}{kT} \right) - 1 \right\} 4\pi r^2 \, dr = -\left(2B_2 - \frac{32\pi}{3} R^3 \right) \quad (8.1.3)$$

When the particle–particle interaction potential $V(r)$ is strongly negative over an appreciable range, K can become large and dimerization can occur at low concentrations. The fraction of dimerized particles relative to monomers is

$$\frac{2[P_2]}{[P]} = 2K[P] \qquad (8.1.4)$$

using eq. 8.1.2. Thus, aggregation is not a major event if

$$2K[P]_{\text{tot}} \ll 1 \qquad (8.1.5)$$

where $[P]_{\text{tot}}$ stands for the total particle concentration. On the other hand, if $\frac{1}{2}K[P]_{\text{tot}} \gtrsim 1$, substantial aggregation occurs in the system and the aggregates are likely to grow. Thus for the solution to be stable we require that the second virial coefficient B_2 should be substantially smaller in magnitude than the volume per particle; that is $[P]_{\text{tot}}^{-1} \gg -B_2$.

8.1.2 Particles Deformed upon Aggregation Change Their Effective Interaction Potential

In the interaction potentials shown in Figure 8.1, the independent cordinate represents the distance between the particles. We implicitly assumed that no major internal rearrangements had altered the particle structure. However, at short range, particularly in the primary minimum, the interparticle interactions conceivably might be strong enough to affect the shape and nature of the particle.

Whether such changes of the particles occur provides a demarcation line between the sol systems to be discussed in Chapter 9 and the emulsions and foams discussed in Chapter 11. A solid particle rearranges at high temperatures only in a sintering process. On the other hand, liquid drops or bubbles deform easily. They flatten on contact and coalesce without much effort.

A more intricate question of particle integrity arises when a surfactant or polymer molecular film covers the surface. The final outcome of an aggregation process essentially depends on how the film responds to the perturbation of the interparticle contact. In vesicle systems, the result may be fusion or flocculation; for emulsions, the response determines whether coalescence accompanies coagulation; for sterically stabilized sols, the properties of the surface film regulate the degree of coagulation versus coalescence.

Clearly, the result of the colloidal aggregation process in these cases depends on a crucial balance between interparticle and intraparticle interactions. If we seek a detailed understanding, we must consider both types of interaction and the interplay between them. However, in this chapter we will concentrate on the first step in the aggregation process and ignore the complications associated with the internal rearrangements of the particles until we discuss emulsions in Chapter 11.

8.2 The DLVO Theory Provides Our Basic Framework for Thinking About Colloidal Interactions

8.2.1 Competition Between Attractive van der Waals and Repulsive Double-Layer Forces Determines the Stability or Instability of Many Colloidal Systems

A major advance in colloid science occurred during the 1940s when two groups of scientists—Boris Derjaguin and the renowned physicist Lev Landau in the Soviet Union and Evert Verwey and Theo Overbeek in the Netherlands—independently published a quantitative theoretical analysis of the problem of colloidal stability. The theory they proposed became known by the initial letters of their names: DLVO.

DLVO theory assumes that the more long-ranged interparticle interactions mainly control colloidal stability. Two types of force are considered. A long range van der Waals force operates irrespective of the chemical nature of the particles or the medium. If the particles are similar, this force is always attractive (see Section 5.2). Furthermore, most colloidal particles acquire a charge either from surface charge groups or by specific ion adsorption from the solution. For similar particles this charge leads to a repulsive double-layer force.

Thus, we can write the total interaction potential as

$$V(h) = V_A(h) + V_R(h) \qquad (8.2.1a)$$

and the corresponding force equals

$$F = -\frac{dV}{dh} = -\frac{dV_A}{dh} - \frac{dV_R}{dh} \qquad (8.2.1b)$$

Depending on the relative strength of the attractive and repulsive terms, we can generate the interaction potential versus distance curves shown in Figure 8.1, which also illustrates the corresponding force curves. We should realize that interaction potentials have the character of free energies and contain both energetic and entropic contributions.

Let us illustrate the typical properties of the interaction potentials and the corresponding force by considering the interaction between two identical half-planes separated by a distance h. According to eq. 5.2.7, the attractive van der Waals interaction V_A is

$$\frac{V_A(h)}{area} = \frac{-H_{121}}{12\pi h^2} \qquad (8.2.2)$$

where H_{121} represents the Hamaker constant for medium 1 separated by a liquid 2.

At large separations, the double-layer interaction force is

$$\frac{F}{area} \simeq 32(kT)^2 \frac{\varepsilon_r \varepsilon_o}{z^2 e^2} \kappa^2 \Gamma_0^2 \exp(-\kappa h)$$

$$= 64kTc_o^* \Gamma_0^2 \exp(-\kappa h) \qquad (5.1.31)$$

as derived in Section 5.1. We obtain the corresponding potential by integrating eq. 5.1.31 and

$$\frac{V_R(h)}{area} = \frac{64kTc_o^* \Gamma_0^2}{\kappa} \exp(-\kappa h) \qquad (8.2.3)$$

Here Γ_0, defined as $\tanh(ze\Phi_0/4kT)$ in eq. 3.8.14, depends on the potential at the surface, but the distance dependence is solely determined by the Debye screening in the solution.

Now we can write eqs. 8.2.1 as

$$\frac{V(h)}{area} = -\frac{H_{121}}{12\pi h^2} + \frac{64kTc_o^* \Gamma_0^2}{\kappa} \exp(-\kappa h) \qquad (8.2.4a)$$

and

$$\frac{F}{area} = -\frac{H_{121}}{(6\pi h^3)} + 64kTc_o^* \Gamma_0^2 \exp(-\kappa h) \qquad (8.2.4b)$$

Figure 8.2
Potential energy of interaction between two planar surfaces according to the DLVO theory, eqs. 8.2.4. The curves $V(1)$ and $V(2)$ are calculated for two different values of the Debye length with $\kappa_1 < \kappa_2$, where κ_2 is chosen so that the maximum equals zero. (D.J. Shaw, *Introduction to Colloid and Surface Chemistry*, 3rd ed., Butterworth-Heinemann, London, 1980, p. 192.)

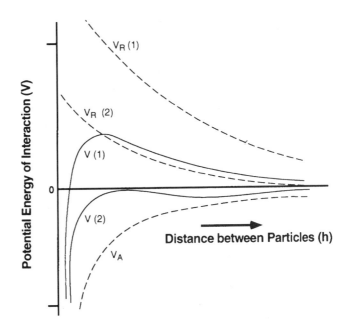

The attractive term $V_A(h)$ dominates the repulsive term $V_R(h)$ when h is very large or very small. At intermediate separations the double-layer force gives rise to a potential energy barrier if the surface is sufficiently charged and if the electrolyte ions do not screen too much. Figure 8.2 illustrates the resulting total potential for two choices of κ.

When the low electrolyte concentration makes κ small, a substantial energy barrier exists in the total $V(h)$ and the colloidal system is typically stable. On the other hand, for large κ and short screening length κ^{-1}, the barrier is small or nonexistent and the colloidal system is unstable.

To understand the approximate location of the barrier, let us consider the special case when the barrier is so small that the maximum value of $V(h)$ is zero. (See Figure 8.2, curve $V(2)$.) At the maximum

$$\frac{\partial V(h)}{\partial h} = \frac{\partial V_A}{\partial h} + \frac{\partial V_R}{\partial h} = 0 \tag{8.2.5}$$

and differentiating eq. 8.2.4a yields

$$\frac{\partial V(h)}{\partial h} = -\kappa V_R - \frac{2}{h} V_A = 0 \tag{8.2.6}$$

Combined with the condition $V(h_0) = 0$, that is, $V_R = -V_A$, we obtain simply

$$h_0 = \frac{2}{\kappa} \tag{8.2.7}$$

where h_0 is the separation at the top of the barrier.

The barrier moves to larger separations when $V(h_0) > 0$, but when the barrier is lower, the colloidal system coagulates rapidly. Thus, as the asymptotic expressions in eqs. 8.2.2 and 8.2.3 are valid for $h > \kappa^{-1}$, they are useful for predicting stability/instability.

8.2.2 The Critical Coagulation Concentration Is Sensitive to Counterion Valency

In the DLVO theory, colloidal stability is caused by a balance between double-layer repulsions and van der Waals attraction. Therefore, the simplest means of controlling it is to change the nature and concentration of the electrolyte. Experimentally we observe a rather abrupt change from stability to instability on changing the salt concentration. Such studies measure the critical coagulation (electrolyte) concentration, CCC.

Starting from a DLVO potential and viscosities of the solvent, we can calculate the expected association rate for any electrolyte concentration. However, a simpler, more qualitative, approach is to specify a potential $V(h)$, which approximately represents the transition from stability to instability in practical terms. A mathematically convenient choice is to let $V(h) = 0$ at the barrier maximum. In this case increasing the electrolyte concentration would yield an unstable system, or, slightly decreasing the electrolyte concentration would achieve long-term stability. Later in the chapter, we will analyze the relation between kinetics of coagulation and interaction potential more closely.

Let us apply the model to a system of identical spherical particles, in which the radii R are large enough for the Derjaguin approximation to hold. We can obtain the force by combining eq. 5.2.33, which relates the force between spherical particles to the interaction potential between flat surfaces, with eqs. 8.2.2 and 8.2.3

$$F(h) = \pi R \left\{ -\frac{H_{121}}{12\pi} \frac{1}{h^2} + \frac{64 k T c_o^* \Gamma_0^2}{\kappa} \exp(-\kappa h) \right\} \quad (8.2.8)$$

When integrated with $dV(h) = -F(h)d(h)$ this equation yields

$$V(h) = \pi R \left\{ -\frac{H_{121}}{12\pi} \frac{1}{h} + \frac{64 k T c_o^* \Gamma_0^2}{\kappa^2} \exp(-\kappa h) \right\} \quad (8.2.9)$$

In Figure 8.3 we show how the total interaction potential $V(h)$ varies with the screening length κ^{-1}. Thus, we seek a relation between κ and h_0 when the maximum of the potential curve ($F(h_0) = 0$) occurs for an interaction potential of zero ($V(h_0) = 0$). Solving eqs. 8.2.8 and 8.2.9 for the distance h_0 at the top of the barrier, we find

$$h_0 = \frac{1}{\kappa} \quad (8.2.10)$$

Figure 8.3
The influence of electrolyte concentration κ on the total potential energy of interaction of two spherical particles calculated using eq. 8.2.9. Parameters: $R = 10^{-7}$ m; $T = 298$ K; $z = 1$; $H_{121} = 6 \times 10^{-20}$ J; $\varepsilon = 78.5$; $\Phi_0 = 50$ mV $\approx 2kT/e$. (D.J. Shaw, *Introduction to Colloid and Surface Chemistry*, 3rd ed., Butterworth-Heinemann, London, 1980, p. 193.)

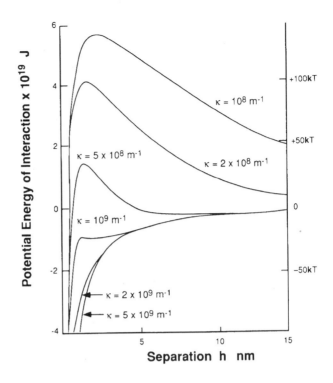

in close analogy with eq. 8.2.7. For spherical particles this maximum occurs at half the distance required for flat surfaces.

We can estimate the critical coagulation concentration by inserting $h_0 = 1/\kappa$ into eq. 8.2.9 and solving for c_0^*, noting that κ also involves c_0^*, as in eq. 3.8.13. This substitution yields

$$\text{CCC} \simeq 1 \times 10^5 \frac{(\varepsilon_r \varepsilon_o)^3 (kT)^5}{z^6 e^6 H_{121}^2} \Gamma_0^4 \qquad (8.2.11)$$

Although this formula is numerically approximate, it clarifies the CCC's strong dependence on the parameters.

For a highly charged surface, Γ_0 approaches a value of unity see Section 3.8) and the CCC depends on the sixth power of the ion valency. The strong valency dependence of the CCC was observed empirically around 1900 and became known as the Schulze–Hardy rule. Providing a simple theoretical derivation of this rule was one of the successes of the DLVO theory.

At lower charge densities, where we can linearize the Poisson–Boltzmann equation, Γ_0 takes the value $ze\Phi_0/4kT$ and eq. 8.2.11 becomes

$$\text{CCC} \simeq 4 \times 10^2 \frac{(\varepsilon_r \varepsilon_o)^3 kT \Phi_0^4}{(zeH_{121})^2} \qquad (8.2.12)$$

Thus for low surface potentials Φ_0 the CCC depends on the ion valency to the inverse second power.

Problem: Derive the CCC for constant surface charge density in the linear regime of the Poisson–Boltzmann equation.

Solution: In the linear regime $\sigma = \varepsilon_r \varepsilon_o \kappa \Phi_0$, eq. 3.8.20 and $\kappa^{-1} = (\varepsilon_r \varepsilon_o kT/2z^2 e^2 c_o^*)^{1/2}$ from eq. 3.8.13. From eq. 8.2.12

$$
\text{CCC} \simeq \frac{4 \times 10^2 (\varepsilon_r \varepsilon_o)^3 kT}{(zeH_{121})^2} \frac{\sigma^4}{(\varepsilon_r \varepsilon_o)^4 \kappa^4}
$$

$$
= \frac{4 \times 10^2 kT\sigma^4}{(zeH_{121})^2 (\varepsilon_r \varepsilon_o)} \frac{(\varepsilon_r \varepsilon_o kT)^2}{4 \cdot (ze)^4 \text{CCC}^2}
$$

$$
\rightarrow \text{CCC}^3 \simeq \frac{10^2 (kT)^3 \sigma^4 (\varepsilon_r \varepsilon_o)}{(ze)^6 H_{121}^2}
$$

$$
\rightarrow \text{CCC} \simeq \frac{5 kT (\varepsilon_r \varepsilon_o)^{1/3} \sigma^{4/3}}{(ze)^2 H_{121}^{2/3}}
$$

In eqs. 8.2.11 to 8.2.13 we consistently use SI units, and the CCC is given in molecules per volume.

For mathematical convenience, the derivations have been made for a symmetrical electrolyte. In practice, this turns out not to be a major problem, provided we obey some simple rules when applying the results to unsymmetrical electrolyte systems. For example, in highly charged systems eq. 8.2.11 still applies if we interpret z as the valency of the counterion.

In the linear regime, the true parameter in the problem is κ, and by using the general formula for κ given in eq. 3.8.13, we can remake the derivation for an arbitrary mixture of electrolyte. The result will deviate somewhat from those given in eq. 8.2.12.

A notable feature of the equations for the CCC is that they do not depend on the particle radius. This independence arises from our somewhat arbitrary condition that $V(h_0) = 0$ at the CCC. In reality, a slightly higher $V(h_0)$ is required to obtain long-term stability, so that an R-dependence enters the calculation through a nonzero left-hand side of eq. 8.2.9.

For many colloidal systems we do not know either the surface potential Φ_0 or the surface charge density σ. In practice we often must resort to indirect methods to characterize the electrical properties of a particle. Perhaps the most convenient way to do this is to measure the migration of single particles in an electric field, a method that enables us to deduce the so-called zeta potential. This concept will be discussed in more detail in Section 8.4 on electrokinetic phenomena.

Until this point we have discussed colloidal stability simply as a question of reaching the primary minimum. The further fate of the aggregated particles depends on the interaction at short range and here chemical specificity enters the problem. The usefulness of the DLVO theory is based on the fact that the van der Waals and electrical double-layer forces are long-ranged, with a known asymptotic form. Thus, we can hope that the DLVO potential is accurate up to the location of the energy barrier, even without a good theoretical or experimental basis for assuming that this potential is accurate up to contact between the surfaces. At short range all the complications discussed in Sections 5.3–5.6 can enter the problem.

Thus, the properties of the aggregated state depend in a sensitive way on the nature of these short range forces as well as on the detailed geometry of the particles. In general, the DLVO potential cannot be used to discuss this aspect of the problem of colloidal particle aggregation.

8.2.3 A Colloidal Suspension Can Be Stabilized by Adsorbing Surfactants or Polymers

In the DLVO theory description, colloidal stability emerges for charged particles in a medium with low electrolyte concentration. In practice, there are many other circumstances when we want to achieve stability of a suspension. The most efficient way to accomplish this is to add an additional component like an amphiphile or a polymer which adsorbs on the colloidal particles and changes their surface properties. With a judicious choice of adsorbent one can obtain stability for uncharged systems, such as particles in an apolar medium, and also for aqueous high salt conditions. There are many applications where we want to be able to control the properties of the suspension so that it is stable up to a certain point at which instability may be introduced by some simple means. Polymer and surfactant stabilized systems provide a greater flexibility in this respect.

Surfactants adsorb readily from an aqueous solution on the surface of a (slightly) hydrophobic particle and also on hydrophilic surfaces in an apolar medium. They self-assemble on the surface to form either micellar type aggregates or a monomolecular film. The latter is more effective in inducing colloidal stability. An adsorbed film of surfactant typically eliminates the charges on the particle itself; the resulting repulsive force between the particles is mainly determined by the interaction between the surfactant monolayers. There is a contribution to the force from the dispersion attraction between the particles themselves, but due to the substantial thickness of the adsorbed layers this dispersion interaction is only operating at a relatively long range, substantially reducing its importance.

Problem: Calculate the dispersion interaction between two similar spherical particles of radius $R = 1 \times 10^{-7}$ m separated by two monolayers of phospholipids (2.0 nm per monolayer) and water region of thickness 2.0 nm. Assume a Hamaker constant of 6×10^{-20} J and neglect the contribution from the lipid layers.

Solution: From eq. 5.2.34

$$U_{\text{sphere}} = -\frac{RH_{121}}{12h} = \frac{1 \times 10^{-7}(\text{m})6 \times 10^{-20}(\text{J})}{12(2+2+2)10^{-9}(\text{m})} = 6.25 \times 10^{20}\,(\text{J})$$

corresponding to $15kT$.

In an apolar medium the apolar chains of the surfactant will be exposed to the medium and provide stability like a short polymer brush if the medium is a good solvent for the chains. In an aqueous suspension the stability with repulsive force will depend on the nature of the polar head group. For charged surfactants the long range electrostatic repulsion will operate effectively, and even at high salt concentrations stability can prevail under favorable conditions through a more short range interaction since the strength of the van der Waals force has been reduced.

One disadvantage in using surfactants to stabilize a dispersion is that the required monomer concentration in the bulk is approximately equal to the CMC. For dilute suspensions, in particular, this can consume large amounts counted per particle.

It was noted in the previous chapter that polymers readily adsorb on surfaces and that this can occur even at very low bulk concentrations. In a good solvent the adsorbed polymers generate a repulsive steric entropic force with a range determined by the radius of gyration of the polymer (cf. Figure 7.18). Since for a typical polymer R_g should exceed 10 nm, the attractive dispersion force between the particles is virtually eliminated. However, an attraction can be introduced by changing to bad solvent conditions (cf. Table 7.3 and Figure 7.19).

With a polyelectrolyte on the particle surface one has the combined effect of an electrostatic and a steric entropic repulsion. As long as the polymer remains on the surface a suspension is stable even for high electrolyte concentrations due to the steric entropic repulsion. In this limit the charges on the polymer have the effect of ensuring good solvent conditions in an aqueous medium.

It is an interesting question whether the stabilizing effect of an adsorbing surfactant or polymer is kinetic or thermodynamic. As we will discuss in Chapter 11, both possibilities are found in emulsion systems. With colloidal particles of a solid character the shapes are given by the preparation and the question of thermodynamic stability seems to depend strongly on particle shape. For perfect spheres there is not much direct molecular contact in the adhered state and the particle–particle interaction might not be strong enough to push away an adsorbed surfactant or polymer layer. If the particles, on the other hand, have large planar faces, which easily occur for particles with crystalline atomic order, the direct particle–particle contact is more effective in giving a strong adhesion and the adsorbed layers can be pushed out. Molecularly rough particles represent an intermediate case between the perfect sphere and the locally planar particles.

8.3 Kinetics of Aggregation Allow Us to Predict How Fast Colloidal Systems Will Coagulate

8.3.1 We Can Determine the Binary Rate Constant for Rapid Aggregation from the Diffusional Motion

The initial rate of coagulation involves the bimolecular association between two identical colloidal particles P to form a dimer and

$$-\frac{d}{dt}[P] = k_r[P]^2 \qquad (8.3.1)$$

where k_r represents the second-order rate constant for rapid coagulation, which we assume is irreversible. In the equations that follow, concentrations are given as particles per volume. When molar concentration units are used, the rate constant should be multiplied by Avogadro's number, N_{Av}.

To evaluate k_r, we can consider a system of uniform spherical particles of radius R, which are undergoing Brownian motion. We assume that the spheres interact only upon contact and adhere to form a doublet. This assumption is equivalent to replacing the potential energy curve in Figure 8.1a by a square well potential with an interaction distance equal to $2R$ as shown in Figure 8.4. Because our model ignores the attractive interac-

Figure 8.4
Simplified square well potential with an interaction distance equal to the sum of the particle radius used to model the kinetics of coagulation. At $h > 2R$, particles do not interact, but a strong attractive interaction occurs with $h \leqslant 2R$.

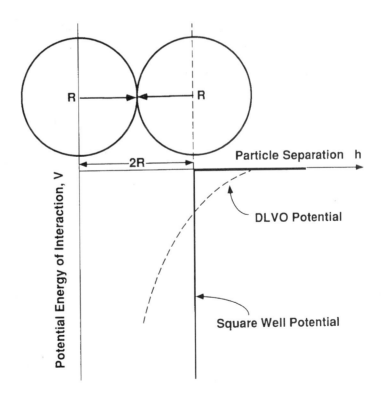

tions at distances greater than 2R, it simplifies reality. However, it does provide a simple analytical expression and can be used to develop more realistic models. The relation between the rate of coagulation and particle diffusion was established by Marian Smoluchowski in 1917.

We focus our attention on one central stationary particle i and calculate the number of particles that diffuse toward it. The flux $\vec{J}$ of particles crossing the unit area is given by Fick's first law

$$\vec{J} = -D_j \vec{\nabla}[P] \tag{8.3.2}$$

where D_j represents the diffusion coefficient of the particles. The total flux of particles crossing a spherical surface at radius r and area $A = 4\pi r^2$, surrounding the central particle i, is

$$(J_r A)_i = -(4\pi r^2) D_j \frac{d[P]}{dr} \tag{8.3.3}$$

At steady state, the flux remains constant, and we can integrate this equation using the boundary conditions $[P] = [P]_b$ at $r = \infty$ and $[P] = 0$ at $2R$ to give

$$(J_r A)_i = -8\pi R D_j [P]_b \tag{8.3.4}$$

where $[P]_b$ equals the bulk particle concentration.

Equation 8.3.4 gives the number of particles that move toward the fixed particle per unit time, that is, the association rate. However, the central particle also undergoes Brownian motion. We can take this effect on the rate of coagulation into account by replacing the diffusion coefficient for a single particle, D_j, with a relative diffusion coefficient for two particles, D_{ij}. Using Einstein's equation

$$D = \frac{\overline{X^2}}{2t} \tag{8.3.5}$$

which relates the diffusion coefficient D to the distance X that a particle moves in time t. Assuming that the Brownian motions of particles i and j are uncorrelated,

$$D_{ij} = \frac{\overline{(X_i - X_j)^2}}{2t}$$

$$= \frac{\overline{X_i^2} - 2\overline{X_i X_j} + \overline{X_j^2}}{2t} \tag{8.3.6}$$

$$= D_i + D_j$$

because $\overline{X_i X_j} = \overline{X_i} \cdot \overline{X_j} = 0$ for uncorrelated Brownian motion. Therefore, when the diffusing and the target particles are equivalent, $D_{ij} = 2D_i$. If we include hydrodynamic interactions, the analysis of the relative Brownian motion becomes more complex.

Equation 8.3.4 gives the flux toward one central particle. Because N such particles exist in the system, we must sum over all of them and divide by 2 to avoid counting the association of particle i with particle j twice. Substituting eq. 8.3.6 into 8.3.4 yields

$$-\frac{d[P]}{dt} = -\frac{1}{2}(J_r A)_i [P] = 8\pi R D_i [P]^2 \qquad (8.3.7)$$

An identification with eq. 8.3.1 yields an expression

$$k_r = 8\pi R D_i \qquad (8.3.8)$$

for the binary association rate constant of two similar particles. To account for a heteroassociation between two different particles, we write

$$k_r = 4\pi(R_i + R_j)(D_i + D_j) \qquad (8.3.9)$$

If $R_i \approx R_j$ and $D_i \approx D_j$, the two equations differ by a factor of 2. A difference in the rate equation (eq. 8.3.1) causes this difference when it is applied to distinguishable particles.

For uniform spheres, Stokes's law (eq. 5.7.2) gives a relation between the diffusion coefficient, particle size, and solvent viscosity η

$$D = \frac{kT}{6\pi\eta R} \qquad (8.3.10)$$

When substituted into eq. 8.3.8 this equation yields

$$k_r = \frac{4kT}{3\eta} \qquad (8.3.11)$$

The rate constant of coagulation of identical spherical particles depends only on the viscosity of the solutions and is independent of particle size. This cancellation is due to the increased probability of large particles colliding because of their size (as measued by R); which compensates for the increased probability of small particles colliding because of their greater mobility (as measured by D). In water at 20°C, the diffusion-controlled binary association rate constant k_r equals $0.54 \times 10^{-17}\,\text{m}^3$ per particle per second, which equals $3.3 \times 10^9\,\text{M}^{-1}\,\text{s}^{-1}$. For the heteroassociation, we should multiply these numbers by 2.

Use of eq. 8.3.1 to evaluate the rate of coagulation is limited to measurements that only determine the early stage of coagulation. We also have assumed that the association is irreversible so that $K[P] \gg 1$ in eq. 8.1.4. As the dimers increase in concentration, they in turn will react to form trimers, tetramers, and so on. A kinetic scheme describing the complete course of coagulation appears in the next section.

8.3.2 We Can Calculate Complete Aggregation Kinetics If We Assume That Rate Constants Are Practically Independent of Particle Size

In a situation involving rapid coagulation, the association process does not stop with dimers but continues to trimers, tetramers, and so forth until macroscopic particles precipitate. Following the argument of the preceding section, we can obtain the rate constant k_{ij} for the diffusion-controlled association of an i-mer P_i with a j-mer P_j by combining eqs. 8.3.9 and 8.3.10

$$k_{ij} = 4\pi(D_i + D_j)(R_i + R_j)$$
$$= \frac{2}{3}\frac{kT}{\eta}\left(\frac{1}{R_i} + \frac{1}{R_j}\right)(R_i + R_j) \tag{8.3.12}$$

The rate constant carriers only a weak dependence on the size of the particles because the factor $(1/R_i + 1/R_j)(R_i + R_j)$ varies between 4 (when $R_i = R_j$) and R_i/R_j (when $R_i \gg R_j$). Thus, we can create a reasonable approximation of reality by making all rate constants equal and setting $(1/R_i + 1/R_j)(R_i + R_j)$ to 4 so that

$$k_{ij} \simeq \frac{8}{3}\frac{kT}{\eta} \equiv k \tag{8.3.13}$$

becomes a universal rate constant that we can insert in the kinetic scheme of the total aggregation process. Because $k_{ij} \gtrsim k$, we will underestimate the rate somewhat.

For the concentration of species P_j the rate equation is

$$\frac{d[P_j]}{dt} = k\left\{\frac{1}{2}\sum_{i<j}[P_i][P_{j-i}] - [P_j]\sum_i[P_i]\right\} \tag{8.3.14}$$

When j goes from 1 to j_{max}, eq. 8.3.14 represents j_{max} coupled differential rate equations. In fact, we can solve this system of equations.

Let us start by considering the change in the total particle concentration,

$$\sum_j[P_j]$$

This quantity is changed by all associations, and because the rate constant is the same, we can draw an analogy with eq. 8.3.1

$$-\frac{d}{dt}\sum_j[P_j] = \frac{k}{2}\left(\sum[P_j]\right)^2 \tag{8.3.15}$$

where we divide by 2 to correct for the double counting of each association.

With the initial condition of total particle concentration

$$[P]_0^{tot} = \sum_j[P_j]_0 \quad \text{at } t = 0$$

eq. 8.3.15 has the solution

$$\sum_j [P_j] = \frac{[P]_0^{tot}}{1 + t/\tau} \qquad (8.3.16)$$

where $\tau = 2/(k[P]_0^{tot})$. This solution shows that for short times, $t < \tau$, the decrease in total number of particles is linear in time, while for long times, $t > \tau$, the decrease is inversely proportional to t. The characteristic time τ separating the two regimes decreases with increasing total particle concentration. Note that τ equals the time required to reduce the total number of particles in solution to one-half their initial number.

Now we can integrate eq. 8.3.14 in an iterative way by noting that

$$\sum_i [P_i]$$

in the last term is given explicitly by eq. 8.3.16. For $j = 1$, the first term on the right-hand side of eq. 8.3.14 vanishes and

$$\frac{d[P_1]}{dt} = -k[P_1]\frac{[P]_0^{tot}}{1 + t/\tau} \qquad (8.3.17)$$

where $[P_1]$ remains the only unknown. A straightforward integration yields

$$[P_1] = [P_1]_0 \left(1 + \frac{t}{\tau}\right)^{-2} \qquad (8.3.18)$$

From this equation we can construct an uncoupled equation for $k = 2$ and so on. The general solution for the initial condition of only monomers at $t = 0$ is

$$[P_k] = [P]_0^{tot} \left(\frac{t}{\tau}\right)^{k-1} \left(1 + \frac{t}{\tau}\right)^{-k-1} \qquad (8.3.19)$$

Figure 8.5 illustrates the predicted change in concentration of monomers, dimers, and so forth, where $[P_k]/[P]_0^{tot}$ is plotted versus t/τ. The concentration of monomers decreases continuously, but the concentration of all other species initially increases, then goes through a maximum, and finally decreases.

These equations provide several predictions that we can test directly:

1. If eq. 8.3.16 describes the total course of coagulation, then the value of τ should be independent of the stage of coagulation. Table 8.1 provides an example of this.

2. According to eq. 8.3.16, the product, $\tau[P]_0^{tot}$, should be independent of the initial concentration as shown in Table 8.2.

3. Above the CCC, τ should become independent of the added electrolyte concentration. In Table 8.3 the measured values of τ do become almost constant above this point.

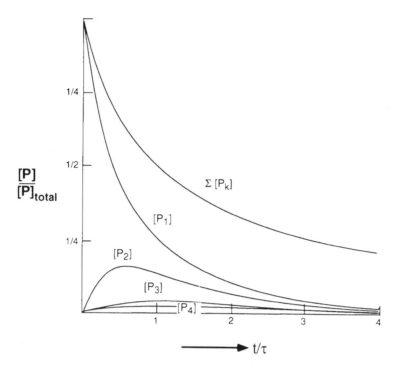

$$\frac{[P]}{[P]_{total}}$$

$\Sigma\,[P_k]$

$[P_1]$

$[P_2]$

$[P_3]$

$[P_4]$

t/τ

Figure 8.5
Plot of the relative
concentrations $[P_1]/[P]_{total}$
of monomers ($k = 1$),
dimers ($k = 2$), and so on
(eq. 8.3.19), and total
number of particles, $\Sigma\,[P_1]/$
$[P]_{total}$, as a function of
time t/τ, where τ, the time
of coagulation, equals the
time required to reduce the
particle concentration to
half its initial value. (H.R.
Kruyt, *Colloid Science*,
Vol. 1, Elsevier, New York,
1952, p. 282.)

Several features of the coagulation process deserve additional
comments.

Effect of Stirring on the Rate of Flocculation

As we saw in Section 4.4, fluid flow can have a pronounced
effect on the rate at which a process involving mass transfer

TABLE 8.1 Illustration of the Validity of Eq. 8.3.16 for the
Coagulation of a Gold Sol

	Gold Sol (number of particles $\times 10^{-14}\,m^{-3}$)		
t	Obs.	Calc ($\tau = 79$ s)	$\tau(s)^a$
0	20.20	20.20	
30	14.70	14.40	80
60	10.80	11.19	69
120	8.25	7.74	83
240	4.89	4.78	76
490	3.03	2.71	85
			$\tau_{av} = 79$

Data from T. Tuorila, *Kolloid Chem. Beih.* **22**, 191 (1926).
[a]See eq. 8.3.16 =

$$\tau = t\left(\frac{\Sigma\,[P]_0^{tot}}{\Sigma\,[P_j]} - 1\right)^{-1}$$

TABLE 8.2 Illustration of the Independence of the Rate Constant $k = 2(\tau[P]_0^{\text{tot}})^{-1}$ on Initial Condition

Dilution Solution	Mean $\tau(s)$	$\tau[P]_0^{\text{tot}}$ (sm^{-3})
1:1	22.7	7.1×10^{17}
1:2	44.0	7.0×10^{17}
1:10	213.4	6.9×10^{17}

Data from H.R. Kruyt and A.E. van Arkel, *Rec. Trav. Chim.* **39**, 656 (1920).

TABLE 8.3 When the Critical Coagulation Concentration of a Selenium Sol Is Exceeded, τ Should Be Independent of Electrolyte

Salt	Concentration of Salt (mM)	Number of Particles per m^3	τ (s)
KCl	20	33.5×10^{15}	10^7
	30	33.5×10^{15}	6×10^6
	65	33.5×10^{15}	360
	80	33.5×10^{15}	19.5
	100	33.5×10^{15}	20.3
	200	33.5×10^{15}	20.7
BaCl_1	2	33×10^{13}	2×10^5
	10	33×10^{13}	1700
	40	33×10^{13}	1000
	100	33×10^{13}	1400

Data from H.R. Kruyt and A.E. van Arkel, *Rec. Trav. Chim.* **39**, 656 (1920).

occurs. We can express the ratio for the rates of coagulation without and with forced convection as

$$\frac{J_{\text{stirring}}}{J_{\text{Brownian motion}}} = \frac{\eta R_{ij}^3 (dv/dx)}{2kT} \tag{8.3.20}$$

where dv/dx is the rate of stirring. The ratio of the rates involves the sum of the particle radii, $R_{ij} = R_i + R_j$, to the third power. For water, if the shear rate $dv/dx = 1\,\text{s}^{-1}$, then $J_s/J_B \simeq 1$ when $R_{ij} = 2\,\mu\text{m}$ or 125 when $R_{ij} = 10$. Thus, stirring has no effect at the beginning of a coagulation process when the particles are small. However, stirring a quiescent, still clear solution some time after coagulation has begun and the particle size has enlarged immediately increases turbidity.

Nonspherical Particles

All the discussion and experimental results cited above involve spherical particles. The product of RD of the effective radius R_{eff} and the diffusion coefficient D is always larger for any non-spherical shape than it is for a sphere of the identical mass. This result occurs because the former particles can also meet through rotational diffusion. Thus, nonspherical particles flocculate more rapidly than spherical ones.

8.3.3 Kinetics of Slow Flocculation Depends Critically on Barrier Height

When the potential energy curve V has a barrier appreciably larger than kT, the rate of coagulation can become orders of magnitude slower than that calculated by eq. 8.3.16. Table 8.3 provides examples of such slow coagulation rates. We can derive rate expression for the initial stages of coagulation by generalizing eq. 8.3.2 to allow for an external potential. In a more general form, Fick's first law says that the flux $\vec{J}$ is proportional to a friction factor D/kT, a concentration term $[P]$, and a driving force, the gradient in chemical potential $\vec{\nabla}\mu$, so that (cf. eq. 1.6.10)

$$\vec{J} = -\frac{D}{kT}[P]\vec{\nabla}\mu \qquad (8.3.21)$$

The chemical potential is in an external potential, $V(\vec{r})$

$$\mu = \mu^\theta + kT\ln[P] + V(\vec{r}) \qquad (8.3.22)$$

Combining eqs. 8.3.21 and 8.3.22 leads to

$$\vec{J} = -D\vec{\nabla}[P] - \frac{D}{kT}[P]\vec{\nabla}V \qquad (8.3.23)$$

which in comparison with eq. 8.3.3 has an extra potential-dependent term.

We can solve eq. 8.3.23 using the same procedure we used to calculate rapid coagulation in Section 8.3.1. If the central particle is located at the origin of the coordinate system and the particle is a spherically symmetrical, the total flux through a shell of radius r is

$$(JA)_i = -8\pi r^2 D\left\{\frac{d[P]}{dr} + [P]\frac{d(V/kT)}{dr}\right\} = \text{constant} \qquad (8.3.24)$$

at steady state. In this equation we have invoked the approximation that the relative diffusion coefficient is twice the individual D for similar particles.

We can integrate eq. 8.3.24 by noting that $\exp[V/kT]$ is an integrating factor. The concentration profile around the central particle is

$$[P] = -\frac{(JA)_i}{8\pi D} \int_r^\infty r'^{-2} \exp\left(\frac{V(r')}{kT}\right) dr' \exp\left(-\frac{V}{kT}\right) + [P]_b \exp\left(-\frac{V}{kT}\right) \quad (8.3.25)$$

where we have used the boundary condition that $[P]$ should tend to the bulk value $[P]_b$ for large r. At a radius $r < 2R$, we consider a dimer to have formed. Thus, a second boundary condition is $[P] = 0$ for $r = 2R$, as stated earlier. We can now solve for the flux $(JA)_i$ in eq. 8.3.25 and

$$(JA)_i = -8\pi D[P]_b \left\{ \int_{2R}^\infty r^{-2} \exp\left(\frac{V}{kT}\right) dr \right\}^{-1} \quad (8.3.26)$$

Using the same argument that leads from eq. 8.3.4 to eq. 8.3.8, the rate constant k_s, in the presence of a barrier, becomes

$$k_s = 8\pi D \left\{ \int_{2R}^\infty r^{-2} \exp\left(\frac{V}{kT}\right) dr \right\}^{-1} \equiv \frac{k_r}{W} \quad (8.3.27)$$

Thus, an energy barrier decreases the rate of coagulation by an amount equal to the stability ratio W

$$W = 2R \int_{2R}^\infty \exp\left(\frac{V}{kT}\right) r^{-2} dr \quad (8.3.28)$$

If we can express the interaction potential energy as a function of particle separation, we can calculate W. Usually this integral must be evaluated numerically.

Verwey and Overbeek developed an approximate expression for eq. 8.3.28 that permits direct comparison with experimental data. They showed that W is determined almost entirely by the value of V at the maximum, V_{max}. In other words, the function $(r^{-2}) \exp V/kT$ has its maximum at V_{max} and drops off rapidly for $V < V_{max}$. Because the exponential is more important than r^{-2}, we can write $r^{-2} \approx r_{max}^{-2} (r_{max} \simeq 2R)$. V expanded in a Taylor series around V_{max} gives

$$V = V_{max} + (r - 2R)\left(\frac{\partial V}{\partial r}\right)_{max} + \frac{(r - 2R)^2}{2}\left(\frac{\partial^2 V}{\partial r^2}\right)_{max} \quad (8.3.29)$$

in which $(\partial V/\partial r)_{max} = 0$. Substituting into eq. 8.3.28 yields

$$\begin{aligned}
W &= \frac{2R}{4R^2} \exp\left(\frac{V_{max}}{kT}\right) \int_{2R}^\infty \exp\left(\frac{(\partial^2 V/\partial r^2)_{max}(r - 2R)^2}{2kT}\right) dr \\
&= \frac{1}{2R} \exp\left(\frac{V_{max}}{kT}\right) \int_0^\infty \exp(-p^2 r'^2) dr' \quad (8.3.30)
\end{aligned}$$

where $p^2 = -(\partial^2 V/\partial r^2)_{max}/2kT$ and the variable changes to r' with appropriately adjusted limits. Because we are at a maximum, $(\partial^2 V/\partial r^2)_{max}$ is negative. The integral in eq. 8.3.30 is

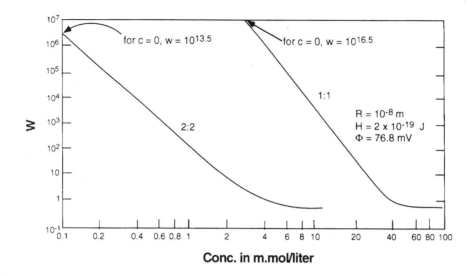

Figure 8.6
Plot of the stability ratio W (eq. 8.3.31) as a function of electrolyte concentration. W represents the ratio k_r/k_s, where k_r equals the rapid diffusion-controlled coagulation rate given by eq. 8.3.11 and k_s is the measured coagulation rate when the electrolyte concentration is less than the CCC, as in eq. 8.2.11. (H.R. Kruyt, *Colloid Science*, Vol. 1, Elsevier, New York, 1952, p. 286.)

$\pi^{0.5}/2p$ and

$$W = \frac{\pi^{0.5}}{4Rp} \exp \frac{V_{max}}{kT} \qquad (8.3.31)$$

When we calculate V_{max} using DLVO theory eq. 8.3.31, provides a relationship between W, the electrolyte concentration c, and the surface potential Φ_0.

Figure 8.6 shows a sample calculation of how W depends on the electrolyte concentration for a 1:1 and a 2:2 electrolyte. For $W \lesssim 1$, the theory ceases to be relevant. Experimentally one often finds a less dramatic change in W with changes in the electrolyte concentration. It is not clear what precisely causes this discrepancy, but one could note that most deviations from the idealized model will have the qualitative effect of producing a less strong dependence of W on c.

8.3.4 Aggregates of Colloidal Particles Can Show Fractal Properties

The final outcome of an aggregation of colloidal particles depends also on processes occurring after the initial association. As discussed in Section 8.1, rearrangements are likely to occur once two particles have come into molecular contact. However, for an irreversible association of solid particles, the system is often kinetically trapped in its initial configuration. As the association progresses, this fact has important consequences for the nature of the aggregates and also for the kinetics of the association.

The equilibrium state of the aggregate should always be compact, but when the aggregate is locked into a configuration set by the initial contact, a much less dense system is formed. The analysis of the kinetics made in Section 8.3.2 no longer applies. It has been shown that aggregates grow in size as a so-called fractal object. In contrast to a compact aggregate, in which the mass M and the radius R are related as $M \sim R^3$, the radius of a fractal object depends more strongly on M. By definition, if M and R are related through

$$M \sim R^{d_f} \tag{8.3.32}$$

we are dealing with a fractal object. The exponent d_f represents the fractal dimension, which is a rational number less than 3; the smaller the d_f, the less compact the aggregate.

In the previous chapter we discussed the polymer coil, which in fact shows fractal properties. By inverting eqs. 7.1.16 and 7.1.17 we find $d_f = 2$ for a coil in a theta solvent while $d_f = 1.67$ in a good solvent.

Figure 8.7 shows a transmission electron micrograph of an aggregate of colloidal gold particles. The open nature of the aggregate is apparent, and we can see clearly how the density of particles decreases away from the center. This effect is a necessary consequence of eq. 8.3.32.

Fractal dimensionality depends on the nature of the aggregation process. The least dense structure appears for a diffusion-limited aggregation in which particles stick irreversibly and permanently at contact. Computer simulations have shown that $d_f = 1.8$ in this case. For the experiments with colloidal gold shown in Figure 8.7, the estimated value of $d_f = 1.75 \pm 0.05$ agreed well with theory.

When a potential barrier must be overcome so that $W \gg 1$ in eq. 8.3.28, an incoming particle will bounce against the

Figure 8.7
Fractal cluster of gold particles prepared by diffusion-limited aggregation. (D.A. Weitz, M.Y. Lin, and J.S. Huang, in *Physics of Complex and Supramolecular Fluids*, S.A. Safran and N.A. Clark, eds., Wiley, New York, 1987, p. 518.)

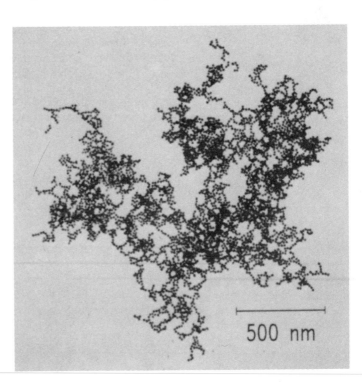

500 nm

aggregate several times. In that case, it is more likely to reach the interior of the aggregate and stick there, leading to a more compact structure and a larger d_f is observed experimentally.

8.4 Electrokinetic Phenomena Are Used to Determine Zeta Potentials of Charged Surfaces and Particles

To use the DLVO theory, we need to know the potential energy of attraction and repulsion between colloidal particles. As we have seen, we can estimate the attraction contributions in terms of the Hamaker constant. However, determining the repulsive double-layer force is more problematic. In many practical situations, it is difficult to obtain a reliable estimate of a colloidal particle's surface potential. Under these circumstances, an alternative strategy is to use electrokinetic measurements, which we can interpret in terms of the zeta potential, ζ. Experience has shown that we can correlate colloid stability with this readily accessible experimental quantity: Table 8.4 correlates critical coagulation concentrations with zeta potentials. Consequently, it is important to see how electokinetic techniques can be used to determine the zeta potential.

Four electrokinetic phenomena play a role in colloid science (see Table 8.5). Electrophoresis and sedimentation potential involve the motion of charged particles in quiescent liquid. Streaming potential and electro-osmosis involve the flow of fluid past a stationary charged surface. We can also classify these phenomena in terms of driving force and response. In electrophoresis and electro-osmosis experiments, we apply an electrical field and generate particle or fluid flow. In sedimentation potential or streaming potential experiments, we impose an external pressure gradient or an acceleration force and generate an electrical potential. Our main interest lies in

TABLE 8.4 Comparison of Critical Coagulation Concentration (CCC) and Critical Zeta Potential

	As_2S_3 Sol($-$)			Fe_2O_3 Sol($+$)	
Electrolyte	CCC (meq/L)	Critical Zeta Potential (mV)	Electrolyte	CCC (meq/L)	Critical Zeta Potential (mV)
KCl	40.0	44	KCl	100.0	33.7
$BaCl_2$	1.0	26	NaOH	7.5	31.5
$AlCl_3$	0.15	25	$CaSO_4$	6.6	32.5
$Th(NO_3)_4$	0.20	27	K_2CrO_4	6.6	32.5
	0.28	26	$K_2Fe(CN)_6$	0.65	30.2
	0.40	24			

TABLE 8.5 Summary of Electrokinetic Effects

Potential	Stationary (wall)	Moving (colloid particles)
Applied	Electro-osmosis	Electrophoresis
Induced	Streaming potential	Sedimentation potential

electrophoresis, but we will consider streaming potential and electro-osmosis as well.

There are also many examples of electrokinetic effects in technical processes and in biological systems. The flow of liquids containing even a small amount of charge can result in an induced streaming potential. This in turn can trigger corrosion processes in pumps and valves. It can even lead to damaging electrical discharges. Electrophoresis is used in coating processes, while electro-osmosis provides a mechanism for inducing flow in thin capillary tubes.

8.4.1 We Can Relate the Electrophoretic Velocity of a Colloidal Particle to the Electrical Potential at the Slip Plane

In an electrophoresis experiment, we impose an electrical field across a colloidal solution and measure the colloidal particle velocity, v. Often colloidal particles are large enough to permit the use of a light or a dark field microscope to follow their motion directly. We simply measure distance traveled versus time to obtain v.

The term *electrophoresis* implies the interplay between electrical phenomena and motion. We know that a viscous fluid adjacent to a solid surface moves with the same velocity as the surface, yet at a distance the surface does not affect the fluid's velocity. Therefore, if we induce motion either of a charged solid particle or in the adjacent solution, we expect to find relative motion between the particle and the counterions. The point in the electrical double layer δ where the liquid begins to flow defines the location of the zeta potential. Although we expect the distance δ to be on the order of a few molecular diameters, we do not know its exact position.

When we first apply an electrical field, the charged particles experience a force that causes them to accelerate. However, the opposing viscous force increases linearly with velocity so that the particle almost instantaneously reaches a steady state velocity in which electrical force and viscous force are equal. In that situation we can estimate the velocity using Stokes's law (cf. eqs. 5.7.1, 5.7.2) to yield

$$\vec{v} = \frac{\vec{F}}{6\pi\eta R} \tag{8.4.1}$$

in which $\vec{F}$ is the external force experienced by the particle.

In general, we do not know the force acting on the particle caused by the external electrical field E. For small fields the force varies linearly with the field, so that $\vec{F} \sim \vec{E}$. We can then define the electrophoretic mobility u as

$$u = \frac{\vec{v}}{\vec{E}} \qquad (8.4.2)$$

This quantity is the primary value determined in an electrophoretic experiment.

For the simple case of an isolated particle with charge q in a field $\vec{E}$, the force is $q\vec{E}$, and by combining eqs. 8.4.1 and 8.4.2 we have

$$u = \frac{q}{6\pi\eta R} \qquad (8.4.3)$$

Equation 8.4.3 is of limited use for a colloidal system because we cannot know the extent to which the particle moves independently of the surrounding electrolyte ions. However, we can rephrase the problem and consider the (spherical) particle plus some counterions as one unit with effective charge q' and effective radius R'.

According to eq. 3.8.25, in a system for which the Debye–Hückel theory is valid, the potential is

$$\Phi(r) = \frac{q'}{4\pi\varepsilon_r\varepsilon_o} \frac{e^{-\kappa r}}{r} \frac{e^{\kappa R'}}{1 + \kappa R'} \qquad (8.4.4)$$

Here we can identify the factor q'/R' (see also eq. 8.4.3). By choosing $r = R'$, we define

$$\zeta \equiv \Phi(R') = \frac{(q'/R')}{4\pi\varepsilon_r\varepsilon_o} (1 + \kappa R')^{-1} \qquad (8.4.5)$$

where ζ is called the zeta potential. Furthermore, if we assume $\kappa R' \ll 1$, eq. 8.4.3 provides a simple relation between the electrophoretic mobility and the zeta potential

$$\zeta = \frac{3}{2} \frac{\eta u}{\varepsilon_r\varepsilon_o} \qquad (8.4.6)$$

This relationship is known as the Hückel equation. We have subsumed the uncertainty in both charge q' and radius R' into one quantity, the zeta potential. Even though absolute numbers are uncertain, we should expect the measured ζ to increase if we increase the true charge q, or if we decrease the true radius R. As a result, we can use ζ measurements for comparative studies.

In the electrophoretic measurement, we primarily measure the balance between a viscous drag force and an electric pulling force. Our discussion used the Stokes equation for the viscous part and the Debye–Hückel theory for the electric force. Furthermore, we restricted our treatment to small particles through the condition $\kappa R \ll 1$. Now we approach the problem from a slightly different angle. Consider instead the limit of large particles so that $\kappa R \gg 1$. In such a case it suffices to consider a planar charged surface exposed to a liquid and to balance the viscous and electrical forces in this system (see Figure 8.8).

Consider a slice of the liquid with a thickness dx, at a distance x from the surface. The viscous force on either side of the slice is proportional to the gradient dv/dx in the velocity of the liquid. The net viscous force F_v on the slice is proportional to the difference in the gradients and

$$\frac{F_v}{\text{area}} = \eta \frac{d^2 v}{dx^2} dx \qquad (8.4.7)$$

The electrical force F_E equals the field E times the charge in the slice

$$F_E = E\rho \, \text{area} \cdot dx \qquad (8.4.8)$$

in which we can obtain the charge density ρ from the Poisson equation (eq. 3.5.13) so that

$$\rho = -\varepsilon_r \varepsilon_o \frac{d^2 \Phi}{dx^2} \qquad (8.4.9)$$

Figure 8.8
Electrophoresis of a large particle ($\kappa R \gg 1$). At a distance δ from the surface, the liquid velocity equals that of the particle, while a velocity gradient exists for larger x values. Across the AB sheet the viscous force is

$$\eta \left\{ \frac{dv}{dx}\bigg|_{x+dx} - \frac{dv}{dx}\bigg|_{x} \right\}$$

$$= \eta \frac{d^2 v}{dx^2} dx$$

while the balancing electric force is $E\rho(x) \cdot dx$.

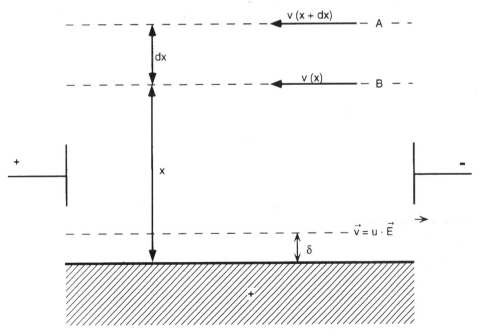

At steady state the forces should balance at all x and

$$\eta \frac{d^2v}{dx^2} = \varepsilon_r \varepsilon_o E \frac{d^2\Phi}{dx^2} \tag{8.4.10}$$

The general solution of eq. 8.4.10 is

$$\eta v(x) = \varepsilon_r \varepsilon_o E\Phi(x) + C_1 x + C_2 \tag{8.4.11}$$

because $v(x)$ and $\Phi(x)$ have proportional second derivatives. Now $v(x)$ and $\Phi(x)$ cannot increase without limit as x increases, so $C_1 = 0$. Furthermore, far from the surface both $v(x)$ and $\Phi(x)$ should remain constant, and we assign zero for both quantities so that $C_2 = 0$.

In the electrical field, the particle will have a velocity $\vec{v} = u\vec{E}$. At a small distance δ from the surface we apply the hydrodynamic stick boundary condition, so that the velocity is $\vec{v}(\delta) = u\vec{E}$. Writing $\Phi(\delta) \equiv \zeta$, we can express eq. 8.4.11 as

$$\zeta = \frac{\eta}{\varepsilon_r \varepsilon_o} u \tag{8.4.12}$$

which is called the Helmholtz–Smoluchowski equation.

The Hückel equation (eq. 8.4.6) and the Helmholtz–Smoluchowski equation (eq. 8.4.12), which are valid in the two limits $\kappa R \ll 1$ and $\kappa R \gg 1$, respectively, differ by a factor of 1.5. These two limits seldom apply strictly in colloidal applications.

Figure 8.9 gives a graphical representation of $\zeta(\varepsilon_r \varepsilon_o)/(\eta\mu)$ as a function of κR. It shows a smooth transition from the Hückel to the Helmholtz–Smoluchowski equation for low zeta potentials, but a marked dip occurs in the range $1 < \kappa R < 10$ for high potential ($\zeta e/kT > 1$).

Figure 8.9
Graphical representation of $C = (\varepsilon_r \varepsilon_o)\zeta/\eta u$ relating electrophoretic mobility and zeta potential for spherical colloidal particles in 1:1 electrolyte solutions. The solid line shows the Henry equation valid for low zeta potentials. The dashed curves represents numerical solutions for $e\zeta/kT = 1, 2, 3,$ and 4 (i.e., $\zeta = 25.7, 51.4, 77.7,$ and 102.8 mV at 25°C). In the regime where $e\zeta/kT > 1$, the curves are sensitive to the valency of the counterions and coions. (P.H. Weirsma, A.L. Loeb, and J.T.G. Overbeek, *J. Colloid Interface Sci.* **22**, 78 (1966).)

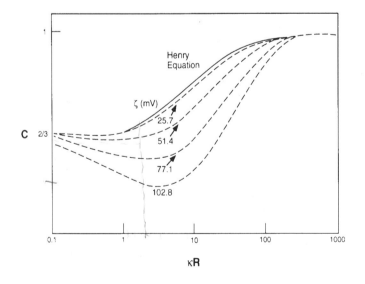

Figure 8.10
Electrophoresis cell in which particle mobility is measured directly using a microscope. (D.J. Shaw, *Introduction to Colloid and Surface Chemistry*, 3rd ed., Butterworth-Heinemann, London, 1980, p. 164.)

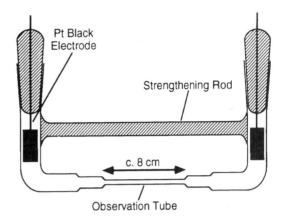

(Internal diameter: c. 2 mm)
(Wall thickness: c. 0.05 mm)

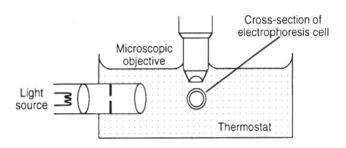

Our analysis assumed that the liquid flow or the external electric field scarcely perturbs the electrical double layer outside a charged particle. Full analysis of the dynamic effects occurring in an electrophoresis experiment of a colloidal system presents a major challenge. For example, a more detailed description of the electrophoretic rate of the particles also depends on the mobilities of the counterions and on local viscosities close to the particle. However, in most cases the simple equations 8.4.6 and 8.4.12, as well as the results of Figure 8.9, provide a reasonable estimate of the zeta potential.

Figure 8.10 shows a schematic of a microelectrophoresis cell that consists of a horizontal glass tube of either rectangular or circular cross section with an electrode at each end. The inlet and outlet taps are used for cleaning and filling the cell. During experiments one side of the cell is closed and the other left open to the atmosphere.

We determine electrophoretic velocity by measuring the time required for the particles to move a fixed distance ($<100 \, \mu m$) on a calibrated eyepiece. The field strength is adjusted to give measurement time on the order of 10 seconds. Faster times result in unacceptably high timing errors: slower times increase the unacceptable error caused by Brownian motion.

8.4.2 We Can Determine the Zeta Potential for a Surface by Measuring the Streaming Potential

In a streaming potential measurement, we cause a solution to flow through a capillary tube by applying a pressure gradient across it (Figure 8.11). This movement displaces part of the double layer in the mobile phase so that a charge accumulates in one flask relative to the other flask. The flow of counterions constitutes a current and generates a potential difference between the two flasks. However, the potential does not increase without limit because the induced potential generates a back-flow of current. As a result, the potential difference rapidly reaches a steady state Φ_{st}. Usually the compensating current also goes through the solution, but if the solid surface is conducting, a current can occur in the substrate. In practical situations this process can lead to corrosion problems.

The fluid flow through the capillary follows Poiseuille's law, and as we can see in Figure 8.12a, the fluid velocity at radius r in a capillary of radius R and length l is

$$v = \frac{\Delta p^2}{4\eta l}(R^2 - r^2) \tag{8.4.13}$$

in which Δp equals the pressure drop.

The electrical current dI associated with the fluid flow in a shell of area $dA = 2\pi r\, dr$ is

$$dI = \rho v\, dA = \frac{\rho\Delta p}{4\eta l}(R^2 - r^2)2\pi r\, dr \tag{8.4.14}$$

where ρ stands for the charge density.

If we replace r by a distance measured from the surface of shear, $x = R - r$, we can write

$$dI = \frac{\rho\Delta p}{4\eta l}(2Rx - x^2)2\pi(R - x)\, dx$$

$$\simeq -\frac{\rho\pi\Delta p}{\eta l}R^2 x\, dx \tag{8.4.15}$$

Figure 8.11
Apparatus for measuring the streaming potential. A pressure gradient applied across the capillary tubes causes fluid to flow through from right to left. The two electrodes measure the resulting potential difference, which then can be related to the zeta potential by eq. 8.4.18. (H.R. Kruyt, *Colloid Science*, Vol. 1, Elsevier, New York, 1952, p. 196.)

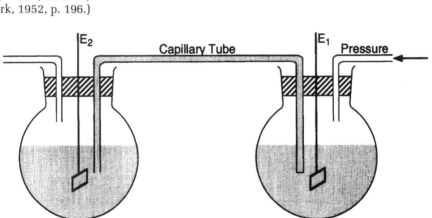

Figure 8.12
Flow profiles in a capillary tube encountered during (A) streaming potential measurements (Poiseuille flow, eq. 8.4.13), (B) electro-osmosis measurements, and (C) electrophoretic measurements in a capillary with a closed end. In electrophoresis experiments, the particle velocity varies as a function of position in the capillary because electro-osmosis also occurs. The true particle velocity, which we obtain at the stationary level where the net fluid velocity is zero, occurs at $0.146R$.

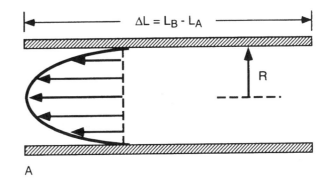

A

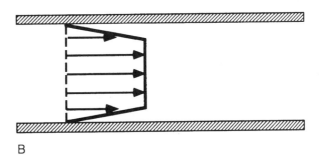

B

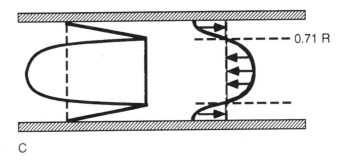

C

We obtain the second equation by using the approximation $x \ll R$. From the point of view of the electrical double layer, the surface of the capillary is flat ($\kappa R \gg 1$). Using the Poisson equation (eq. 3.5.13) for ρ yields

$$dI = \frac{\pi \varepsilon_r \varepsilon_o \Delta p R^2}{\eta l} \frac{d^2\Phi}{dx^2} x \, dx \qquad (8.4.16)$$

Through an integration by parts

$$I = \frac{\pi \varepsilon_r \varepsilon_o \Delta p R^2}{\eta l} \left(x \frac{d\Phi}{dx}\bigg|_0^R - \int_0^R \frac{d\Phi}{dx} \, dx \right) = \frac{\pi \varepsilon_r \varepsilon_o \Delta p R^2 \zeta}{\eta l} \qquad (8.4.17)$$

using $\Phi = \zeta$ at $x = 0$ and $\Phi = d\Phi/dx = 0$ at $x = R$.

The current in eq. 8.4.17 is balanced by a current in the bulk liquid of conductivity, k_b. The resistance in the tube is $(\pi R^2 k_b)^{-1}$ per unit length to give

$$\Phi_{st} = \frac{\varepsilon_r \varepsilon_o \zeta}{\eta} \frac{\Delta p}{k_b} \qquad (8.4.18)$$

using eq. 8.4.17 for the current. We can determine the zeta potential by measuring Φ_{st} versus Δp and knowing ε_r, η, and k_b of the bulk.

8.4.3 Electro-osmosis Provides Another Way To Measure the Zeta Potential

In an electro-osmosis experiment, we apply a potential Φ_{eo} across two reversible electrodes connected by a capillary tube of length L as shown in Figure 8.11 or across a porous plug that can be approximated as a bundle of capillaries. For charged surfaces the electrical field exerts a force on the counterions in the double layer. This force causes the counterions to move, and they carry the solution in the tube along with them (see Figure 8.12B). Because the tube's radius R is on the order of millimeters the double layer is virtually planar, $1/\kappa \ll R$.

In electrophoresis, we considered a moving particle in a fixed solvent, but in electro-osmosis, the converse is true. However, the relative motion between surface and bulk solution is the same, and because $\kappa R \gg 1$, the Helmholtz–Smoluchowski equation applies. We obtain the volume flow of liquid dV/dt as the product of the velocity and the area of the tube. The velocity is related to the mobility u and the electric field, $E = \Phi_{eo}/L$, by equation 8.4.2 so that

$$\frac{dV}{dt} = vA = uEA = \frac{\varepsilon_r \varepsilon_o \zeta \pi R^2 \Phi_{eo}}{\eta L} \qquad (8.4.19)$$

Sometimes it is more convenient to measure the current I than the potential difference Φ_{eo}. The current is carried by the bulk with conductance k_b, and sometimes also by currents on the surface with surface conductance k_s. The total conductance Λ along the tube is then

$$\Lambda = L(\pi R^2 k_b + 2\pi R k_s) \qquad (8.4.20)$$

We can measure the total flow of liquid dV/dt and determine the zeta potential using the relation

$$\frac{dV}{dt} = \frac{\varepsilon_r \varepsilon_o I \zeta}{\eta(k_b + 2k_s/R)} \qquad (8.4.21)$$

The importance of surface conductance can be checked by changing the radius of the capillary.

Electrophoretic measurements by the microscope method are complicated by the simultaneous occurrence of electro-osmosis. The internal glass surfaces of the cell are usually charged, which causes an electro-osmotic flow of liquid near to the tube walls. Because the cell is closed (as we can see in Figure 8.12C) the charged glass surfaces also cause a compensating return flow of liquid with maximum velocity at the center of the tube. This combination results in a parabolic distribution of liquid speeds with depth, and we can observe the true electrophoretic velocity only at locations in the tube where the electro-osmotic flow and the return flow of the liquid cancel each other. For a cylindrical cell, this "stationary level" is located at 0.146 of the internal diameter from the cell wall. For a flat cell, "stationry levels" are located at fractions of about 0.2 and 0.8 of the total depth, the exact locations depending on the ratio of width to depth. If the particle and cell surfaces have the same zeta potential, the velocity of particles at the center of the cell is twice their true electrophoretic velocity in a cylindrical cell and 1.5 time their true electrophoretic velocity in a flat cell.

Exercises

8.1. Consider two identically charged spherical colloidal particles with a surface potential of 25 mV dispersed in an aqueous medium. The effective Hamaker constant is 7×10^{-21} J. Determine the separation h_0 for which the interaction potential goes through zero between the barrier and the primary minimum. Assume $h_0 \ll \kappa^{-1}$.

8.2. Calculate the interaction potential between two similar charged spheres of radius 10^{-7} m and a surface potential of -50 mV. The NaCl concentration of the aqueous medium is 10 mM and the Hamaker constant 5×10^{-20} J.

8.3. Calculate the quantity

$$-\left(\frac{\partial^2 V}{\partial h^2}\right)_{h = h_{max}}$$

for a system of spherical colloidal particles of radius 100 nm and surface potential 35 mV in a 2×10^{-3} M KCl solution using the Hamaker constant $H = 5 \times 10^{-20}$ J and $T = 300$ K. The second derivative of the interaction potential should be evaluated at the primary maximum.

8.4. Tuorila (*Kolloid Chem. Beih.* **22**, 191 (1926)) measured the rapid coagulation of a gold sol, radius 51 nm, and obtained the following values:

$t(s)$	Number of Particles/m$^3 \times 10^{-14}$
0	20.2
30	14.7
60	11.2
120	7.74
240	4.78
480	2.71

Determine the association rate constant and compare with the theoretical value for a diffusion-controlled reaction.

8.5. Estimate the critical coagulation concentration in water at $T = 300\,K$ for highly charged spherical colloidal particles for counterion valencies $z = 1, 2, 3$ ($H_{121} = 1.0 \times 10^{-20}$ J).

8.6. Calculate the expected value of the stability ratio W for two spheres of radius 100 nm in a 2×10^{-3} M aqueous KCl solution at $T = 300\,K$ and with surface potential 35 mV and Hamaker constant 5×10^{-20} J.

8.7. Polymer chains in a solvent can be viewed as fractal objects. What is the fractal dimension (a) at theta conditions and (b) in a good solvent?

8.8. Spherical particles of radius 100 nm aggregate by a diffusion limited process. Estimate the size of a cluster of (a) 100 particles, and (b) 1000 particles, assuming a fractal aggregate of dimensionality 1.8.

8.9. Dunstant and Saville (*J. Chem. Soc. Faraday Trans.* **88**, 2031 (1992)) measured the electrophoretic mobility of intrinsically uncharged alkane prticles, average radius 500 nm, in aqueous salt solutions and found the following electrophoretic mobilities, u:

Electrolyte	Concentration (M)	$u \times 10^8$ ($m^2\,s^{-1}\,V^{-1}$)
$AlCl_3$	10^{-3}	3.1
	10^{-2}	3.0
HCl	10^{-3}	-0.9
	10^{-2}	± 0
KCl	10^{-3}	-2.7
	10^{-2}	-2.1
$CaCl_2$	10^{-3}	-2.1
	10^{-2}	-1.8

Determine the zeta potential for these cases. Estimate the minimum magnitude of the surface charge. Discuss the molecular origin of the charge. If the surface charge arises from impurities, what is the lower value of concentration of the impurity if the particles are present at a volume fraction of 0.001? Using a more general theory, Dunstan and Saville estimated the following zeta potentials (mV): 40, 35, -15, 0, -45, -30, -35, -25, in the order given in the table. How do these values deviate from the ones you have calculated?

8.10. Stigter and Mysels (*J. Phys. Chem.* **59**, 45 (1955)) measured the electrophoretic mobility of SDS micelles as a function of added NaCl. Their results are summarized as follows.

NaCl (m/L)	CMC (mM/L)	N	$10^8 u$ $(m^2 V^{-1} s^{-1})$	$10^{10} R_{mic}$ (m)	κR_{mic}
0	7.12	80	4.55	21.5	0.61
0.01	5.29	89	4.26	22.1	0.86
0.03	3.13	100	3.84	23.0	1.32
0.05	2.27	105	3.63	23.4	1.69
0.1	1.46	112	3.42	24.0	2.40

(a) Calculate the zeta potential using the Smoluchowski equation. Estimate the value (using the information in Figure 8.9) when the Henry equation is employed.

(b) Using the data in the table, compare the values of Φ_0 determined from the dressed micelle model and the calculated zeta potential. Discuss the physical reason for the differences in these values.

8.11. A spherical colloid with a radius of 50 nm has a mobility of $4.0 \times 10^{-8} m^2 V^{-1} s^{-1}$ in a 0.001 M KCl solution. Calculate the zeta potential using the Smoluchowski equation and then the Henry correction.

8.12. Calculate the electro-osmotic transport of an aqueous solution through a glass capillary 5 cm long with a 0.5 mm internal diameter under a potential difference of 100 V when the zeta potential is 80 mV. The viscosity of the solution is 0.01 poise ($1.0 \times 10^{-3} kg \cdot m^{-1} s^{-1}$) and its dielectric constant ε_r is 78.5.

8.13. Calculate the expected streaming potential and streaming current (i.e., charge transported by convection) if the zeta potential is -100 mV and the solution conductance is $10^{-5} \Omega^{-1} cm^{-1}$. The pressure difference is 0.1 atm across the capillary tube, which is 30 cm long and 1 mm in diameter. Also calculate the resistance of the capillary ($\eta = 1 \times 10^{-3} kg \cdot m^{-1} s^{-1}$ and $\varepsilon_r = 78.5$). Repeat the calculation for the flow of a hydrocarbon (gasoline) through the capillary under the same conditions; specific conductance equals $10^{-8} \Omega^{-1} m^{-1}$ and $\varepsilon_r = 2$. Do you see why the first thing that is done when delivering fuel to an airplane via a meters-long fueling line is to connect a grounding wire between the airplane and the pumping track?

8.14. In an electro-osmosis experiment, the flow rate was $2 \times 10^{-5} m^3/s$, the current through the cell 0.5 mA, and the conductance of the solution $10^{-3} \Omega^{-1} m^{-1}$. Given $\eta = 0.01$ poise and $\varepsilon_r = 78.5$, calculate the zeta potential of the porous plug. (You can neglect the pressure difference on both sides of the plug.) (1 poise $= 0.1 kg \cdot m^{-1} s^{-1}$.) If the effective radius of the porous plug is 0.5 mm and its length 15 cm, what pressure difference across the plug would result in zero net flow of solution?

8.15. In an electrophoresis experiment, we visually measure the distance traveled by individual particles over a set period of time and calculate an average displacement from a number of observations. The distance an individual particle moves reflects a combination of its motion under the influence of an electrical field and the random movement resulting from Brownian motion. Suppose that you determine an electrophoretic mobility of $40 \times 10^{-8}\,\mathrm{m^2\,V^{-1}\,s^{-1}}$ for a colloidal sol in water.

(a) How far does the particle move under an electric field of 4000 V/m in 20 seconds?

(b) Calculate the maximum and minimum distance value you might expect when the effect of Brownian motion is taken into account for particles with a radius of 5000 Å. Repeat the calculation for particles $R = 50\,\text{Å}$.

Literature

References[a]	Section 8.1 Overview: Kinetics and Thermodynamics of Colloids	Section 8.2: DLVO Theory	Section 8.3: Kinetics of Aggregation	Section 8.4: Electrokinetics
Adamson & Gast				CHAPTER V 8–9 Zeta potentials
Davis		CHAPTER 12.3 DLVO theory		
Hiemenz & Rajagopalan		CHAPTER 11 DLVO theory	CHAPTER 13 Slow and fast coagulation	CHAPTER 12 Properties and measurements
			CHAPTER 12.11 Coagulation kinetics of charged systems	
Hunter (I)		CHAPTER 9.1—9.4 DLVO theory	CHAPTER 9.5—9.6 Kinetics of coagulation	CHAPTER 8 Zeta potential
Hunter (II)	CHAPTER 1.5 Electrical charge and colloidal stability	CHAPTER 7.6–7.7 DLVO and Schulze–Hardy rules	CHAPTER 7.8 Slow and fast coagulation	CHAPTER 9.11 Electrokinetic effects
	CHAPTER 3.3–3.6 Sol size distribution and determination			
Israelachvili		CHAPTER 12.17–12.20 DLVO		
Kruyt	CHAPTER II.5 Stability of sols	CHAPTER VI DLVO Interactions between sols	CHAPTER VII Kinetics of flocculation	CHAPTER V.8 Electrokinetic phenomena

Meger	CHAPTER 10 Colloidal stability		CHAPTER 5 Electrokinetics
Shaw	CHAPTER 8 Stability	CHAPTER 8 Coagulation	CHAPTER 7 Theory and phenomena
Vold and Vold	CHAPTER 8.2 DLVO	CHAPTER 8.3 Flocculation kinetics	CHAPTER 6.6 Zeta potentials

[a]For complete reference citations, see alphabetic listing, page xxxi.

9

COLLOIDAL SOLS

CONCEPT MAP

Preparation of Colloidal Particles

In Chapter 8 we described the general principles that govern colloidal stability; here we consider specific colloidal sols that demonstrate typical properties and illustrate some of the complexities encountered in practical applications.

Two basic strategies for forming colloidal particles involve (1) dispersion in which large particles are broken into small ones, and (2) condensation, in which exceeding an equilibrium condition, such as a solubility limit, leads to the formation of a new phase (examples in eqs. 9.1.1–9.1.5). Usually these processes lead to heterogeneous sols that possess a wide range of particle sizes.

To prepare monodisperse colloids, we must separate and control nucleation and growth, as illustrated in Figure 9.1. The idea is to induce nucleation in a short burst by briefly exceeding the nucleation concentration and then promote growth by maintaining the reactant concentration below that leading to nucleation, but above the equilibrium concentration.

Potential-Determining Ions

We prepare AgI sols by exceeding the solubility product $(c_{Ag^+})(c_{I^-}) = 7.5 \times 10^{-17} \, M^2$.

Zeta potential measurements show that:

- The point of zero charge pzc occurs when $c_{Ag^+}(pzc) = 3.0 \times 10^{-6} \, M$.

- When $c_{Ag^+} > c(pzc)$, the particles are positively charged

- ζ varies linearly (Figure 9.3) with $\log(c_{Ag^+})$ or $\log(c_{I^-})$.

- These observations establish that iodide ions adsorb preferentially on AgI particles and that the concentration of Ag^+ (or I^-) sets the potential on the AgI particles.

With such systems, we divide ions into two categories: (1) indifferent ions (KNO_3), which effect interaction by changing κ, and (2) potential-determining ions (Ag^+ and I^-), which directly change Φ_0 and thus can have a profound effect on colloidal stability, particularly near the c(pzc).

With proteins and latex spheres containing carboxyl and/or amino groups, H^+ ions are potential determining, while with a sol like $Al(OH)_3Al^{3+}$, OH^- plays the same role; in either case the surface potential depends on pH.

Potential-Determining Ions Lead to Complex Sol Behavior

Some of the complexities encountered in colloidal systems are illustrated by the interactions between the negatively charged AgI sol and Al^{3+}, which forms a variety of complexes whose electrical charge is set by concentration and pH (Figure 9.4).

Clays

Clays are platelike structures containing multilayers held together by van der Waals forces. The individual layers are comprised of two-dimensional tetrahedral arrays containing either Si^{4+} and O^{2-} or octahedral arrays containing Al^{3+}, Mg^{2+}, O^{2-}, and OH^- (Figures 9.6 and 9.7).

Atomic substitutions ($Al^{3+} \rightarrow Mg^{2+}$) lead to negatively charged clays like the montmorillonites (eq. 9.3.1), which adsorb water between their layers and swell, or the illites like mica, which do not swell.

Surface force apparatus experiments directly measure forces between mica surfaces (Figures 9.8–9.10), illustrating the role of ion exchange in controlling forces between clay colloidal particles (eq. 9.3.4).

Latex Spheres

Neutron scattering experiments on monodisperse latex spheres provide values of the form factor $P(Q)$ which give particle size (Figure 9.13) and the structure factor $S(Q)$, which measures particle–particle interactions (Figure 9.14). We can express $S(Q)$ in terms of the pair distribution function $g(r)$ (eq. 9.4.4).

Neutron scattering experiments permit us to follow the transformation of latex particles from gaslike independent behavior (Figure 9.16) through liquidlike behavior to solidlike order (Figure 9.17) with changes in latex volume fraction, temperature, or salt concentration.

By matching the refractive index of the particles and solvent, and by adding trace amounts of a chemically different but equal-sized particle, we can obtain diffusion coefficients for isolated particles in a latex matrix.

Homocoagulation Versus Heterocoagulation

Many practical colloidal systems, such as paints, contain mixtures of sols, while many colloidal processes, like mineral flotation, involve heterogeneous surfaces. In such systems van der Waals interactions can be attractive or repulsive and electrical double-layer interactions between likes and unlikes can differ significantly.

Separate characterizaton of the homocoagulation behavior of three sols (an aluminum hydrous oxide sol, a cationic PS latex, and an anionic PS latex: Figure 9.24) provides a basis for interpreting the heterocoagulation between mixtures of the two (Figures 9.25 and 9.26).

By writing the total rate of coagulation v_T (eq. 9.5.2) as a sum of independent coagulation rates, v_{11}, v_{22}, and v_{12}, we can relate the measured total stability ratio W_T to three individual stability ratios (eq. 9.5.7) and obtain predictions in accord with measurements.

Aerosols

Aerosols containing solid or liquid particles dispersed in gas (air) play important roles in many medical, technological, and pollution processes.

In a gaseous medium, gravitational and van der Waals forces dominate particle interactions except when particles become charged. Then extraordinary long range repulsive forces operate.

Hetero- and homocoagulation are equally likely to occur, leading to the formation of open fractallike aggregates.

9

During the past century, the study of small solid particles dispersed in liquid has played a pivotal role in defining the colloidal domain. In addition, analysis of these sols has helped to establish fundamental relationships in nature. For example, both the kinetic nature of matter as manifested by Brownian motion and the first determination of Avogadro's number involved colloidal sols. More recently, characterization by neutron scattering and video enhanced microscopy of polymer colloids such as latex spheres has provided insights into phase transformations.

Colloidal sols form the basis of many industrial processes and products. Processing of clay–water dispersions produces ceramics ranging from delicate china figurines to toilet bowls and masonry bricks. Applying modern colloid technology to silica sols yield tough, fracture-resistant ceramics that eventually find application in high temperature automobile engines, in rocket nose cones, and as longer lasting medical prostheses.

Virtually all coating systems employ colloidal materials, and most industrial products are coated for either protective or decorative reasons. Paints are colloidal suspensions containing titanium dioxide or zinc oxide and latex particles. Drilling muds, which are indispensable as lubricants and rheology control agents in oil exploration, are complex colloidal materials. Papermaking involves producing a meshwork from colloidal cellulose fibers in which clay particles are used as filler to improve capacity and to produce a pleasing surface texture. Inks used in ordinary ballpoint pens, xerography, and high speed printing presses are colloidal liquids or paste. Toothpaste, scouring powders, and other heavy-duty cleaning agents contain colloidal materials such as pumice.

Control of colloidal systems is also a key issue in dealing with a host of environmental problems associated with our heavy use of technology. Smog consists of colloidal-sized particles generated by an atmospheric photochemical reaction involving petroleum as well as natural products. Controlling the fine powders and other colloidal debris associated with processing at wood pulping plants, mineral flotation sites, coal grinding processing units, and asbestos plants requires application of colloidal chemical techniques. Some of the key steps in the purification of water and the treatment of sewage also depend on the adsorptive capacity of colloidal materials.

In this chapter we explain methods for preparing colloidal sols and describe several colloidal systems. The examples we have chosen illustrate important principles and provide useful insights into the world in which we live.

9.1 Colloidal Sols Formed by Dispersion or Condensation Processes Usually Are Heterogeneous

Two basic strategies for preparing colloidal sols are dispersion, which involves breaking large particles into small ones, and condensation, in which exceeding an equilibrium condition (e.g., the solubility limit) leads to the formation of a new phase.

In the dispersion method, a mechanical process breaks bulk material down to colloidal dimensions. For example, the titanium dioxide used in paints is processed in a ball mill. In fact, we can view clays as colloidal materials that originally were crystallized from silicon, aluminum, and oxygen compounds and subsequently dispersed into colloidal-size particles by natural forces.

Dispersion produces particles of limited size because small particles tend to agglomerate during the grinding process and to resist redispersion in subsequent processing steps. Use of a suitable dispersing agent can control agglomeration during grinding. For example, if we want to grind sulfur in a ball mill, we can add sucrose to coat the surface of the particles and prevent them from agglomerating. We can remove the sugar later by dialysis.

Another dispersion method involves passing an electric arc between two plates submerged in a liquid. This process tears small particles from the metal surfaces. If we perform the dispersion in the absence of oxygen, we obtain a metal sol. Otherwise the sol's furface is covered by metal oxide.

Condensation methods are much more widely used and more versatile than dispersion methods. They include chemical reaction, condensation from the vapor, and dissolution and reprecipitation.

The reaction

$$KI + AgNO_3 \rightarrow AgI(sol) + K^+, I^-, NO_3^- \qquad (9.1.1)$$

exemplifies how we can form a sol by exceeding the solubility limit of AgI. To stabilize the sol, we must reduce the concentration of unprecipitated ions by dialysis.

Condensation processes can form new colloidal phases of a variety of materials. For example, reduction of the gold chloride complex by hydrogen peroxide

$$2Au(Cl)_4^- + 3H_2O_2 \rightarrow 2Au(sol) + 6HCl + 3O_2 + 2Cl^- \quad (9.1.2)$$

forms colloidal gold. Oxidation of hydrogen sulfide with oxygen

$$2H_2S + O_2 \rightarrow 2S(sol) + 2H_2O \quad (9.1.3)$$

forms the sulfur sol. Heating an aqueous solution of $Fe(Cl)_3$

$$Fe(Cl)_2 \rightarrow Fe(OH)_3(sol) + 3HCl \quad (9.1.4)$$

forms the azeotrope of HCl and simultaneously the $Fe(OH)_3$ sol.

The dissolution–reprecipitation method involves, as an example, rapidly injecting a solution of steric acid in ethanol into a large volume of water

$$C_{17}H_{35}COOH_{EtOH} + H_2O \rightarrow fatty\ acid\ (sol) \quad (9.1.5)$$

Because all the processes just described involve concurrent nucleation and growth, they lead to colloidal particles with a wide range of particle sizes. For example, when we simply mix together a solution of $AgNO_3$ and KI, we create small regions of solution in which the concentration of Ag^+ and I^- ions exceeds the solubility product. This condition initiates the nucleation process and growth follows. As mixing continues, nucleation begins in other regions of the solution. Sequential nucleation, accompanied by continuous growth, results in sols with a wide distribution of particle sizes. Although such sols have many uses, monodisperse colloids are desirable in many scientific investigations and also in many refined applications.

9.1.1 Controlling Nucleation and Growth Steps Can Produce Monodisperse Sols

The key to forming monodisperse sols is to separate and control the nucleation and growth processes. As we saw in Chapter 2, nucleation often occurs at concentrations or vapor pressures that exceed the equilibrium value by a considerable amount. By analogy, precipitation of colloidal material also can begin at a concentration of reactants that considerably exceeds the solubility product.

Figure 9.1 illustrates the strategy used to produce monodisperse colloids. Two important concentrations are involved: the nucleation concentration at which growth begins and the solubility limit or saturation concentration at which growth stops. The idea is to adjust the temperature or concentration so that the initial concentration uniformly exceeds the nucleation concentration in the sample, with the result that nucleation occurs in a single short burst that causes the concentration of reactants to fall below the nucleation threshold. We can control subsequent growth by adjusting reaction conditions to remain below the nucleation threshold, but above the equilibrium line,

Figure 9.1
Producing a monodisperse sol requires control and separation of nucleation and growth processes. The number of nuclei is set by increasing the concentration of reactants above the nucleation concentration for a brief period. Particle growth is controlled by maintaining the reactant concentration below the nucleation concentration but above the equilibrium concentration long enough to obtain the desired particle size. Arrows indicate small additions of reactants after the nucleation step.

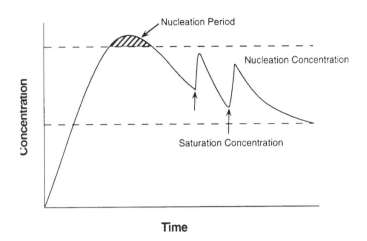

by consecutive small additions of reactants or by changing the temperature. The number of nuclei formed in the initial stage determines the number of particles; the length of the growth period determines their size.

Formation of monodisperse gold sols illustrates how we can separate nucleation and growth entirely. In the nucleation step, small gold particles are formed by the reaction

$$Au(Cl)_4^- + P(red) \rightarrow Au(nuclei) + reduction\ products \qquad (9.1.6)$$

The growth step involves adding a mild reducing agent (such as formaldehyde) along with more $Au(Cl)_4^-$ under conditions that allow no new nuclei to be formed

$$Au(nuclei) + Au(Cl)_4^- + H_2CO \rightarrow Au(sol) + reduction\ products \qquad (9.1.7)$$

Egon Matijević and his co-workers have prepared a number of homogeneous metal oxide or hydroxide sols using a controlled hydrolysis technique. Their scheme consists of heating an acidic solution containing a transition metal complexed with an anion, such as chloride, sulfate, or phosphate. If the anion is sulfate, complexes with the general formula $M_a(OH)_b(SO_4)_c^{\{(za-b-2c)+\}}$ are formed. Increasing the temperature causes acceleration of the rate of deprotonation of coordinated water. This generates the intermediates and leads to precipitation. Manipulating the rate of heating, the concentration and purity of reactants, the growth temperature, and the time, controls the nucleation bursts and the extent of subsequent particle growth. Because transition metal compounds have a marked tendency to form polynuclear complexes of varying mass and composition, the exact mechanism for forming these sols is usually difficult to specify. The concentration and chemical nature of the complexing counterions—Cl^-, SO_4^{2-}, PO_4^{3-}—also play an important role in determining the final morphology of such metal oxide sols. Figure 9.2 shows electron micrographs of two such sols.

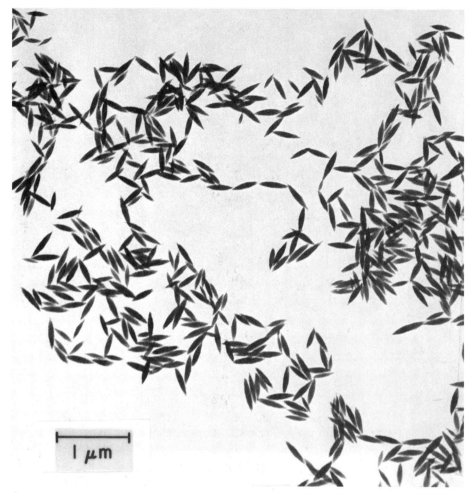

A

Figure 9.2
Examples of monodisperse
colloids prepared using
the controlled hydrolysis
technique. (A) Hematite
($\beta - Fe_2O_3$) particles. Very
small variations in
experimental nucleation
and growth conditions can
lead to very different
particle morphologies.

Several factors associated with the growth step actually
increase the homogeneous nature of the sol. Studies on monodisperse sulfur sols have demonstrated that when diffusion of
material to the particle surface, rather than surface incorporation, is the rate-determining step, the rate at which the surface
area changes over time is the same for all particles, although the
rate itself may vary. Overbeek extended and generalized these
observations by showing that under such circumstances size
distribution becomes narrower with time. In addition, even if the
incorporation step is rate determining, a narrower distribution
results if the rate of incorporation is the same for all particles or
is proportional to the surface area of the particle. Only in the
unlikely event that the rate is proportional to the particle's
volume does narrowing of the distribution fail to occur, and even
then the size distribution remains constant. Thus, the growth
process usually leads to a narrowing of the particle size distribution. One exception is the Ostwald ripening (see Chapter 2),
where large particles grow and small ones decrease in size.

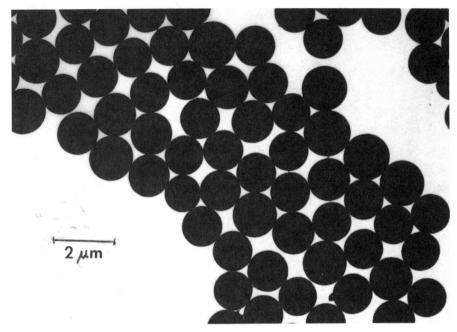

B

Figure 9.2
(B) Cadmium sulphite
spherulites.

The process of preparing organic polymers as monodisperse spheres by emulsion polymerization starts with an emulsion of a monomer, such as styrene, which is usually stabilized by a surfactant, such as SDS. We can adjust the concentration of SDS so that the emulsion also includes mixed micelles containing some solubilized monomer. Then we add a water-soluble initiator, such as potassium persulfate. Statistically, polymerization begins in the micelles rather than in the emulsion droplets. As the number of monomers in the micelles is depleted, more monomer diffuses into them from the emulsion. Thus, the number of activated micelles determines the number of particles, but the amount of monomer present in the system determines their size.

9.2 The Concentration of Silver and Iodide Ions Determines the Surface Potential of Silver Iodide Sols

The AgI sol, one of the most thoroughly characterized colloidal systems, has played an important role in developing our understanding of colloidal suspensions.

The solubility product of AgI is $7.5 \times 10^{-17} \, M^2$ at 25°C. Under conditions in which $c_{Ag^+} = c_{I^-} = 8.7 \times 10^{-9} \, M$, electrophoretic measurements on AgI colloidal particles show that the particles carry a negative charge. By varying the concentration of Ag^+ (or I^-) by adding an appropriate salt, we can determine that when $c_{Ag^+} > 3.0 \times 10^{-6} M$, the particles are positively charged; below this value they bear a negative charge. Thus, the point of zero charge pzc, occurs when $c_{Ag^+} = 3.0 \times 10^{-6} M$. Furthermore, the potential on the particles (as determined by electrophoretic measurements) varies linearly as a function of the logarithm of added Ag^+ (or I^-) ions. This means that the relative concentrations of Ag^+ and I^- ions in solution determine the potential on the surface of the AgI particles.

We can understand this behavior in terms of the ion binding model of section 3.6. At equilibrium the chemical potentials of the Ag^+ and I^- ions in solution equal those on the colloidal particle, and on the surface the chemical potential of the ions vary with the electrostatic surface potential. The point of zero charge occurs when the ions are adsorbed equally, and the fact that $c_{I^-} = 2.5 \times 10^{-11} M$ is much smaller than $c_{Ag^+} = 3.0 \times 10^{-6} M$ simply reflects readier adsorption of I^- ions on the AgI surface.

Away from the point of zero charge, a nonzero potential Φ_0 exists at the surface of the sol particle. For $c_{Ag^+} > 3.0 \times 10^{-6} M$, this potential is positive, while it is negative at lower concentrations of Ag^+. We can understand the quantitative relation between ion concentration and surface potential by considering the sol particle as a giant ion, P^{m+}, of charge me, where m can have either sign.

In the solution we have the equilibria

$$P^{m+} + Ag^+ \rightleftharpoons P^{(m+1)+} \tag{9.2.1a}$$

$$P^{m+} + I^- \rightleftharpoons P^{(m-1)+} \tag{9.2.1b}$$

The free energy change of adsorbing a silver ion is

$$\Delta G_{Ag} = \Delta\mu^\theta_{Ag} + (m+1)e\Phi_0 - kT\ln c_{Ag^+} - me\Phi_0 \tag{9.2.2}$$

neglecting the small entropy of mixing terms for the particles. Similarly for the addition of an iodide ion

$$\Delta G_I = \Delta\mu^\theta_I + (m-1)e\Phi_0 - kT\ln c_{I^-} - me\Phi_0 \tag{9.2.3}$$

The free energy changes ΔG_{AG} and ΔG_I are both zero at equilibrium. In particular, at the point of zero charge (pzc), where $\Phi_0 = 0$, we find

$$\Delta\mu^\theta_{Ag} - \Delta\mu^\theta_I = kT\ln\left\{\frac{c_{Ag^+}(\text{pzc})}{c_{I^-}(\text{pzc})}\right\} \tag{9.2.4}$$

By combining eqs. 9.2.2–9.2.4 we can eliminate $\Delta\mu^\theta_{Ag}$ and $\Delta\mu^\theta_I$ to find the surface potential

$$\Phi_0 = \frac{kT}{2e}\left(\ln\left\{\frac{c_{Ag^+}}{c_{Ag^+}(pzc)}\right\} + \ln\left\{\frac{c_{I^-}(pzc)}{c_{I^-}}\right\}\right)$$

$$= 0.059\log\left(\frac{c_{Ag^+}}{3.0\times10^{-6}}\right) = -0.059\log\left(\frac{c_{I^-}}{2.5\times10^{-11}}\right)$$

(9.2.5)

This is sometimes called the Nernst equation.

The concentrations, or more specifically the chemical potentials, of the Ag^+ and I^- in the solution set the surface potential on the AgI particle. Figure 9.3 shows how the zeta potential varies with $-\log c_{Ag^+} = pAg$. We see two types of deviation from the ideal behavior described by eq. 9.2.5. Around the point of zero charge at $pA = 5.5$, a linear relation exists between the zeta potential and pAg, but the slope is smaller than in eq. 9.2.5. This deviation could be explained by the difference between the surface potential Φ_0 and the zeta potential.

Furthermore, at high and low values of pAg, the zeta potential seems to saturate. It may reflect a deviation from the surface homogeneity implicitly assumed in the derivation of eq. 9.2.5.

In a system like the AgI sol, we divide ions into two categories, potential-determining ions and indifferent ions. For example, KNO_3 functions as an indifferent electrolyte. Indifferent electrolytes affect the interaction between colloidal particles mainly by changing the Debye length, $1/\kappa$. They also change the activity of potential-determining ions through the Debye—Hückel screening, but that effect is generally a small, secondary one. Potential-determining ions directly change Φ_0. Consequently, minute changes in the concentration of potential-determining ions can produce dramatic effects on the stability of colloidal systems. For example, as we saw in Table 8.4, most colloids undergo rapid flocculation when the zeta potential

Figure 9.3
Zeta potentials ζ for silver iodide sol as a function of $-\log[Ag^+]$. As the silver ion concentration decreases, ζ changes from a positive to a negative value; the point of zero charge occurs at 3×10^{-6} M. The prediction of the Nernst equation, $\Phi_0 = 59 (5.5-pAg)\,mV$ (eq. 9.2.5), is shown for comparison purposes. (H.R. Kruyt, ed., *Colloid Science*, Vol. 1, Elsevier, New York, 1952, p. 231.)

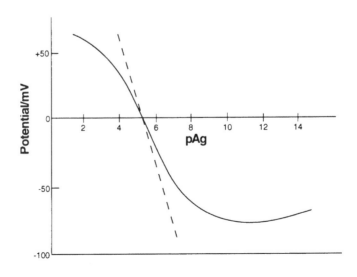

decreases to 25–35 mV. When the concentration of Ag^+ ion varies from 3×10^{-7} to 3×10^{-5} M, the potential changes from -59 to 59 mV according to eq. 9.2.5.

9.2.1 Potential-Determining Ions Play an Important Role in Controlling Stability

Potential-determining ions play a key role in a large number of important colloidal systems. For example, hydroxide and hydronium ions serve as potential-determining ions for colloidal molecules or particles containing weak acidic or basic groups or both. This category includes ionic polymers, latex spheres, and amphiphilic aggregates containing carboxyl and amino groups as well as almost all proteins. Chloride ion is a potential-determining ion for the gold sol because it forms a particularly strong complex. Trivalent ions like Al^{3+} and Fe^{3+} form a variety of complexes whose electrical charge is set by the ion concentrations and the pH. Because the aluminum system is so important in water and sewage treatment, we will consider its behavior.

Like many other multivalent cations, aluminum can form a bewildering variety of complexes whose stability depends on metal ion concentration and pH. At low pH (<4) $Al^{3+}(H_2O)_6$ is the dominating species while at high pH ($\gtrsim 10$) $Al(OH)_4^-(H_2O)_2$ predominates. Around neutral pH the aluminum hydroxide $Al(OH)_3$ precipitates at equilibrium but due to the loose network structure, kinetics is slow. In intermediate pH regimes the aluminum ions associate into oligomeric and polymeric structures which can be seen as precursors of the precipitating hydroxide. Below pH 7 these oligomers/polymers should be positively charged while at higher pH values they are predominately negatively charged. We know that the complexes are important in controlling colloid stability in the aluminum system, and we illustrate it by considering the effect of $Al(NO_3)_3$ on the stability of the AgI sol.

Figure 9.4 displays the critical coagulation concentration (CCC) of an AgI sol prepared by mixing 1×10^{-4} M $AgNO_3$ with a 4×10^{-4} M KI solution as a function of $Al(NO_3)_3$ and pH. In the absence of aluminum, these sols will be highly negatively charged according to eq. 9.2.5, and the high charge ensures colloidal stability. The presence of aluminum has a dramatic effect on this stability in a way that depends strongly on pH. It is possible to understand the pH effects on the basis of the solution of the aluminum ions:

1. When the pH is less than 4, the CCC is constant and equal to 2.5×10^{-5} M, a value characteristic for destabilization of the colloid by the hydrated trivalent Al^{3+} species. Zeta-potential measurements show that the AgI sol remains negatively charged in acidic solutions even at very high $Al(NO_3)_3$ concentrations.

2. At pH 4, the CCC drops sharply to 7×10^{-6} M and remains almost constant up to pH 7. This fact implies the formation of a hydrolyzed species with a charge greater

Figure 9.4
Coagulation behavior of a
silver iodide sol as a
function of pH and
aluminum nitrate
concentration. The AA′
curve represents the CCC
and the BB′ curve the
stability limit due to charge
reversal resulting from
adsorption of positively
charged aluminum
complexes. (E. Matijević,
K.L. Mittal, R.H. Ottewill,
and M. Kerker, *J. Phys.
Chem.* **65**, 826 (1961).)

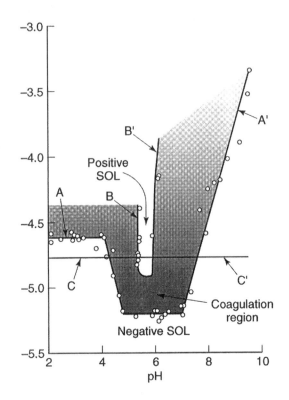

than $+3$. In this regime we have a range of oligomeric and
even polymeric aluminum species that have a charge
per Al of less than $+3e$ but where the total charge can
reach higher values. The highly charged species will act
like a polyelectrolyte and associate with the negatively
charged surfaces of the particles. At low concentrations of
aluminum this will result in an electrostatic attraction
like the bridging type discussed in Chapter 7 for
polyelectrolytes.

3. When the pH increases above 7, the CCC rises strongly.
Oligomers and polymers are here progressively more and
more negatively charged and they also decrease in
size. This leads to a transition toward the behavior of a
1:1 electrolyte solution. At not too high pH it is probable
that negatively charged particle surfaces have an
influence on the charge of the bound aluminum
complexes.

4. Following the *BB′* path: between pH 5 and 6 and at
sufficiently high concentration of $Al(NO_3)_3$, zeta-potential
measurements show that the AgI sol becomes positively
charged. This charge reversal results from the adsorption
of multinuclear Al complex and reflects the preferential
adsorption of highly charged complexes rather than
the hydrated metal ions themselves. In Section 7.4 we
discussed a similar recharging of surfaces by adsorption of
a polyelectrolyte on a surface.

5. Following the *CC′* path: by rapidly increasing the pH from
2 to 5.8, it is possible to go from a stable negatively
charged sol to a stable positively charged sol. Increasing

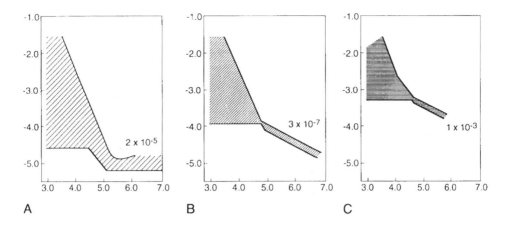

A   B   C

Figure 9.5
The stability of AgBr sol as a function of $Al(ClO_4)_3$ concentration and pH in the presence of various amounts of KF. Hatching denotes unstable regions (A) No KF, (B) 1×10^{-4} M KF, and (C) 1×10^{-3} M KF. (Adapted from E. Matijević, S. Krathvil, and J. Sticher, *J. Phys. Chem.* **75**, 564 (1969).)

the pH to 8 results in the reappearance of stable negatively charged sol.

In many practical systems, the interplay between potential-determining ions, pH, and colloidal stability is even more intricate, but often we can understand this behavior by analyzing the chemistry involved. For example, Figure 9.5 shows how the stability behavior of the AgBr sol depends on pH and the concentrations of $Al(ClO_4)_3$ and KF. In this case, the interactions between Al^{3+} and F^- ions lead to the formation of six complexes, ranging from AlF^{2+} to $Al(F)_6^{3-}$. We can use their equilibrium constants to calculate the chemical potential of Al^{3+} which, in turn, controls the formation of pure oligomeric aluminum complexes. As shown in Figure 9.4 it is these that determine the stability of the AgBr sol. This manifestation of the Schulze—Hardy rule demonstrates once more that the stability of a colloidal sol depends very delicately on the concentration of highly charged counterions.

9.3 Clays Are Colloidal Sols Whose Surface Charge Density Reflects the Chemistry of Their Crystal Structure

The building blocks of clays are two-dimensional arrays either of silicon and aluminum tetrahedrally coordinated with oxygen or of aluminum and magnesium octahedrally coordinated. Such clay crystals are composed of sheets consisting of either three stacked arrays (Figure 9.6) or two arrays (Figure 9.7). The forces between sheets are much weaker than within them. As a consequence the clay minerals easily break up into platelets.

Depending on the relative amounts of Si, Al, Mg, and O they contain, the sheets may be charged or neutral. Most commonly some oxygen atoms coordinate covalently bound hydrogens to achieve an electroneutral system.

In many clays, atom substitution occurs in either the tetrahedral or the octahedral layers. In the tetrahedral sheet,

Three Layers

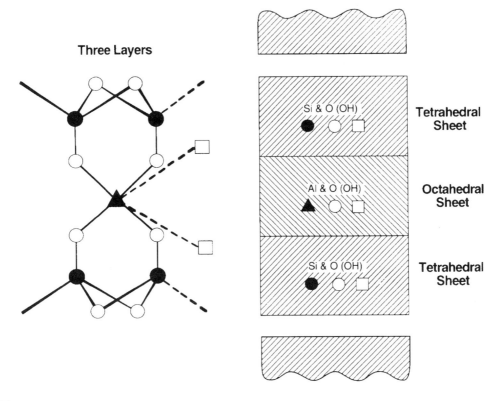

Figure 9.6
Atomic arrangements in
the unit cell of a typical
three-layer clay. The
tetrahedral sheet consists
of Si bonded to four atoms,
two of which connect to
the octahedral sheet
containing aluminum,
oxygen, and hydroxyl
atoms. Clay minerals
consist of stacks of the two-
or three-layered structures.
No covalent bonding exists
between the stacks. Clays
can be neutral as shown or
charged as a result of lattice
substitutions of Al for Si
or Mg for Al. (Adapted
from H. van Olphen,
*Introduction to Clay
Colloid Chemistry*, Wiley-
Interscience, New York,
1963, p. 65.)

trivalent Al sometimes partly replaces tetravalent Si. In the
octahedral sheet, Mg replaces trivalent Al without completely
filling the third vacant octahedral position. When an atom of
lower valency takes the place of one of higher valency, the
lattice acquires a negative charge although the structure does
not change in any other way. This excess negative charge results
in the adsorption of positive ions like Na^+, K^+, or Ca^{2+} that are
too large to be accommodated in the interior of the lattice.
Instead the cations are located between sheets.

Two classes of charged three-layer clays are the montmorillonites, which expand in the presence of water, an the illites,
which do not. We will focus first on the properties of isolated
clay particles and then consider the colloidal interactions between them.

A typical montmorillonite clay has a unit cell formula

$$(Si_8)(Al_{3.33}Mg_{0.67})O_{20}(OH)_4Na_{0.67} \tag{9.3.1}$$

Its formula weight of 734 g means that this amount of clay
contains Avogadro's number of unit cells. From x-ray measurements, we know that the in plane area per unit cell is
$5.15\,Å \times 8.9\,Å$.

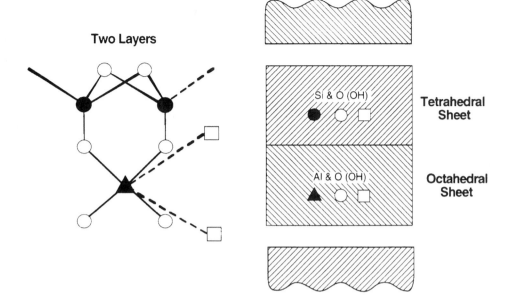

Two Layers

Si & O (OH)

Tetrahedral Sheet

Al & O (OH)

Octahedral Sheet

Figure 9.7
Atomic arrangements in the unit cell of a typical two-layer clay. The tetrahedral sheet consists of Si bonded to four atoms, two of which connects to the octahedral sheet containing aluminum, oxygen, and hydroxyl atoms. (Adapted from H. van Olphen, *Introduction to Clay Colloid Chemistry*, Wiley-Interscience, New York, 1963, p. 64.)

Problem: Calculate the total surface area per kg of the montmorillomite clay if the material is dispersed into individual sheets.

Solution: With a unit cell molecular weight of 734, 1 kg contains 1000/734 unit cells each with an exposed area of $2.5 \times 5.15 \times 8.9 \times 10^{-20}\,\mathrm{m^2}$. The total area is

$$\frac{10^3}{734} \times N_{\mathrm{Av}} \times 2 \times 5.15 \times 8.9 \cdot 10^{-20} = 7.5 \times 10^5\,\mathrm{m^2}$$

The surface area per univalent cation equals

$$\frac{2 \times 5.15\,\text{Å} \times 8.9\,\text{Å}}{0.67} = 136\,\text{Å}^2/\text{univalent cation} \qquad (9.3.2)$$

A charge per every $136\,\text{Å}^2$ corresponds to a charge density of $0.12\,\mathrm{C/m^2}$.

When montmorillonite clays come into contact with water, the water molecules penetrate between the unit layers. This interlayer swelling increases the clay's basal spacing, which can double the volume of the particle. The extent of swelling depends on the type of clay and the cation, but we can observe definite values for the interlayer spacing ranging from $12.5\,\text{Å}$ to $20\,\text{Å}$. More or less stable configurations containing one to four monomolecular layers of water between the unit layers can be obtained.

As a consequence of the clay's swelling, its cations can undergo exchange of counterions. For the montmorillonite clay described in eq. 9.3.1, the cation-exchange capacity (CEC) is 95.1 meq/100 g clay. The original water softeners employed such clays as ion exchangers. A bizarre application of clays as ion exchangers occurred after the Chernobyl accident, when reindeer in northern Scandinavia before slaughter were force-fed clays to reduce the content of radioactive cesium.

In illite (nonswelling) clays such as muscovite mica, lattice substitutions occur mainly in the tetrahedral sheet. The counterions are usually potassium. Because the unit layers do not expand upon addition of water, only the counterions on the external surfaces can be exchanged.

We begin our discussion of the complex colloidal properties of clay minerals by considering the interaction forces between two mica surfaces because this is the best understood system.

9.3.1 Directly Measurable Interaction Forces Between Two Mica Surfaces Provide Insight into the Complexities of Colloidal Systems

Atomic substitutions that lead to the formation of charged clays are locked into the clay's lattice as it takes shape. For each clay mineral, these substitutions lead to a characteristic surface density of ions that can exchange or dissociate. Mica has one charge per every $50\,\text{Å}^2$, which corresponds to a charge density σ of $0.32\,\text{C/m}^2$. Using the Gouy–Chapman equation, we can estimate the mica's surface potential as a function of the electrolyte concentration if all ions dissociate into the solution. For a 1:1 electrolyte, the calculated values of -355, -240, and $-130\,\text{mV}$ at 10^{-5}, 10^{-3}, and $10^{-1}\,\text{M}$ salt, respectively, are much larger than the measured values we will describe.

The surface forces apparatus (SFA) permits us to determine the forces of interaction between two mica sheets as a direct function of distance. Consequently, we can determine surface potentials and effective charge densities as well. Figure 9.8A shows the force curve measured in 1.2 mM HCl and Figure 9.8B shows the effect of adding KOH to the HCl solution to produce a $2.3\,\text{mM}\,\text{K}^+$ concentration at pH 11.1. The dashed lines represent predictions from the DLVO theory using a Hamaker constant for the mica–water system of $2.2 \times 10^{-20}\,\text{J}$. The lower theoretical line reflects the assumption that the surfaces remain at constant potential as they approach each other, and the upper line projects constant charge density. We see that at any range ($d > 10\,\text{nm}$) there is an excellent agreement between calculated and observed force with the surface potential as the only fitting parameter. In particular the slope of the force curve in the log-linear plot matches the Debye screening length calculated on the basis of the known composition of the electrolyte concentration.

There are two basic differences between the force curves in Figures 9.8A and B, the fitted surface potentials are not the same and there is a qualitative discrepancy in the behavior at short range.

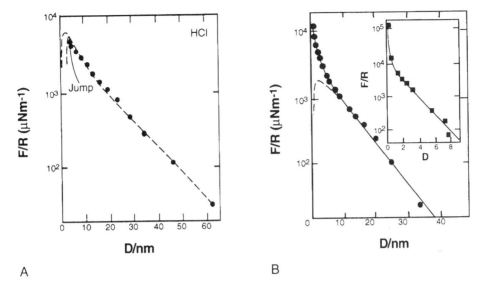

Figure 9.8
(A) Measured forces between mica sheets in 1.2 mM HCl; dashed line shows fit to DLVO theory using $\Phi_0 = 130$ mV. (B) Measured forces in a solution containing 1.2 mM HCl and 3.5 mM KOH ($\Phi_0 = 67$ mV). Inset: repulsive force at short range. (R.M. Pashley, *J. Colloid Interface Sci.* **80**, 153 (1981).)

We can understand the differences in these force curves from the following analysis. With HCl the fitted surface charge density is 1.8×10^{-2} C/m², while in water a value of 2.3×10^{-3} C/m² is measured. These charge densities correspond to one charge per 7000 Å² in water and 900 Å² in 1.2×10^{-3} M HCl. These values are considerably lower than the charge density of one charge per 50 Å² associated with charged sites on mica. This discrepancy emphasizes that the complex adsorption processes discussed in Section 3.7 occur at the surface and influence the interaction between colloidal particles.

What is the dominating cation on the surface? In CO_2-saturated water at pH 6.5, the concentration of H^+ is 3.2×10^{-6} M. The mica surfaces are about 2 cm² and the volume of the SFA is 30 mL. Therefore, if all the potassium ions are exchanged off the mica, the resulting concentration of K^+ in the SFA's aqueous solution is $\approx 10^{-10}$ M. Thus, protons are the dominant cation in the solution, and we also could expect them to dominate at the surface, where they could associate covalently with an oxygen atom. When HCl is added to water, the proton chemical potential is increased and so is κ, facilitating the ionization of the surface (see Section 3.7).

With KOH the surface potential required to fit the curve corresponds to a relatively low surface charge density of 0.012 C/m². A new feature is that we also can detect a strong short range repulsive force at small distances of separation between the two mica surfaces, which represents a significant deviation from the DLVO theory as discussed in Section 5.5.

Force curves for different alkali metal counterions show a specificity with respect to the onset of the departure from DLVO behavior. For example, the force curves for LiCl solutions of varying concentration shown in Figure 9.9 reveal that we observe only DLVO-type behavior when the alkali metal ion concentration is low. However, with increasing salt concentra-

Figure 9.9
Forces measured between
mica surfaces in LiCl
solutions of 1×10^{-4} M,
1×10^{-3} M, 1×10^{-2} M,
and 6×10^{-2} M. Inset: a
short range repulsive force
is observed first at a
concentration of
6×10^{-2} M LiCl. (R.M.
Pashley, *J. Colloid Interface
Sci.* **83**, 531 (1981).)

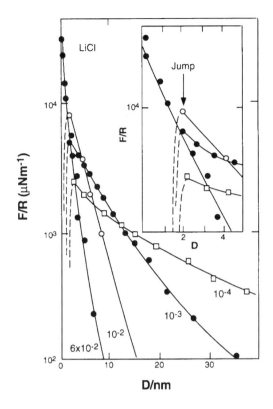

tion the observed departures depend on the pH of the solution.
For example, at pH 5.7, the presence of a short range repulsion
becomes evident above concentrations of $K^+ > 4 \times 10^{-5}$, $Cs^+ >$
10^{-3}, $Na^+ > 10^{-2}$, and $Li^+ > 6 \times 10^{-2}$ M. This observation suggests that competitive adsorption of hydrogen ion and alkali
metal ions on the mica surface controls the interaction force at
short range.

We can take competitive adsorption into account by analyzing the association of the alkali metal ions M^+ and the
protons H^+ with surface groups S^-, as discussed in Section 3.7,
and calculating the surface concentrations of the alkali metal
ions and protons on mica as a function of concentration at
pH 5.6 once the surface potentials and the equilibrium constants are known. Figure 9.10 shows the results for Li^+, Na^+,
K^+, and Cs^+ counterions. The surface binding constants follow
the sequence $K_K \approx K_{Cs} > K_{Na} > K_{Li}$.

These ideas suggest that at concentrations permitting both
M^+ and H^+ to be adsorbed on the surface, ion exchange can
perform an important role in controlling the interaction forces
as the mica surfaces move together. Analyzing the mica system
illustrates how chemical events at a uniform smooth surface
influence the interaction between surfaces. Most systems of
practical importance possess less homogeneous surfaces and are
chemically as complex as the mica surface. As a result we
should expect a rich behavior, particularly at small separations

Figure 9.10
Estimates of surface-adsorbed H^+, Li^+, Na^+, K^+, and Cs^+ ion densities on mica based on force measurements. Shaded region shows concentrations in which a short range repulsion prevents adhesive contact. (R.M. Pashley, *J. Colloid Interface Sci.* **80**, 153 (1981).)

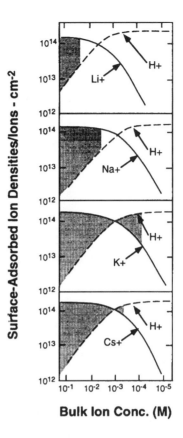

between surfaces, where the properties may exhibit a critical dependence on the concentration of different components in the bulk solution.

9.3.2 Coagulated Structures Complicate the Colloidal Stability of Clay Sols

When montmorillonite clays come into contact with water, they can either hydrate to take up one to four water monolayers or swell extensively. The latter behavior is typical of the chemically impure clays in ordinary soil, which can swell to many times their original volume. This continued swelling results from the osmotic pressure created by the high concentration of ions in the double-layer regions between the colloidal clay particles and the low concentrations of ions in the bulk solution as discussed in Chapter 5.

The behavior of clays is more complex than we have implied so far. Clays contain surfaces of two types: those associated with platelike surfaces, which we have already discussed, and those associated with edges. At the edges of the plates, the tetrahedral silica and the octahedral alumina sheets are disrupted and the primary bonds are broken. Clearly, such edges display very different surface properties. Figure 9.11 shows an electron micrograph obtained from a kaolinite clay

Figure 9.11
Transmission electron microscope image of clay particles mixed with a negatively charged gold sol showing that the gold particles adsorb only onto the positively charged edges of the clay particles, not onto the negatively charged surfaces (P.A. Thiessen, *Z. Anorg. Chem.* **253**, 162 (1947).)

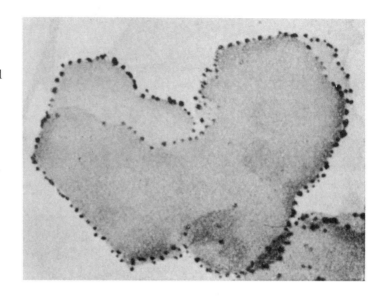

mixed with a negative gold sol. The gold particles are located exclusively on the edges, which suggests that the edges are positively charged.

We can understand the chemical basis for this affinity in terms of the behavior of the amphoteric properties of aluminum ions. Because clays are slightly soluble, aluminum ions present in solution can be adsorbed onto the edges. As we have seen, protons and hydroxide ions serve as potential-determining ions for aluminum ions. Therefore, in acidic solution aluminum ions will adsorb onto the edges of the clay particles, rendering them positively charged. In basic solution, on the other hand, the edges remain uncharged. Thus, although a clay particle's net charge is always negative, it can possess both negatively charged platelike surfaces and positively charged edges.

Positively and negatively charged regions and their associated double layers have a pronounced effect on the flocculation properties of clays. In addition to the fact-to-face (FF) association described in the SFA studies, clay particles can associate to form edge-to-face (EF) and edge-to-edge (EE) structures like those shown in Figure 9.12. Because the double-layer interactions differ for the three structures, the potential energy curves of interaction for the three modes of association also differ. Consequently, the type of aggregate that forms can change in response to variations in the solution's properties.

FF association leads to thicker and sometimes larger aggregates, while EF association leads to voluminous three-dimensional "house of cards" structures. Although these structures interconvert if we add or change the clay concentration, the interplay between attractive and repulsive forces is subtle. For example, the viscosity of a montmorillonite clay suspension initially increases upon addition of salt. We can interpret this behavior in terms of the EF structures that form in salt-free solution and disengage when small amounts of electrolyte are

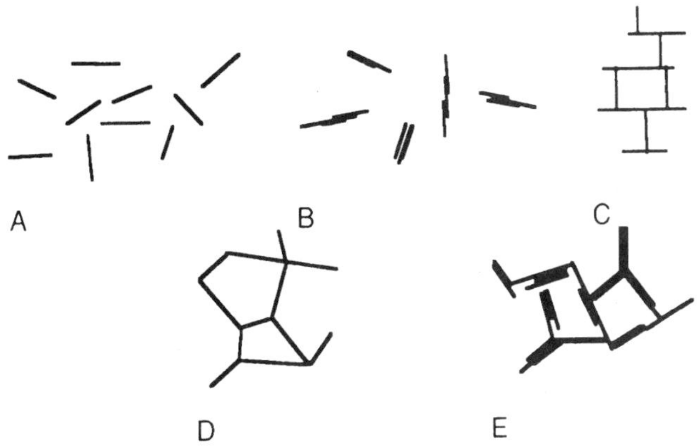

Figure 9.12
Modes of particle association in clay suspensions: (A) dispersed state, (B) face-to-face association, (C) edge-to-face association, (D) edge-to-edge association, and (E) combinations of structures. Particle concentration and presence of various electrolytes control the type of association and lead to very different mechanical properties. (Adapted from H. van Olphen, *Introduction to Clay Colloid Chemistry*, Wiley-Interscience, New York, 1963, p. 94.)

added. Adding more electrolyte results in the reestablishment of aggregates and conversion to FF structures.

Highly concentrated clay suspensions with low viscosities are desirable in many practical applications, such as drilling muds. We can achieve such properties by adding a peptization agent to reverse the charge on the edges of the clay particles. For example, polymetaphosphates chemisorb onto the clay's edges, forming complexes with the aluminum ions. By causing the charge on the edges to become negative, small amounts of such material can transform rigid gels into freely flowing solutions.

9.4 Monodisperse Latex Spheres Can Model Various States of Matter as Well as the Phase Transformations Between Them

Latex spheres find widespread industrial applications. For example, they are found in water-based paints, adhesives, and fabric coatings, and they form a major constituent in automobile tires.

We also can use these polymer colloids to simulate many features of molecular systems. By varying sol and salt concentrations and temperature, we can make latex systems behave in a manner analogous to the liquid, crystalline, and glassy states of matter. Because colloidal particles possess high mass compared to solvent molecules, various processes (such as phase changes) are slowed down to the point where we can follow them in real time.

Three conditions must be met before such studies can be done. First, the latex spheres must be homogeneous in size. If they are not, phase transformations will be badly defined and the crystalline region inaccessible. Emulsion polymerization provides the means to achieve this goal. Second, the level of salt in the system must be reduced to a very low value, preferably almost to zero. We can achieve this by performing dialysis and passing the latex system through a highly purified mixed-bed ion-exchange resin. Third, we must employ experimental techniques that permit us to determine interparticle structure in concentrated colloidal dispersions. Neutrons, which have a short wavelength of ~ 10 Å and are scattered by atomic nuclei, are particularly useful for this purpose. Neutron scattering has the advantage of minimizing the multiple scattering effects that create acute problems when we try to measure concentrated latex dispersions with light.

Our discussion of latex spheres focuses first on charged latex systems, in which interactions are long-ranged and dominated by electrostatics, and then on sterically stabilized systems, which behave more like hard-sphere systems. We can demonstrate how to use the scattering equations given in Chapter 7 to obtain information about the structure of latex colloidal polymer solutions. As we already have seen (Section 7.1.6), we can write the sample scattering function as a product

$$I(\vec{Q}) = P(\vec{Q})S(\vec{Q}) \tag{9.4.1}$$

where $P(\vec{Q})$ represents the scattering of individual particles and $S(\vec{Q})$ the structure factor.

For an ensemble of noninteracting spherical particles irradiated with neutrons, the scattering intensity is

$$\text{intensity} = Ac_p^* V_p^2(\rho_p - \rho_m)^2 P(\vec{Q}) \tag{9.4.2}$$

in which A is the instrument calibration factor, V_p equals the volume of each spherical particle ($V_p = 4\pi R^3/3$), and c_p^* is the number of particles per unit volume. Coherent neutron scattering densities are represented by ρ_p for the particle and ρ_m for the dispersion medium.

The scattering densities ρ_i can be modified by isotope substitution: in particular, by exchanging protons for deuterium molecules in both the medium and the particle. This modification is called contrast variation. For isotropic samples and unpolarized neutrons only the magnitude Q of the scattering vector $\vec{Q}$ matters.

9.4.1 Long Range Electrostatic Repulsions Dominate the Solution Behavior of Ionic Latex Spheres

In dilute solutions of ionic latex spheres with an appropriate concentration of a screening electrolyte, the interaction between particles is small. Consequently, we can ignore the interparticle interference, and under these conditions, $S(Q)$ is constant.

Figure 9.13a plots $I(Q)$ versus Q for a latex system with a volume fraction equal to 0.011 and 1.3×10^{-2} M NaCl. Using eqs. 9.4.2 and 7.1.29 for the form factor and introducing a size polydispersity produces a fit to the experimental data as shown in the solid line. The result, $R = 157$ Å, agrees with values determined by electron microscopy and light scattering. This

Figure 9.13
Neutron scattering intensity $I(\vec{Q})$ versus scattering vector $\vec{Q}$, for a polystyrene latex. (a) In 1.3×10^{-2} N NaCl, volume fraction = 0.011. Under these conditions particle interactions are so small that they can be ignored and $S(\vec{Q})$ is constant. The data can be fitted using eqs. 9.4.2, and 7.1.25 (solid line) to give a particle radius $R = 157$ Å, in agreement with independent measurements. (b) In 10^{-4} M NaCl, volume fraction = 0.13. The change in shape of the curves in (a) and (b) illustrates the effect of particle interactions on $I(\vec{Q})$. (R.H. Ottewill, *Langmuir* **5**, 4 (1989).)

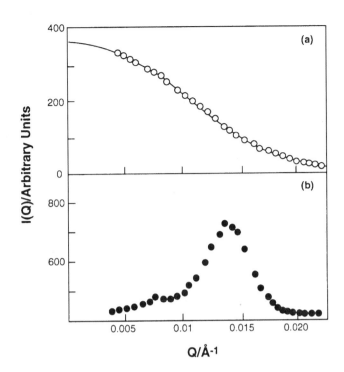

quantitative agreement between theory and data confirms the spherical shape of the particles.

If the concentration of latex increases or the salt concentration decreases, particle interactions become important. They cause increased correlations between particles and a substantial change in the $I(Q)$ curves, as shown in Figure 9.13b for 10^{-4} M NaCl and particle volume fraction, $\Phi = 0.13$.

A more convenient way to follow particle–particle interactions with changing solution conditions is to plot $S(Q)$ versus Q as shown in Figure 9.14. Since we can obtain the value of $P(Q)$ from measurements in dilute solution (cf. Figure 9.13), we can use eq. 9.4.1 to calculate $S(Q)$.

Changes in the shape and position of the $S(Q)$ curves with volume fraction and salt follow definable trends. At constant salt but increasing Φ, a peak develops in the $S(Q)$ curve and its maximum moves toward larger Q, which is consistent with a decrease in the interparticle spacing. For small values of Q, the $S(Q)$ decreases with increasing Φ.

At constant Φ but increasing salt concentration, the $S(Q)$ curve becomes broader, which is consistent with a decrease in the interparticle repulsive interactions and a correspondingly smaller correlation between the particles. The increase of $S(Q)$ as Q goes to zero signals a decrease in the osmotic compressibility of the system as its ionic strength increases.

We can make these qualitative observations more quantitative by relating the particle–particle interaction to the pair distribution function $g(r)$. For particles with a small charge $Z_p e$, we can use the Debye–Hückel theory (eq. 3.8.25) to calculate

Figure 9.14
Plots of the structure factor $S(Q)$, versus Q showing how interparticle interactions change as a function of latex volume fraction and salt concentration. (a) 10^{-3} M NaCl at various latex volume fractions: ○, 0.01; ●, 0.04; ▲, 0.08; □, 0.13. (b) Volume fraction = 0.04 at various NaCl concentrations (M): □, ion exchanged; ▲, 10^{-4}; ●, 10^{-3}; ○, 5×10^{-3}. The $S(\vec{Q})$ data points shown in (a) and (b) are obtained by using $I(\vec{Q})$ of the particles from Figure 9.13 and eq. 9.4.1. (R.H. Ottewill, *Langmuir* **5**, 4 (1989).)

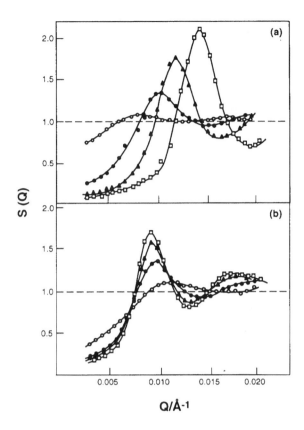

the interaction at long range and

$$V(r) \simeq \Phi_{\text{DH}}(r)Z_p e = \frac{(Z_p e)^2}{4\pi\varepsilon_r\varepsilon_o} \frac{\exp[-\kappa(r-R)]}{r(1+\kappa R)}$$

$$= 4\pi\varepsilon_r\varepsilon_o R^2(1+\kappa R)\Phi_s^2 \exp(\kappa R)\exp(-\kappa r)/r$$

$$\simeq 4\pi\varepsilon_r\varepsilon_o R^2\Phi_s^2 \exp(2\kappa R)\exp(-\kappa r)/r \qquad (9.4.3)$$

where the third equality introduces the surface potential Φ_s through evaluating eq. 3.8.25 at $r = R$, while the last equality is a numerical approximation, $(1+\kappa R) \simeq \exp(\kappa R)$, valid for $\kappa R \leqslant 1$. It turns out that even for $\kappa R > 1$, this approximation partly accounts for the deficiency in the Debye–Hückel potential arising from the large size of the particles.

For a well-defined monodisperse polystyrene latex, we know all the parameters in eq. 9.4.3 except Φ_s. Adding the condition that $V(r)$ is infinite for $r < 2R$ makes it possible to calculate the pair distribution function $g(r)$, which is related to the observed structure factor $S(Q)$ by a Fourier transformation. Figure 9.15 compares calculated curves for $S(Q)$ to data points. The calculation is based on a procedure called the mean spherical approximation (MSA), which is essentially the Debye—Hückel theory corrected for the finite size of the particles. The fitted values of Φ_s agree reasonably well with zeta potentials

Figure 9.15
Fit of $S(Q)$ versus Q for a polystyrene latex at a volume fraction of 0.04 and various salt concentrations c_s, using eq. 9.4.3 to fit the electrostatic surface potential Φ_0: (a) $c_s = 5 \times 10^{-3}$, $|\Phi_0| = 47\,\mathrm{mV}$; (b) $c_s = 10^{-3}\,\mathrm{M}$, $|\Phi_0| = 49\,\mathrm{mV}$; and (c) $c_s = 10^{-4}\,\mathrm{M}$, $|\Phi_s| = 51\,\mathrm{mV}$. The surface potentials are consistent with zeta-potential measurements. (R.H. Ottewill, *Langmuir* **5**, 4 (1989).)

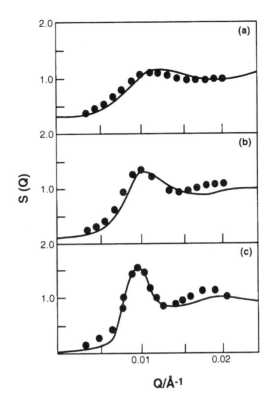

obtained on the same latex sample, but at much lower volume fractions. These results substantiate the idea that the diffuse double layer controls particle–particle interactions and that eq. 9.4.3 satisfactorily represents the electrostatic interactions between particles at the distance involved in these studies.

We can follow the transformation from gaslike-independent behavior to solidlike-ordered behavior by expressing $S(Q)$ in terms of the pair distribution function $g(r)$. This process requires very good data because we must perform a Fourier transformation of eq. 7.1.28 to yield

$$g(r) = 1 + \frac{2R^3}{2\pi^2 r \Phi} \int_0^\infty [S(Q) - 1] Q \sin Qr \, dQ \qquad (9.4.4)$$

Figure 9.16 shows the curves of $g(r)$ versus r for a polystyrene latex ($D = 310\,\text{Å}$ for $10^{-4}\,\mathrm{M}$ NaCl) at two volume fractions. The curve for $\Phi = 0.01$ indicates that the particles scarcely interact at distances greater than $1000\,\text{Å}$. However, the interactions become more repulsive with decreasing distance and show an excluded volume effect. At $\Phi = 0.13$, the $g(r)$ curve displays liquidlike properties with short range order but long range disorder.

Solidlike behavior has been observed with a latex of $0.1\,\mu\mathrm{m}$ radius and a volume fraction of 0.08. In this experiment, the latex was passed through a mixed-bed ion exchange to

Figure 9.16
Plots of the pair correlation function $g(r)$ obtained from eq. 9.4.4, versus r for polystyrene latex particles in 10^{-4} M NaCl at two different volume fractions: 0.01 (solid line, which shows gaslike correlation) and 0.13 (dashed line, which shows liquidlike correlation). (R.H. Ottewill, *Langmuir* **5**, 4 (1989).)

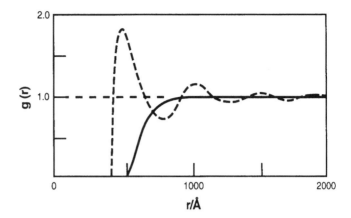

remove any spurious electrolyte. Figure 9.17 indicates Bragg peaks in the sample's intensity pattern. As the spot pattern index shows, particles formed a face-centered-cubic crystalline array with the 111 plane oriented against the face of the cell. The unit cell dimension in this latex is 6400 Å.

9.4.2 Sterically Stabilized Latex Spheres Show Only Short Range Interactions and Form Structured Solutions Only at Higher Concentrations

Figure 9.17
Three-dimensional plot of intensity for ion-exchanged latex with a radius of 0.1 μm at a volume fraction of 0.08. Analysis of the Bragg diffraction peaks shows a cubic crystalline array. (R.H. Ottewill, *Langmuir* **5**, 4 (1989).)

We also can use scattering techniques to study interaction of sterically stabilized latex particles. Figure 9.18 shows an example of such a latex particle. It consists of a spherical core of polymethyl methacrylate (PMMA) surrounded by a shell of poly(12-hydroxystearic acid) (PHS), grafted to the PMMA core covalently.

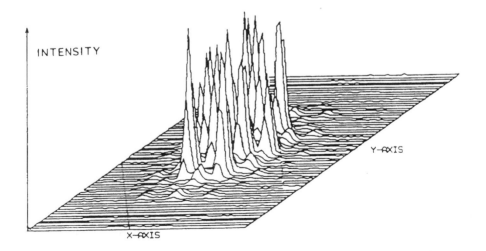

Figure 9.18
Schematic of a latex with a
polymethyl methacrylate
(PMMA) core and shell
composed of a grafted layer
of poly(12-hydroxystearic
acid) (PHS). (R.H. Ottewill,
Langmuir **5**, 4 (1989).)

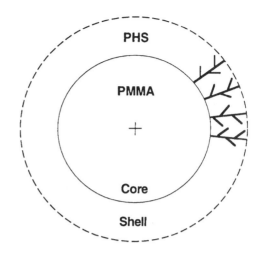

In such sterically stabilized systems, interaction occurs only when the stabilizing shells belonging to two particles approach closely or touch. Thus, we should expect to find a potential energy function like that shown in Figure 9.19, in which the energy of repulsion rises sharply to infinity on close approach.

Figure 9.20 gives an example that plots $g(r)$ versus r as a function of volume fraction for PMMA–PHS lattices with a core of 310 Å. Comparison with similar plots for ionic lattices shows that nonionic systems require much higher volume fractions to produce a significant structure.

These sterically stabilized latex systems also lend themselves to using photon correlation spectroscopy to measure particle mobility. The basic idea is to create a system in which the latex particles match the solvent so that they have the same refractive index. We then add a tracer particle that is identical

Figure 9.19
Diagrams of potential
energy against distance r
for (a) hard-sphere
interactions and (b) soft-
sphere interactions.
Dashed line shows DLVO-
type interaction without
protecting shell. (R.H.
Ottewill, *Langmuir* **5**, 4
(1989).)

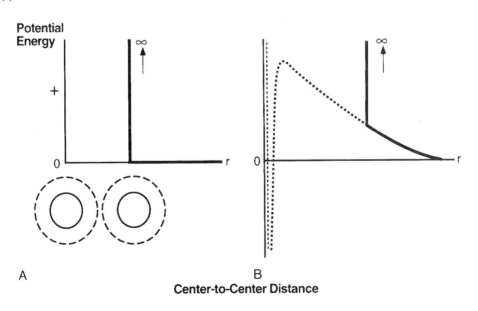

Figure 9.20
Plots of $g(r)$ versus r for a PMMA–PHS particle (core diameter = 310 Å) in dodecane at various volume fractions: —, 0.23; —·—, 0.28; ···, 0.36; and ---, 0.42. Comparison with Figure 9.16 for an ionic latex shows that much higher volume fractions are required for nonionic latex to produce significant pairwise interactions. (R.H. Ottewill, *Langmuir* **5**, 4 (1989).)

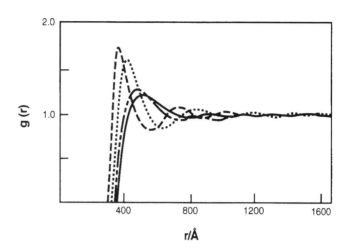

in size but has a different refractive index. Under these circumstances, we can follow the tracer particle motion in a lattice provided by the other particles.

For example, PMMA–PHS has a refractive index of 1.471, which we can match optically with a mixture of *cis*-decalin ($n = 1.481$) and *trans*-decalin ($n = 1.469$). We can use a polyvinyl acetate (PVAc) core stabilized by PHS and obtain particles with a refractive index of 1.483.

The hydrodynamic radius of the tracer PVAc–PHS is 83 ± 2 nm and that of the PMMA–PHS is 85 ± 2 nm. Because we use the same stabilizing moieties and the particle sizes are very similar, the repulsive steric interaction should be almost identical for these systems.

Photon correlation spectroscopy can determine the diffusion coefficient of the tracer particles in the latex system. We can make these measurements over a wide Q range, and by varying the correlation delay time, we can obtain diffusion coefficients that sample particle interactions on varying time scales. If the measurement times are long compared to the time required for the particle to move several particle diameters, we sample the colloidal and hydrodynamic interaction of the tracer with other particles. Alternatively, if the time scale is sufficiently short, we can measure a self-diffusion coefficient that corresponds to the particle moving only a small fraction of the particle diameter and thus obtain information on the interaction of the tracer with particles that constitute a solvent cage.

Figure 9.21 shows the diffusion coefficient of the PVAc–PHS tracer as a function of the PMMA–PHS volume fraction. It is expressed in normalized form—that is, divided by the free diffusion coefficient of the particle. The long-time diffusion coefficient decreases with volume fraction and tends to a very small value as the volume fraction approaches 0.5.

For hard-sphere systems, a liquid–solid coexistence is found in the range $0.49 \lesssim \Phi \lesssim 0.54$. (See Chapter 10 for a discussion of the phenomenon.) For nearly hard spheres of equal

Figure 9.21
Ratio of observed self-diffusion coefficient D to the diffusion coefficient at infinite dilution D_0 versus volume fraction of particles: ●, measurements yield interaction of nearest neighbors, and △, measurements yield a bulk diffusion coefficient (R.H. Ottewill, *Langmuir* **5**, 4 (1989).)

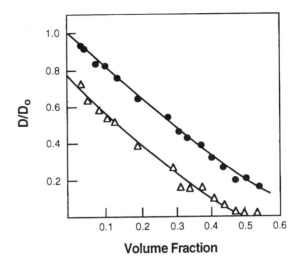

size, the long-time D, which essentially represents a movement through the lattice, appears to cease at the freezing transition. In contrast, the short-time D is still finite at $\Phi = 0.5$, showing that some particles continue to move within their cage of surrounding particles.

These results also provide insight into questions regarding the formation of crystalline or glassy states. If the particles at $\Phi = 0.5$ slowly sediment, permitting diffusion processes to continue, crystals can form in the lower part of a container. When $\Phi > 0.5$, a glassy state forms, presumably as a consequence of the inhibition of diffusion, and a disordered state follows. Shearing such a sample induces crystallization.

If we graft a sterically stabilized latex with polar polymer chains, such as polyethylene glycol, to a polystyrene core, we can obtain aqueous systems that display properties like those observed in nonaqueous solvent. In general, this means that we can form highly organized, high density systems by using particles with short range repulsive barriers and that crystallization or glass formation will occur in solutions of monodisperse particles with a volume fraction above 0.5.

9.5 Homocoagulation and Heterocoagulation Occur Simultaneously in Many Colloidal Systems

Many colloidal systems contain particles with different sizes, shapes, Hamaker constants, and surface potentials. Practical examples include paint formulation, water purification, filtration, flotation, and corrosion. In these systems particles of different types can associate, leading to so-called heterocoagulation rather than homocoagulation. However, heterocoagulation is not limited to colloidal sols. This effect also occurs in microflotation, which involves particle–bubble interactions, and adhesion, in which particles attach to planar surfaces. This area of study is complex, because we simultaneously must consider both homocoagulation between like particles and heterocoagulation between unlike particles. It is clearly a more demanding task to design a system of several types of colloidal particle that are nevertheless stable. When should we expect heterocoagulation to be a significant process? Within the framework of the DLVO theory, stability is determined by a balance between attractive van der Waals and repulsive double-layer forces. The van der Waals force always gives a preference for like associating with like (see Chapter 5), while the repulsive electrostatic force is always strongest when likes interact.

If the different types of particle have opposite charge, both the van der Waals and the electrostatic forces are attractive at long range and flocculation or coagulation should occur. Note that the electrostatic force can turn repulsive at short range as discussed in Section 5.1.5.

When all types of particle have the same sign of the charge (although with different magnitude), a delicate situation can arise when the difference in the attractive force between different pairs of particles is nearly balanced by a similar difference in the repulsive force. The three parameters that mainly govern this balance are the surface potentials, the Hamaker constants, and the particle sizes.

When the particles with the smallest magnitude of the surface potential also have the highest Hamaker constant and the largest size, homocoagulation of these particles will be the first event when the sol concentration is increased gradually. In other cases the expected outcome is determined from a detailed analysis of the forces. As an illustration of the phenomenon, we focus on how to characterize stability behavior of three nearly monodisperse colloidal sols and describe experiments that examine these mixtures.

We consider three sols:

An aluminum hydrous oxide sol prepared by aging a 5.0×10^{-4} M solution of $Al_2(SO_4)_3$ at 125°C for 20 hours. Light scattering measurements determined a particle diameter of 0.57 μm.

A cationic polystyrene latex containing 2% aminoethyl methacrylate–HCl synthesized by emulsion polymerization, with a particle diameter of 0.38 μm.

An anionic polystyrene latex synthesized without an emulsifier using $K_2S_2O_8$, with a particle diameter of 0.38 μm.

Electrophoretic mobilities, measured as a function of pH in a electrophoresis cell, establish the point of zero charge. Inspection of Figure 9.22 shows that mobility is pH dependent for all three sols. Corresponding zeta and surface potentials change sign at the point of zero charge.

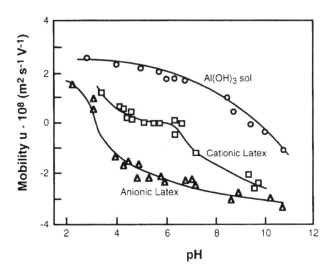

Figure 9.22
Electrokinetic mobilities μ ($m^2 s^{-1} V^{-1}$) as a function of pH for aluminum hydrous oxide, cationic polystyrene (PS) latex, and anionic PS latex sols at a constant ionic strength (0.01 M KNO_3). (H. Sasaki, E. Matijević, and E. Barouch, *J. Colloid Interface Sci.* **76**, 319 (1980).)

Changes in light scattering intensity with time, measured with a differential light scattering photometer, determined the rate of coagulation. These measurements were made at a scattering angle of 5° to minimize multiple scattering effect. The coagulation rate constant k was determined from the expression

$$\frac{I(t)}{I(0)} = 1 + 2kc_o t \qquad (9.5.1)$$

in which $I(t)$ and $I(0)$ give the scattering intensities at time t and at $t = 0$, and c_o represents the initial particle concentration. For all three systems, the quantity $I(t)/I(0) - 1$ changed linearly with time, and the scattering intensity was directly proportional to the particle number concentration over the range studied.

Rate data were obtained by dividing the slopes of the intensity–time lines at different conditions with the same slopes when the systems undergo rapid coagulation determining the stability ratio $W = k_r/k_s$ (eq. 8.3.27). Figure 9.23 gives values for W for homocoagulation of the anionic polystyrene latex at pH 3.5. The critical coagulation concentration strongly depends on the counterion valency in accordance with the Schulze–Hardy rule.

Since the zeta potential (and by implication the surface potential) of all three systems reflects a strong pH dependence, the stability ratio W for homocoagulation should also vary strongly with pH. Figure 9.24, which summarizes values of W for the three systems as a function of pH at fixed electrolyte concentration, illustrates this point. The curves are drawn so that $\log W$ is taken as zero at the isoelectric points. Rapid coagulation takes place close to the point of zero charge, as expected.

Rate constants calculated using eq. 9.5.1 on the basis of scattering intensity–time curves displaying the maximum slope give values for the three systems of 3.6×10^{-18} (aluminum hydroxide), 3.4×10^{-18} (anionic PS latex), and $3.3 \times 10^{-18} m^1/s$

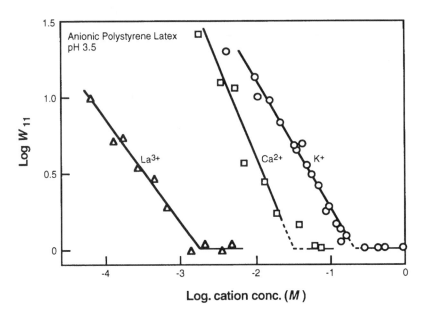

Figure 9.23
Stability ratio W_{11} of the anionic PS latex at pH 3.5 in the presence of various concentrations of $La(NO_3)_3$, $Ca(NO_3)_2$, and KNO_3. Total particle number concentration: 3×10^{14} per cubic meter. (H. Sasaki, E. Matijević, and E. Barouch, *J. Colloid Interface Sci.* **76**, 319 (1980).)

Figure 9.24
Stability ratios W_{11} for the aluminum hydrous oxide sol, the anionic PS latex, and cationic PS latex as a function of pH at a constant ionic strength (0.01 M KNO_3). (H. Sasaki, E. Matijević, and E. Barouch, *J. Colloid Interface Sci.* **76**, 319 (1980).)

hydroxide), 3.4×10^{-18} (anionic PS latex), and $3.3 \times 10^{-18} \, m^1/s$ (cationic PS latex). These rate constants represent 60% of the value for a diffusion-controlled reaction in water (see eq. 8.3.11).

For the mixed aluminum hydroxide sol–anionic polystyrene latex, the measured total stability ratio W_T is shown in

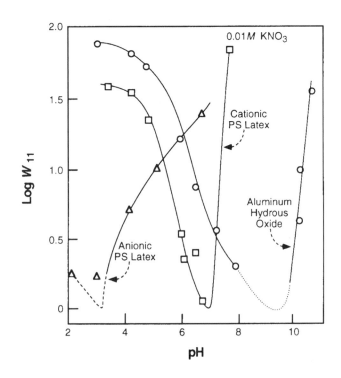

Figure 9.25
Overall stability ratio W_T for the mixture of anionic PS latex (particle 1) and aluminum hydroxide (particle 2) as a function of pH at a constant ionic strength (0.010 M KNO$_3$). Particle number fraction of anionic PS latex: 0.75 (●), 0.50 (□), 0.25 (○). Total particle number concentration: 1.12×10^{14} per cubic centimeter. (H. Sasaki, E. Matijević, and E. Barouch, *J. Colloid Interface Sci.* **76**, 319 (1980).)

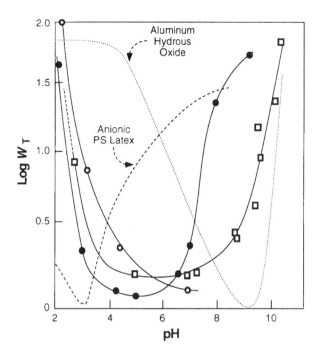

Figure 9.25. A high coagulation rate occurs in the whole pH range between the respective points of zero charge for the types of particle.

In an elementary analysis of these curves, we first assume the homocoagulation process occurs with the same rate as in a pure system of the same concentration. Similarly, we assume that heterocoagulation occurs independently of homocoagulation.

The total rate of the coagulation v_T is built up from three independent coagulation rates; the two homocoagulation rates v_{11} and v_{22}, and the heterocoagulation, v_{12},

$$v_T = v_{11} + v_{22} + v_{12} \tag{9.5.2}$$

where each process has its characteristic rate at the given pH. The rate constant for the total process, k_S, represents the weighted average over the individual coagulation events. We have

$$v_{ij} = k_{ij}c_i c_j \tag{9.5.3}$$

and

$$
\begin{aligned}
v_T &= k_{11}c_1^2 + k_{22}c_2^2 + k_{12}c_1 c_2 \\
&= (k_{11}p_1^2 + k_{22}p_2^2 + k_{12}p_1 p_2)c_{\text{tot}}^2
\end{aligned}
\tag{9.5.4}
$$

where c_i is the concentration of component i and $p_i = c_i/c_{\text{tot}}$ its fraction. For two components $p_2 = 1 - p_1$. Thus, the total observed rate constant is

$$k_S = p_1^2 k_{11} + p_2^2 k_{22} + p_1 p_2 k_{12} \qquad (9.5.5)$$

The rate constants k_{11} and k_{22} for the pure system were measured in Figure 9.24, and k_{12} is the only unknown. We can make the simple assumption that heterocoagulation occurs as a diffusion-limited aggregation if the particles have opposite charge and that otherwise the rate is negligibly small. Then

$$k_{12} = 2k_r \qquad (9.5.6)$$

because the diffusion-controlled rate constant is twice that for a heterocoagulation relative to a homocoagulation (see discussion of eq. 8.3.11).

Combining eqs. 8.3.27, 9.5.5, and 9.5.6, the stability ratio W_T is

$$W_T = (p_1^2 W_1^{-1} + p_2^2 W_2^{-1} + 2p_1 p_2)^{-1} \qquad (9.5.7)$$

in the pH range between the minima of the stability ratios, W_1 and W_2, for the respective pure systems. Figure 9.26 shows the calculated W_T for the mixed anionic polystyrene latex–aluminum hydrous oxide system. The main features of the experimental curves for $\log W_T$ versus pH in Figure 9.25 are reproduced. For example, we find an almost constant value of $W_T = 1.8$ in the pH range of 4–8 both for the experiments and in the calculation. Furthermore, when one type of particle is in

Figure 9.26
Calculated stability ratio W_T using eq. 9.5.7 for the aluminum hydroxide–anionic polystyrene system of Figure 9.25.

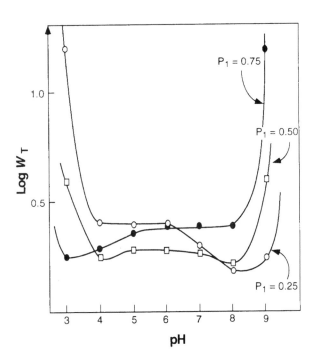

excess, the stability ratio has its minimum closer to the pH of zero charge for the more abundant particle.

The agreement between experiments and the predictions of eq. 9.5.7 shows that heterocoagulation is the dominant process when particles carry opposite sign, while it occurs to a lesser degree for particles of similar charge.

9.6 Aerosols Involve Particles in the Gas Phase

9.6.1 Some Aerosols Occur Naturally, But Many Others Are Produced in Technical Processes

Aerosols are systems in which colloidal solid particles or liquid drops are dispersed in a gaseous medium, typically air. They are used in a number of technical processes, ranging from the synthesis of sulfuric acid to aerosol coating processes and spray drying. Both liquid and solid aerosols also are used in nose sprays and to administer asthma drugs. However, the most important aerosols occur naturally or are inadvertently created.

Clouds are magnificiently shaped collections of small water droplets. Trees, grasses, and other plants cross-fertilize through airborne pollen particles. An intensely red evening sky can owe its color to volcanic ash particles generated on the other side of the globe. The Smoky Mountains get their name from aerosols formed by organic compounds emitted by coniferous trees growing on the slopes. During the great fires in Yellowstone National Park in the summer of 1988, afternoon skies were redish-gray to viewers more than a thousand kilometers from the scene.

These days we usually connect aerosols with air pollution problems. Urban smog is a complex mixture of particles containing water and inorganic and organic compounds. Through the action of sunlight and some highly reactive pollutants, the chemical composition of these particles undergoes a continuing change. Recently it has become clear that in a delicate chemical way, aerosols are involved in another major environmental hazard. The depletion of the ozone layer over the polar caps during the dark period of the year is catalyzed not only by Freons used (among other things) to create aerosols for coating processes, but by a chain of reactions that also depends on the presence of ice particles contaminated by inorganic compounds.

Drugs delivered in the form of aerosols help asthma patients and other allergy sufferers, but in fact, aerosols in the form of pollens, mold spores, and other particles also cause the ailment. Some infections, such as the common cold, are transmitted by viruses carried in the form of aersols.

9.6.2 Aerosol Properties Differ Quantitatively from Those of Other Colloidal Dispersions in Three Respects

Because aerosols are dispersed in a gas, their colloidal properties differ from those of the dispersions discussed thus far in this chapter. No compensating buoyancy exists in the medium, so substantial gravitational force is always acting on aerosol particles. Gravitational potential energy is $m_p gh$ and changes by one kT per meter for a particle with a mass

$$m_p = \frac{kT}{g} = 4 \times 10^{-22}\,\text{kg}$$

which corresponds to a radius of $\sim 50\,\text{Å}$ for a water droplet. At equilibrium, therefore, all but the very smallest aerosol particles should settle to the bottom of a container.

A second difference between aerosols and dispersions in liquids emerges when we consider the kinetics of particle motion, for example, in gravitational settling. The steady state velocity of a drop with radius R and mass density ρ under the gravitational force $F = mg$ is

$$v = \frac{mg}{6\pi R\eta} = \frac{2R^2 g\rho}{9\eta} \tag{9.6.1}$$

using the Stokes equation (8.3.10). At $p = 1$ atm. and $T = 20°C$, the viscosity, η, of air is $1.82 \times 10^{-5}\,\text{kg/m}\cdot\text{s}$, which is 50 times lower than the value in water. Thus, the settling rate is much higher in air than in a liquid at the same gravitational force.

Problem: Calculate the settling rate in air (20°C, 1 atm.) for a water droplet of radius $R = 1\,\mu\text{m}$.
Solution:

$$v = \frac{2(10^{-6}\,\text{m})^2 9.81\,\text{ms}^{-2} 1 \times 10^3 (\text{kg/m}^3)}{9 \times 1.82 \times 10^{-5}(\text{kg/ms})}$$

$$= 1.2 \times 10^{-4}\,\text{m/s}$$

A velocity of $1 \times 10^{-4}\,\text{m/s}$ found in the problem is low in relation to a macroscopic scale and it takes a substantial time for micron size small particles to settle in a gravitational field. More importantly, air velocities of $10^{-4}\,\text{m/s}$ occur readily by convection so that, in practice, gravitational equilibrium is hardly ever reached for small aerosol particles.

Interparticle interactions constitute a third special feature of aerosols. The gaseous medium does not effectively screen interactions, and for practical purposes we can neglect the effect of the medium. Thus, Hamaker constants are large, electrostatic interactions derived in the first sections of Chapter 3, and we need not screen from a dielectric medium nor from free charges.

The Debye screening length is effectively infinite, except for the contribution from the aerosol particles themselves.

In the three examples given above, aerosol dispersion has the same quantitative properties as any other colloidal dispersion. The quantitative differences are substantial, however, and their combined effect cause aerosols to show qualitatively different colloidal behavior. For example, they can be studied only under conditions that are dilute from a colloidal dispersion point of view. Even then stability is a major concern in the laboratory. Particles not only coagulate and settle, but they also combine with the container walls. Particle trajectories are strongly influenced by convective motions, which occur much more readily in a gas than in a liquid.

9.6.3 Aerosol Particles Interact by van der Waals Forces as Well as Electrostatically and Hydrodynamically

Aerosol particles always interact through an attractive van der Waals force. The screening effect exerted by the intervening medium is negligible, and Hamaker constants are large with typical values in the range of $5–20 \times 10^{-20}$ J (see Table 5.1). Whatever the nature of the particles, the van der Waals force is strong, so heterocoagulation occurs as readily as homocoagulation. Van der Waals attraction also operates between an aerosol particle and a macroscopic body.

When liquid aerosol droplets coagulate, they also coalesce to form a single larger droplet. Solid aerosol particles will stick at the point of contact. When a third particle is added to the aggregate, it also will stick at contact. If no subsequent rearrangements take place, this aggregation process leads to the formation of loose aggregates whose fractal appearance resembles that shown in Figure 8.7. Household dust particles provide one example of such aggregated aerosol particles.

The van der Waals interaction energy U_{vdW} between two spheres of radii R_1 and R_2 is (eq. 5.2.35)

$$U_{vdW} = -\left(\frac{4}{3}\right)^2 \frac{H_{12}R_1^3R_2^3}{R^6} \qquad (9.6.2)$$

when the separation R is much larger than the radii. At short distances, the interaction changes to (eq. 5.2.34)

$$U_{vdW} = -H_{12}\frac{R_1R_2}{6(R_1 + R_2)}(R - R_1 - R_2)^{-1} \qquad (9.6.3)$$

If we consider the ambient atmosphere, $kT \simeq 4 \times 10^{-21}$ J, typical Hamaker constants are a factor of 10–50 larger. For eq. 9.6.2 to apply, the separation R must be at least three times larger than the radii R_1 and R_2, and it then follows that $U_{vdW} \ll kT$ at these separations. On the other hand, for small separations R, in which eq. 9.6.3 is valid, the interaction energy is always larger than kT. Thus, the van der Waals interaction has the same magnitude as the average kinetic energy at particle surface–surface separations of order the radii of the particles.

In practice most aerosol particles acquire a net electrical charge, so they interact also by electrostatic forces. A range of mechanisms result in charged particles. When they are detached from a bulk phase their electrolyte ions may not obey total charge neutralization or the particle can acquire excess electrons. It is an everyday experience that friction between two different fabric materials can induce such an electron transfer. Once formed, aerosol particles collide with ions and electrons generated in the atmosphere by, for example, ionizing radiation.

Ambient urban air contains on the order of $1-4 \times 10^8$ small ions of either sign per cubic meter. When these ions collide with aerosol particles, they stick and change the particle's net charge. One way of estimating the distribution of charge over the particles is to assume that a Boltzmann equilibrium is established.

In Section 3.6 we used the Born model to calculate the solvation energy for an ion in a bulk phase. This energy is smaller in magnitude in a spherical drop of radius R_1 because the electrical field is that of a vacuum for distances from the ion larger than R_1. For a single charge q in the drop the solvation energy is (see eq. 3.6.3)

$$\mathbf{A}_{\text{solv}} = \frac{-q^2}{8\pi\varepsilon_o} \frac{1}{R_{\text{ion}}} \left[1 - \frac{1}{\varepsilon_r}\left(1 - \frac{R_{\text{ion}}}{R_1}\right) - \frac{R_{\text{ion}}}{R_1} \right] \quad (9.6.4)$$

For a polar medium like water we can neglect the term that depends on ε_r and write

$$\mathbf{A}_{\text{solv}} = \text{const} + \frac{q^2}{8\pi\varepsilon_o} \frac{1}{R_1} \quad (9.6.5)$$

For a collection of drops of the same radius, the Boltzmann distribution implies that the probability $P(ze)$ for having a charge $q = ze$ on the particle is

$$P(ze) = \frac{\exp[-z^2 e^2/(8\pi\varepsilon_o \mathbf{k}TR_1)]}{\sum\limits_{z=-\infty}^{\infty} P(ze)} \quad (9.6.6)$$

Problem: Determine the radius R_1 for which the factor in the exponent reaches the value -1 for $z = 1$ and at $T = 300$ K.
Solution:

$$\frac{e^2}{8\pi\varepsilon_o \mathbf{k}TR_1} = -1 \rightarrow R_1 = \frac{e^2}{8\pi\varepsilon_o \mathbf{k}T}$$

$$= \frac{(1.6 \times 10^{-19}\,\text{C})^2}{8\pi 8.85 \times 10^{-12}(\text{C/Vm})1.38 \times 10^{-23}(\text{JK}) \times 300\text{K}}$$

$$= 28 \times 10^{-9}\,\text{m}$$

For $R = 28$ nm, the ratio between particles with a single charge and the neutral ones is thus $2 \times e^{-1}/1 = 0.74$ at equilibrium. Evaluating the sum over all particle charges in eq. 9.6.6 shows that for $R_1 \gtrsim 35$ nm the majority of particles will carry a net charge under equilibrium conditions. A similar factor $(ze)^2/4\pi\varepsilon_o R$ enters into the direct Coulomb interaction between two particles. When eq. 9.6.6 applies, therefore, the electrostatic interaction between drops is smaller than kT away from contact and does not influence the coagulation kinetics in any major way.

An important consequence of the particle charge is that particles migrate in an electrical field as they do in an electrophoresis experiment. Low conductance of air makes it possible to achieve a high electrical field over macroscopic distances. By applying a voltage between two large electrode surfaces and letting air slowly flow by, we can collect aerosol particles on the electrodes.

The electrical force on a particle is the charge ze times the field $\vec{E}$. At steady state the particle has a drift velocity v, determined by a balance between the electric force and the viscous drag force $6\pi R_1 \eta v$, assuming that Stokes's law (eqs. 5.7.1, 5.7.4) applies.

For air at room temperature the viscosity $\eta = 1.8 \times 10^{-5}$ kg/m·s. Assuming $R_1 = 10^{-6}$ m and field of 10^6 V/m, the velocity of a particle of charge e is

$$v = \frac{e \cdot E}{6\pi R \eta} = 5 \times 10^{-4} \, \text{m/s}$$

If the particle spends a minute in an electrode gap 2 cm wide, it will reach the electrode surface.

Such a depositing of aerosol particles on electrodes is widely used. Charged aerosol pigmenting particles can selectively coat metallic surfaces. Smoke from a combustion process can be purified with respect to soot particles, or aerosol particles can be removed from ambient air in rooms occupied by allergic persons.

The hydrodynamic interaction discussed in Section 5.7 causes an important long range correlation between aerosol particles. Consider a cloud of aerosol particles either settling under gravity or blown away by the wind. A particle moving at the front of the cloud will cause a counterflow of air around itself. This flow extends several diameters away from the particle. A second particle located behind the first not only will sense the external force, but it also will be pulled by the air flowing into the space vacated by the first particle. Thus, the two particles tend to move collectively. Operating throughout the cloud, this mechanism keeps the cloud intact.

9.6.4 Aerosol Particles Possess Three Motion Regimes

Several times in the two preceding sections we have used the Stokes equation when describing the dynamics of aerosol particles. This equation is derived from the Navier–Stokes hydrodynamic equation, assuming that viscous forces predominate over inertial ones. This assumption covers most cases of interest for a colloidal particle in a liquid, but we need to be cautious for aerosols.

The viscosity of air is much ($\sim 10^{-2}$) lower than that of liquids. Consequently, viscous forces are small, and we are likely to encounter conditions under which inertial forces play a role. We can measure the relative importance of inertial and viscous effects by the dimensionless Reynolds number Re, which we encountered in Section 4.4.3. For a sphere of radius R_1 and density ρ, moving with a velocity v in a medium of viscosity η,

$$\text{Re} = \frac{\text{inertial force}}{\text{viscous force}} = \frac{2vR_1\rho}{\eta} \qquad (9.6.7)$$

When Re < 1, the fluid flow around the moving particle is laminar and the Stokes equation applies. For Re $> 10^3$, on the other hand, the fluid flow is fully turbulent. In this regime the resistance coming from the medium is the drag force

$$F_D \simeq 0.2\pi\rho_m R_1^2 v^2 \qquad (9.6.8)$$

proportional to the velocity squared, the medium mass density ρ_m, and the cross-sectional area πR_1^2 of the particle. In the regime $1 < \text{Re} < 1000$, a gradual transition from the Stokes regime takes place in which the drag force is proportional to the velocity and the turbulent regime, where $F_D \sim v^2$. Raindrops, for example, move under turbulent flow conditions so that eq. 9.6.8 describes the drag force.

A second regime in which the Stokes equation fails is found when the gas can no longer be described as a homogeneous medium. In deriving the hydrodynamic equations, we assume that all relaxation processes in the medium occur rapidly relative to the motion of the particle. The local equilibrium is established through molecular collisions in the medium.

The mean free path length λ measures the average distance traveled between collisions. For hard-sphere particles of radius R_1 at a number density ρ_N, we have (cf eq. 1.6.2)

$$\lambda = \left(\sqrt{8}\pi\rho_N R_1^2\right)^{-1} \qquad (9.6.9)$$

For a nitrogen molecule in air at 20°C and $p = 1$ atm, the estimated mean free path length is 70 nm. We can expect Stokes' law only to apply to aerosol particles longer than this size.

We can generalize Stokes's law to include a so-called Cunningham correction factor C_c, which has the effect of reducing the drag force

$$F_D = \frac{6\pi\eta R_1 v}{C_c}$$

$$\simeq 6\pi\eta R_1 v \left\{ 1 + \frac{\lambda}{R_1} \left[1.26 + 0.40 \exp\left(\frac{-1.1 R_1}{\lambda} \right) \right] \right\}^{-1}$$

(9.6.10)

Equation 9.6.10 should be used for particles in ambient air with a diameter less than 1×10^{-6} m. This correction becomes even more important in the upper atmosphere, where the mean free path length is longer.

Exercises

9.1. The following critical coagulation concentrations, CCC were found for an $Fe[OH]_3$ sol:

Electrolyte	CCC (mM)
NaCl	9.25
KBr	12.2
$BaCl_2$	4.8
K_2SO_4	0.20
$MgSO_4$	0.22

What qualitative conclusions can you draw from these observations? Try also to estimate the surface potential.

9.2. The CCC for a 0.25% (w/w) suspension of a montmorillonite clay was found to be 14 mM in a NaCl solution and 2.8 mM in a $CaCl_2$ solution. Interpret this observation.

9.3. Homogeneous AgI particles of radius 1 μm in a 2×10^{-2} M $NaNO_3$ solution are at a concentration of 10^{18} particles per cubic meter. The silver ion concentration is adjusted to 3×10^{-7} M.
   (a) What is the surface potential?
   (b) Estimate the net charge on one colloidal AgI particle.
   (c) How much $AgNO_3$ must be added to the sol to cause rapid coagulation? Assume a Hamaker constant of 5×10^{-20} J.
   (d) How much KI must be added to cause rapid coagulation?
   (e) Estimate the second-order rate constant for rapid coagulation.
   (f) When rapid coagulation occurs, how much time is needed to reduce the number of particles by (i) a factor of 2 and (ii) a factor of 8?

9.4. Horn et al. (R.G. Horn, D.R. Clarke, and M.T. Clarkson, *J. Mater. Res.* **3**, 413 (1988)) measured the interaction between two sapphire surfaces in 1×10^{-3} M NaCl at different pH values as follows:

Separation h (nm)	Energy ($\times 10^5$ J/m^2)	
	pH 6.7	pH 11
30	~0	0.40
25	−0.15	0.55
20	−0.23	1.27
17.5	−0.32	1.8
15	−0.40	2.4
12.5	−0.55	2.9
10	< −0.55	4.3
8.0	< −0.55	4.9
6.0		< 4.9

The value of the Hamakar constant is 6.7×10^{-20} J. Do the data fit the DLVO theory? Estimate the magnitude of the surface potential for the two cases. What is its sign?

9.5. De Kruif et al. (G.C. de Kruif, J.W. Jansen and A. Vrij, in *Physics of Complex and Supermolecular Fluids*, Safran and Clark, eds., Wiley, New York, 1985, p. 315) measured the structure factor $S(Q = 0)$ by light scattering in a sterically stabilized solution of monodispersed silica particles:

$S(Q = 0)$	Concentration (kg/m^3)
0.91	25
0.83	50
0.67	75
0.59	100
0.50	125
0.43	150
0.31	200

The density of the particles is 1.35×10^3 kg/m^3 and their radius 34 nm. Determine the second virial coefficient Make a quantitative interpretation of the value.

9.6. Kihira et al. (H. Kihira, N. Ryde, and E. Matijevic, *J. Chem. Soc. Faraday Trans.* **88**, 2379 (1992)) determined the stability ratio W of polystyrene latex spheres $R = 65$ nm as a function of colloid KNO$_3$. The following stability ratios and zeta potentials were found:

C KNO$_3$ (mM)	W	ξ(mV)
10	1000	-55
20	500	-40
40	100	-33
100	10	-29
300	1	-22
500	1	18

Determine the CCC and the zeta potential at this point. Equating the surface potential and zeta potential and using a Hamaker constant of 1.0×10^{-20} J, does the CCC agree with the estimate of eq. 8.2.11? Estimate theoretically the value of W at 0.1 M KNO$_3$ and 0.01 M KNO$_3$. How do these estimates agree with the observations?

9.7. If equal numbers of equal-sized AgI particles ($\Phi_0 = 50$ mV) and Au particles ($\Phi_0 = -50$ mV) are mixed, a rapid coagulation is observed. However, if the relative concentrations differ by a factor of 4, extensive coagulation does not occur. Suggest a mechanim for this behavior.

9.8. A water droplet contains four monovalent positive ions. The water is gradually evaporated so the drop is shrinking. At what size is it favorable for the drop to split into two parts containing two ions each?

9.9. A water drop contains a single monovalent ion. What is the equilibrium drop size in a relative humidity of 95% at $T = 25°C$?

9.10. Calculate the Debye screening length for air with a density of monovalent ions of 2×10^8 per cubic meter.

9.11. A raindrop with a radius of 1 mm is falling. What is the steady state velocity?

9.12. Water droplets are formed as the vapor from a cup of hot coffee meets the cool morning air. The heat from the cup also generates a convective upward flow of air, which we assume is 10^{-2} m/s. Which water drops will rise and which will fall?

9.13. Use eq. 5.4.8 to estimate the hydrodynamic force between two aerosol particles of radius $R = 10^{-5}$ m separated by 10^{-4} m and moving with a relative radial velocity of 1 m/s. Compare with the gravitational force.

9.14. A water drop of radius 1×10^{-4} m is falling in air (20°C, $p = 1$ atm.) with a rate of 0.25 m/s. What is the Reynolds number?

9.15. A silica sphere of radius 2×10^{-7} m is settling in air (20°C, $p = 1$ atm.). Determine the velocity at steady state. How long does it take to fall 1 m?

Literature

References[a]	Section 9.1: Formation of Colloidal Sols	Section 9.2: Systems with Potential-Determining Ions	Section 9.3: Clays	Section 9.4: Latex Spheres	Section 9.5: Heterocoagulation	Section 9.6: Aerosols
Hiemenz & Rajagopalan		CHAPTER 11.0–11.3 Potential ions				
Hunter (II)	CHAPTER 1.5 Preparation of monodispersed sols CHAPTER 1.6–1.7 Impurities & purification	CHAPTER 6.7–6.8 Potential-determining ions	CHAPTER 1.5 Clay structure CHAPTER 7.7.3 Swelling CHAPTER 7.4 SFA measurements on mica	CHAPTER 14 Neutron scattering, concentrated, dispersions	CHAPTER 7.4.1 Theory of heterocoagulation	
Kruyt	CHAPTER II 1–4 Preparation and purification of sols					
Meyers						CHAPTER 13 Aerosols
Vold and Vold	CHAPTER 9 Preparation of sols					CHAPTER 11 Aerosols

[a]For complete reference citations, see alphabetic listing, page xxxi.

10

PHASE EQUILIBRIA, PHASE DIAGRAMS, AND THEIR APPLICATION

CONCEPT MAP

Phase Diagrams

Phase diagrams constitute the most essential part of the thermodynamic description of a colloidal system. Analysis of their features provides links between macroscopic behavior and intermolecular interactions.

Phase diagrams for colloidal systems display a rich morphology, which includes:

- The usual states of matter—**crystalline solid, isotropic liquid**, and **gas**.

- **Plastic crystals** in which molecules remain organized as a lattice, but rotate—characteristic of long chain amphiphiles.

- **Amorphous solids** and **glasses** in which molecules simultaneously display liquidlike molecular arrangements and local dynamics of solids—typical of polymers.

- **Liquid crystals** and **gels**, some of which were described in Chapter 6.

Phase Equilibria

The Gibbs phase rule (eq. 10.1.3) gives the number of intensive variables f (T, P, etc.) that can be varied independently—given the number of components c and phases p in a system.

In a multicomponent system with two phases in equilibrium, the chemical potential μ_i of each component is the same in both phases. In a two-component system, the Gibbs–Duhem equation permits us to calculate the chemical potential of one component by measuring the concentration dependence of the other.

Phase diagrams provide powerful ways to visualize phase behavior:

- Two-component systems are often represented by plotting temperature versus composition (Figure 10.5) at fixed pressure. In two-phase regions, phase composition is given by the lever rule (eq. 10.1.12).

- Three-component systems use triangular diagrams (Figure 10.6) by fixing two variables (T and P).

Constructing phase diagrams involves preparing samples and determining the number, nature, and composition of the phases by using visual examination and calorimetric, scattering, and spectroscopic measurements.

Examples of Multiple-Phase Colloidal Systems

If two phases are in equilibrium, one is usually more ordered—possessing low energy and entropy—while the other is less ordered, with higher energy and entropy.

In colloidal systems, interparticle interaction energies are $\geqslant kT$, while ΔS_{mix} depends only on X_A and X_B. This fact promotes formation of multiple ordered phases, particularly in self-assembled systems that involve an intricate interplay between inter- and intramolecular interactions. Examples illustrating this include:

- Monodisperse hard spheres display a liquid–solid transition at a volume fraction of $\Phi = 0.49$ to 0.54.

- Ionic surfactants in which inter- and intraaggregate electrostatic interactions lead to multiple phases (Figures 10.9 and 10.10) that are almost independent of temperature, yet can be changed by addition of a cosurfactant such as an alcohol.

- Nonionic surfactants in which phase equilibria are very sensitive to changes in temperature that change solvent–head group interactions.

- Block copolymers whose amphiphilic behavior (micelles–inverted micelles) in an appropriate solvent is set by the relative volumes of the two subblocks.

- Monomolecular films whose rich phase behavior reflects the multiple ways of packing parallel hydrocarbon chains at the air–water interface.

Calculation of Phase Diagrams

By calculating phase equilibria, we can relate microscopic interaction to macroscopic phase behavior. The strategy involves developing expressions for free energy or chemical potentials and requires careful — often tedious — manipulation of thermodynamics expressions, as illustrated by the following examples:

- Liquid-liquid separations modeled using regular solution theory show that when $w/RT = 3$, phase separation occurs, but a value of $w/RT = 2$ and $X_A = X_B$ corresponds to a critical point at which two coexisting phases have just merged.

- Two-component systems can be modeled in which A and B form an ideal liquid although solid solutions do not exist, and three-phase coexistence regions can be established using the intercept method (Figure 10.15).

- Impurities create a special type of two-component system. Unequal distributions of the impurity between two phases leads to creation of a two-phase area in a phase diagram, as illustrated in Figure 10.20.

- Calculation of a phase diagram (Figure 10.22) showing how a short range stabilizing force determines lamellar–gel equilibria in lipid–H_2O systems.

Continuous Versus Discontinuous Phase Transformations

First-order phase transformations involve a discontinuous change in all dependent thermodynamic variables except free energy, and at a molecular level they reflect transformations from an ordered to an disordered state (or some other drastic structural change) or vice versa. Melting of a crystalline solid is an example.

In continuous phase transformations, dependent variables — such as heat capacity, compressibility, and surface tension—diverge/vanish as the independent variable, for example T, approaches a critical value, T_c,

The transformation of a gas–liquid two-phase system to a supercritical fluid at the critical point provides an example. As the physical differences between the two phases diminish and merge at T_c, long range molecular correlations become important.

Continuous transformations can be characterized by power law expressions with characteristic exponents (eqs. 10.4.1–10.4.4) that can be predicted from statistical mechanics. Surprisingly, the same set of critical exponents describes many seemingly diverse physical systems.

We can illustrate many features of continuous phase transitions using regular solution theory. The critical exponents are inaccurate, reflecting the failure of mean field theory to correctly account for long range molecular correlations.

Other Sources of Continuous Transformations

So far we have focused on macroscopic systems possessing a few well-defined components. Other sources of apparently continuous phase transformations include:

- Systems in which finite size effects play a dominant role, such as in the formation of small micelles leading to a broad CMC.

- Systems containing impurities or inhomogeneities, which broaden phase transformations.

- Systems in which equilibrium times are long compared to observation times, as in glass transitions for polymers.

10

In colloidal systems, interparticle or interaggregate interactions can easily be strong enough to promote macroscopic ordering that can lead to the formation of a sequence of phases. Slight variations in conditions may be enough to trigger phase changes that in turn induce major changes in macroscopic properties. Although surfactants and biological lipids display particularly rich behavior, the phenomenon is also important in other colloidal systems.

The study of phase equilibria fulfills a dual purpose. As the most important factor in the thermodynamic characterization of a system, phase behavior often provides an essential clue to macroscopic behavior. In addition, we can link phase behavior in an intimate but nontrivial way with molecular or particle interactions. If we understand this connection properly, we can control phase behavior by changing interactions on a molecular level. Conversely, we can study interaggregate interactions by observing phase behavior.

10.1 Phase Diagrams Depicting Colloidal Systems Are Generally Richer Than Those for Molecular Systems

10.1.1 Several Uncommon Aggregation States Appear in Colloidal Systems

Colloidal systems show a greater diversity in aggregation states than we find in molecular systems. Reviewing the general terminology used to describe different states and phases can help us appreciate this fact, but we must remember that whenever we attempt to systematize a complex physical reality, we encounter borderline cases that may not fit a given classification system.

At ambient temperatures, matter, exists in a solid, liquid, or gaseous state or in a mixture of these. Dynamic properties at the molecular level provide the basis for characterizing aggregation states. In a **solid**, a molecule's translational motion is slow to very slow. In a **liquid**, molecules (or colloidal particles) diffuse quite rapidly in space and interact constantly with their neighbors. In a **gas**, a molecule may travel unperturbed over a distance relatively large compared to its size before colliding with another molecule.

These three main aggregation states display a range of phase types, each with its own specific characteristics. For example, the most common form of solid is a **crystal**, in which molecules are arranged with long range positional and orientational order. Closer to the melting point, a crystal can undergo a phase transition, becoming noticeably softer. It maintains full positional order in this **plastic crystal** state, but it loses some orientational order and molecules rotate in their lattice sites. This behavior commonly occurs for compounds with long hydrocarbon chains.

In **amorphous solids**, molecules (or atoms) are not located in precisely determined lattice sites, as they are in crystals. An important example—found commonly in solid polymers but also seen in other colloidal systems—is a **glass**. The molecular arrangements we observe in glasses are very similar to those found in liquids, while the local dynamic properties are typical of solids. Amorphous solids often represent nonequilibrium structures, but because of the slow dynamics in the solid state, they can remain unchanged for indefinite times.

Although molecules diffuse rapidly in a liquid, in principle, a long range order can be maintained. However, for ordinary **isotropic liquids** only a short range order exists between neighboring molecules, and no correlations exist between molecules farther away. Most substances melt from a crystalline solid to an isotropic liquid. This change implies a transition from a very ordered state to a totally disordered one.

However, a number of compounds show a more gradual melting behavior, with one or more transitions between the crystalline solid and the isotropic liquid. These intermediate phases or **liquid crystals**, retain some, but not all, of the macroscopic order present in crystal, while there is a high molecular mobility.

Liquid crystalline phases that are induced by temperature changes are denoted as thermotropic liquid crystals. There is a considerable chemical diversity among the substances forming thermotropic liquid crystals. The most well known type is the one containing a stiff elongated aromatic backbone with a short alkyl chain at the ends. However, compounds like cholesterol esters, pure soaps, and some polymers can also form liquid crystals. In colloidal systems we are normally dealing with two or multicomponent systems, with particles or aggregates in a solvent. In this case, one uses the term lyotropic liquid crystals. The thermotropic and lyotropic liquid crystals differ somewhat in the chemical interactions causing the formation of the ordered phases, but the thermodynamic properties are basically the same.

In a **smectic** liquid crystal, molecules show long range (macroscopic) orientational order and also long range translational order in one or two directions. The term **smectic**, which is derived from the Greek word for soap, sometimes refers specifically to a layered structure like those found in the lamellar liquid crystals discussed in Chapter 6, but other forms of positional ordering are also possible like the hexagonal or cubic phases we have mentioned in Chapter 1. Figure 10.1 illustrates a typical molecular ordering in a thermotropic smectic phase.

When a smectic phase is heated, it can melt into an isotropic liquid. More often, however, it transforms into a phase in which the long range positional order has disappeared, but long range orientational order remains. The nature of this orientational order depends on whether the molecules are optically active (chiral).

A nonchiral molecule produces a **nematic** liquid crystalline phase, as illustrated in Figure 10.2, in which the preferred orientation is homogeneous throughout the system. Because

Figure 10.1
Illustration of molecular order in a smectic phase. The molecules are preferentially oriented along the direction n. A careful examination also reveals a layering in the direction of n. When the layering and the orientation lie along the same direction, n, the phase is called smectic A, otherwise it is called a smectic C phase. (A) The orientational distribution function $f(\theta)$ with maxima at $\theta = 0$, π and a minimum at $\theta = \pi/2$. (B) The center-of-mass density along the director showing a periodic variation defining the layer thickness.

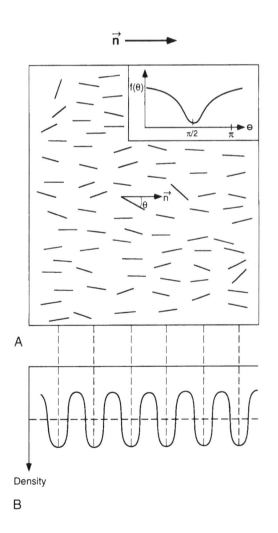

Figure 10.2
Molecular order in a
nematic phase. (A) The
orientational distribution
function $f(\theta)$ having the
same qualitative properties
as in the smectic phase.
(B) The center-of-mass
density, which is uniform.

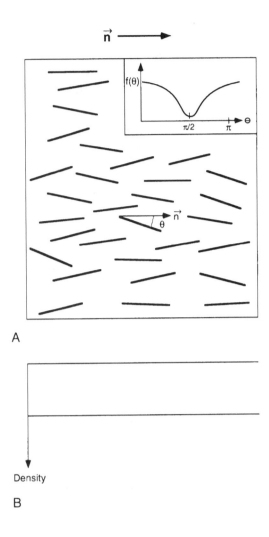

A

B

Density

nematic phases readily orient in magnetic and electric fields,
they can be used to control the optical properties of a macroscopic sample. This property is one reason for the popularity of
liquid crystals in optical displays.

In chiral systems, the molecules have a nonuniform orientational order, and the preferred orientation changes direction
in a screwlike fashion (Figure 10.3). This behavior leads to a
cholesteric also called **chiral nematic** phase.

The pitch of the screw typically lies in the range 300 nm
to 30 μm, depending on the nature of the molecules and on
temperature. When this screwlike order is present on length
scales matching the wavelength of visible light, it leads to
spectacular optical properties.

A familiar use of cholesteric liquid crystals is in thermom-
esters with color indicators. The screw pitch is very temperature sensitive, being determined by a compromise between
energy and entropy, so that changing light-scattering properties
create an apparent color change. Cholesteric liquid crystals are

Figure 10.3
In a cholesteric phase, the preferred orientation rotates through the sample. Each plane shown corresponds to the situation in the nematic phase illustrated in Figure 10.2.

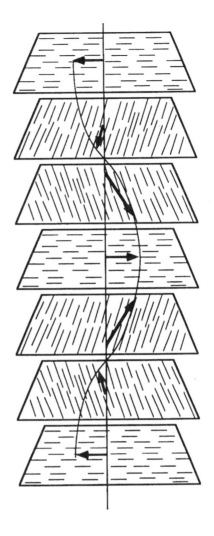

transparent to light except for wavelengths that are integer multiples of the pitch multiplied by the refractive index.

In systems with two or more components, phases may show solidlike behavior with respect to some components, but liquidlike behavior with respect to others. Usually we call such phases **gels**, although some authors attach a broader meaning to this term. In a typical gel, one component—a crystalline lipid bilayer or a crosslinked polymer—has low molecular mobility, while the other—usually a solvent—moves as in a liquid. Both components should form domains of an infinite extension.

The term gel has been used rather nebulously to describe anything with a jelly-like consistency, sometimes with the specification that there should be a non-zero yield stress. This is a phenomenological classification criterion which, in our opinion, should be avoided when possible. Thus, liquid crystalline phases are sometimes denoted as gels and also foams could easily classify as gels. Such an inconsistent nomenclature is bound to cause confusion.

The weakly interacting molecules found in the gas phase do not show the complex behavior displayed in condensed phases. In the colloidal literature, the term **gas** sometimes designates a dilute solution of particles. This terminology is based on a close analogy between the equilibrium statistical mechanics of a two-component incompressible liquid and a one-component compressible gas. We also used the term for a dilute surface film in Chapter 2, where the gaseous state implies that the area per molecule is so large that surface molecules rarely meet. From a molecular point of view, both these cases involve a liquid system, and the use of the term gas can be somewhat misleading. Aerosols, on the other hand, are true gaseous colloidal systems.

10.1.2 The Gibbs Phase Rule Guides the Thermodynamic Description of Phase Equilibria

Thermodynamics provides some general and very useful rules for phase equilibria. The most important of these is the Gibbs phase rule. To derive this rule, let us first recall some basic thermodynamic concepts.

We can write the equation of state for a one-component system as $F(V,n,p,T) = 0$, as for example, in $pV/(nRT) - 1 = 0$. This equation shows that three independent variables uniquely determine the equilibrium state. We can now distinguish between extensive variables like volume V and amount n that are proportional to the size of the system, and intensive variables like pressure p and temperature T that are independent of system size.

Of the three independent variables, we can choose one as extensive, characterizing the size of the system, and two as intensive. If we now take a dependent intensive variable—for example, the density $\rho = n/V$—we can generally write

$$\rho = f_\rho(n,p,T) = f_\rho(p,T) \tag{10.1.1}$$

where the second equality follows from our conclusion that intensive variables are independent of the system size. Thus, by this formal argument we have shown the intuitively obvious fact that the density of a gas, for example, depends only on temperature and pressure; the amount of the gas is irrelevant. The general statement of eq. 10.1.1 is that all intensive quantities in a one-component system are a function solely of two independent intensive variables.

We can now introduce the concept of **number of degrees of freedom f**, representing the number of independent intensive variables that remain after we have taken all possible constraints into account. For a one-component system with no additional constraints, $f = 2$. If we add the condition that two phases, α and β, are at equilibrium, an additional equation, $\mathbf{G}_\alpha^0(T,p) = \mathbf{G}_\beta^0(T,p)$, states that the two phases possess equal molar free energies. This equation reduces f to one, since T and

p are no longer independent and, in fact,

$$\frac{dp}{dT} = \frac{\Delta \mathbf{S}}{\Delta V} \qquad (10.1.2)$$

for two phases in equilibrium according to the Clausius–Clapeyron equation.

For a two-component system, we must add another intensive variable, which we can represent conveniently as the mole fraction of the second component, $X_2 = n_2/(n_1 + n_2)$. Thus, with three independent intensive variables and no constraints, $f = 3$.

We can generalize these observations to apply to systems with an arbitrary number of components c. The **Gibbs phase rule** states that

$$f + p = c + 2 \qquad (10.1.3)$$

where p is the number of phases.

Sometimes determining the number of components presents a nontrivial task. For example, we usually consider H_2O to be a one-component system, but this is true only as long as dissociation into H_2 and O_2 remains negligible. In the presence of a dissociation, water becomes a two-component system. The constraint that H_2 and O_2 occurs in the molar ratio 2:1 in the total system affects as extensive variable, not an intensive one.

A mixture of two salts, such as NaCl + KBr, in which all four ions are different, is a three-component system. For example, a particular phase may well be rich in NaBr. The requirement that the total system contain equal molar amounts of Na and Cl again constrains an extensive variable, not an intensive one.

Polydispersity in the molecular weight always characterizes a polymer system, and according to strict thermodynamics the number of components is large. However, the components are so similar that we can normally consider the polymer as one component if it is chemically homogeneous.

10.1.3 In a Multicomponent System with Two Phases in Equilibrium, the Chemical Potential of Each Component Is the Same in Both Phases

Equilibrium at constant temperature and pressure occurs when the Gibbs free energy $\mathbf{G}$ has a minimum value. For a system with two components, 1 and 2, with two phases, α and β,

$$\mathbf{G} = \mathbf{G}^\alpha + \mathbf{G}^\beta \rightarrow d\mathbf{G} = d\mathbf{G}^\alpha + d\mathbf{G}^\beta \qquad (10.1.4)$$

but the change in free energy as molecules are added to or subtracted from the phase is determined by the chemical potentials μ_1 and μ_2, and

$$d\mathbf{G}^\alpha = \mu_1^\alpha dn_1^\alpha + \mu_2^\alpha dn_2^\alpha \quad \text{and} \quad d\mathbf{G}^\beta = \mu_1^\beta dn_1^\beta + \mu_2^\beta dn_2^\beta \qquad (10.1.5)$$

The total amount of matter is constant so that

$$dn_1^\beta = -dn_1^\alpha \quad \text{and} \quad dn_2^\beta = -dn_2^\alpha \qquad (10.1.6)$$

Thus,

$$d\mathbf{G} = (\mu_1^\alpha - \mu_1^\beta)dn_1^\alpha + (\mu_2^\alpha - \mu_2^\beta)dn_2^\alpha \qquad (10.1.7)$$

and at the minimum, $d\mathbf{G} = 0$, that can only occur for

$$\mu_1^\alpha = \mu_1^\beta \quad \text{and} \quad \mu_2^\alpha = \mu_2^\beta \qquad (10.1.8)$$

because n_1^α and n_2^α can vary independently. In the general case,

$$\mu_i^\alpha(X^\alpha) = \mu^\beta(X^\beta) \qquad (10.1.9)$$

for all components i, where X^α and X^B denote the compositions of the two phases α and β, respectively.

By definition

$$\mu_i = \left(\frac{\partial \mathbf{G}}{\partial n_i}\right)_{T,p,nj} \qquad (10.1.10)$$

so we can calculate the chemical potential if we know $\mathbf{G}$ as a function of the composition. The intercept method provides a convenient way to graphically determine μ_1 and μ_2 for two-component systems. It is particularly useful for determining qualitative features.

Figure 10.4 illustrates the method through a plot of $\mathbf{G}/(n_1 + n_2)$ versus X_1 (or X_2). We obtain the chemical potentials

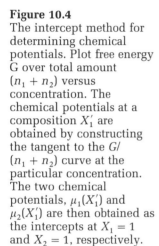

Figure 10.4
The intercept method for determining chemical potentials. Plot free energy G over total amount $(n_1 + n_2)$ versus concentration. The chemical potentials at a composition X_1' are obtained by constructing the tangent to the $G/(n_1 + n_2)$ curve at the particular concentration. The two chemical potentials, $\mu_1(X_1')$ and $\mu_2(X_1')$ are then obtained as the intercepts at $X_1 = 1$ and $X_2 = 1$, respectively.

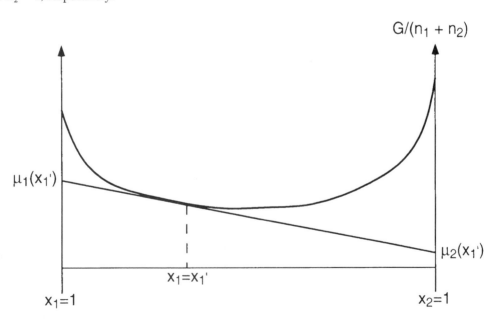

at an arbitrary composition X_1' by taking the tangent of the curve, $\mathbf{G}/(n_1 + n_2)$, at X_1'. The intercepts of this tangent at $X_1 = 1$ and $X_2 = 1$ are the chemical potentials $\mu_1(X_1')$ and $\mu_2(X_1')$. (For proof see any text on physical chemistry.)

With two components, there are three independent intensive variables. At fixed T and p, only one is left. Now the chemical potentials, μ_1, and μ_2, are both intensive. Thus, they cannot be varied independently at constant T and p. The interdependence between the two is expressed in the Gibbs–Duhem equation

$$n_1 d\mu_1 = -n_2 d\mu_2 \qquad (10.1.11)$$

If we measure the concentration dependence of the chemical potential of one component, we can calculate the concentration dependence of the other by using eq. 10.1.11. This explains, for example, how we can use a measurement of the osmotic pressure—that is, the chemical potential of the solvent—to study interactions between solute molecules or colloidal particles.

10.1.4 Phase Diagrams Conveniently Represent Phase Equilibria

Phase diagrams provide us with a useful visual way to represent phase behavior. In one diagram we can summarize a large body of experimental data, and we also can use the phase diagram for predictive purposes.

A two-component system may have up to three degrees of freedom, but we can display only two dimensions conveniently in a graphic representation. Pressure usually influences condensed phases only slightly. In such a case, we decrease the number of degrees of freedom by one by holding the pressure constant (e.g., at 1 atm), while we vary temperature T and composition X.

Figure 10.5 illustrates some general features of T–X (p fixed) diagrams through a generic case of two components A and B, which form separate solid phases, yet show miscibility in the liquid state l. For one-phase regions with two degrees of freedom ($f = 2$), T and X vary independently, and the phase remains the same, although it exhibits continuously varying properties.

When two phases are in equilibrium, $f = 1$. In the diagram, these regions also have a two-dimensional character, but changes in total composition do not change the individual composition of the two phases in equilibrium. Their only effect is to change the relative amount of each phase. For example, a system containing one mole at composition X and temperature T (such as the one illustrated in Figure 10.5) will separate a moles of solid A and b moles of solid B in such a way that $a + b = 1$. Conservation of component 1 implies

$$X_2' = aX_2^A + bX_2^B \rightarrow a(X_2^A - X_2') = b(X_2' - X_2^B) \quad (10.1.12)$$

which is called the **lever rule**.

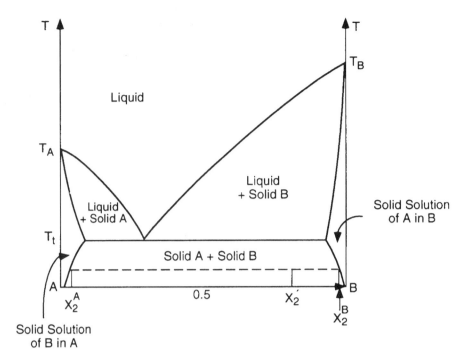

Figure 10.5
Binary T–X phase diagram showing one-phase and two-phase areas and a three-phase line at $T \simeq T_t$. Use of the lever rule in a two-phase area is also shown. The diagram represents the system Ag(A)–Cu(B).

The horizontal straight line at temperature T_t is called a three-phase line. The three phases, A, B, and l, are in equilibrium at this temperature, and consequently $f = 0$. No degree of freedom exists, and both the temperature and the compositions of the phase in equilibrium are unique.

Systems with three components may have as many as four degrees of freedom. To represent the phase behavior of such a system in a two-dimensional diagram, we must fix two degrees of freedom, usually T and p. Then we can represent the composition in a triangular diagram. As an example, Figure 10.6 shows the phase equilibria for a water–potassium decanoate ($C_9H_{19}COOK$)–octanol system.

Figure 10.7 demonstrates how to represent the composition by means of such a diagram. Each corner represents a pure component, while the opposite base of the triangle represents the binary system of the other two components. We arrive at the relative amount of component A at an arbitrary point P by drawing a line parallel to the A base. As shown in Figure 10.7, the intersection of the parallel line with the sides of the triangle determines the fraction of A.

The phase diagram divides the triangle into three types of area. One-phase areas have two degrees of freedom and can take any shape. Two-phase areas may have four corners joined by two opposing straight lines and two opposing curved lines, or we can replace the straight line(s) by concave lines ending in a critical point. These phases have only one degree of freedom, which shows that variations in one direction do not change intensive variables. In a complete phase diagram, we mark the

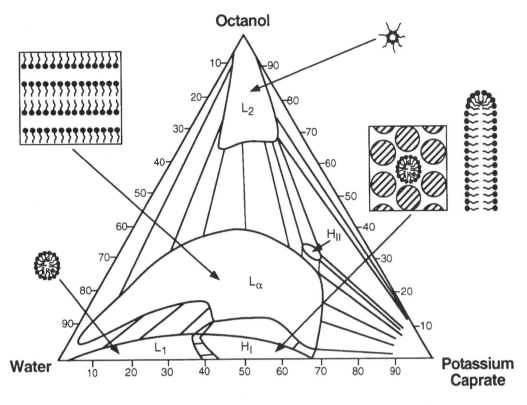

Figure 10.6
Phase diagram of the ternary system water–potassium decanoate (captrate)–octanol at constant pressure (1 atm.) and constant temperature (20°C). L_1 indicates a micellar and L_2 a reversed micellar solution. H_I is a normal hexagonal and H_{II} a reversed hexagonal liquid crystalline phase; L_α is a lamellar phase. (B. Jönsson and H. Wennerström, *J. Phys. Chem.* **91**, 338 (1987).)

invariant direction by so-called tie lines, which connect points on the two opposing curved lines as exemplified in Figure 10.6. Tie lines connect one-phase points in equilibrium. Only the proportions of the two phases change for points along the same tie line, and the lever rule (eq. 10.1.12) applies.

Finally, for three phases in equilibrium, $f = 0$. This situation occurs in so-called three-phase triangles. For a point within such a triangle the three phases represented by the corners of the triangles coexist in equilibrium. The relative amounts are determined in the same way as explained in connection with Figure 10.7, except that we are no longer dealing with an equilateral triangle.

10.1.5 Determining Phase Equilibria Is a Demanding Task

Determining the phase equilibria or phase diagram is usually the most important characterization of the macroscopic properties of a colloidal system. In general, determination of a complete phase diagram involves a lot of work. The basic principle is to mix the components and observe the number and nature of the phases. However, this can be done in a number of ways, depending on the particular characteristics of the sample.

The most obvious, direct, and often the most versatile method, is direct visual observation, be it by the naked eye or

Figure 10.7
Evaluating the composition at a point P in a triangular diagram. The relative amounts of A, B, and C are 0.6, 0.1, and 0.3, respectively.

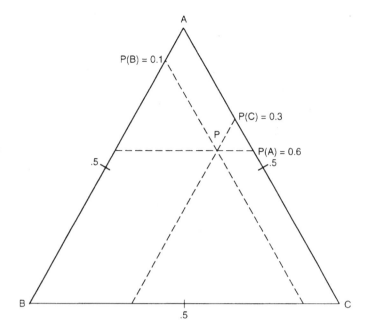

in a polarizing microscope. Isotropic liquid phases typically phase separate readily, and mixing at a given composition and analyzing the amount and composition of the different phases in equilibrium can make it possible to determine a tie line. So when phases readily separate macroscopically, determining the phase diagram is a straightforward task.

However, colloidal systems that abound in liquid crystals and other viscous phases require a long time for achieving a macroscopic separation of the sample into two distinct regions with a single planar interface. It is more common that phases in equilibrium are dispersed in one another with domain sizes of tens or hundreds of micrometers. For each prepared sample, we face three questions:

How many phases are present?

What is the nature of the phases (liquid, liquid crystal, or solid) and what is the symmetry?

What are the compositions of the phases (if more than one phase is present)?

If macroscopic phase separation does not exist, we are forced to perform measurements on the whole sample and obtain some sort of macroscopic average. The three most common approaches are calorimetry, (x-ray) scattering, and spectroscopic (NMR, ESR, fluorescence) measurements.

Calorimetric measurements are useful only for making a T–X diagram. Differential scanning calorimetry (DSC) readily measures the heat capacity as a function of the temperature, T. By repeating the measurement for a range of compositions, we can localize the temperatures of three-phase lines and of melting transitions. (See, e.g., Figure 6.18). However, phase changes that involve only small changes in enthalpic properties may remain undetected.

Small-angle x-ray or neutron scattering is sensitive to the structure of the phase. Thus, it is very useful for determining the nature of a phase, although it is more difficult, but still possible, to use the method when several phases are present.

Spectroscopic methods rely on the fact that spectroscopic properties are often sensitive to the local molecular environment. Thus, the same molecule in different phases will give different spectroscopic responses. In a multiphase sample, this behavior can be very useful, making it possible to count the number of phases in a straightforward way. Figure 10.8 shows as an example a series of deuterium NMR spectra used to construct the phase diagram of the system dipalmitoylphosphatidylcholine–water shown later (in Figure 10.21). By measuring the number of signals, their intensity, and the magnitude of the quadruple splittings, the combined information more than suffices to construct the phase diagram.

We conclude this section by noting that some occasions require a rapid method for determining phase behavior. It is no coincidence that such methods have been developed at the scientific laboratories of the two leading detergent producers in the world, Procter & Gamble and Unilever. A common principle is to take a surfactant crystal, place it in a microscope between coverslips, and let water penetrate into the crystal from one side by a diffusion process. During the diffusion process, a concentration gradient is generated and, in principle, the whole compositional range can be sampled across the coverslip. By focusing the (polarizing) microscope at different positions along the concentration gradient, one can determine the sequence of phases. Phase boundaries appear as sharp lines, making it relatively easy to detect a change in phase structure.

10.2 Examples Illustrate the Importance of Phase Equilibria for Colloidal Systems

A phase transition or an equilibrium between two phases typically occurs when a balance exists between two macroscopic states: one more ordered (with a low energy and entropy) and one less ordered (with a higher energy and entropy). In colloidal systems, interactions between particles can easily be of order kT and higher, yet entropies of mixing N_A and N_B objects are the same irrespective of the size of the objects. Consequently, ordered phases occur more readily in colloidal than in small molecule systems. This observation is particularly true for self-assembling systems whose intricate interplay between inter- and intra-aggregate interactions results in a particularly rich phase behavior. We will discuss a number of colloidal systems in which phase behavior constitutes an important aspect of the properties of the system in general.

Figure 10.8
Representative deuterium
NMR spectra of the binary
dipalmitoylphospha-
tidylcholine (DPPC)–D_2O
system. (a) Molar ratio
D_2O/DPPC = 13,
$t = 49.3°C$. Lamellar liquid
crystalline phase L_α. (b)
Molar ratio 10.5,
$t = 40.2°C$. Rippled phase
$P_{\beta'}$. (c) Molar ratio 7.0,
$t = 29.2°C$. Gel phase $L_{\beta'}$.
(d) Molar ratio 7.0,
$t = 49.3°C$. Two phase
$L_\alpha + L_{\beta'}$, since this
spectrum is a superposition
of spectra (a) and (c). (e)
Molar ratio 9.2, $t = 40.1°C$.
Two phase $L_\alpha + P_{\beta'}$, since
the spectrum is a
superposition of (a) and (b).
(f) Molar ratio 10.7,
$t = 43.0°C$. P_β phase. (g)
Molar ratio 24.6,
$t = 45.6°C$. $L_\alpha + D_2O$. (h)
Molar ratio 15.5, $t = 30.0°C$
$L_{\beta'} + D_2O$. (J. Ulmius, H.
Wennerström, B.
Lindblom, and G.
Arvidson, *Biochemistry* **16**,
5742 (1977).)

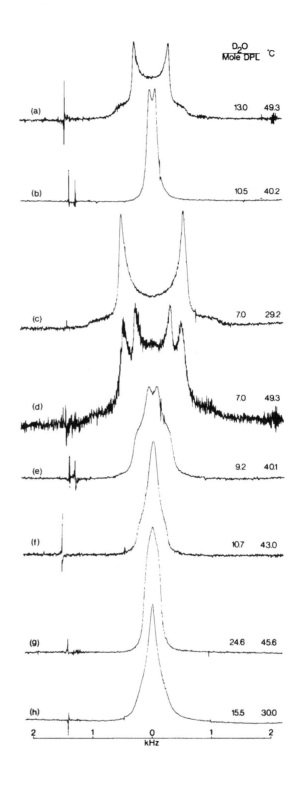

10.2.1 Purely Repulsive Interactions Can Promote the Formation of Ordered Phases

One of the early successes in computer simulation studies of liquids was the demonstration that even for a hard-sphere system (such as the system of equal-sized spheres that cannot overlap but otherwise are noninteracting, discussed in Section 3.1), an ordered (solid) phase can coexist with a disordered (fluid) phase. The system's ordered phase is stable as a result of the close packing limit of volume fraction $\Phi = 0.74$ down to $\Phi \simeq 0.54$, while the fluid state is stable below $\Phi \simeq 0.49$. Ordering occurs at high values of Φ because the increased entropy created by the rattling of the hard spheres at their sites in the lattice more than compensates for decrease in the entropy associated with changing from long-range disorder to long range order.

The hard-sphere order–disorder transition would remain an interesting theoretical observation if we could not design colloidal particles that behave like hard spheres. The recipe for doing so is first to make spherical silica or polymer particles that are as monodisperse as possible. Second, the chemistry of the particle surfaces is controlled to avoid charges. Then the particles are dispersed in a solvent whose refractive index nearly matches that of the particle. This match is important because according to eq. 5.2.36, it reduces the effective Hamaker constant to a negligibly low value. The resulting particle system exhibits neither double-layer repulsion nor a van der Waals attraction at long range. At molecular contact, some soft repulsive interaction might occur, but because the range of this interaction is much smaller than the particle radius, the system is effectively a hard-sphere one.

These systems have been prepared, and their agreement with the hard-sphere model is surprisingly good. Because the lattice spacings in the ordered phase are of the same magnitude as the wavelength of visible light, we can see spectacular optical effects caused by Bragg reflections. The optical properties of opals, which are formed by crystallized spherical silica particles of colloidal size, have the same origin.

If we prepare the silica or polymer latex particles under such conditions that they acquire an electrical charge, a dominant repulsive electrostatic interaction will result. If we reduce the level of electrolyte, an ordered phase can appear at low volume fractions. Here the ordering also is caused by repulsive interaction. Repulsion is minimized by having the particles localized on sites, and this can compensate for the entropy decrease in the ordering process.

The system responds very sensitively to electrolyte, and this sensitivity can be used to test for the involvement of electrostatic double-layer forces.

10.2.2 Ionic Surfactants Self-Assemble into a Multiplicity of Isotroic and Liquid Crystalline Phases

When surfactant molecules self-assemble in a dilute system,

local interactions determine the preferred aggregate shape as measured by the surfactant number discussed in Section 1.6. However, if the concentration increases, the interaction between aggregates becomes important, and this effect can lead either to aggregates ordering relative to one another or interactions becoming strong enough to change the aggregate shape. The compromise between these two effects leads to very rich phase behavior. Figure 10.9 shows the phase diagram for a binary cationic soap–water system. In addition to the micellar solution, it exhibits a number of different liquid crystalline phases. The most abundant is the normal hexagonal phase H_I, formed by infinite cylindrical aggregates packed in a two-dimensional hexagonal array. A schematic picture of this phase was shown in Figure 10.6. The hexagonal phase was called the middle phase by the soap boilers. The other common phase is the lamellar phase L_α, which soap boilers called the neat phase. Figure 10.9 also shows two different cubic phases. The one between the micellar solution and the hexagonal phase contains finite, short, rodlike aggregates packed in a cubic array as shown in Figure 10.10. The other is a bicontinuous phase whose structure resembles the cubic structure shown in Figure 1.6e.

In fact, a range of different cubic phases have been found in addition to so-called intermediate phases with symmetries lower than the two-dimensional hexagonal and the three-di-

Figure 10.9
$T - X$ phase diagram of the binary system water–dodecyltrimethyl-ammonium chloride. Recent studies show that intermediate phases are present between the hexagonal and cubic phases. (R. R. Balmbra, J. S. Clunie, and J. F. Goodman, *Nature* **222**, 1159 (1969).)

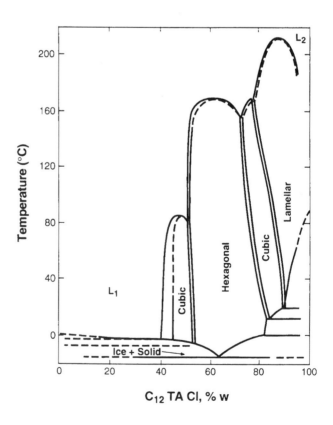

Figure 10.10
Schematized structure in a cubic phase containing ordered ellipsoidal micellar aggregates. Two inequivalent positions exist as well as a certain degree of orientational disorder. Dashed lines represent projections of unit cell boundaries. (K. Fontell, C. Fox, and E. Hansson, *Mol. Cryst. Liq. Cryst.* **1**, 9 (1985).)

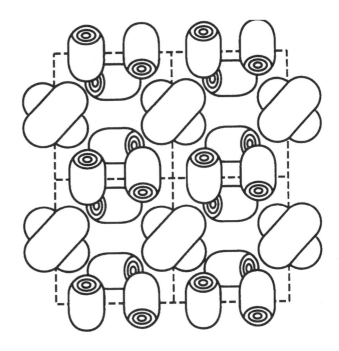

mensional cubic. Detailed structures still remain to be established for a number of these phases.

Another noticeable feature of the phase diagram in Figure 10.9 is that the phase boundaries between the different liquid crystalline phases are surprisingly insensitive to temperature changes. This behavior is typical for ionic surfactant systems, and we can trace it back to the fact that all the important free energy contributions have approximately the same temperature dependence.

Figure 10.6 shows a ternary phase diagram for ionic surfactant–alcohol–water. In such cases the alcohol is often called a cosurfactant. Because it is not charged, it reduces the charge density on aggregate surfaces and also the optimal area per head group, thus changing the effective surfactant number N_s. These effects explain why adding octanol induces the formation of a lamellar phase. New phases in this diagram are the reversed micellar solution L_2, which consists of small water surfactant aggregates in a continuous octanol phase, and the reversed hexagonal phase H_{II}, which contains small water channels packed in a two-dimensional hexagonal array.

10.2.3 Temperature Changes Dramatically Affect Phase Equilibria for Nonionic Surfactants

Surfactants with a polar group consisting of an oligoethylene oxide chain with the general formula $CH_3(CH_2)_{n-1}(OC_2H_4)_mOH$ also display a range of different isotropic and liquid crystalline phases. In spite of profound differences in the inter- and intra-aggregate forces relative to those of ionic surfactant systems, phases of the same types occur.

Figure 10.11
$T - X$ phase diagram of the binary system H_2O–$C_{12}H_{25}(OC_2H_4)_5OH$. To show more clearly the behavior at low surfactant concentration, the composition scale is nonlinear. L_1 and L_2 denote normal and reversed micellar solutions respectively. H_1 is a hexagonal, V_1 a cubic, and L_α a lamellar liquid crystal. L_3 is an isotropic solution containing bilayers, and S denotes the solid surfactant. Horizontal lines are tie lines. The critical demixing temperature $T_c = 31.9°C$. Above $54°C$ the action of undulation forces causes the lamellar phase to swell extensively to at most 99% H_2O (w/w). (R. Strey, R. Schomäcker, D. Roux, F. Nallet, and U. Olsson, *J. Chem. Soc. Faraday Trans.* **86**, 2253, (1990).)

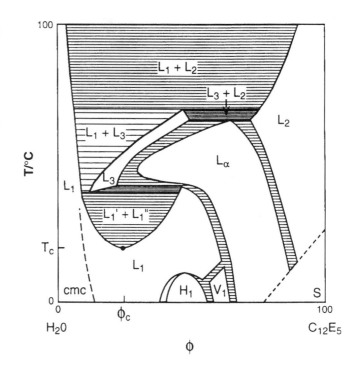

Figure 10.11 shows a $T - X$ diagram of the binary system water–$C_{12}E_5$, in which the notation refers to $n = 12$ and $m = 5$ on the general formula given in the preceding paragraph. At low temperatures ($t \lesssim 20°C$), the sequence of phases resembles the one in Figure 10.9. However, increasing the temperature has a profound effect that is qualitatively similar to adding octanol to a potassium decanoate–water mixture as in Figure 10.6.

Increase in temperature does not lead primarily to melting of aggregate order; the decrease in area per polar group, which leads to a change in the surfactant number, is more significant. The basic molecular reason for the decrease in area of the polar group with increasing T has not been unequivocally established, but it is generally observed that water becomes a relatively less efficient solvent of polar groups as temperature increases. So in the language of polymer solutions, water becomes a less good solvent for the C_2H_4O groups at higher T, leading to a partial dehydration of the polar group.

10.2.4 Block Copolymers Show as Rich a Phase Behavior as Surfactants

In Chapter 7 we showed that block copolymers self-assemble into micellar aggregates in dilute solution, according to the same principles that apply to surfactants. It should come as no surprise that when the polymer concentration is increased, ordered phases readily appear. Depending on the selectivity of the solvent and on the relative size of the two blocks, hexagonal, lamellar, and bicontinuous cubic phases are formed. Even the pure polymer can show this range of phase as the relative size of the blocks changes.

Let us illustrate the general behavior through the polystyrene (PS)–polyisoprene (PI) system:

$$H-(CH_2-CH)_n-(CH_2-\underset{\underset{\underset{CH_2}{\parallel}}{\underset{CH}{|}}}{\overset{\overset{CH_3}{|}}{C}})_m$$

As the ratio between PS and PI is changed, the polymer melt goes through a sequence of structures that is very similar to the ones found for surfactant–water systems.

Four of these are illustrated in Figure 10.12. To rationalize the trend in the structural changes, we could clearly use a parameter similar to the surfactant number introduced in Chapter 1.

The PS and the PI units virtually do not mix at all, and the samples consist of pure PS and PI regions. Because PS and PI parts are joined by covalent bonds, some sort of regular pattern must arise. One important factor in determining the structure is the relative volume of each of the two subblocks. In a homogeneous polymer melt, the chains fold as an ideal random coil, and the dimension of this coil is proportional to $\sqrt{N}\, l_p$, where l_p stands for the persistence length and N the degree of polymerization. Because the persistence lengths of the polystyrene and polybutadiene are of similar magnitude and have identical polymer backbones, phase behavior is symmetrical with respect to the volume fraction of the two components.

Figure 10.12
Transmission electron micrographs of a sequence of morphologies with increasing homopolymer concentration in blends of polystyrene (PS) and styrene-*b*-isoprene diblock copolymer. (a) 10% PS, lamellae, (b) 30% PS, the ordered bicontinuous cubic morphology, (c) 50% PS, cylinders on a hexagonal lattice, and (d) 70% PS, spheres on a cubic lattice. (K. I. Winey, E. L. Thomas, and L. J. Fetter, *J. Chem. Phys.* **95**, 9367 (1991).)

Figure 10.13
Experimentally determined
phase diagram for the block
copolymer system
polyisoprene–polystyrene.
The fraction of styrene
is represented by f_s, and χN
is the product of the χ
parameter and the number
N of monomer units per
block. bcc denotes a body-
centered-cubic phase of
finite aggregates of blocks
of the less abundant
component in a matrix of
the more abundant
component. If the blocks
are too short, ordered
phases do not form, and
ODT denotes the curve for
the order–disorder
transition. (F. Bates,
Science **251**, 898 (1991).)

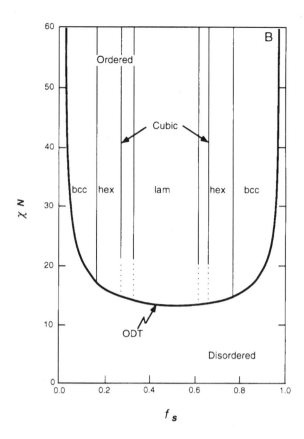

Thus, PI forms spherical centers in PS up to 18%, while above
80% PS forms spheres in a PI matrix. A lamellar state is
preferred around equal volume fractions, and a two-dimensional hexagonal packing occupies the space between the lamellar packing and the spheres. Between the hexagonal and the
lamellar phase, a bicontinuous cubic phase appears as shown in
the phase diagram of Figure 10.13.

10.2.5 Monomolecular Films Show a Rich Phase Behavior

Chapter 2 briefly discussed the relationship between surface
pressure Π_s and the area per molecules A as measured in a
Langmuir balance. This relationship corresponds to a p–V
diagram for a bulk system, but here the Π_s–A curves show much
more structure. One reason is that the change from three-
dimensional to two-dimensional molecular packing profoundly
influences the cooperativity in a molecular association process.
The number of nearest-neighbor molecules is lower in the film
than in bulk.

In the Π_s–A curves, we can observe two types of transition. At a certain pressure, the curve appears flat, showing that
the area can be changed without affecting the pressure. This flat
curve is typical of a normal first-order phase transition. In

Figure 10.14
Phase diagram showing the different phases formed in a two-dimensional film of n-docosanoic acid ($C_{21}COOH$) at the air–water interface. Depending on surface pressure and temperature, seven phases are formed. The molecular arrangement in the different phases has not been identified, but the higher the pressure and the lower the temperature, the more the film becomes like a solid. Solid lines denote first-order transitions and dashed lines continuous transitions. The dotted line at pressures around $50 \times 10^{-3} J/m^2$ denotes a mechanical instability of the film. (M. Lundquist, in *New Concepts in Lipid Research*, R. T. Holman, ed., Pergamon Press, New York, 1978, p. 101.)

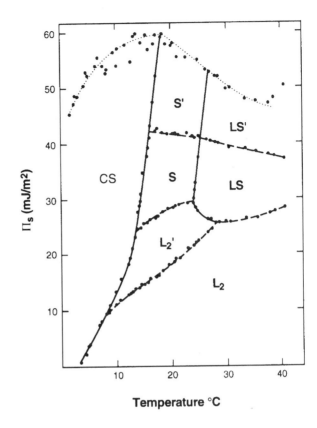

another case, we observe a distinct kink in the Π_s–A curve with different but nonzero slopes on either side of it. This kink indicates a phase transition of the second kind (also called a continuous transition), which we will discuss in Section 10.4. As an illustration of the complexity found for high surface pressures or small areas per molecule, Figure 10.14 shows the phase T–Π_s diagram for the saturated n-docosanoic acid of 22 carbon atoms at the interface between air and an aqueous hydrochloric acid solution. In the temperature range of 10–40°C, this diagram displays a sequence of two to four transitions of both the first-order type and the continuous type. All the different phases shown represent dense films whose the area varies from between 25 and 22 Å^2 per molecule at low pressures to approximately 18 Å^2 at the collapse of the film.

Detailed molecular organizations present in the individual phases in this example are not fully understood, but the rich phase behavior (which is not particularly sensitive to the nature of the polar head group) observed signals the existence of many possibilities for varying the packing of approximately parallel hydrocarbon chains. In a bulk phase we usually see a distinct transition from a solid state low molecular mobility to a liquid state whose molecular mobility is high. In the surface film, the transition between these two general states seem to be more gradual and to involve a series of phase transitions.

At the highest values of Π_s and at lower temperatures, the molecules are packed with a long range positional and orientational order along the monolayer. Decreasing Π_s or increasing T results in less ordered structures. Several of these show long-range positional order and molecules in a hexagonal arrangement, but with some disorder in alkyl chains. Furthermore, these chains can show a long range tilt order. The combination of these effects leads to a multitude of possible phases, many of which are observed.

10.3 We Obtain an Understanding of the Factors That Determine Phase Equilibria by Calculating Phase Diagrams

To understand the relationship between microscopic interactions and macroscopic phase behavior, we need to develop methods for calculating phase equilibria. The general principle requires us to find an expression of the free energy or its derivative, the chemical potential, and calculate the phase behavior from this. Such a calculation can be accomplished in many ways, however, and the choice depends on purpose and circumstances. Because thermodynamic phase behavior shows a certain degree of universality, different molecular mechanisms can lead to essentially the same macroscopic properties.

We will illustrate several ways to calculate phase diagrams using different methods.

10.3.1 The Regular Solution Model Illustrates Liquid–Liquid Phase Separation

In Section 1.5 we showed that we can calculate the chemical potential for a binary mixture of liquids A and B using the regular solution model (see eq. 1.5.7) as

$$\mu_A = \mu_A^{\theta} + \boldsymbol{R}T \ln X_A + wX_B^2$$
$$\mu_B = \mu_B^{\theta} + \boldsymbol{R}T \ln X_B + wX_A^2 \tag{10.3.1}$$

where w represents the effective interaction parameter between molecules A and B (eq. 1.4.16). If w is positive, the A–B interaction is unfavorable, and from this point of view the system would prefer to separate into a pure A and a pure B phase. However, the entropy of mixing, represented by the term $\boldsymbol{R}T \ln X$ in eq. 10.3.1, promotes mixing of the liquids. Clearly, when w is sufficiently large, the entropy of mixing is not strong enough to prevent phase separation.

Let us assume that the system separates into two liquid phases, α and β, with a composition X_A^{α} and X_A^{β}. The chemical potentials in these two liquids should be equal, and combining equations 10.1.9 and 10.3.1, we have the two equations

$$\boldsymbol{R}T \ln X_A^{\alpha} + w(1 - X_A^{\alpha})^2 = \boldsymbol{R}T \ln X_A^{\beta} + w(1 - X_A^{\beta})^2$$
$$\boldsymbol{R}T \ln(1 - X_A^{\alpha}) + w(X_A^{\alpha})^2 = \boldsymbol{R}T \ln(1 - X_A^{\beta}) + w(X_A^{\beta})^2 \tag{10.3.2}$$

which we can solve for the two unknowns, X_A^α and X_A^β. For what values of w/RT can we obtain a solution where $X_A^\alpha \neq X_A^\beta$? We can easily solve the system of equations 10.3.2 numerically, but to illustrate a graphical method, let us use eq. 10.3.1 directly. The parameter that determines whether a phase separation occurs is w/RT. Now we can solve for equal chemical potentials for both components by plotting $\mu_B - \mu_B^\theta$ versus $\mu_A - \mu_A^\theta$. In practice, this plot is done for a given w/RT by choosing a number of compositions X_A and calculating both μ_A and μ_B.

In Figure 10.15 we show such plots for the three cases, $w/RT = 1$, 2, and 3. When $w/RT = 1$, the μ_B versus μ_A plot simply provides a monotonic curve and the chemical potentials are never equal at different compositions. On the other hand, for $w/RT = 3$ the curve displays a loop, and at its intersection point the chemical potentials are equal for both components at two different compositions. This means that phase separation has occurred. By going back and determining the compositions at the crossing, we can obtain the phase boundaries. The case $w/RT = 2$ represents the transition point between monotonic and loop behavior, which signals that the two coexisting phases have just merged into one phase and we have reached the critical point.

Figure 10.16 shows the $T–X$ phase diagram calculated using the regular solution model. The critical point occurs at $T = w/2R$ and $X_B = X_A = 0.5$. In the coexistence region, the dashed line signifies the compositions at the cusps in the loop in Figure 10.15C, and it follows that between the two cusps $\partial\mu_A/\partial n_A$ and $\partial\mu_B/\partial n_B < 0$—that is, the chemical potential

Figure 10.15
Plots of μ_s versus μ_A for the regular solution model. (A) At $w/RT = 1$ the curve is monotonic, implying miscibility at all compositions. (B) At $w/RT = 2$ a kink in the curve signals a critical point. (C) At $w/RT = 3$ the curve makes a closed loop. At the self-intersection μ_A and μ_B are equal at two different compositions, $X_A = 0.93$ and $X_A = 0.07$. For compositions intermediate between these two, corresponding to the closed loop shown, the system phase separates. Between the kinks at $X_A = 0.79$ and $X_A = 0.21$ the liquid is locally unstable and would show spinodal decomposition. Arrows indicate increasing X_A.

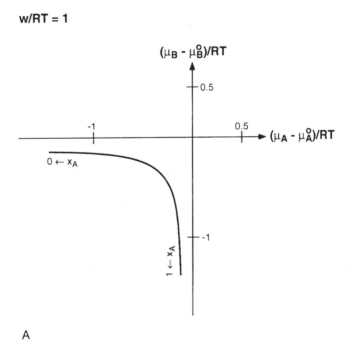

w/RT = 1

A

Figure 10.15
Continued

w/RT = 2

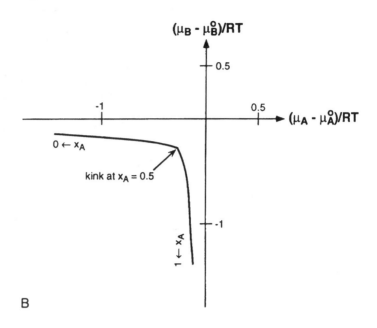

B

w/RT = 3

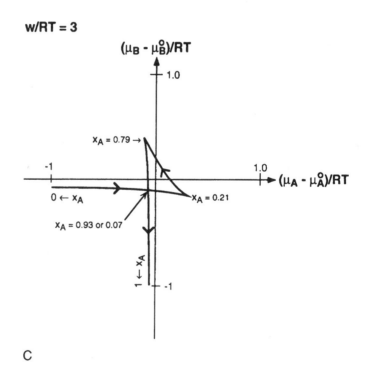

C

Figure 10.16
Phase diagram calculated for two liquids using the regular solution theory. The dashed line is the calculated spinodal line. A phase quenched to below this line is locally unstable, and the components segregate spontaneously without requiring a nucleation process.

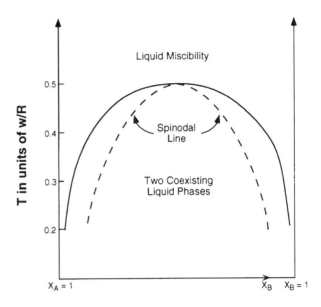

decreases when the component is added. In this case, the system becomes locally unstable so that spontaneous fluctuations in the concentration will tend to grow. On the other side of the dashed line in Figure 10.16, which is called the spinodal line, the system is locally stable and requires a nucleation process for the equilibrium phases to form.

Particularly for polymer and colloidal systems, in which equilibration processes can be inherently slow, we can observe intricate domain formations if a system is quickly brought from the isotropic solution into the unstable region. In such a spinodal decomposition, characteristic metastable concentration patterns can develop.

10.3.2 Liquid State Miscibility and Solid State Demixing Lead to a Characteristic Phase Diagram

Figure 10.5 showed a typical phase diagram for a binary A—B system with a common liquid phase but separated solid phases. We can rationalize this phase behavior within a simple model, again using the regular solution model.

If we assume the melted (liquid) phases of A and B mix readily, and for the sake of simplicity that they form an ideal solution, the free energy per mole of substance becomes

$$\frac{\mathbf{G}_l}{n_A + n_B} = X_A(\mu_A^\theta(l) + \mathbf{R}T \ln X_A) + X_B(\mu_B^\theta(l) + \mathbf{R}T \ln X_B) \tag{10.3.3}$$

In the solid phases, which place much higher constraints on the packing of a B molecule in an A lattice and vice versa, the effective interaction parameter in the regular solution theory is high and positive. A solid solution does not form readily. For the solid phase

$$\frac{\mathbf{G}_s}{n_A + n_B} = X_A(\mu_A^\theta(s) + \mathbf{R}T \ln X_A + wX_B^2) + X_B(\mu_B^\theta(s) + \mathbf{R}T \ln X_B + wX_A^2) \qquad (10.3.4)$$

where we expect $w/\mathbf{R}T \gg 1$.

The standard chemical potentials are temperature dependent, but at the melting point T of A, we have

$$\mu_A^\theta(T_A, s) = \mu_A^\theta(T_A, l)$$

but for $T < T_A$

$$\mu_A^\theta(T, s) < \mu_A^\theta(T, l)$$

and vice versa for $T > T_A$. The same behavior occurs for B $\mu_B^\theta(T_B, s) = \mu_B^\theta(T_B, l)$ at its melting point T_B. Furthermore, $\mu_B^\theta(T, l)$ decreases more rapidly with increasing T than $\mu_B^\theta(T, s)$. To find the general phase behavior, we need not specify the parameters quantitatively. It suffices to note that the standard chemical potentials display the most significant temperature dependence.

Let us again use a graphical method for determining the phase diagram. In Figure 10.4 we saw how the intercept method can be used to find the chemical potentials once the free energy is known. To solve eq. 10.1.9 for coexistence of two phases, we need to find a common tangent at two points. The two compositions X_A^α and X_A^β in which a common tangent touches the free energy curve, $\mathbf{G}/(n_A + n_B)$, represent the coexisting phases.

As Figure 10.17 shows, we have two free energy curves: one for the solid phase, the other for the liquid. Above the melting point of both compounds at temperatures $T > T_A, T_B$, the free energy of the liquid is lowest at all compositions, and the liquid is the stable phase irrespective of the value of X_A (Figure 10.17A). At low temperatures $T \ll T_A, T_B$, the solid phases are more stable than the liquid ones. As a result of the high $w/\mathbf{R}T$ value, however, a solid–solid phase separation occurs, which can be determined by finding the common tangent in Figure 10.17B. Now as T increases from a low value, the free energy of the liquid phase comes closer and closer to that of the solid system. At a particular temperature T_t, the tangent joining the two solid phases also touches the liquid free energy curve as shown in Figure 10.17C. Here three phases manifest equal chemical potential of the two components. Thus, at T_t we have a three-phase line. For $T_t < T < T_A, T_B$, we can find two common tangents: one joining the solid A phase and the liquid and the other joining the liquid and the solid B phase (Figure 10.17D).

By collecting the behavior we have found in the different temperature regimes, we arrive at the result shown in Figure 10.18. This type of reasoning, for example, is very useful for understanding the phase behavior of mixtures of phospholipids

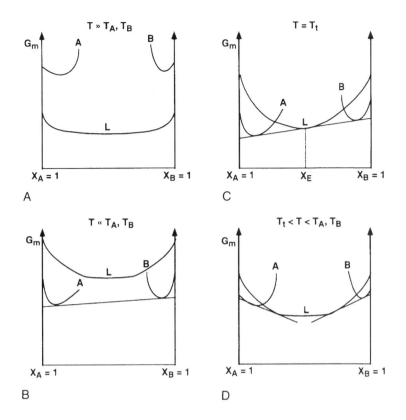

Figure 10.17
Plots of concentration dependences of the free energy, $G_m = G/(n_A + n_B)$, per mole of a binary A–B mixture with one liquid L and two solid phases, A and B. (A) The temperature T is higher than both the melting temperatures, T_A and T_B, of the pure solids, A and B, $T \gg T_A$, T_B. (B) Temperature much lower than the melting temperatures, $T \ll T_A$, T_B. (C) Temperature $T = T_t$, where a common tangent to the free energy curves of all three phases exists. (D) Temperature intermediate between T_t and the melting temperatures T_A and T_B, $T_t < T < T_A$, T_B.

that form bilayer structures. Here the liquid phase is really a liquid crystalline phase and the solid is the gel phase with solid alkyl chains. Depending on the similarity between the two components, a range of behavior can be observed, from that shown in Figure 10.18 to that shown in Figure 10.19.

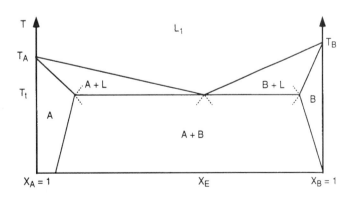

Figure 10.18
Phase diagram deduced from the curves of Figure 10.17. Dotted boundaries show metastable equilibria and X_E is the composition at the eutectic point (i.e., the composition of the liquid phase that emerges first when a two-phase sample is heated).

Figure 10.19
Experimental phase
diagram for the binary
mixture of dimyristoyl-
phosphatidylcholine
(DMPC) and dipalmitoyl-
phosphatidylcholine
(DPPC) in the presence of
excess water. A liquid
crystalline, L_α, or a gel, P_β,
lamellar phase (see Chapter
6) or both, are present
in the system. The dashed
line shows the calculated
phase diagram. (S. Marley
and J. M. Sturtevant, *Proc.
Natl. Acad. Sci. USA* **73**,
3862 (1976).)

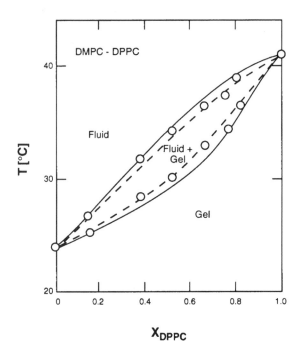

10.3.3 Two Lipids That Exhibit Different Melting Points But Ideal Mixing in Both the Gel and Liquid Crystalline Phases Produce a Simple Phase Diagram

Figure 10.19 shows an experimentally determined phase diagram for a mixture of two different lecithins with slightly different alkyl chain lengths in excess water. We can generate a model that reproduces the qualitative features of Figure 10.19 simply by assuming that the two components mix ideally in both lamellar phases: gel (*g*) and liquid crystal (*lc*).

The chemical potentials for the two components are in the gel

$$\mu_1^g(T, X^g) = \mu_1^{\theta,g}(T) + \boldsymbol{R}T \ln X^g \qquad (10.3.5a)$$

$$\mu_2^g(T, X^g) = \mu_2^{\theta,g}(T) + \boldsymbol{R}T \ln (1 - X^g) \qquad (10.3.5b)$$

where X stands for the mole fraction of component 1. The same expressions apply for the liquid crystal with a change of superscript *g* to *lc*. Equilibrium at equal chemical potentials implies

$$\mu_i^g(T, X^g) = \mu_i^{lc}(T, X^{lc}) \quad i = 1,2 \qquad (10.3.6)$$

We can solve for the two unknows, $X^g(T)$ and $X^{lc}(T)$ in terms of the difference in standard chemical potential

$$\Delta\mu_i^\theta(T) = \mu_i^{\theta,g}(T) - \mu_i^{\theta,lc}(T) \qquad (10.3.7)$$

Denote $\Delta\mu_1^\theta(T)/RT \equiv A$; $\Delta\mu_2^\theta(T)/RT \equiv B$. Then we can write the solution to eq. 10.3.6 as

$$X^g(T) = \frac{1 - e^B}{e^A - e^B} \tag{10.3.8a}$$

$$X^{lc}(T) = \frac{e^A(1 - e^B)}{e^A - e^B} \tag{10.3.8b}$$

Problem: Show that $X^i(T)$ of eq. 10.3.8 solves eqs. 10.3.6 with μ_i given by eq. 10.3.5.

Solution: The equality of the chemical potentials of component 1 yields

$$A = \ln(X^{lc}/X^g)$$

the r.h.s. equals using 10.3.8

$$\ln\left\{\frac{e^A(1 - e^B)}{e^A - e^B}\frac{(e^A - e^B)}{1 - e^B}\right\} = \ln e^A = A = \text{l.h.s.}$$

For the second component

$$B = \ln\left\{\frac{(1 - X^{lc})}{(1 - X^g)}\right\}$$

the r.h.s. equals

$$\ln\left\{\left[1 - \frac{e^A(1 - e^B)}{e^A - e^B}\right]\Big/\left[1 - \frac{1 - e^B}{e^A - e^B}\right]\right\}$$

$$= \ln\{[e^A - e^B - e^A + e^{A+B}]/[e^A - e^B - 1 + e^B]\}$$

$$= \ln\{e^B(e^A - 1)/(e^A - 1)\} = B = \text{l.h.s.}$$

The mole fractions $X^g(T)$ and $X^{lc}(T)$ are temperature dependent through the temperature dependence of the standard chemical potentials $\Delta\mu_i^\theta(T)$, which are properties of the puresystems. At the melting points T_{oi}, we have equilibrium between the two phases, and for the pure compounds $\Delta\mu_i^\theta(T_{oi}) = \Delta H_i^\theta - T\Delta S_i^\theta = 0$. We can expand $\Delta\mu_i^\theta(T)$ around the points T_{oi} in a Taylor series, using the relation $\partial\mu/\partial T = \mathbf{S}$ to obtain

$$\Delta\mu_i^\theta(T) \simeq (T - T_{oi})\frac{\partial}{\partial T}\Delta\mu_i^\theta = (T - T_{oi})\Delta S_i^\theta = \left(\frac{T}{T_{oi}} - 1\right)\Delta H_i^\theta \tag{10.3.9}$$

where ΔH_i^θ represents the measured transition enthalpies for the pure lipids. From this equation we find the temperature dependence of the two parameters A and B to be

$$A = -\frac{\Delta \mathbf{H}_i^\theta}{\mathbf{R}T}\left(\frac{T}{T_{01}} - 1\right) \quad \text{and} \quad B = \frac{\Delta \mathbf{H}_2^\theta}{\mathbf{R}T}\left(\frac{T}{T_{02}} - 1\right).$$

Figure 10.19 also shows the calculated phase diagram as a dashed line. It has the same shape as the experimental one, but its two-phase area is narrower. A simple extension of this model would introduce a nonideality connection of a regular solution type, which we would expect to be larger in the gel than in the liquid crystal.

10.3.4 The Presence of an Impurity Broadens a Phase Transition by Introducing a Two-Phase Area

The presence of an excessive amount of one component creates a special type of two-component system. Understanding how an impurity influences the phase behavior relative to that of a pure one-component system is often of practical importance.

Consider the case in which the pure abundant component (1) exhibits a phase transition $\alpha \rightarrow \beta$ at a temperature T_c. The presence of impurity (2) will somewhat change the properties of this phase transition.

A sufficient dilution, the entropy of the mixing term, $\mathbf{R}T \ln X_2$, dominates the chemical potential of the rare component. We can see this, for example, from the discussion of the regular solution theory in Section 1.5, in which $\mathbf{R}T \ln X_2$ dominates $w(1 - X_2)^2$ at small X_2 values even when the interaction parameter w is very large.

In phase i, the chemical potential of the impurity is

$$\mu_2^i = \mu_2^\theta(T, i) + \mathbf{R}T \ln X_2^i \qquad (10.3.10)$$

at sufficient dilution. If both phases α and β are present, the impurity will be distributed between the phases according to the distribution equilibrium

$$\frac{X_2^\beta}{X_2^\alpha} = \exp\left[\frac{\mu_2^\theta(T, \alpha) - \mu_2^\theta(T, \beta)}{\mathbf{R}T}\right] \equiv K_d \qquad (10.3.11)$$

where K_d represents the distribution coefficient of component 2 between phases α and β.

Equation 10.3.11 gives the ratio of the concentrations. If we also want to obtain the absolute value of X_2, we must consider the chemical potential of component 1. For small X_2, this equals

$$\mu_1^i \cong \mu_1^\theta(T, i) + \mathbf{R}T \ln(1 - X_2^i) \cong \mu_i^\theta(T, i) - \mathbf{R}T X_2^i \quad (10.3.12)$$

As in the preceding example, we can relate the temperature dependence of the chemical potential to the transition enthalpy

$\Delta\mathbf{H}$ through eq. 10.3.8 as

$$\mu_1^\theta(T, \alpha) - \mu_1^\theta(T, \beta) = \Delta\mu^\theta = \left(\frac{T}{T_c} - 1\right)\Delta\mathbf{H} \qquad (10.3.13)$$

Solving for equal chemical potential of component 1, we find

$$X_2^\alpha - X_2^\beta = \left(\frac{T}{T_c} - 1\right)\frac{\Delta\mathbf{H}}{\mathbf{R}T} \qquad (10.3.14)$$

and using eq. 10.3.11,

$$X_2^\beta = \frac{(T_c - T)\Delta\mathbf{H}}{T_c T \mathbf{R}(1 - K_d^{-1})} \qquad (10.3.15)$$

If $K_d > 1$ (i.e., the impurity dissolves better in the high temperature phase), a solution exists for $T < T_c$. Otherwise X_2^β is negative. For the other phase,

$$X_2^\alpha = \frac{(T_c - T)\Delta\mathbf{H}}{T_c T \mathbf{R}(K_d - 1)} \qquad (10.3.16)$$

The constructed phase diagram of Figure 10.20 shows that straight lines initially border the two-phase area and that an impurity introduces a two-phase area into the diagram. Increasing the amount of impurity widens the two-phase area, but width also depends on the magnitude of the distribution coefficient K_d. The larger the K_d, the wider the two-phase area.

Normally, a high temperature phase provides a better solvent than a low temperature phase, so the two-phase area slopes down in temperature, as we see in Figure 10.20. However, exceptions to the rule clearly exist. Figure 10.19 shows one such exception. Although statements in the literature often

Figure 10.20
Calculated effect of an impurity (2) on the equilibrium between phases α and β. For the pure component 1, the transition temperature occurs at $T = T_c$. The slope of the lines is determined by the transition enthalpy $\Delta\mathrm{H}$ and the partition coefficient K_d for the impurity between the two phases.

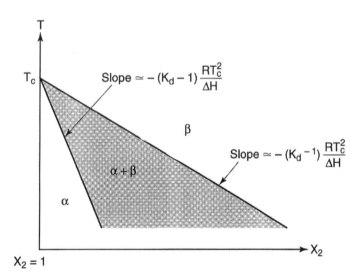

assert that an added component broadens a phase transition by decreasing the cooperativity of the transition, in fact, the typical observation is the introduction of a two-phase coexistence as shown in Figure 10.20. In this case, we have no basis for drawing conclusions about the actual cooperativity of the phase transitions as such.

10.3.5 The Short Range Stabilizing Force Influences the Equilibrium Between Liquid Crystalline and Gel Phases in Lecithin–Water Systems

We can study another example of phase equilibria between lamellar liquid crystal and gel phases of lecithins in the phase diagram of dipalmitoylphosphatidylcholine (DPPC) and water illustrated in Figure 10.21. In contrast to the example discussed in Section 10.3.3, water is not necessarily present in excess, and lowering the water content can trigger phase changes between liquid crystal and gel.

As we saw in Section 5.5, phospholipid bilayers interact across water through a short range repulsive force. The lower the water content, the stronger this interaction becomes, and interaggregate interaction can cause phase changes. In the osmotic stress technique, we primarily measure how the composition of the phase changes with chemical potential of the solvent (water). Usually such data are transformed into a force, but when calculating phase equilibria, we are interested in the chemical potentials themselves.

To establish the existence of a short range repulsive force, we can use the osmotic stress technique to measure repeat distance (which we can convert to water content) as a function of water chemical potential (see Chapter 6). This means that we know $\mu_{H_2O}(X) - \mu_{H_2O}^{\theta}$ experimentally for a number of X, T points

Figure 10.21
Experimentally determined phase diagram for the binary system dipalmitoyl-phosphatidylcholine and water. Concentration is measured as the number of water molecules n_1 per molecules of lipid n_2. The diagram includes a lamellar liquid crystalline phase L_α, a rippled gel phase P_β, and a tilted gel phase $L_{\beta'}$. See Figure 6.6 for the structure of the phases. (J. Ulmius, H. Wennerström, B. Lindblom, and G. Arvidson, *Biochemistry* **16**, 5742 (1977).)

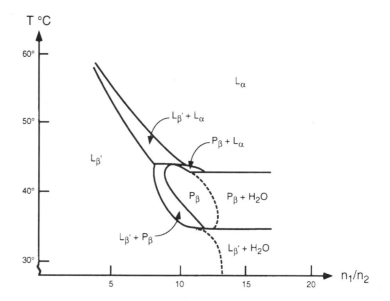

within the stability range of the respective phases. In this calculation, X represents the molar ratio n_{H_2O}/n_{lipid}, not the mole fraction.

We need to know the chemical potentials of both components to calculate phase equilibria. However, at fixed temperature and pressure, we have only one independent intensive variable, and $\mu_{H_2O}(X)$ and $\mu_{DPPC}(X)$ cannot be independent. This interdependence constitutes the essence of the Gibbs–Duhem relation

$$n_{DPPC} d\mu_{DPPC} = -n_{H_2O} d\mu_{H_2O} \qquad (10.3.17)$$

which we can integrate to yield

$$\mu_{DPPC}(X) - \mu_{DPPC}(X_0) = -\int_{\mu_{H_2O}(X_0)}^{\mu_{H_2O}(X)} X \, d\mu_{H_2O}$$

$$= X_0 \mu_{H_2O}(X_0) - X\mu_{H_2O}(X) + \int_{X_0}^{X} \mu_{H_2O}(X) \, dX \qquad (10.3.18)$$

The last equality follows from a partial integration. In eq. 10.3.18, X_0 stands for an arbitrary reference point. It is convenient to choose X_0 as the composition at the swelling limit at which the lamellar system is in equilibrium with pure water so that $\mu_{H_2O}(X_0) = \mu_{H_2O}^{\theta}$. This reference state is the same one we used in Section 10.3.3 to discuss the mixed DMPC–DPPC system. The temperature dependence of the difference in standard chemical potential

$$\mu_{DPPC}^{\theta}(X_0^i, T, i) - \mu_{DPPC}^{\theta}(X_0^i, T_0, j)$$

between two phases, i and j, then obeys eq. 10.3.8. Thus, from the measured transition enthalpies in excess water, we can estimate the temperature dependence of the lipid chemical potential.

For the water, we can obtain the concentration dependence of the chemical potentials directly from experiments, while for the lipid, we use the Gibbs–Duhem relation of eq. 10.3.17. However, to perform calculations we fit the experimental data to a functional form in which we assume that the water chemical potential is determined by a repulsive force that varies exponentially with thickness h of the aqueous layer and by an attractive van der Waals force proportional to the inverse third power of this thickness (see Chapter 6). The resulting expression for the water chemical potential in phase i is

$$\mu_{H_2O}(X, T, i) = \mu_{H_2O}^{\theta}(T) - \left(A_i \exp\left(\frac{-h}{\lambda_i}\right) + \frac{H}{6\pi} h^{-3} \right) v_0$$

$$= \mu_{H_2O}^{\theta}(T) - v_0 A_i \exp\left(\frac{-2Xv_0}{a_0^i \lambda_i}\right) + \frac{H}{6\pi} \frac{(a_0^i)^3}{8v_0^2} \frac{1}{X^3} \qquad (10.3.19)$$

where we have taken the thickness of the aqueous sheath to equal $2Xv_0(a_0^i)^{-1}$, where v_0 represents the volume per water molecule and a_0^i is the area per lipid in the bilayer. Here H is

TABLE 10.1 Parameters Used to Calculate Phase Diagram in Figure 10.22

H	Hamaker Constant	$6 \times 10^{-21}\,\mathrm{J}$
A_{L_β}	Amplitude of repulsive force	$6.8 \times 10^8\,\mathrm{N/m^2}$
A_{P_β}		$1.2 \times 10^9\,\mathrm{N/m^2}$
A_{L_α}		$1.3 \times 10^9\,\mathrm{N/m^2}$
λ_{L_β}	Decay length of repulsive	$2.0\,\mathrm{\AA}$
λ_{P_β}	force	$2.1\,\mathrm{\AA}$
λ_{L_α}		$2.2\,\mathrm{\AA}$
$a_0^{L_\beta}$	Area per molecule	$49\,\mathrm{\AA^2}$
$a_0^{P_\beta}$		$49\,\mathrm{\AA^2}$
$a_0^{L_\alpha}$		$57\,\mathrm{\AA^2}$
$\Delta \mathbf{H}_{L_\beta \to P_\beta}$	Transition enthalpy	$7.5\,\mathrm{kJ/mol}$
$\Delta \mathbf{H}_{P_\beta \to L_\alpha}$		$36.4\,\mathrm{kJ/mol}$
$T_{0 L_\beta \to P_\beta}$	Transition temperature	$34°\mathrm{C}$
$T_{0 P_\beta \to L_\alpha}$		$41°\mathrm{C}$

From L. Guldbrand, B. Jönsson, and H. Wennerström, *J. Colloid Interface Sci.* **89**, 532 (1982).

the hydrocarbon–water–hydrocarbon Hamaker constant, and we determine the parameters A_i, a_0^i, and λ_i by fitting to the x-ray diffraction and force measurements in te respective phases.

By integrating eq. 10.3.18, we can find the concentration dependence of the lipid chemical potential.

With these expressions for the concentration dependence of the chemical potentials and eq. 10.3.8 for the temperature dependence of the chemical potentials in the reference state, we can solve eq. 10.1.9 numerically for varying temperatures.

Table 10.1 summarizes the constants that enter eqs. 10.3.8, 10.3.19, and 10.3.20. Figure 10.22 shows the calculated phase diagram based on a numerical solution of the coupled equations. This calculation reproduces all the main features of the experimental phase diagram and demonstrates the crucial role of the repulsive short range force in inducing the phase changes. At first sight, the complexities of the phase equilibria present a confusing picture, yet these calculations serve as an important basis for understanding the intimate relation between interbilayer forces and the phase behavior. As the water content is decreased, conditions become more and more favorable for the bilayer to transform into a state in which the repulsive interaction is weaker. At some stage this interbilayer force compensates for the difference in intrabilayer interactions.

10.3.6 The Isotropic to Nematic Transition Can Be Caused by an Orientation Dependent Excluded Volume

A nematic phase with long range orientational correlations is typically found in one-component systems when the molecules

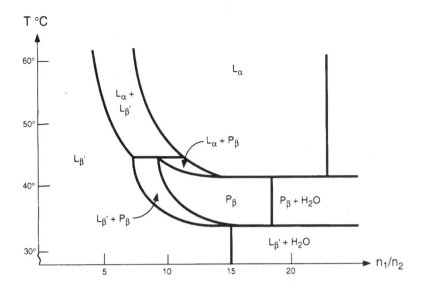

Figure 10.22
Calculated phase diagram
for the system shown in
Figure 10.21. (L.
Guldbrand, B. Jönsson, and
H. Wennerström, *J. Colloid
Interface Sci.* **89**, 532
(1982).)

are rather stiff and elongated. In colloidal systems, nematic
phases occur with stiff rodlike particles that do not attract one
another. In fact, the first theoretical description of an isotropic
to nematic transition was developed by Lars Onsager in 1949 to
explain the observation of a nematic phase in suspensions of
Tobacco Mosaic Virus, TMV. He modelled the TMV as stiff rods
of length L and diameter D, and they were supposed to interact
by a hard rod potential stating that rods cannot overlap. Onsager
realized that there was also an electrostatic interaction but left
that complication for future work.

In Section 10.2.1 we discussed the formation of an ordered phase in a hard-sphere system. In the hard rod system the
isotropic to nematic transition has a similar cause of a balance
between two competing entropy contributions. In the ordered
phase the entropy of the orientational degrees of freedom is
decreased. What could compensate for that loss of entropy?

In the isotropic phase the particles adopt all orientations,
and neglecting correlations the rotating rods exclude a volume
of order L^2D, which for large L/D ratios is substantially larger
than the actual particle volume. On the other hand, for orienta-
tionally correlated particles the excluded volume approaches a
value of four times the particle volume. The total excluded
volume increases linearly with particle concentration so in the
oriented phase the particles have more accessible volume than
in the isotropic phase and the relative difference is higher with
increased concentration. Thus, for the oriented phase there is a
higher value of the entropy associated with the translational
degrees of freedom. Thus, a dilute solution should be isotropic,
while there is a possibility that there is a transition to an
ordered phase as the concentration is increased. Onsager was
able to capture these qualitative arguments in a quantitative
description.

We have a two-component system with solute particles in
a solvent. In Section 1.5 we pointed out that it is often conveni-

ent to base the thermodynamic description of such a system in terms of the osmotic pressure and its virial expansion

$$\Pi = kTc^* + kTB_2c^{*2} + \cdots \qquad (10.3.20)$$

where c^* is the solute concentration and B_2 the second virial coefficient. If we truncate at second-order the chemical potential of the solvent (subscript 1, molecular volume V_1)

$$\mu_1 = \mu_1^{\theta} - \Pi V_1 = \mu_1^{\theta} - V_1 kT(c^* + B_2c^{*2}) \qquad (10.3.21)$$

Specifying the chemical potential of one component fixes the free energy of the two-component system through the Gibbs–Duhem relation, $d\mu_2 = (-n_1/n_2)d\mu_1$, and from eq. 10.3.21 one can determine the free energy to be

$$\mathbf{G} = n_1\mu_1 + n_2\mu_2 = n_1\mu_1^{\theta} + n_2\mu_2^{\theta} + kTn_2 \ln c^* + kTn_2c^*B_2 \qquad (10.3.22)$$

Problem: Show that eqs. 10.3.21 and 10.3.22 are consistent.
Solution: Differentiating 10.3.22 gives

$$\mu_1 = \left(\frac{\partial \mathbf{G}}{\partial n_1}\right)_{n_2} = \mu_1^{\theta} + kTn_2 \frac{1}{c^*}\left(\frac{\partial c^*}{\partial n_1}\right)_{n_2} + kTn_2B_2\left(\frac{\partial c^*}{\partial n_1}\right)_{n_2}$$

since $c^* = n_2/(n_1V_1 + n_2V_2)$

$$\left(\frac{\partial c^*}{\partial n_1}\right)_{n_2} = \frac{\partial}{\partial n_1}\left(\frac{n_2}{n_1V_1 + n_2V_2}\right) = -\frac{n_2}{(n_1V_2 + n_2V_2)^2}\cdot V_2$$

$$= -\frac{c^*V_2}{(n_1V_2 + n - 2V_2)}$$

which inserted into the expression for μ_1 results in

$$\mu_1 = \mu_1^{\theta} - kTn_2\frac{V_1}{(n_1V_1 + n_2V_2)}\cdot\frac{c^*}{c^*} - kTn_2B_2\frac{c^*V_1}{n_1V_1 + n_2V_2}$$

$$= \mu_1^{\theta} - kTV_1c^* - kTV_1c^{*2}B_2$$

as in eq. 10.3.21.

By direct differentiation of the free energy in eq. 10.3.22 we find the chemical potential of the solute

$$\mu_2 = \mu_2^{\theta} + kT\ln c^* + kT + kTc^*(2B_2 - V_2) \qquad (10.3.23)$$

We are considering an isotropic and an oriented phase and they differ in the magnitude of the second virial coefficient as well as in the standard chemical potential since the oriented phase has a smaller orientational entropy. The general expression for the virial coefficient for cylindrically symmetrical molecules including the possibility of an orientational potential $V(\theta)$ is

$$B_2 = \frac{1}{2V \cdot N_\theta^2} \iint \left\{ \exp\left(\frac{-V_{12}}{kT}\right) - \exp\left(\frac{-V(\theta_1)}{kT}\right) \exp\left(\frac{-V(\theta_2)}{kT}\right) \right\} d\vec{r}_1 d\vec{r}_2 d\cos\theta_1 d\cos\theta_2 \qquad (10.3.24)$$

where $N_\theta = \int \exp(V(\theta)/kT) d\cos\theta$ and θ_i denotes the orientation of the molecules with respect to a reference direction. For the case where there is no orientational potential, $V(\theta) = 0$, and the pair interaction V_{12} only depends on the distance r_{12} between the molecules, we can integrate eq. 10.3.24 with respect to all variables except r_{12} and regain the expression 1.5.15 for the virial coefficient in this more special case. For long, hard rods $L \gg D$ the integral over the center of mass coordinates can be approximately performed reducing the expression for B_2 to

$$B_2 = \frac{L^2 D}{N_\theta^2} \iint \exp\left(\frac{-V(\theta_1)}{kT}\right) \exp\left(\frac{-V(\theta_2)}{kT}\right) |\sin\gamma| d\cos\theta_2 d\cos\theta_2 \qquad (10.3.25)$$

where γ is the angle between the two rods. For the isotropic phase $V(\theta = 0)$ and the integral can be calculated in a straightforward way. For the oriented phase, on the other hand, the orientational distribution has to be determined by minimizing the free energy since $V(\theta)$ is an effective potential caused by the interaction between the rods. We encountered a similar self-consistency criterion in the Poisson–Boltzmann equation where the ion distribution is calculated using an electrostatic potential generated by the ions themselves. In the present case the orientational distribution takes the same role as the ion concentration profile in the electrostatic case and the expression for the entropy is

$$S = k \int f(\theta) \ln f(\theta) d\cos\theta \qquad (10.3.26)$$

in analogy with eq. 3.8.9. Here

$$f(\theta) \equiv \exp(V(\theta)/kT)/N_\theta \qquad (10.3.27)$$

is the orientational distribution function.

Minimizing the free energy with respect to variations in the normalized distribution function $f(\theta)$ leads using the calculus of variation to the nonlinear integral equation for $f(\theta)$

$$\ln f(\theta) = -1 - 2c^* L^2 D \int f(\theta') |\sin\gamma(\theta, \theta')| d\cos\theta' + \lambda \qquad (10.3.28)$$

where λ is a so-called Lagrangian multiplier introduced to ensure that $f(\theta)$ remains normalized during the variation. One solution to this equation is $f(\theta) = 1/2$. To arrive at a physically more interesting solution corresponding to an unoriented phase Onsager introduced as a simplifying assumption that the distribution function could be parameterized as

$$f(\theta, \{\alpha\}) = \text{const} \times \cosh(\alpha \cos \theta)$$

where α is a parameter to be determined from eq. 10.3.28. Note that the $\cosh(x)$ function is chosen on physical grounds so that orientations $\theta = 0$ and π are equally probable with a monotonic decay toward $\theta = \pi/2$. For each value of c^* the equation provides one value of α through a numerical solution. By repeating this solution for a range of concentrations c^* Onsager obtained the free energy and the chemical potentials at a series of concentrations. Solving for phase coexistence using eq. 10.2.3 resulted in a single isotropic phase for $c^* < 4.2/DL^2$, while the isotropic and oriented (nematic) phases coexist in the concentration range $4.2 /DL^2 < c^* < 7.0/ DL^2$. The volume of the rods is $\pi D^2 L/4$ so the larger the L/D ratio the lower the volume fraction at which the oriented phase appears.

Take as an example a system of rods with $D = 4.0 \, \text{nm}$ and $L = 50 \, \text{nm}$, which could be values typical of rod shaped micelles. Since the volume of the rod is $\pi D^2 L/4$ the isotropic phase is stable up to a volume fraction $\Phi = (4.2/DL^2)\pi D^2 L/ 4 = 0.26$ and the nematic phase appears at $\Phi = 0.44$. In the typical surfactant system one observes an isotropic to hexagonal transition in this regime. However, the hexagonal phase also has a positional ordering and belongs to the general class of smectic phases. Even though the excluded volume effect is one mechanism that drives the isotropic to hexagonal transition there are also additional effects that cause concomitant orientational and translational ordering. In the case of the surfactant system the growth of the aggregates provides an additional degree of freedom and the growth favors the formation of the smectic phase rather than a nematic phase.

The competition between nematic and smectic order is a common feature in practice. By elaborating the rod model by making a distinction between cap–cap and cap–center interactions between two rods, we could introduce a mechanism for the formation of a smectic system. When the cap–cap interaction is more favorable than the cap–center, the oriented system will also have a tendency to form layers as illustrated in Figure 10.1. The free energy balance between the nematic and the smectic phase is determined by the strength of the interaction relative to kT. Depending on such details of the interaction between the particles, one can have a sequence isotropic solution–nematic liquid crystal–smectic liquid crystal–solid crystal. In this sequence either of the two middle stages, or both, can be absent.

10.4 Continuous Phase Transitions Can Be Described by Critical Exponents

10.4.1 Phase Changes Can Be Continuous

The phase changes discussed in this chapter have been almost exclusively of a discrete character. At a given temperature or composition, a new phase appears whose properties distinctly differ from those present earlier. Such first-order phase changes are characterized by discontinuous changes in the dependent thermodynamic variables, except for the free energy. They are indeed the most common, but not the only, type of phase transition. The alternative is transitions that are known collectively as second-order transitions, phase transitions of the second kind, or continuous phase transitions. We will adopt the latter term, since it is most descriptive.

In a continuous phase transition, some of the dependent thermodynamic variables diverge as the independent variable say the temperature T, reaches the critical or transition value, T_c. As an example, Figure 10.23 shows how heat capacity can vary at a continuous phase transition. By integration we find that the enthalpy varies strongly but smoothly around T_c. In a first-order transition, on the other hand, enthalpy changes discontinuously at T_c and consequently, C_p is unlimited at T_c.

Figure 10.23
Schematic comparison of a first-order (a), (c) and a continuous (b), (d) phase transition. In the continuous transition, enthalpy varies continuously but strongly in the region of the transition temperature T_c, while for a first-order transition a jump in H occurs. This jump also implies an infinite heat capacity for the first-order transition, while in a continuous transition the heat capacity can diverge gradually. For some types of continuous transition it can also stay finite.

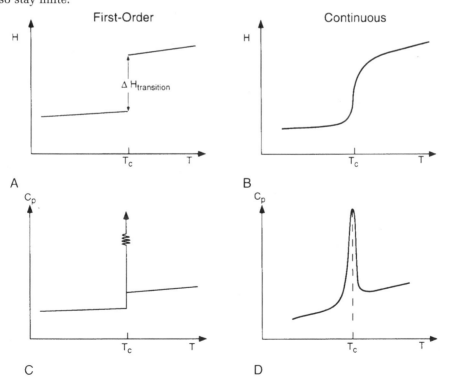

The different thermodynamic characteristics of first-order and continuous transitions have their source in the molecular events that occur as the transition point is approached. Melting of a crystal to a liquid is a typical example of a first-order transition. From a molecular point of view, an abrupt change occurs at the melting point from a well-ordered crystalline state to the much more disordered liquid state. Clearly, discontinuity in the thermodynamic variables reflects a discontinuous change on the molecular level.

The change from gas–liquid coexistence to a supercritical fluid at the critical point serves as an illustration of a continuous transition. At temperatures below the critical point, a gas and a liquid phase coexist. As the temperature is raised following the critical isochore, the physical difference between the two coexisting phases diminishes until they merge at the critical point. Above the critical temperature, there is a single fluid phase. In this transition the molecular conditions change in a smooth way and discontinuities in the thermodynamic variables are not necessarily present. However, in the example, we are dealing with a change from a two-phase to a one-phase situation, so we should expect the thermodynamics to exhibit some signs of this change.

10.4.2 Continuous Transitions Are Characterized by the Values of Critical Exponents

For a continuous transition, a limited number of thermodynamic variables diverge or alternatively become zero. For these variables power laws with characteristic critical exponents can describe the behavior around the transition point. Determining the values of these critical exponents for a particular transition has challenged both theory and experiment.

The heat capacity around the transition temperature can be described by

$$C_p = \begin{cases} \text{constant } (T - T_c)^{-\alpha} & T > T_c \\ \text{constant } (T_c - T)^{-\alpha} & T < T_c \end{cases} \qquad (10.4.1)$$

where the same critical exponent α has been found to apply at either side of the transition point.

For the gas–liquid system approaching the critical point, the difference in density, $\rho(\text{liquid}) - \rho(\text{gas})$, approaches zero in a characteristic way described by a critical exponent β

$$\rho(\text{liquid}) - \rho(\text{gas}) = \text{constant}(T_c - T)^{\beta} \qquad (10.4.2)$$

The density difference in this example is a particular case of a more general concept called the order parameter of the transition. For a general continuous transition, the exponent, β, describes how the order parameter approaches a constant value (zero in eq. 10.4.2). A first step in any analysis of a continuous phase transition is to identify the relevant order parameter(s).

For a liquid–liquid phase separation the density difference is the order parameter, while in a smectic to nematic transition the order parameter is the amplitude of the density variation illustrated in Figure 10.1.

Large fluctuations in density or more generally large fluctuations in the order parameter are found above the critical point. We can relate the long-wavelength part of these fluctuations to the compressibility, $(1/V)(\partial V/\partial p)_T$, for the fluid. Compressibility diverges at the critical point as

$$-\frac{1}{V}\left(\frac{\partial V}{\partial p}\right)_T = \text{const}(T - T_c)^{-\gamma} \quad T > T_c \qquad (10.4.3)$$

Equations 10.4.1–10.4.3 summarize the thermodynamic behavior around a critical point. What we observe at the molecular level is an increase in the range of molecular correlations in the system as we approach the critical point. The correlation length diverges at the critical point, and in our example the correlations have resulted in the formation of two separate liquid and gas phases below the critical point. Above the critical point we can picture either small drops of liquid in a gas or vice versa. These drops become larger and larger as we approach the critical point, but the density contrast relative to the bulk becomes smaller and smaller. Just on the other side of the critical point, the drop has reached a macroscopic size, while the density difference relative to the other phase is negligibly small.

For applications to colloid systems, it is particularly important to realize that the surface tension between two phases also goes to zero at the critical point, and the width of the interface diverges. The decrease in surface tension follows naturally from the diminution of the molecular difference between the two coexisting phases on approaching the critical point. The surface tension γ follows the general equation

$$\gamma = \gamma_0 \left(1 - \frac{T}{T_c}\right)^\mu \qquad (10.4.4)$$

where μ is another critical exponent.

For a long time the theoretical calculation of critical exponents was a major challenge in statistical mechanics. The basic problem was to account properly for the long range correlations. This problem was solved by introducing the so-called renormalization group theory. To describe even the essence of this theory would lead far outside the scope of this text, but one important qualitative result is the explanation of the so-called universality of critical behavior. Systems that superficially appear to be physically very different can have the same set of critical exponents. In fact only a limited number of sets of critical exponents have been observed. The exponents at both the gas–liquid interface for one-component systems and the liquid–liquid interface for two-component systems display so-called Ising-like values with $\alpha \simeq 0.1$, $\beta \simeq 0.32$, and $\gamma \simeq 1.25$, while the surface tension exponent is $\mu \simeq 1.26$.

10.4.3 We Can Use the Regular Solution Theory to Illustrate the Ising Model and to Calculate Mean Field Critical Exponents

The Ising model has been of fundamental importance in building up our understanding of continuous phase transitions. Originally the model was introduced to describe magnetic systems with two discrete states of each spin. However, we can establish the model by referring to the regular solution model of liquid mixtures discussed in Chapter 1.

In a lattice model of a liquid mixture of A and B (see Figure 1.9), each site can be occupied by either A or B. The energy associated with the site depends on the occupancy of the z near-neighbor sites, where an A—B pair has an effective interaction energy of w/z (see eq. 1.4.16).

We can solve for the most probable configurations and the thermodynamic variables in this model using exact statistical mechanics. Near the critical point of the mixture, this approach ultimately would lead to the complexities of the renormalization group theory. As an illustrative alternative, we adopt the simplifying assumptions of the regular solution theory that lead to so-called mean field exponents.

In the calculations of Section 10.3.1, we found that for sufficiently high temperature, $w/RT < 2$, A and B will mix at all proportions. For $w/RT = 2$, a critical point occurs at $X_A = X_B = 1/2$. The transition at $X_A = X_B = 1/2$ and T from slightly above to slightly below $w/2R$ is continuous.

Using the regular solution theory, we can calculate the critical exponents. For example, the width of the coexistence region is described by the exponent β. Combining eq. 10.3.1 for the chemical potential with the condition for equilibrium between two phases, eq. 10.3.2, implies that

$$RT \ln\left(\frac{1}{2} - \Delta X\right) + w\left(\frac{1}{2} + \Delta X\right)^2 = RT \ln\left(\frac{1}{2} + \Delta X\right) + w\left(\frac{1}{2} - \Delta X\right)^2 \qquad (10.4.5)$$

Here the compositions of the two phases are denoted by $X_A^\alpha = \frac{1}{2} - \Delta X$ and $X_A^\beta = \frac{1}{2} + \Delta X$, using the fact that the system is symmetric with respect to exchange of A and B. Writing $T = (T - T_c) + w/2R$, we find after expanding the logarithm to the third order that

$$\Delta X = \left(\frac{3}{4T_c}\right)^{1/2} (T_c - T)^{1/2} \qquad (10.4.6)$$

Comparing this result with eq. 10.4.2, we find that the mean field critical exponent β is 0.5 rather than the experimentally found value of 0.32.

In eq. 10.4.3 we saw that compressibility diverges at the critical point of a fluid. For a binary mixture, osmotic compressibility $(\partial \Pi_{osm}/\partial c_2)^{-1}$ exhibits a corresponding divergence. In Chapter 4 we saw that this quality determines the intensity of light scattering, and we explained the critical opalescence as "cloud point" behavior observed close to critical points. To determine the mean field value of the critical exponent γ, we can again use the regular solution theory. Because there is no basis for designating one component as solvent, we replace $-V_A \pi$ with the chemical potential of the A component

$$\left(\frac{\partial \Pi_{\text{osm}}}{\partial c_2}\right)^{-1} \simeq -\left(\frac{\partial \mu_A}{\partial c_B}\right)^{-1} V_A \simeq -\left(\frac{\partial \mu_A}{\partial X_B}\right)^{-1} \quad (10.4.7)$$

where V_A represents the molar volume of A (and B). Using eq. 10.3.1, we find

$$\frac{\partial \mu_A}{\partial X_B} = -\frac{RT}{1 - X_B} + 2wX_B \quad (10.4.8)$$

At the critical point, $T_c = w/2R$ and $X_B = 0.5$

$$-\left(\frac{\partial \mu_A}{\partial X_B}\right)^{-1} = \frac{1}{2R}(T - T_c)^{-1} \quad (10.4.9)$$

Thus, the mean field value of the exponent γ becomes 1 instead of the true value $\simeq 1.25$.

Mean field values of the exponents differ from the exact values because the critical fluctuations have not been properly accounted for. Theories much more more elaborate than the regular solution theory also misrepresent the critical exponents if they do not treat the critical correlations properly.

10.4.4 Nonionic Surfactants Show a Critical Demixing When the Temperature Increases

Many nonionic surfactants, particularly those based on ethylene oxide polar groups, show a liquid–liquid phase separation when the temperature is increased. (c.f. Figures 4.14 and 10.11.) Miscibility at low temperatures and phase separation at high temperatures is a priori an unexpected finding. As Figure 10.16 shows, normal behavior is the reverse. In general terms we can correlate the observed anomaly with the fact that water becomes a less good solvent for polyethylene oxide, and polar groups in general, as temperatures increases. (The fundamental reason for this effect can be found in eq. 3.5.16.) The detailed molecular mechanism that drives the phase separation was discussed in Section 4.3.5.

For nonionics of the type $C_m E_n$ (see Section 10.2.3), the critical behavior has been carefully studied. By measuring the locus of the two-phase area, we can determine the value of the exponent β. We find $\beta = 0.35$, close to the value expected for an Ising-like system.

The critical point for nonionic systems is often referred to as the cloud point because we can observe a strong increase in the scattered light upon increasing the temperature of the micellar solution. Early studies interpreted this phenomenon as being caused solely by aggregate growth. However, at a critical point we expect a divergence of the light scattering, and the cloudy appearance of the solutions is caused by critical concentration fluctuations. Careful studies reveal that the light scattering intensity diverges with a critical exponent $\gamma = 1.25$, consistent with an Ising-like behavior.

Because critical fluctuations dominate the light scattering, use of scattering behavior to measure aggregate size becomes problematical in the vicinity of the critical point. Studies made primarily with NMR techniques have revealed that micelles do grow in size as the cloud point is approached, but the extent of the growth depends on the ratio of n to m. Thus, micelle growth is not the inherent cause of the liquid–liquid phase separation. Instead, it is the result of an attractive micelle–micelle interaction.

10.4.5 The Term Continuous Phase Transition Sometimes Characterizes Less Well-Defined Phase Changes

We have discussed first-order and continuous phase changes using a thermodynamic point of view. The systems have been considered to consist of a few (one, two, or three), well-defined components. In principle, we assumed that samples were indefinitely large, and true equilibrium was always implied. Many real physical systems provide examples that deviate from the idealized models, although we can identify many features of a phase transition in them.

In a system of finite size, the variables cannot show true discontinuities or divergences. The phase change occurs more smoothly. However, for the finite size effect to be observable with typical experimental methods, we must go down to a colloidal scale. As an example, consider the formation of micelles. With some effort, we can see that the micellization of a cooperative aggregation of 100 monomers takes place over a small but finite concentration range. However, if we increase the aggregation number to 1000 monomers the transition is so sharp that its width lies below the typical detection limit.

Sample inhomogeneity also broadens a phase transition. In a solution of colloidal particles or in a synthetic polymer system, it is virtually impossible to obtain a system of perfect homogeneous composition. We always find a size distribution and usually also a certain chemical heterogeneity. Thermodynamically we are dealing with a truly multicomponent system, and in any phase transition the different components will prefer, however slightly, one phase over another relative to the mean behavior. This preference leads to a certain broadening of the transition.

Another common effect in colloidal systems is slow equilibration. Two particles may easily be stuck in a particular configuration for times longer than are experimentally available for obtaining equilibrium. Even though thermodynamics would predict a sharp phase transition, the slow kinetics precludes this from happening. Instead, we observe a somewhat more gradual change into a partly nonequilibrium state. The glass transition often seen in pure polymer systems provides an example of such nonequilibrium behavior.

Exercises

10.1. The phase diagram of Figure 10.9 for the water–dodecyltrimethylammonium chloride (DTAC) is a $T-X$ diagram. Consider a composition of 20% H_2O and 80% DTAC. Give the sequence of phases when such a sample is heated from 0°C to 180°C with the stability range for each phase.

10.2. Using the phase diagram of Figure 10.6, determine the nature of phases and their composition for the following total composition in grams:

H_2O	Octanol	Potassium Caprate
4	2	4
1	12	2
16	1.2	2.8
2.5	10	7.5
1	5	4.5

10.3. Figure 10.19 shows the calculated phase diagram of the DMPC–DPPC binary system. Calculate the temperature dependence of the heat capacity for an equimolar mixture. Assume that the enthalpy change in going from gel to liquid crystal is the same as in the pure compounds. The heat capacity is 1.3 kJ mol^{-1}K^{-1} for DMPC and 1.6 kJ mol^{-1}K^{-1} for DPPC.

10.4. Calculate the ternary phase diagram for a mixture of three liquids, A, B, and C, at 25°C. Use regular solution theory with the parameters
(a) $w_{AB} = 7$ kJ/mol, $w_{AC} = w_{BC} = 0$
(b) $w_{AB} = 7$ kJ/mol; $w_{AC} = 6$ kJ/mol; $w_{BC} = 0$.

10.5. For Exercise 7.8 we found that the free energy of mixing of two polymers of equal size is

$$\mathbf{G}_{\text{mix}} = kT \{n_A \ln \Phi_A + n_B \ln \Phi_B + N_p(n_A + n_B)\}\chi\Phi_A\Phi_B$$

Calculate the combinations of χ and of polymerization, N_p, for which the polymers are miscible at all compositions.

10.6. A small membrane protein is added to a DPPC bilayer system. The protein is much more soluble in the liquid crystalline than in the gel phase. Calculate the heat capacity C_p for the temperature range of 35–45°C for protein concentrations of 0.01 and 0.02 in mole fraction units. Plot the result as C_p versus T as it would be measured in a differential scanning calorimetry experiment. Assume that the contribution to the heat capacity from the protein is constant.

10.7. Extend the model calculations leading to the calculated phase diagram of Figure 10.19 by allowing for a nonideal mixing in the gel state. Assuming regular solution theory, use (a) $w = 1.0$ kJ/mol, and (b) $w = 10.0$ kJ/mol.

10.8. Calculate the phase diagram of Figure 10.18, using the following parameters: $T_A = 40°C$, $T_B = 60°C$; $\Delta H_A = H_A(l) - H_A(s) = 40$ kJ/mol; $\Delta H_B = H_B(l) - H_B(s) = 50$ kJ/mol; $C_p^A(l) = C_p^B(l) = 1.6$ kJ/mol·K, $C_p^A(s) = C_p^B(s) = 1.3$ kJ/mol·K; and $w_{AB}(l) = 0$, $w_{AB}(s) = 10$ kJ/mol).

10.9. Use the Flory–Huggins equation to calculate the critical point for a polymer of N_p units in a solvent.

10.10. The phase diagram of Figure 4.14 shows how a micellar solution of $C_{10}E_4$ splits into two isotropic phases as the temperature is increased above $\sim 21°C$. Determine the critical exponent β. The critical temperature is $20.56°C$.

Literature

A detailed discussion of all aspects of phase behavior in amphiphilic systems is given by Davis in chapters 6 and 14. For examples of phase equilibria (Section 10.2), see Chapter 10 of H. R. Kruyt, *Colloid Science* (New York: Elsevier, 1952).

MICRO- AND MACRO- EMULSIONS

CONCEPT MAP

Emulsions

Emulsions are mixtures of two immiscible liquids, such as oil and water, stabilized by an emulsifier. Thermodynamically stable mixtures (called microemulsions) are always stabilized by surfactants. Metastable mixtures (called macroemulsions) can be stabilized by surfactant (Figure 11.1), polymer, or small particles (Figure 11.22).

Macroemulsions contain either drops of oil in water (O/W) or water in oil (W/O). For microemulsions, the possibility of bicontinuous structure also exists.

The emulsifier protects the drops. An emulsion's properties depend crucially on the static and dynamic properties of the protecting film.

Characterization of Surfactant Films

The energetic properties of a surfactant film at an oil–water interface can be characterized by five phenomenological parameters

- γ, the surface tension at a macroscopic interface.

- K_a, the area expansion elastic modulus, which for phospholipid bilayers is $0.1–1$ J/m^2.

- H_o, the spontaneous curvature — giving the curvature an unconstrained film would adopt.

- κ, the bending rigidity (see Chapter 6 — undulation forces) — typically $1-20\,kT$.

- $\bar{\kappa}$, the saddle splay, modulus, which is usually negative and of order $-kT$.

The Helfrich curvature free energy $\mathbf{G}_f$ (eq. 11.1.6) provides a relation between these parameters and the properties of surfactant films.

- For planar monolayers $\mathbf{G}_f = 2\kappa H_0^2$.

- For spheres of radius R, $\mathbf{G}_f = 8\pi\kappa(1 - RH_o)^2 + 4\pi\bar{H}$.

Microemulsions

Microemulsions are thermodynamically stable isotropic solutions containing oil, water, and surfactants, that are composed of either droplets (Figure 11.4) or cylindrical structures whose size can be calculated from simple geometrical arguments, or bicontinuous structures (Figure 11.5), which occur when $H_o \cong 0$, corresponding to no preferred curvature of the surfactant film.

Triangular phase diagrams (Figure 11.6) at constant T and p are widely used; in addition the χ (Figure 11.7) and fish (Figure 11.9) plots provide useful phase representations.

With nonionic surfactants, temperature controls microemulsion structure and stability, H_o decreases strongly with increasing temperature, so that O/W forms when $T <$ PIT (phase inversion temperature). Around the PIT we find a bicontinuous structure and for $T >$ PIT a W/O microemulsion.

With ionic surfactants we can add salt and cosurfactants to obtain the same type of phase behavior obtainable with nonionic surfactants via variations in temperature.

Microemulsions containing dialkyldimethylammonium surfactants readily form inverted, interconnected W/O cylinders (Figure 11.15), which upon addition of water transform to W/O spheres.

Macroemulsions

Macroemulsions are thermodynamically unstable dispersions continuing two immiscible liquids and an emulsifier.

Emulsions combine stability properties associated with colloidal sols and transformations characteristic of amphiphilic aggregates.

We prepare emulsions by **dispersion**—that is, injecting the mechanical energy required to break up bulk phases and create large interfacial area (Figure 11.19), or by **condensation**—that is, using chemical energy stored in reactants to transform bulk phases via surface-driven instabilities into emulsion droplets.

Macroemulsion Stability

Figure 11.22 depicts the sequence of steps that determine the time evolution of an emulsion:

- Creaming—differences in density between the drop and the medium result in concentration of drops in the top or bottom of the container.

- Flocculation—drops move into the secondary minimum, a process promoted by large drop size.

- Coagulation—drops move into the primary minimum.

- Ostvald ripening—diffusion of droplet molecules through the medium causes small drops to decrease in size and disappear while large drops grow.

- Coalescence—two drops merge into one.

Stabilizing an emulsion involves minimizing these steps; destabilization involves promoting them.

Surfactant Film Properties

Dynamic surfactant film properties play key roles in formation and destabilization of emulsions.

During emulsion formation, the processes of drop formation and coalescence compete, and the fastest-formed proper surfactant film leads to the longest-lived drop.

- While new O/W interface is being created, surfactant interfacial density is nonuniform. Nonuniformity induces Marangoni surface-driven flows, promoting formation of a stabilizing surfactant film.

- Subsequent transport of surfactant from the bulk of the interface via convection promotes a complete surfactant film more effectively than diffuse transport from within a drop. The phase in which the emulsifier is most soluble usually emerges as the continuous phase. This phase correlates with the surfactant number because $N_s < 1$ corresponds to greater water solubility and W/O emulsions, while $N_s > 1$ leads to O/W emulsions.

During coalescence, two fluctuation paths lead to film rupture and fluid flow between drops:

- Formation of oil channels between two O/W droplets via bicontinuous structures (Figure 11.25).

- Formation of transient holes in adjacent monolayers, arising from density fluctuations; when two such holes coincide, fluid squirts from one drop to another (Figure 11.25).

Such fluctuations are minimized when the spontaneous curvature H_o lies between $l_o^{-1} \gg H_o \gg 0$ for O/W emulsions or $-l_o^{-1} \ll H_o \ll 0$ for W/O emulsions, which frustrates formation of flat interfaces.

Destabilization Strategies

We can add salt to break an electrostatically stabilized O/W emulsion; centrifuge to promote creaming; agitate to promote coalescence; or heat if the surfactant is in gel state.

We can change curvature by adding a cosurfactant or an antagonistic surfactant.

Foams

A foam is a dispersion of gas bubbles in a solid or liquid. It is never thermodynamically stable.

A foam is formed in a liquid from:

- Forcing gas through a nozzle.

- A supersaturated solution of gas.

- Superheating a liquid.

- Mechanical agitation of a liquid surface.

A solid foam forms when the liquid of a foam solidifies.

A gas bubble in a liquid rises to the surface by the action of gravitation. There it can form a concentrated foam or froth if the liquid film that separates the bubbles is stabilized by surfactants or polymers. In the concentrated foams the bubbles deform to polyhedral shapes with flat contact areas and plateau borders at the corners.

Foams disintegrate by the action of Ostwald ripening, where gas molecules moves from small bubbles to large ones by a molecular diffusion over the liquid film. The film can also rupture so that two bubbles coalesce.

The properties of the liquid film in a foam can be studied in the surface film balance, which measures film thickness as a function of bulk osmotic pressure. This technique also allows for reproducible measurements of film rupture under osmotic stress.

11

Oil and water do not mix. This observation is equally relevant in the chemical laboratory, the oil field, and the kitchen. However, on many occasions we want oil and water to mix, and we accomplish this by means of an emulsification. For example, the art of making a good sauce often contains a crucial moment of oil and water mixing. In the food industry emulsions have numerous uses in products such as low fat margarines, coffee whiteners, soups, and chocolate drinks. The active components of many drugs are insoluble in water, and emulsification is often an important step in the formulation of medications in the form of pastes or salves. A remarkable application that is in the developing stage is the use of emulsified fluorocarbons as a blood substitute. Emulsions have found an increasing use in paints and in other situations calling for the replacement of organic solvents for reasons of health and conservation. They are also used in formulating machining fluids, in surface coating processes, in making the bitumen composite used in surfacing highways, and in emulsion polymerization and numerous other processes.

Sometimes we need to create very low surface tension between water and oil rather than obtaining mixing. The prime example is tertiary oil recovery, in which an aqueous phase is used to push out the oil remaining in porous rock.

Many living organisms process fatty materials in a basically polar environment. Examples include emulsification of fat droplets in milk or solubilization of fats by lipoprotein particles in blood or by bile in the intestine. Also in these cases the procedure employs surface-active components.

In foams we disperse a gas in a liquid or solid medium. As we will see in the last section of this chapter, foams show many important analogies with emulsions when it comes to the central questions of structure and stability.

From a thermodynamic point of view, we can distinguish two types of emulsions. Systems in a thermodynamically stable state are called microemulsions, while metastable (or unstable) systems are known as macroemulsions. Although the systems have many structural features in common, microemulsions typically involve a smaller length scale than macroemulsions, as the prefixes suggest.

To ensure reasonable stability of the dispersed state, the interface water–oil or air–liquid must be protected by some third component. This chapter focuses, for the most part, on amphiphile-stabilized interfaces, keeping in mind that polymer or particle adsorption to the interface also can stabilize emulsions and foams.

11.1 Surfactants Form a Semiflexible Elastic Film at Interfaces

11.1.1 We Can Characterize the Elastic Properties of a Film Through Five Phenomenological Constants

Micro- and macroemulsions, as well as foams, are stabilized by a surfactant film separating the aqueous and oil (gas) domains. The properties of these systems are largely determined by those of the surfactant film. A fruitful approach to describing the system is to view the surfactant layer as a film characterized by a few phenomenological parameters that depend on intensive thermodynamic variables: temperature, pressure, electrolyte concentration, etc. In the emulsions and foams we have domains of bulk phases (water, oil, air), separated by the film. The free energy of the system can then be seen as the sum of the free energy of the respective bulk phase plus an extra contribution that can be solely attributed to the film. The free energy of the film can, in turn, be seen as a sum of three different contributions. For a given configuration of film a free energy can be associated with the number of molecules per unit area. Furthermore, there is an additional contribution from the curvature of the film. A third free energy contribution comes from the fact that the film can adopt many different configurations. In the discussion of the undulation force in Section 5.5 we already encountered one manifestation of the entropic contribution to the free energy of a system containing surfactant films.

When we consider a film that can stretch and bend there are, to second order, five characteristic parameters. We can define two of these with reference to the simplest geometry of the film: the planar state. The surface tension or surface free energy per unit area, γ, which was introduced in the beginning of this book, represents the first derivative of the film free energy $\mathbf{G}_f$ with respect to area. The area expansion modulus K_a is a second order derivative

$$K_A = \left(\frac{\partial^2 \mathbf{G}_f}{\partial A^2}\right)_{n_s} \frac{A}{2} \ (\text{J/m}^2) \tag{11.1.1}$$

evaluated at constant number n_s of surfactants in the film. Normally, when we change the area of the film, the surfactants will equilibrate between film and bulk, and in this case $\mathbf{G}_f$ is a truly linear function of A. However, in some situations, the constraint of constant amount is relevant. Such cases exist when there is no reservoir of surfactants, when transport of surfactants to the film is slow relative to the time scale of area changes, or when we want to consider concentration fluctuations. For macroscopic interfaces the area expansion modulus K_A is called the Gibbs elasticity.

To illustrate the significance of the remaining three parameters H_o, κ, and $\bar{\kappa}$, we must consider curved surfaces. Figure 11.1 shows an arbitrarily curved film. As discussed in Section 1.3, the curvature at each point on a surface is described by the inverse of the two principal radii of curvature, R_1 and R_2. A flat film has zero curvature since $1/R_1 = 1/R_2 = 0$. Now consider a film curved into a uniform cylinder of radius R. In this case $1/R_1 = 1/R$, while $R_2^{-1} = 0$. We adopt the sign convention that H is positive when the film curves toward oil while it becomes negative when the aqueous medium is inside the cylinder. For a generally curved surface the mean curvature is

$$\mathbf{H} = \tfrac{1}{2}(\text{sign } 1/R_2 + \text{sign } 1/R_2) \qquad (1.3.3)$$

where the $+/-$ of the sign function follows from the convention that the curvature toward oil is positive.

The free energy of the film depends on how much it is curved. The free energy g_f per unit area, which for the present case is $g_f = \mathbf{G}_f/(2\pi RL)$ (where L is cylinder length), is a function of R. Now by physical arguments g_f cannot reach arbitrarily small values while it should reach high values for highly curved surfaces in the limits $R \to 0$. Consequently, a global minimum $g_f(R)$ must exist so that

$$\frac{dg_f}{d1/R} = 0 \qquad (11.1.2)$$

at some value $H_o = \tfrac{1}{2}\,\text{sign } 1/R_o$ of the curvature. This defines the spontaneous curvature H_o of the film. We shall see that this important characteristic of a surfactant film is closely related to the surfactant number N_s introduced in Chapter 1.

We obtain the fourth parameter, the bending rigidity or bending elastic modulus κ of the film, by taking the second

Figure 11.1
Perpendicular cut through an oil–water interface showing the curved surfactant film. Note the tighter chain packing when the film curves toward the oil compared to when it curves toward water.

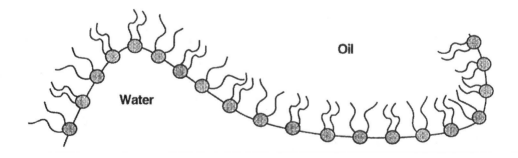

derivative of g_f so that

$$\kappa = \frac{d^2 g_f}{d(1/R)^2} \qquad (11.1.3)$$

It is physically most consistent to evaluate the derivative at $1/R_1 = 2H_o$, although most authors use $1/R_1 = 0$ as the reference point. Furthermore, we choose to consider a situation of constant amount of surfactant per area although one also could evaluate the derivative at constant chemical potential. A third technical complication in the description of curved films of finite thickness occurs because the area of the film depends on the plane in the film along which the area is evaluated. For example, in the cylindrical case, $A = 2\pi R L$, but the value of R changes by an amount corresponding to the film thickness if the area is evaluated on the inside rather than the outside surface. The conventional choice is to consider the so-called neutral surface, the plane that gives the same area per molecule as the planar surface. For surfactants this neutral surface usually is located close to the plane where the polar head group meets the alkyl chains.

We can introduce the fifth elastic constant, the saddle splay modulus $\bar{\kappa}$ by considering a saddle-like deformation of a planar interface such that sign $1/R_1 = -$sign $1/R_2$. Then

$$\bar{\kappa} = -\tfrac{1}{2} \frac{d^2 g_f}{d(1/R_1)^2} \qquad (11.1.4)$$

The first derivative is zero since a substitution sign $1/R_1 \to -$sign $1/R_1$ only amounts to a 90° rotation of the saddle while otherwise the film is unaffected.

In the next section we will discuss the typical magnitudes of these coefficients. Now we merely note that the area expansion modulus is typically so large that we often can assume that the area per molecule remains constant as the film is bent. For this case we can cast $g_f(R_1, R_2)$ in a convenient form by expanding to second order in the curvatures

$$g_f = \gamma + 2\kappa(H - H_o)^2 + \bar{\kappa} K \qquad (11.1.5)$$

where $K = $ sign $1/R_1 \cdot$ sign $1/R_2$ is called the Gaussian curvature.

Problem: Show that eq. 11.1.5 is consistent with eqs. 11.1.2–11.1.4.

Solution: Taking the first derivative of eq. 11.1.5 with respect to $1/R_1$

$$\frac{\partial g_f}{\partial 1/R_1} = 4\kappa(H - H_o)\frac{\partial H}{\partial 1/R_1} + \bar{\kappa}\frac{\partial K}{\partial 1/R_1}$$

$$= 2\kappa\left(\frac{1}{2}\frac{1}{R_1} + \frac{1}{2}\frac{1}{R_2} - H_o\right) + \frac{\bar{\kappa}}{2}\frac{1}{R_2}$$

At $1/R_1 = 2H_o$ and $1/R_2 = 0$, we find

$$\frac{\partial g_f}{\partial 1/R_1} = 0$$

as in eq. 11.1.2. The second derivative is

$$\frac{\partial^2 g_f}{\partial (1/R_1)^2} = \kappa$$

as in eq. 11.1.3. For sign $1/R_2 = -\text{sign } 1/R_1$ so that $H = 0$, eq. 11.1.5 reduces to

$$g_f = \gamma + 2\kappa H_o^2 - \bar{\kappa}(1/R_1)^2$$

and

$$\frac{d^2 g_f}{d(1/R_1)^2} = -2\bar{\kappa}$$

By integrating g_f of eq. 11.1.5 minus γ we obtain the curvature free energy

$$\mathbf{G}_f = \int (g_f - \gamma) \, dA = \int \{2\kappa(H - H_o)^2 + \bar{\kappa}K\} \, dA \quad (11.1.6)$$

This equation, which is called the Helfrich curvature (free) energy after the German physicist Wolfgang Helfrich, has played a pivotal role in establishing our current understanding of monolayer and bilayer systems. This curvature free energy expression implies that for $\bar{\kappa} < 0$ and $2\kappa + \bar{\kappa} > 0$ there exists an optimal state of zero curvature energy when $H = H_o(1 + \bar{\kappa}/2\kappa)^{-1}$ and $1/R_1 = 1/R_2$. As the curvature deviates from these optimal values there is a free energy penalty that depends on the values of κ and κ', respectively. For $\bar{\kappa} > 0$ or $2\kappa + \bar{\kappa} < 0$ additional terms have to be added to the curvature free energy expression to ensure the existence of a stable film configuration.

Problem: Calculate the curvature energy of a film on a sphere of radius R when $H_o = 0$.
Solution: For a sphere $H = \pm 1/R$ and $K = 1/R^2$ so that the curvature energy is

$$\mathbf{G}_f = \int_A \left(2\kappa \frac{1}{R^2} + \frac{\bar{\kappa}}{R^2}\right) dA = 4\pi R^2 \left(\frac{2\kappa}{R^2} + \frac{\kappa}{R^2}\right) = 4\pi(2\kappa + \bar{\kappa})$$

irrespective of the value of R.

A virtue of eq. 11.1.6 is that we can show mathematically in the so-called Gauss–Bonnet theorem, that the integral of the Gaussian curvature depends solely on the topology of the surface. In physical terms this means that deformations of the film leave the integral over K unchanged. This changes value only when the film breaks.

11.1.2 With Some Effort We Can Measure Elastic Constants

In the preceding section we introduced four new phenomenological parameters: the spontaneous curvature H_o, the bending elastic modulus κ, the saddle splay modulus $\bar{\kappa}$, and the area expansion elastic modulus K_a. Later in this chapter we will see that H_o is an important parameter for understanding the structure and stability of microemulsions and is also most relevant for the discussion of emulsion properties. Bending elasticity influences both the static and dynamic properties of microemulsions and emulsions, while the area expansion modulus is particularly relevant in connection with film rupture under tension.

Obtaining precise experimental values for all four constants has turned out to be experimentally difficult, although certain trends have been well established. The curvature H_o depends both on the surfactant and on the composition of the polar and apolar phases it separates. On the apolar side the oil penetrates to some extent between the hydrocarbon tails of the surfactant, as shown in Figure 11.2. The more extensive the penetration, the more curvature is imposed toward the polar side. As a result H_o decreases because we define curvature as positive when it is toward the oil. The longer the oil chains, the less they penetrate the surfactant film for entropic reasons and the smaller the effect on H_o. In addition, raising the temperature induces more gauche conformations in the surfactant chains, which become more coiled, also decreasing H_o.

The nature of the polar head group influences how changes in the conditions of the polar (aqueous) medium affect H_o. For ionic surfactants the bulk electrolyte concentration exerts a strong effect. The higher the salt content, the more

Figure 11.2
Penetration of oil molecules between the apolar chains in the surfactant film. Note the greater extent of penetration when the film curves toward water.

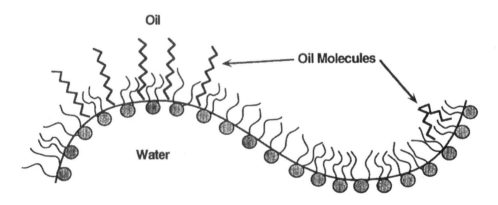

easily the film curves toward the water, leading to a decrease in H_o. We learned in Chapter 3 that the electrostatic free energy is dominantly entropic. Thus, increasing the temperature leads to an increase in the electrostatic free energy and therefore also in H_o.

The combined effects of temperature on the apolar chains and on the electrostatic interactions are competitive. Empirically, the electrostatic term seems to dominate·slightly, so H_o increases weakly with increasing temperature.

For nonionic surfactants of the ethylene oxide type, electrolytes affect H_o only at high concentrations, and the sign of the effect depends on the nature of the ions according to the Hofmeister series discussed in Section 3.7. On the other hand, temperature has a profound effect. As T is increased, water becomes a less good solvent for the ethylene oxide monomers. As a result, water penetrates less into the palisade layer, with a concomitant decrease in H_o. Because the chains on the other side of the film support the same trend in the variation of H_o, increasing temperature for this type of surfactant causes a strong decrease in H_o. This phenomenon explains the strong temperature effects on the phase equilibria of Figure 10.11.

Thus, by varying the polar head group and the length and number of the apolar chains, adding a cosurfactant, or changing external parameters (such as temperature, the nature of the oil, or the electrolyte concentration), we can tune the spontaneous curvature to any value in the range 0.5 to $-0.5 \, \mathrm{nm}^{-1}$.

The bending elasticity modulus κ is expressed in units of energy, and its typical measured values lie in the range of $1-20 kT$ $(T = 300 \, \mathrm{K})$ for amphiphile monolayers and bilayers in the liquid state. Factors that make the film more flexible are short chains, cosurfactants, double-chain surfactants with unequal chains, more oil penetration, and a higher fraction of cis-unsaturated bonds. Other things being equal, the value of κ for a monolayer represents half that of the corresponding bilayer.

Few reliable measurements have been made of the saddle splay modulus $\bar{\kappa}$. The typical trend is that $\bar{\kappa}$ is negative, a situation that emerges naturally considering the minus sign in eq. 11.1.5. The magnitude of $\bar{\kappa}$ is usually smaller than that of κ for the same system.

Bilayers are a symmetric combination of two monolayers. Due to the symmetry $H_o = 0$ we expect that $\kappa(\text{bilayer}) \gtrsim 2\kappa(\text{monolayer})$, while the saddle splay constant of the bilayer is affected by the monolayer spontaneous curvature $H_o(\text{mono})$ and

$$\bar{\kappa}(\text{bilayer}) \simeq 2\bar{\kappa}(\text{monolayer}) - \kappa(\text{monolayer})l_o H_o(\text{mono})$$

Here the second term on the r.h.s. can dominate over the first and cause $\bar{\kappa}(\text{bilayer})$ to change sign relative to that of the monolayer. This effect is important for understanding the stability of sponge phases and some cubic liquid crystals.

The area expansion modulus K_a has been measured for only a few cases of phospholipid bilayer systems. Typical values are in the range of $0.1-1 \, \mathrm{J/m^2}$. Figure 11.3 illustrates an

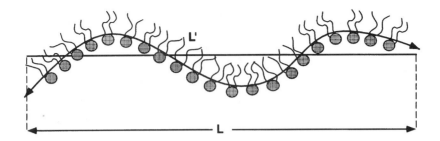

Figure 11.3
In a perpendicular cut, the contour length L' of the film is longer than the projected length L when bending fluctuations are present. For a two-dimensional film the actual area A' would be similarly larger than the projected area A.

important complication found in applying this value to a surfactant film or a bilayer. For a fluctuating interface we can measure the area in two different ways: either along the film or at the projected area at the film's mean location. Experimentally the latter area is most accessible. However, when we stretch a film and change its projected area, the intrinsic area remains virtually the same, while the fluctuations are damped. In such a case, the apparent area expansion modulus is much smaller than K_a until we reach the regime where all the bending fluctuations of the film have ceased.

11.2 Microemulsions Are Thermodynamically Stable Isotropic Solutions That Display a Range of Self-Assembly Structures

11.2.1 Microemulsions Can Contain Spherical Drops or Bicontinuous Structures

A microemulsion is an isotropic solution containing substantial amounts of both a strongly polar component (usually water) and a strongly apolar component (usually oil) that are stabilized thermodynamically by an amphiphilic additive. As we will see there is a substantial structural diversity in microemulsions meeting this thermodynamic criterion.

The most clear-cut examples are the oil-in-water (O/W) and water-in-oil (W/O) microemulsions illustrated in Figure 11.4. In these cases the dispersed medium forms spherical droplets. We can calculate their size using a simple geometrical argument. It is easy to derive the volume fractions Φ_w, Φ_o, and Φ_s (water, oil, and surfactant) from the given composition and densities of the components.

We can see the oil-in-water microemulsion droplet as composed of a hydrocarbon core of radius R_c covered by a layer of surfactant head groups making the overall radius R_H. By simple geometry the radius R_c is related to the ratio between the hydrocarbon volume V_{hc} and the area A_{hc} of the core

$$R_c = 3V_{hc}/A_{hc} \qquad (11.2.1)$$

The hydrocarbon volume is the sum of the oil and a fraction α

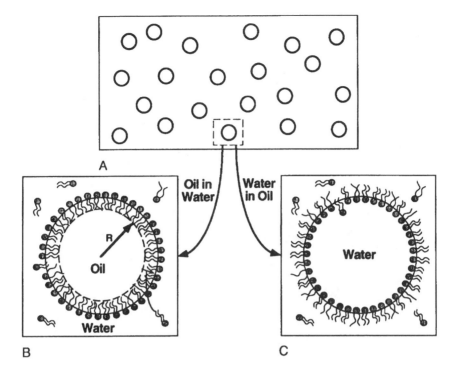

Figure 11.4
A droplet microemulsion
phase (a) can be of the
oil-in-water (b) or the
water-in-oil type (c).
Typically, but not
exclusively, the drops are
nearly spherical and have a
small size polydispersity.
The radii R of the drops are
usually one to four times
the thickness l_o of the
surfactant film. The excess
surfactant is almost
exclusively in the
continuous phase.

of the surfactant volume so that

$$V_{hc} = V_{tot}(\Phi_o + \alpha\Phi_s) \tag{11.2.2}$$

We now assume that the surfactant contributes a fixed area a_o
per molecule making the total core area

$$A_{hc} = \frac{V_{tot}\Phi_s}{v_s} a_o \tag{11.2.3}$$

where v_s is the molecular volume of the surfactant. We can now,
combining eqs. 11.2.1–11.2.3, obtain an expression for the core
radius in terms of experimentally accessible quantities

$$R_c = 3(\Phi_o/\Phi_s + \alpha)(v_s/a_o) \tag{11.2.4}$$

Here α is typically on the order of 0.5 while v_s/a_o lies in the
range 1.5–2 nm and microemulsion droplets mostly have radii
in the interval 5–20 nm.

We could ascribe the same argument equally well to the
W/O microemulsion. An important feature of eq. 11.2.4 is that
we determine the radius R_c solely by the volume fractions.
When the spontaneous curvature H_o of the film is close to the
inverse radius R_c^{-1} of the drop, we can expect a relatively stable
situation with a practically unfrustrated surfactant film. However, when $H_o \gg R_c^{-1}$, that is, for large values of the spontaneous
curvature, the film can relax toward a lower free energy state by

Figure 11.5
Scanning electron micrographs of a bicontinuous microemulsion containing 7 wt % $C_{12}E_5$ and equal amounts of water and octane. The sample was prepared by freeze-fracture and subsequent deposition of a tantalum–tungsten coating. Contrast between the oil and water phases occurs because the coating forms a mottled pattern on the oil regions. (W. Jahn and R. Strey, *J. Phys. Chem.* **92**, 2294 (1988).)

decreasing the drop size, expelling the emulsified oil (or water for negative H_o) to a bulk phase. When $H_o \ll R_c^{-1}$, the film can relax by forming larger nonspherical aggregates.

Ultimately, when $H_o \approx 0$, there is no preference for a O/W relative to a W/O structure. Instead the surfactant film separates two macroscopic oil and water regions in a bicontinuous structure, which represents a disordered version of the bicontinuous cubic phases discussed in Chapters 1 and 6. Figure 11.5 shows a recent series of cryo-scanning electron micrographs of a fractured bicontinuous microemulsion phase. Figures 11.4 and 11.5 illustrate two types of microemulsion structure. Other structures have been found, but these represent the basic intermediate cases in which large finite flexible aggregates form or a bicontinuous structure is obtained even when one of the solvent components is present in excess.

11.2.2 Temperature Controls the Structure and Stability of Nonionic Surfactant Microemulsions

Nonionic surfactants with oligoethylene oxide head groups form stable microemulsions. These types of surfactants are both technically important, because they can be synthesized readily using ethylene oxide, and fundamentally interesting. As we discussed in the preceding section, their behavior is very temperature dependent.

Most microemulsion systems that have been studied contain four or more components. In addition to oil, water, and surfactant, we usually must add a cosurfactant (alcohol) and/or an electrolyte to tune the system to the rather special conditions necessary for microemulsion stability. However, we can tune the nonionic surfactants in a ternary system by adjusting the temperature. Furthermore, the weak interbilayer forces in these systems also favor the stability of the microemulsion relative to the lamellar phase.

Working with a strict three-component system has two virtues. First, it simplifies the representation of the phase diagram. For each temperature we can use a triangular representation like that shown in Figure 10.7. A second virtue of a ternary system becomes apparent in applications. For example, to employ a microemulsion in tertiary oil recovery, the surfactant film's properties must remain intact as the process goes on in time. With more than one amphiphilic component, the loss of surfactant due to surface adsorption inevitably will be selective. In a sense, the process resembles large-scale silica gel chromatography. Changing composition results in changed properties, which usually is an undesirable feature.

Figure 11.6 shows a series of triangular diagrams for the system $C_{12}E_5$–water–tetradecane at three different temperatures. Determining one such diagram is rather tedious work, and the creation of a consistent series represents a major effort. Conversely, the diagrams summarize a great deal of experimental observation, although extracting this information from the diagrams requires some training.

The diagrams show one liquid crystalline phase, a lamellar phase, and for a given temperature, several isotropic liquid phases. These could be identified as a dilute micellar solution in water, a dilute inverted micellar solution in oil also containing monomeric surfactant, a microemulsion phase, and an L_3 or sponge phase. As the temperature changes, these isotropic one-phase regions split in two and merge with others. Through a concomitant change of concentration and temperature, passing from one of these isotropic phases to all the others can be accomplished without ever entering a two-phase area.

Although the thermodynamic nature of the phase does not change in such a process, major structural changes can occur on a more microscopic scale. In the water corner, $\Phi_w \gg \Phi_o$, Φ_s, for example, we expect surfactant micelles to be slightly swollen with oil. Similarly, in the oil corner, $\Phi_o \gg \Phi_w$, Φ_s, the typical structure is an inverted micelle in oil. In this case, we also find some surfactant monomers in the oil medium. For this surfactant,

Figure 11.6
Phase diagrams of the
ternary system water—
tetradecane–$C_{12}E_5$. At
47.8°C. Five distinct
isotropic phases can be
observed: the dilute normal
micellar solution L_1, in
the left-hand corner; the
concentrated micellar
solution L_1, close to the
binary water–$C_{12}E_5$ axes
(cf. Figure 10.11); the
sponge or L_3 phase in the
left-hand corner; the
inverted micellar solution
L_2, which encompasses
the pure oil and pure
surfactant phases; and
finally the microemulsion
μE, in the center of the
diagram. In addition, there
is a large lamellar phase
region L_a. The three corners
of the fan-shaped μE region
connect to three three-
phase triangles. At low
surfactant concentrations
the microemulsion is in
equilibrium with the L_1
and L_2 phases. (B) At
51.2°C. At a temperature
slightly higher than that
shown in (A), the μE corner
of the three-phase triangle
$L_1 + \mu E + L_2$ has moved
toward the oil corner. The
μE region now has a L_3
"nose" pointing toward the
water corner. Furthermore,
the μE region merges
smoothly into the oil-rich
L_2 phase. (C) At 44.5°C. At
the slightly lower
temperature the three-
phase triangle defined by
$L_1 + \mu E + L_2$ has its μE
apex moved toward higher
water contents. In addition,
the upper right-hand
corner of the μE fan has
been drawn out to a long
"nose" encompassing the
inverted L_3 phase
containing a aqueous sheet
in an apolar environment.
(K. Kunieda and K.
Shinoda, *J. Dispension Sci.
Technol.* **3**, 233 (1982).)

A

B

C

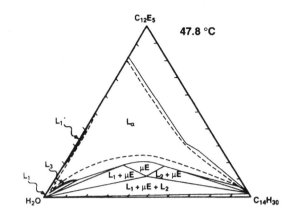

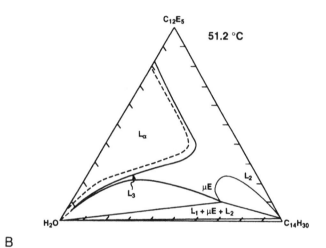

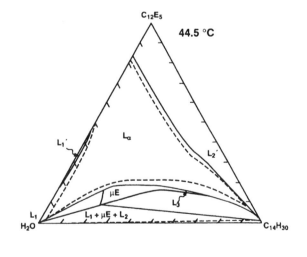

inverted micelles in oil display a substantially higher CMC than normal micelles in water.

For $\Phi_w \approx \Phi_0 \gg \Phi_s$, provided Φ_s is present in nonnegligible amounts, a three-phase equilibrium exists between an aqueous solution, an oil phase, and a microemulsion phase containing the majority of the surfactant. At a specific temperature, $T \simeq$ 47.8°C in Figure 11.6A, the microemulsion in the three-phase triangle contains equal amounts of oil and water. This phase inversion temperature (PIT) will play an important role when we return to its microscopic interpretation.

By increasing the temperature above the PIT, the micro-emulsion of the three-phase triangle becomes increasingly rich in oil, as seen in Figure 11.6B for $T = 51.2$°C. Slightly above 51.2°C (at 51.6°C) the "microemulsion" merges with the in-verted micellar solution and the three-phase triangle disap-pears. Similarly, by lowering the temperature from the PIT, the microemulsion becomes richer in water (see Figure 11.6C) and ultimately merges with the normal micellar solution at a tem-perature slightly below 44.5°C (at 43.8°C). Thus, in the present example, the three-phase triangle appears only in a limited temperature range (43.8–51.6°C).

The phase present between 43.8 and 51.6°C that connects to the three-phase triangle is sometimes called a Winsor III microemulsion. Below 43.8°C the water can still accommo-date substantial amounts of oil and such a phase is called a Winsor I microemulsion. Finally, above 51.6°C the oil can dissolve substantial amounts of water and thus is a Winsor II microemulsion.

This series of ternary phase diagrams provides a nearly complete picture of the phase equilibria in the system. To aid visual analysis, we can make constant composition cuts so that a single diagram can convey various temperature effects. We can make such cuts in many ways. Here we illustrate the two most common ones; the χ cut and the fish cut. The rationale for the names will become apparent.

In the χ cut, we keep the surfactant concentration con-stant at a value slightly higher than that for the microemulsion in the three-phase triangle. In Figure 11.7 the surfactant concen-tration is fixed at 16.6% (w/w). The oil/(oil + water) (w/w) ratio, α, and the temperature are the two independent variables. (Using the volume/volume ratio is more consistent from a conceptual point of view, but experimentally, weights are the measured quantities.

A limitation in using these cuts is that the multiphase areas could contain phases that are not present as pure phases in the cut, while the extent of the one-phase areas emerges clearly. The cut at constant surfactant concentration means a cut parallel to the base of the triangular diagrams in Figures 11.6.

In Figure 11.7, a microemulsion one-phase "channel" goes from the lower left to the upper right corner, representing one leg of a Greek χ. In addition, a narrower one-phase channel goes from the lower right to the upper left corner, representing the other leg of the χ. This channel is a sponge or L_3 phase, and

Figure 11.7
Phase equilibria in the χ cut at a constant amount, 16.6% (w/w), of the surfactant $C_{12}E_5$. The two independent variables are the temperature and the weight fraction oil, $\alpha = (C_{14}H_{30})[(H_2O) + (C_{14}H_{30})]^{-1}$. As both α and temperature increase simultaneously, the channel running from the lower left-hand corner to the upper right defines the isotropic microemulsion phase, which passes from normal surfactant micelles in water to oil droplets in water (O/W) via a bicontinuous structure to water droplets in oil (W/O) and finally reversed micelles in oil. In addition, a narrow L_3 channel goes from the upper left to the lower right-hand corner. Phase notation as in Figure 11.9. (U. Olsson, K. Shinoda, and B. Lindman, *J. Phys. Chem.* **90**, 4083 (1986).)

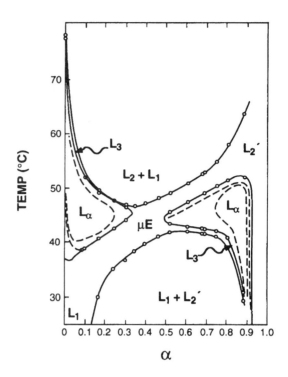

its location at $\alpha = 0$ corresponds — as it should — to the sponge phase in the binary surfactant–water system discussed in Chapter 10.

The PIT is located where the microemulsion and the sponge phase meet at the center of the χ. If we follow the microemulsion channel, we find an uninterrupted transition from a water-continuous medium at low T and α, to a bicontinuous one at intermediate T and α, to an oil-continuous one at high T and α. Figure 11.8 illustrates this transition and shows the variation of self-diffusion coefficients with α plotted as ratios relative to a neat phase of either component. At α around 0.5 both oil and water have reduced diffusion to approximately 60% of the bulk value, while the surfactant has its maximum value. This condition constitutes the signature of a bicontinuous system symmetric with respect to the two continuous media. At either side of this point the surfactant diffusion decreases, while if $\alpha > 0.5$, oil diffusion increases and water diffusion decreases or vice versa if $\alpha < 0.5$.

We can obtain another perspective on the phase equilibria from the fish plot of Figure 11.9. Here the oil/(oil + water) ratio remains constant at 0.5, while the surfactant concentration varies. This plot reveals a striking symmetry with an approximate mirror plane at $T = $ PIT. The three-phase triangle has its largest extension at the PIT. The area where the three phases are in equilibrium adopts the shape of a fish. At higher concentrations the lamellar phase appears. For both the microemulsion and the lamellar phase, the one-phase regions extend to lowest surfactant concentration at the PIT.

Figure 11.8
Variation of the self-diffusion coefficient of water (○) and tetradecane (△) as a function of weight fraction α for the water–oil–surfactant microemulsion system of Figure 11.10. As evident from Figure 11.10, composition and temperature must be varied simultaneously. Diffusion coefficients are plotted relative to the observed values D_0 in the respective neat liquids. At low α values corresponding to a O/W microemulsion, D/D_0 is close to unity for water and small for tetradecane. At high α values, where a W/O microemulsion occurs, the D/D_0 roles are reversed. For α values around 0.5, the water and the oil diffusion are equally affected by the structure, which is the signature of a bicontinuous phase. (U. Olsson, K. Shinoda, and B. Lindman, *J. Phys. Chem.* **90**, 4083 (1986).)

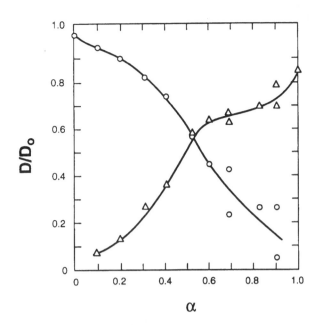

Figures 11.6 to 11.9 summarize experimental observations of a particular three-component microemulsion system, but other nonionic surfactant systems qualitatively behave in the same way. Quantitative changes occur in the PIT and in the surfactant concentration of the Winsor III microemulsion when the surfactant or the oil is changed. However, the characteristic patterns of the χ plot and the fish plot remain unchanged. How can we understand what molecular factors govern this generic phase behavior?

In Figures 11.6, 11.7, and 11.9, $T = $ PIT represents a symmetric state in which water and oil seemingly are equivalent in the sense that the same structure is obtained for α and $1 - α$. Now a microemulsion can be seen as a system composed of two media (polar and apolar) separated by a surfactant film with some spontaneous curvature, H_o. When $H_o = 0$, the film curves as readily toward the oil as toward the water, equating the two media even though chemically they are very different. This equivalence leads to the conclusion that the PIT is the temperature where $H_o = 0$.

At $H_o = 0$ we would expect maximum stability of the lamellar phase because the film then prefers the planar state, and indeed, this can be observed. The microemulsion occurring at the PIT and thus at $H_o = 0$ also should not be of either a W/O or an O/W type, but it should display a structure where the two media have equivalent roles. As the diffusion experiments show, the microemulsion is indeed bicontinuous with equivalent D/D_o for the oil and the water.

If we focus on H_o as the most essential parameter of the system, we can find its temperature dependence by making a Taylor expansion around PIT $= T_o$ so that

$$H_o(T) \simeq \frac{dH_o}{dT}\bigg|_{T_o} (T - T_0) + \cdots \qquad (11.2.6)$$

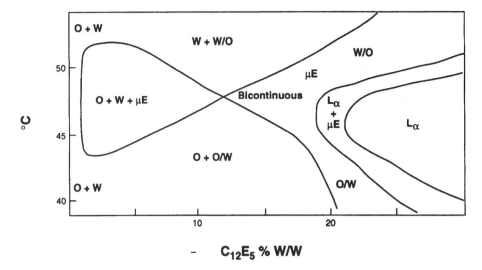

$C_{12}E_5$ % W/W

Figure 11.9
The fish cut. For an equal ratio (W/W) of water and tetradecane ($\alpha = 0.5$) the sequence of phase varies with increasing surfactant concentration at different temperatures around the phase inversion temperature of 47.8°C. The symbol O denotes a tetradecane phase with some surfactant and W an aqueous phase with small amounts of surfactant. Other symbols as in Figure 11.6. The region of the three phases joins the one-phase microemulsion region as in the tail of a fish.

For two temperatures, $T_0 + \Delta T$ and $T_0 - \Delta T$, placed symmetrically relative to T_0 it follows that

$$H_o(T + \Delta T) = -H_o(T - \Delta T) \qquad (11.2.7)$$

At $T + \Delta T$ the film wants to curve toward water as much as it wants to curve toward oil at $T - \Delta T$. This implies that if the film curvature energy provides a dominant contribution to the free energy, we should expect the same phase structure at temperature $T + \Delta T$ and composition α as we find for temperature $T - \Delta T$ and composition $1 - \alpha$. Thus at a change

$$(\alpha, T_0 + \Delta T) \leftrightarrow (1 - \alpha, T_0 - \Delta T) \qquad (11.2.8)$$

the aggregate structure remains the same but the oil and the water exchange places. Thus, at both $T + \Delta T$ and $T - \Delta T$ the spontaneous curvature is H_o toward the medium with the volume fraction α. The film does not notice that the oil and water have exchanged places because it sees only a spontaneous curvature and a volume fraction.

A number of predictions follow from eqs. 11.2.6–11.2.8. First, in the fish diagram $\alpha = 1 - \alpha$ by choice, so that the temperatures PIT $+ \Delta T$ and PIT $- \Delta T$ should give equivalent phases according to eq. 11.2.5. The symmetry apparent in Figure 11.9 supports our assumption.

In addition, the χ plot implies that the partial phase diagram should be invariant with respect to a 180° rotation around the point ($T =$ PIT, $\alpha = 0.5$). Figure 11.7 shows this to be true to an amazing degree.

The simple rationalization of the apparently complex series of phase diagrams represented in Figure 11.6 in terms of a surfactant film with a temperature-dependent spontaneous curvature separating two media of volume fractions Φ_w and Φ_o implies that spontaneous curvature is a main control parameter. When the surfactant volume fraction Φ exceeds either Φ_w or Φ_o, we must refine the discussion.

When $\Phi_s/\Phi_o \gtrsim 1$ the surfactant film is not exposed to an excess of oil and fewer oil molecules penetrate between the apolar chains of the surfactant. This will affect the value of H_o. Similarly, for $\Phi_s/\Phi_w \gtrsim 1$ the polar groups experience an interaction with other polar groups across a thin aqueous region. This affects the spontaneous curvature so that H_o will be concentration dependent under these circumstances.

11.2.3 We Often Need Electrolytes to Obtain Microemulsions for Ionic Surfactants

In Section 11.2.1, we found that bicontinuous microemulsions occur when the spontaneous curvature of the surfactant film is around zero. For nonionic surfactants, we tuned such conditions by changing the temperature. How can we generalize these concepts so that they apply to ionic systems?

For single-chain surfactants, the surfactant number N_s is of the order 0.3 or $H_o \approx 1/l_o$ in terms of curvature, which is far from the optimal condition $H_o \approx 0$ ($N_s \approx 1$). Adding electrolyte can decrease H_o by decreasing the importance of the electrostatic contribution to H_o, but this decrease is insufficient to yield the curvature needed for microemulsion formation.

One solution to the problem of obtaining a balanced microemulsion is to use a cosurfactant, usually a suitable alcohol such as butanol, pentanol, or hexanol. This addition increases the effective surfactant number because the alkyl chains of the alcohols contribute to the hydrocarbon volume, while the alcohol group replaces apolar groups at the polar—apolar interface.

Thus, we can use both cosurfactant/surfactant ratios and electrolyte concentration to control the spontaneous curvature. Clearly, H_o also is a function of temperature as it was for the nonionics, but for reasons discussed in Section 11.1.2, the temperature dependence is weaker. In practice, we can thus use temperature only for fine tuning H_o in ionic surfactant systems.

An obvious alternative to adding cosurfactants to achieve $H_o \approx 0$ conditions is to introduce double-chain surfactants. This approach has been widely used with both anionic and cationic surfactants. We will address the behavior of the cationic surfactants in the following section; here we concentrate mainly on the properties of the anionic microemulsions based on di(ethyl hexyl)sulfosuccinate (AOT).

Figure 11.10 shows the ternary phase diagram AOT–water–isooctane. The binary AOT–water system shows a lamellar phase over a large concentration range. This phase incorporates only small amounts of oil, but an isotropic solution of the W/O microemulsion type appears at high surfactant and low water contents.

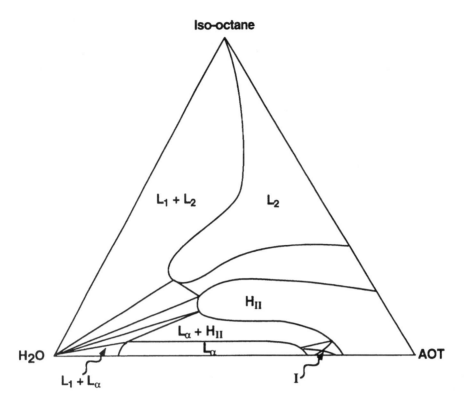

Iso-octane

$L_1 + L_2$

L_2

H_{II}

$L_\alpha + H_{II}$

L_α

H₂O

AOT

$L_1 + L_\alpha$

I

Figure 11.10
Phase diagram of the ternary system H₂O—AOT—isooctane at 20°C. Three liquid crystalline phases are shown: a lamellar phase L_α, a reversed hexagonal phase H_I and a bicontinuous cubic phase I. The two isotropic phases are a dilute aqueous solution L_1, and a W/O microemulsion phase–oil surfactant solution denoted L_2. (K. Fontell, private communication.)

To extend the range of the microemulsion, we must add electrolyte. Its role is twofold: we have already discussed how electrolyte decreases H_o, but here its effect on the intersurface interactions is an even more important factor. As we saw explicitly in the preceding section, both the microemulsion and the lamellar phase are most stable when $H \approx 0$. When there are strong repulsive or attractive interactions between separate segments of the surfactant film, the interaction free energy is minimized for an ordered state with a uniform separation between surfactant films. Thus, this case favors the lamellar phase. When the long-range interactions are weak, as they are for nonionic surfactants and for ionic surfactants at high electrolyte concentrations, a disordered microemulsion phase becomes a more likely event. The higher entropy associated with the disordered structure now provides a relatively more important contribution to the free energy. Thus, for the salt-free AOT system, a lamellar phase rather than a microemulsion forms at $H_o \approx 0$ conditions.

Isotropic phase regions expand when we add electrolyte, and a bicontinuous microemulsion can be found above approximately 0.4% (W/W) of NaCl in the aqueous phase. Figure 11.11 shows a χ plot for a system having a 0.46 (W/W) NaCl in the water. It strongly resembles the χ plot of Figure 11.7, but with the difference that the temperature effect is reversed. For example, at the water-rich end the one-phase region around 60°C is an O/W microemulsion, while the narrow one-phase region at 47–48°C is the sponge phase. In Section 11.1.2 we

Figure 1.11
A χ cut at 5% (w/w) surfactant for the system iso-octane–AOT–H$_2$O (0.46% w/w NaCl). The shape of the diagram is analogous to that in Figure 11.10, but the temperature dependence is reversed. There are no multiphase areas at high oil contents around the phase inversion temperature. However, this part of the diagram was not carefully studied. (J.O. Carnali, A. Ceglie, B. Lindman, and K. Shinoda, *Langmuir* **2**, 417 (1986).)

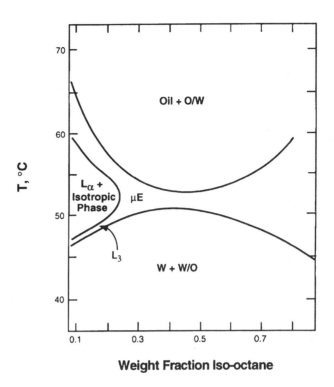

Weight Fraction Iso-octane

gave the molecular reason for this difference between ionic and nonionic systems, namely the temperature variation of the spontaneous curvature of the surfactant film differs in sign.

That the spontaneous curvature increases with increasing temperature for ionic systems is shown even more clearly in the fish plot of Figure 11.12. Increasing the salt content of the aqueous phase leads to a decrease in H_o. Nevertheless, a state of $H_o = 0$ can be found by increasing the temperature. Thus, the three-phase microemulsion equilibrium is found in the range 22–25°C for 0.40% NaCl, 30–37°C for 0.58% NaCl, and 50–65°C for 1.0% NaCl. We see also in Figure 11.12 that the tail of the fish tilts, and the tilt becomes more pronounced, the lower the salt content. The reason is that the electrostatic interactions are affected by all ions in the aqueous phase, including both the added electrolyte and the counterions of the surfactant. Thus, as the AOT is added, the Na$^+$ ion concentration in the aqueous phase increases, leading to a decrease of H_o. The effect of the counterions becomes relatively more important the lower the NaCl concentration.

11.2.4 DDAB Double-Chain Surfactants Show Bicontinuous Inverted Structures

Three-component microemulsions formed with double-chain cationic surfactants, such as diododecyldimethylammonium bromide (DDAB), exhibit behavior that differs substantially

Figure 11.12
Fish cut for the system
n-decane–AOT–H₂O—
NaCl for a mixture of a
constant 1:1 ratio of decane
to brine. Curves are shown
for three different
concentrations of NaCl in
water: $\varepsilon = 1.0$, 0.58, and
0.40% w/w. The higher the
salt content, the higher
the temperature at which
the three-phase triangle
occurs and thus the higher
the point at which $H_o = 0$.
Solid symbols denote the
boundary of the three-
phase equilibrium; open
symbols denote the
boundary of the
microemulsion. (M.
Kahlweit and R. Strey, *J.
Phys. Chem.* **90**, 5239
(1986).)

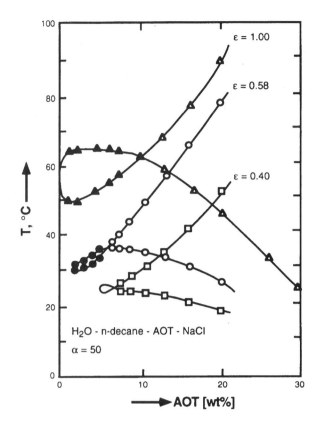

from that seen with surfactants, such as AOT. The major reason
is that optimum curvature of this surfactant in the presence of
oil and water leads to the formation of inverted structures. We
will discuss the properties of these microemulsions after summarizing the characteristic properties of surfactants like DDAB.

1. In water, DDFAB forms two lamellar phases with head
   group areas a_0 of 60 and 68 Å². The volumes of the
   hydrocarbon tails and head groups are estimated to be 704
   and 120 Å³. The chain length of 11.5 Å is 80% of the
   fully extended length.

2. The surfactant is sparingly soluble in water and oil, so that
   its concentration and effective head group area fix the
   interfacial are.

3. Surfactant chains pack flexibly because they are fluid at
   room temperature. This affects H_o by allowing oil to
   penetrate. From studies on bilayer, we know that oil
   uptake is greatest for small alkanes. When the alkane
   chain length exceeds that of the surfactant, oil penetration
   diminishes greatly.

Figure 11.13 shows the single-phase regions for several
DDAB microemulsions at 25°C for a number of oils. We can
summarize the main feature of these microemulsions as follows.

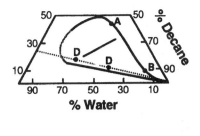

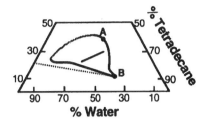

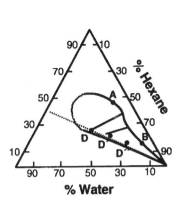

Figure 11.13
Partial phase diagrams (wt %) showing the one-phase didodecyldimethyl-ammonium bromide (DDAB) microemulsion region at 25°C. The *AB* line marks the minimum amount of water required to form the microemulsion region. The shaded line to the left of the single-phase region shows the gradual transition to a viscous, gellike region. The dotted line represents the percolation transition from conducting (low water content) to nonconducting (high water content) microemulsions. (F.D. Blum, S. Pickup, B.W. Ninham, S.-J. Chen, and D.F. Evans, *J. Phys. Chem.* **89**, 711 (1985).)

1. *Oil Specificity.* A high and systematic degree of oil specificity exists. The *AB* boundary at which the microemulsion phase first forms when we add water to an oil–surfactant mixture varies from 6% for hexane to abut 26% for tetradecane. This water volume corresponds to the molar ratio of water to bromide, which ranges from 4:1 to 50:1. For tetradecane, the microemulsion phase does not extend to the oil corner.

2. *Electrical Conductivity Measurements.* All the microemulsions are conducting along the onset line, *AB*. If we add water, conductance decreases for all the oils except tetradecane as we move from the *AB* line towards the water corner of the phase diagram along constant surfactant/oil (S/O) ratio, as shown in Figure 11.14, and we can observe a transition to a nonconducting liquid with a conductivity of half the neat oil. This transition occurs at a point called the percolation threshold. Figure 11.14 also illustrates the percolation thresholds defined by the conductance curves. For a given oil, the percolation concentrations follow a constant surfactant–water ratio, as shown in Figure 11.13. Conductivity also decreases along the *AB* line and displays a percolationlike transition in the region near the point marked *B*. Thus, the microemulsion appears to be bicontinuous along the *AB* line, but becomes water-discontinuous when we add water.

3. *Viscosity Measurements.* For all microemulsions except tetradecane, viscosity decreases when water is added. We can infer from this that the interconnected structure that exists at low water content breaks up when we add water.

4. *NMR Self-Diffusion Measurements.* Self-diffusion coefficients for the surfactant D_{surf} and oil D_{oil} are almost constant throughout the entire single-phase region. Those for the oil are about half as large as those for bulk oil. When we add water D_{H_2O} decreases by a factor of

Figure 11.14
Plot of the logarithm of the reciprocal of specific conductance, that is, measured solution resistance versus water (wt %) for DDAB microemulsions. Adding water to hexane (not shown), octane, decane, and dodecane microemulsions causes the resistance to increase almost to the value for pure oil. Extrapolated dashed lines determine the antipercolation points defined by the dotted lines in Figure 11.16. S/O denotes the ratio of surfactant to oil. Tetradecane microemulsions (solid circles) are always conducting (S/O = 0.343) (S.-J. Chen, D.F. Evans, and B.W. Ninham, *J. Phys. Chem.* **88**, 5855 (1984).)

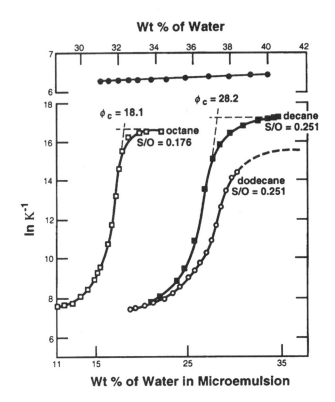

20, which substantiates the conductance measurements. Thus, all the DDAB microemulsions are bicontinuous along the *AB* line and all except those containing tetradecane transform to a water-in-oil droplet structure upon addition of water.

We can deduce a fair amount about the structure of DDAB microemulsions on the basis of these observations if we also take into account the guiding concepts of control of curvature at the oil–water interface and packing constraints.

Curvature is set by a balance between head group repulsion, which favors adminished curvature (such as bilayer-type structures), and oil penetration, which favors increased curvature (such as inverted cylindrical or spherical structures). Interconnected conduits lined with surfactant and filled with water—as shown in Figure 11.15—account for conductance along the minimal water *AB* lines. These microemulsions have a relatively low viscosity, which might suggest that these conduits continually make and break in a dynamic fashion. Another (perhaps more likely) possibility is that liquid flow gradients can exist without extensively perturbing the structure. Contrary to intuitive notions, a bicontinuous structure can sustain a flow gradient since all molecules in the system have full lateral mobility.

Adding water causes the conduits to increase in diameter and to begin to depart from the optimum curvature set by head group repulsion and oil penetration. One way for the structure

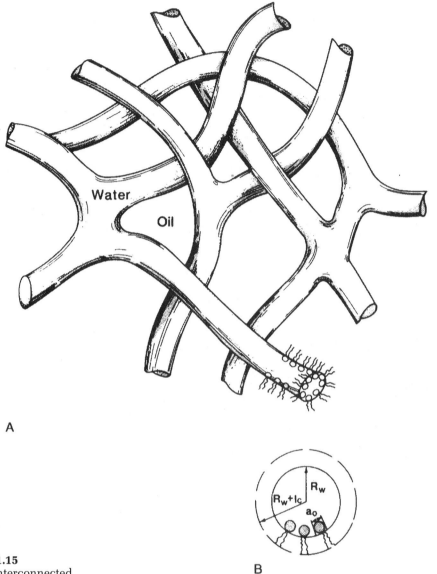

Figure 11.15
(a) The interconnected conduits that comprise the bicontinuous structures in DDAB microemulsions. Upon addition of water, the conduits disconnect and transform to water-in-oil droplets (b). The W/O droplets occur below the dashed lines in Figure 11.14, while the conduits predominate in the *A* region (S.-J. Chen, D.F. Evans, B.W. Ninham, D.J. Mitchell, F.D. Blum, and S. Pickup, *J. Phys. Chem.* **90**, 842 (1986).)

to maintain optimum curvature is to transform from conduits ($H = 1/2R$) to spheres. As the extended conduit structure gradually breaks down, conductance and viscosity decreases, and at the percolation line, the system has converted completely into finite aggregates, probably spheres. Adding more water causes the spheres to grow, but phase separation occurs when the curvature departs too far from its optimum value.

The percolation threshold occurs when the radii of the spheres deduced from eq. 11.2.5 approximately correspond to the spontaneous curvature H_o. As we can see from the dotted lines connecting the percolation thresholds on the phase diagrams (Figure 11.13), the composition at the conducting–nonconducting transition occurs at roughly constant surfactant/

water ratios for a given oil. In each case, this line extends down towards pure oil, which indicates that near the oil corner the microemulsion contains inverse droplets. For the oils ranging from hexane to dodecane, converting cylindrical conduits to spheres requires more water per head group as the molecular weight of the oil increases. Conversion to inverted spheres is the easiest for hexane, since it penetrates the most.

We are now in a position to understand why the behavior of the tetradecane system appears to differ from that of the other oils. If we use the same surfactant/water ratio at the percolation threshold for the tetradecane system as for the dodecane system, we observe that this threshold lies just below the single-phase microemulsion region. (See Figure 11.13.) Therefore, the tetradecane system probably retains a bicontinuous structure.

This conclusion suggests several testable hypotheses. For example, adding a long-chain cosurfactant, such as an alcohol, to the tetradecane microemulsion should increase curvature and induce percolation behavior. In Figure 11.16 plots of the change in conductance against water for several DDAB/dodecanol ratios illustrate this result. Alcohols have relatively little effect on the head group area, and in this case the cosurfactant simply makes H_o more negative.

Alternately, adding a single-chain surfactant like dodecyltrimethylammonium bromide (DTAB) should decrease the curvature until the system ceases to show percolation behavior. In fact, the plot of conductance against added water for the decane system in Figure 11.17 demonstrates how percolation behavior disappears as the ratio of DTAB to DDAB increases.

Thus, we can understand many features of these three-component microemulsions in terms of a concept involving optimum curvature at the oil–water interface. We can gain

Figure 11.16
Adding a cosurfactant, dodecanol, to a conducting DDAB–tetradecane microemulsion induces percolation behavior as indicated by conductance measurements. The cosurfactant is tethered to the oil–water interface, and its hydrocarbon chain induces curvature toward the polar water phase. (S.-J. Chen, D.F. Evans, B.W. Ninham, D.J. Mitchell, F.D. Blum, and S. Pickup, *J. Phys. Chem.* **90**, 842 (1986).)

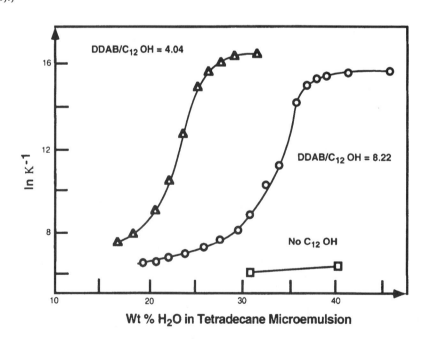

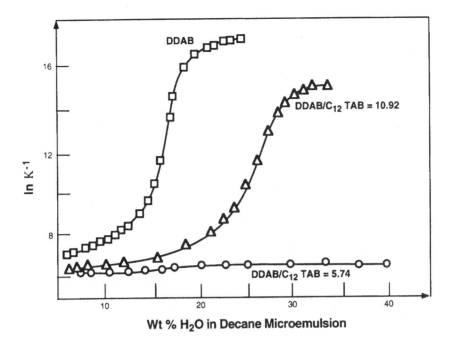

Figure 11.17
Adding the single-chain surfactant C$_{12}$TAB (C$_{12}$H$_{25}$N(CH$_3$)$_3$Br) to a DDAB–decane microemulsion obliterates percolation behavior. The head groups of C$_{12}$TAB and DDAB are equivalent, but the hydrocarbon volume differs by a factor of 2. The effect is to decrease the negative curvature at the oil–water interface. (S.-J. Chen, D.F. Evans, B.W. Ninham, D.J. Mitchell, F.D. Blum, and S. Pickup, *J. Phys. Chem.* **90**, 842 (1986).)

further insight into these systems by considering how packing constraints define the phase boundaries.

Because water-filled cylinders and spheres appear to represent the two extreme structures associated with the DDAB microemulsion, they provide an appropriate focus for our outline of the general strategy used to calculate the constraints imposed by packing. We can obtain the radii of water-filled cylinders and spheres from the following relationships:

$$R_{w(cyl)} = \frac{2V_{aq}}{\text{area}} \qquad (11.2.9)$$

$$R_{w(sph)} = \frac{3V_{aq}}{\text{area}} \qquad (11.2.10)$$

In these equations V_{aq} = (number of DDAB molecules) × (120 Å) + volume of water, and

$$\text{area} = (\text{number of DDAB molecules}) (a_o) \qquad (11.2.11)$$

The radius of the basic packing unit is then

$$R = R_w + l_o \qquad (11.2.12)$$

where l_o is set equal to 11.5 Å.

Oil fills the voids between the surfactant chains within the packing units: excess oil is located between these units. The upper limits the packing units can occupy on close packing are 90.7 and 74.05 vol % for cylinders and spheres. As Figure 11.18 shows, the compositions that satisfy these packing constraints

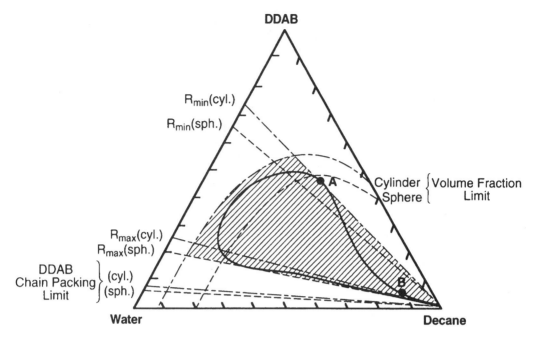

Figure 11.18
Boundaries allowed by geometric constraints due to volume fraction and surfactant chain packing for DDAB with decane as the oil and $a_0 = 60\,\text{Å}^2$. The actual phase boundary for DDAB–decane–water microemulsion systems is shown (solid line). R_{min} and R_{max} were derived from the onset at A and the percolation line (as the limiting radius of curvature for cylinders), respectively. (V. Chen, G.W. Warr, D.F. Evans, and F.G. Prendergast, *J. Phys. Chem.* **92**, 768 (1988).)

lie below the "volume limit" lines. For comparison purposes, we have superimposed the phase diagrams for decane in Figures 11.18.

These geometric constraints define the allowable single-phase region solely on the basis of packing constraints for cylinders and spheres. The W/S limit occurs because the total volume of each packing unit, minus the volume of the water core, cannot be less than that of the surfactant hydrocarbon tails. This limit depends to a large degree on the value of a_0. For $a_o = 60\,\text{Å}^2$, the maximum W/S ratio excludes about 50% of the phase diagram.

The phase envelope we calculate extends over to the surfactant–oil axis, where the conduits and spheres are filled only with head groups. Clearly, such a situation is physically impossible. Therefore, the minimum amount of water required to form the microemulsion reflects the smallest amount of water reuired to solvate the charged head groups and counterions for a given degree of oil penetration. Our discussion thus shows that curvature imposed by the balance between head group repulsion and oil penetration and the limits set by packing constraints largely determines the properties of three-component microemulsions employing DDAB as the surfactant.

11.3 Macroemulsions Consist of Drops of One Liquid in Another

Macroemulsions are thermodynamically unstable dispersions containing two immiscible liquids and an emulsifier. They find wide commercial use because in many instances we want to process a macroscopically homogeneous system containing both water and oil. At times, however, a technical process results in an emulsion, but we desire a separation into two bulk phases. This problem is common in oil recovery.

Emulsions combine properties associated with colloidal sols and amphiphilic aggregates. Because macroemulsions and sols are thermodynamically unstable, we use similar strategies to prepare them. Their kinetic stability as colloidal entities also has a common basis. However, emulsion droplets can transform in size, shape, and number, and the systems exhibit many properties associated with amphiphilic systems. This combination of features makes emulsions particularly complex and challenging to understand. Our present scientific understanding of emulsions clearly does not match their technical importance.

We can state our general goals for forming, stabilizing, and breaking of an emulsion in a simple way. Preparation requires the creation of a metastable state containing either oil in water (O/W) or water in oil (W/O) droplets. Bicontinuous macroemulsions have not been observed, presumably because any perturbation in the system immediately leads to its decomposition. Stabilization demands that we keep the droplet shape and size intact as long as possible, while the strategy for breaking an emulsion requires opening up a pathway for destabilization.

The final fate of an emulsion is clear; it will separate into two or more equilibrium phases. In large part, the way we think about emulsion stability depends on our final goal. If long shelf life or durability during a process is necessary, we want to stabilize the phases. In other cases, emulsions are unwanted products in a manufacturing process. Also, we sometimes want emulsions to be stable up to a certain point in a process, after which phase separation becomes crucial. Thus, the dichotomy between stability and instability so typical of colloidal systems in general also characterizes emulsions.

11.3.1 Forming Macroemulsions Usually Requires Mechanical or Chemical Energy

Two basic kinetic schemes are used for preparing emulsions just as they are for colloidal sols. We can use mechanical energy to mix all the emulsion ingredients (dispersion) or use the chemical energy that is stored in the ingredients and liberated when they come into contact.

The free energy $\Delta \mathbf{G}_{em}$ needed to disperse a liquid of volume V with drops of radius R in a solvent is

$$\Delta \mathbf{G}_{em} = \gamma \frac{3V}{R} \qquad (11.3.1)$$

where γ represents the interfacial tensions. For example, dispersing water in oil ($\gamma \simeq 50 \times 10^{-3} \, \mathrm{J/m^2}$) to form drops of $R = 100 \, \mathrm{nm}$ requires a free energy of 27 J/mol. From a chemical point of view, this is a modest amount of energy. Making the drops larger and reducing γ by adding a surfactant lowers the value of ΔG_{em} even more.

In practice, however, we must use considerably more energy to reach the dispersed state for two reasons. First, directing the input energy solely to forming droplets is technically difficult because any mechanical agitation converts much of the energy to heat through viscous dissipation. Similarly, if we start from a chemical nonequilibrium state, much of the free energy converts to heat or is dissipated in the process of diffusional and convective mixing.

Emulsion formation also requires relatively large energy input for a microscopic reason. The energy barrier against the coalescence of drops that stabilizes the emulsion implies the existence of a barrier against forming two droplets from a bigger one. External perturbation must be large enough to overcome this barrier.

Because an emulsion is a nonequilibrium state, its properties depend not only on the variables of the state, such as temperature and composition, but also on the method of preparation. This fact lends an extra level of complexity to these systems when it comes to scientific experimental studies as well as to practical applications. An example cited by Vold and Vold (p. 400, unnumbered footnote) illustrates the quixotic nature of emulsions:

> Both the scientific literature and the folklore of emulsion preparation are full of examples of reproducible, seemingly capricious behavior. For examples (sic): (1) 20 mL 1% aqueous sodium oleate and 80 mL benzene, shaken vigorously for three minutes, separate again quickly into two clear layers, while a few single gentle swirls suffice to produce a quite durable, milky O/W emulsion. (2) Bubbling air through a 50% oil–water mixture in a glass vessel gave an O/W emulsion, whereas the same mixture in a plastic vessel not wet by water gave a W/O emulsion. (3) The present authors find it entertaining that it seemed appropriate to report that the direction of stirring was without effect.

At least one exception seems to accompany every generalization about emulsions.

Although establishing consistently valid rules for the preparation of emulsions presents real difficulties, we can identify a number of important factors that influence the final properties of the mixed system. Both oil and water droplets generally form in a mechanical agitation process. What determines whether the final product is a W/O or O/W emulsion? The fate of the droplets depends on two competing processes: migration of the emulsifier to the droplet's interface (which stabilizes them) and coalescence (which destroys them). The phase that coalesces most rapidly — for whatever reason — will become the continuous phase.

Another factor that determines the final form of the emulsion is the order in which we mix the ingredients. We can dissolve the emulsifier directly in either the water or the oil phase. Alternatively, we can dissolve components of the emulsifier in both phases. We can carry out the emulsification by simply pouring all the ingredients into a container and mixing them, or we can add the aqueous phase bit by bit to the oil–emulsifier phase (or vice versa).

11.3.2 Turbulent Flow during the Mixing Process Governs Droplet Size

Figure 11.19 illustrates how average droplet size varies with energy input in the case of several emulsification techniques. We can obtain a qualitative description of the droplet formation process using arguments drawn mainly from dimensional analysis.

As we discussed in Chapter 2, a spherical drop exhibits a Laplace pressure

$$p(\text{Laplace}) = \frac{2\gamma}{r} \tag{11.3.2}$$

To tear such a drop into two parts, we need a pressure gradient of magnitude

$$\frac{\Delta p}{\Delta x} \simeq \frac{2\gamma}{r^2} \tag{11.3.3}$$

Figure 11.19
Average droplet size $\bar{R}$ as a function of energy input ($P_1 \times$ time) per volume for diluted paraffin oil-in-water emulsions produced in various machines. Approximate (average) results. (P. Walstra, in *Encyclopedia of Emulsion Technology*, Vol. I. P. Becher, ed., Dekker, New York, 1983, Chap. 2.)

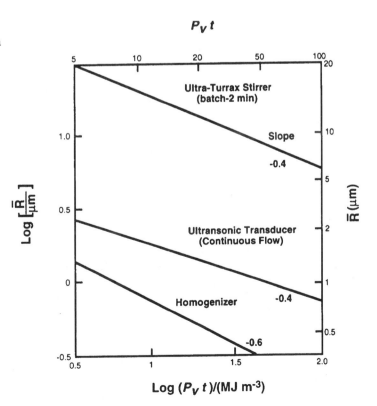

The smaller the drop size, the larger pressure gradient we need. Moreover, eq. 11.3.3 shows that lowering the surface tension γ decreases not only the minimum energy input given by eq. 11.3.1 but also the actual energy input, because a smaller pressure gradient is needed to achieve a certain drop size.

In mixing processes the pressure gradients ($\Delta p/\Delta x$) that exert shearing forces on liquid droplets are associated with the eddies present in turbulent flow. These eddies are local regions of chaotic flow characterized by instantaneous velocities that fluctuate around an average velocity. Eddies have a characteristic size l_e, which measures the distance over which the instantaneous velocity changes appreciably. While we can develop equations that relate eddy velocity, energy, and size to the pressure gradients required for emulsification, they are difficult to apply in a quantitative way. Although our alternative approach is a qualitative one based on dimensional analysis, it provides many of the salient features we need to understand emulsification.

First we consider a pressure difference associated with an eddy of size l_e where a difference in velocity, Δv, exists. According to Bernoulli's equation, the difference in energy density $\rho(\Delta v)^2/2$ relates to the difference in pressure Δp. If this pressure exceeds the Laplace pressure, then a droplet of diameter $d \geqslant d_{min}$ contained in the eddy region will break up. If we assume that this breakup diameter depends on the power input per unit volume P_v, density ρ, and surface tension, we can use dimensional analysis to argue that with a method based on turbulent flow the size distribution function $F(\mathscr{R})$ is uniquely determined by the dimensionless quantity

$$\mathscr{R} = R \left(\frac{P_v}{\rho}\right)^{2/5} \left(\frac{\rho}{\gamma}\right)^{3/5}$$

where R is the radius of the emulsion drop.

Problem: Show that $\mathscr{R}$ is indeed dimensionless.
Solution: $R(m)$; P_v ($Jm^{-3}s^{-1}$); ρ ($kg\,m^{-3} = J\,s^2 m^{-3}$); γ (Jm^{-2})

$$\rightarrow \dim[\mathscr{R}] = m \left(\frac{Jm^2}{m^3 s\, Js^2}\right)^{2/5} \left(\frac{Js^2\, m^2}{m^{-5}J}\right)^{3/5}$$

$$= m \left(\frac{m^2}{s^3}\right)^{2/5} \left(\frac{s^2}{m^3}\right)^{3/5} = m \cdot \frac{m^{4/5}}{s^{6/5}} \frac{s^{6/5}}{m^{9/5}} = 1$$

We thus expect the average radius $\bar{R}$ to depend on the power input P_v as

$$\bar{R} \sim P_v^{-0.4} \tag{11.3.4}$$

which is verified for the upper curve in Figure 11.19. Analysis of the pressure gradients generated by cavitation in ultrasonic irradiation also leads to an energy dependence of -0.4, while for a high pressure homogenizer, the dependence is closer to -0.6, as shown in Figure 11.19.

A whole spectrum of eddy sizes exists in turbulent flow, and so we expect a spectrum of droplet sizes. Mechanically prepared emulsions generally are characterized by a wide particle distribution. However, the difficulty of counting submicrometer particles limits our ability to determine its exact nature. Several studies on emulsions suggest a log normal distribution

$$F(R) = \frac{1}{R\sigma(2\pi)^{1/2}} \exp\left[-\frac{(\ln R - \ln R_o)^2}{2\sigma^2}\right] \qquad (11.3.5)$$

which is based on a normal Gaussian distribution with $\ln R$ as the independent variable, $\ln R_o$ as the mean value and σ^2 as the variance. In eq. 11.3.5, R itself is considered the independent variable so that $\int_0^\infty F(R)\,dR = 1$ and $F(R)$ is a distribution skewed toward large R-values but still specified by only two parameters $(\ln R_o, \sigma)$. This distribution was first derived by the renowned Russian mathematician Kolmagorov when considering the size distribution resulting from grinding particles.

11.3.3 A Chemical Nonequilibrium State Can Induce Emulsification

We can use chemical energy to form emulsions in several ways. Three examples illustrate the principles involved.

1. *Generation of emulsion droplets in the oil—water interfacial region driven by local surface tension gradients and/or chemical reactions.* For example, bringing Nujol mineral oil containing oleic acid into contact with aqueous alkali creates an emulsified oil in the lower phase. Bringing water containing sodium oleate solution into contact with Nujol produces no such effect. Presumably, neutralization of the oleic acid by base provides the energy for the emulsification process. This behavior manifests the Marangoni effect, that is, surface transport of material resulting from transient changes in the surface tension. As the surface tension approaches zero, small thermal fluctuations become sufficient to induce emulsification.

2. *Formation of emulsion droplets at the oil—water interface driven by coupled diffusion and stranding.* For example, when we bring a solution of ethanol in toluene saturated with water into contact with water, the interface becomes quite agitated and emulsions of toluene in water and of water in toluene form. As the ethanol diffuses, it leaves behind the toluene, which coalesces into unstable emulsion drops. The free energy for this process comes from the mixing of ethanol with water.

3. *Formation of emulsion droplets driven by differences in osmotic pressures.* This strategy involves preparing a W/O microemulsion and contacting it with water under

conditions that lead to an inversion process and production of an O/W emulsion. A specific example employs Hercolyn D (methyl ester of partially hydrogenated rosin acid: Figure 11.20) as the oil and potassium rosinate obtained by partial hydrolysis of Hercolyn D as the emulsifier. This rather stable emulsion remains unchanged for at least 5 years.

To understand how a non-equilibrium initial state can evolve into a stable emulsion, let us consider this last example in detail. At low water content, the initial microemulsion is nonconducting, consistent with the formation of inverted W/O spheres. Provided the initial water content is less than 1.8 wt %, the microemulsion undergoes phase inversion to give an O/W emulsion with an unusually uniform droplet diameter of 225 nm when exposed to a water phase. Bringing microemulsions containing more than 1.8 wt % water into contact with water results in unstable emulsions.

If we titrate the 1.8 wt % microemulsion with water, the conductance initially increases, goes through a maximum, and then decreases. This reaction parallels behavior observed with the DDAB microemulsions and suggests that adding water introduces a structural change to interconnected conduits that later transform into spheres.

We can understand these observations as a result of the high concentration of counterions, which considerably increases the osmotic pressure inside the inverted micelles contained in the microemulsion phase compared to that of the contacting water. Consequently, the inverted micelles imbibe

Figure 11.20
Video enhanced microscope images of spontaneous emulsification of Hercolyn D–potassium rosinate–water microemulsions with water: (a) approximately 25 seconds after contact, (b) approximately 40 seconds after contact, (c) 3–5 minutes after contact (explosive injection of oil phase droplets into water); and (d) 15 minutes after contact, showing nearly uniform spheres of oil phase before injection into water.

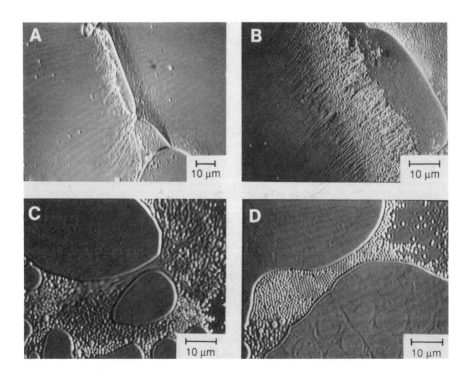

water and grow. This process is equivalent to the titration path described earlier, but on a size scale confined to the interfacial region adjoining the two phases. As a result of high viscosity of the medium and packing constraints, the micelles remain fixed as they grow, interconnect, and eventually invert. Thus, initial microemulsion structure sets the constraints that determine the small, uniform size of the resulting emulsion drops. Immediately after the inversion process has taken place, the emulsion droplets (stabilized by an anionic surfactant) behave as a concentrated colloidal dispersion. Electrostatic repulsion between droplets drives them apart, and they move into the adjacent water phase.

If we replace the contacting water by a 2 M sucrose solution, we completely inhibit the emulsification process. This result is consistent with the role played by the difference in osmotic pressure. When we increase the water content of the microemulsion to the point at which interconnected conduits are present, water from the aqueous phase simply fluxes into the oil phase and causes the formation of a coarse heterogeneous emulsion.

In this example, the difference in osmotic pressure provides the chemical energy to drive the inversion process leading to emulsion formation. Although these processes sometimes are called "spontaneous" emulsification because they do not require mechanical energy, the terminology is misleading. The process involves going from a state of higher free energy (associated with the initial form of emulsion ingredients) to a metastable one (associated with the emulsion) rather than proceeding directly to the final equilibrium state involving several phases.

11.3.4 A Number of Different Mechanisms Affect the Evolution of an Emulsion

A freshly prepared emulsion changes its properties with time owing to a series of different events that occur on a microscopic scale. Depending on the particular molecular characteristics involved, these initial events rapidly lead either to the overall equilibrium state with macroscopically separated phases ("unstable emulsion") or to a metastable state ("stable emulsion"). Clearly, the distinction between stable and unstable emulsion is qualitative and depends partly on the context.

A number of microscopic events contribute to the destabilization of an emulsion. As with the sol systems discussed in Chapter 9, the basis for stability is the existence of a repulsive regime for the force between drops. We can use the general DLVO terminology from Chapter 8 to describe the qualitative features of the droplet–droplet interaction, bearing in mind that the forces involved have a different origin, particularly for W/O emulsions.

Emulsion droplets rather easily **flocculate** into the secondary minimum. This effect is more significant for dilute emulsions ($\lesssim 25\%$) than for more concentrated ones. Since when the mean distance between drops is less than the separation at the secondary minimum, the system experiences no initial energy gain in bringing drops closer together.

Whether flocculated or not, drops can migrate in the gravitational field to increase the concentration at either the top or the bottom of the vessel. The best-known manifestation of the phenomenon gives this effect the name **creaming**.

Once droplets have been brought into proximity by flocculation and/or creaming, they can **coagulate** into the primary minimum while still keeping their identity as separate droplets. This stage of destabilization is final for colloidal sols, but a further process occurs for emulsions.

Coalescence, in which two droplets merge into a bigger one, constitutes the final stage in the life of an emulsion droplet. It requires reorganization of the stabilizing surface layer surrounding the drops. The surfactant film changes its topology during the coalescence process, and this change usually is connected with passing a substantial free energy barrier.

The general DLVO picture only applies for certain emulsion systems. In other cases the interaction curve for two intact droplets exhibits a single energy minimum. In such a case, the distinction between flocculation and coagulation becomes tenuous, and the appropriate term depends on the location and depth of the minimum in the interaction curve. Figure 11.21 summarizes the different steps in the time evolution of an emulsion.

In addition to the sequence of flocculation/creaming, coagulation, and coalescence, a more molecular path contributes to destabilizing the emulsion. When the dispersed sub-

Figure 11.21
Schematic illustration of the different steps in the development/evolution from a freshly prepared emulsion to a final separation into two macroscopic phases.

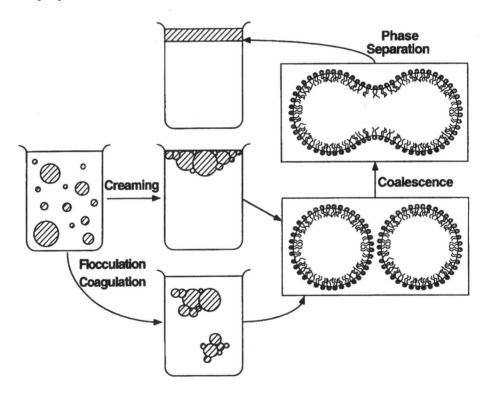

stance has a nonnegligible solubility in the solvent, droplet sizes can change by molecular diffusion. In this Ostwald ripening process, small drops decrease in size and finally disappear, while large drops incorporate material through diffusional transfer.

The surface free energy contribution to the chemical potential is larger in small drops than in large ones. Molecules in the drops are in dynamic equilibrium with molecules dissolved in the medium. The influence of surface free energy causes local concentration to be slightly higher adjacent to small drops than around larger ones. Thus, concentration gradients exist in the medium with concomitant diffusional flows (as discussed in Chapter 2).

We can use the Ostwald ripening process to generate monodisperse drops in an elegant way. If we initially produce an emulsion in the presence of monodisperse polymer so that no drops contain more than one polymer molecule, the presence of the polymer affects the chemical potential of the dispersed medium. Ostwald ripening leads to a quasi-equilibrium state in which all droplets containing a polymer are of a uniform size. Droplets without polymer will disappear irrespective of their size. This process has been used to produce large monodisperse latex particles by emulsion polymerization.

This example suggests an effective way to stabilize an emulsion by quenching the Ostwald ripening mechanism. The basic strategy is to include a solute so that its solubility in the continuous phase is lower than that of the dispersed solvent itself. Obvious choices of solutes are polymers for oil; and electrolyte or sugars for water. As the molecules of the dispersed phase diffuse from the smaller droplets to the larger droplets the solute concentrations increase in the smaller droplets and decrease in the larger ones. Thus, a situation is rapidly established in which the surface tension balances the direct osmotic contributions to give a uniform chemical potential of the dispersed solvent in drops of different sizes.

11.3.5 To Stabilize an Emulsion, the Dispersed Phase in Different Drops Should be Prevented from Reaching Molecular Contact

In practice three different strategies constitute the chief means for achieving emulsion stability. One possibility is to use small solid particles to stabilize the emulsion drops (Figure 11.22a). Particles are usually surface active and adsorb at interfaces. If the dispersed medium does not wet the particle—that is, if the particle prefers the continuous phase—it will stick out from the surface of the drop and prevent direct contact with another drop. This mechanism contributes to emulsion stability in margarine and butter. In this important case, water droplets are stabilized by small fat crystals that adsorb on the drops but are wetted by the apolar continuous medium.

A second method of generating emulsion stability is by adding a polymer. If the monomer unit is slightly surface active,

Figure 11.22
(A) Particles adsorb on the interface between the emulsion drops and the medium. If the particle—liquid surface free energy is lower for the continuous medium than for the emulsified phase, the particles will stick out into the medium and prevent drop–drop contact. (B) A polymer protects an emulsion droplet most efficiently if the continuous medium is a better solvent than the emulsified liquid.

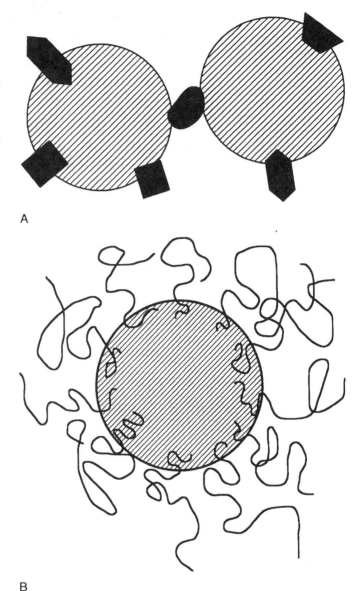

A

B

the polymer will adsorb on the polar–apolar interface and the chain will make excursions into both media (Figure 11.22B). To be effective as a stabilizer, the polymer should prefer the continuous medium, because having the largest fraction of the chain on the outside of the drop generates the strongest repulsive force between drops.

A variation of this stabilization mechanism can be observed for some protein systems. Globular proteins can be surface active even in their native state so that they adsorb on a polar–polar interface. If the major part of the molecule remains in the continuous medium, a surface film of protein

Figure 11.22
(C) A dense film of adsorbed protein molecules also can stabilize an emulsion drop. (D) If an emulsion droplet is prepared in a three-phase system oil–water–lamellar liquid crystal, it can be protected by a few bilayers in addition to a monolayer film.

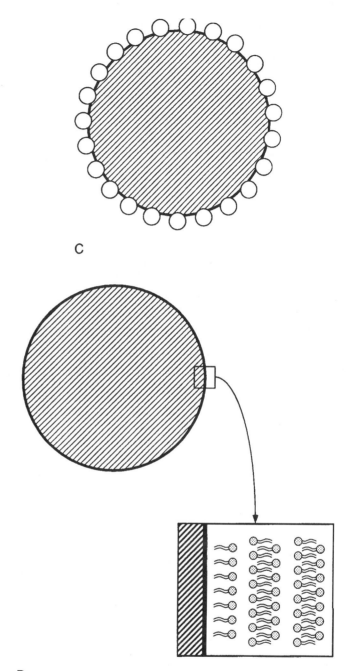

C

D

molecules can stabilize an emulsion drop, as illustrated in Figure 11.22C.

The most common method for obtaining a stable emulsion is by using a surfactant or an amphiphile. Typically they will form a dense monolayer at the polar–apolar interface which acts to prevent direct contact between the dispersed medium of two drops. How successfully the film prevents this contact depends in a sensitive way on the properties of the film itself and on the interaction between films.

For amphiphiles, a monolayer film is not the only possible arrangement that may occur at the polar–apolar interface. Stig Friberg and co-workers convincingly demonstrated that easily prepared and yet stable emulsions can be obtained if the equilibrium state of the emulsified system is a three-phase sample composed of oil, water, and a lamellar liquid crystal containing the majority of the surfactant (see Figure 11.29, below).

In this three-phase equilibrium the liquid crystal phase will wet the oil–water interface; that is, the interfacial tensions between oil–liquid crystal and water–liquid crystal are much smaller than the oil–water tension. Thus, when water is dispersed in oil, or vice versa, the lamellar liquid crystal can remain as a protective sheath around drops, as illustrated in Figure 11.22D. Such a multibilayer arrangement clearly can provide a very strong protection for the emulsion drops.

11.3.6 Emulsion Structure and Stability Depend on the Properties of the Surfactant Film

Section 11.1 discussed the elastic properties of surfactant films in some detail. We can now use ths knowledge to understand how the film responds in an emulsion.

The first queston is, What type of emulsion will we obtain if we mix oil, water, and surfactant and emulsify the mixture mechanically? At the early stage of the emulsification process when a drop has just been formed, there is a critical moment. An optimal surfactant film has not had time to develop at the newly formed surface, and coalescence is very likely to occur if a second drop is encountered. Thus, the more rapidly a proper surfactant film forms, the more long-lived the drop.

When a new drop emerges by budding from a larger one, it has some surfactant at the interface, but because its area has been increased, the mean surfactant density is low and initially nonuniform. At this stage an important phenomenon called the Marangoni effect comes into play. Nonuniform distribution of surfactants creates nonuniform interfacial tension. This non-uniformity results in a mechanical instability in the surface, and surfactant will flow toward the area of low surfactant concentration and high surface tension (Figure 11.23). This process is faster than a surface diffusional equilibration of the concentration. Therefore, the action of the Marangoni effect quickly establishes a uniform but thin surfactant film around the drop.

The second stage of the process leading to a relaxed state of the drop is the transport of surfactant to the surface. In this step, the preferential location of the surfactant becomes important. If it concentrates in the dispersed medium inside the drop, transport mainly occurs by diffusion; but if it concentrates in the continuous medium, which is mechanically agitated, convection dominates and diffusion is less important. As a result, the newly formed film heals more rapidly if the surfactant is found in the continuous phase, and rapid healing leads to a

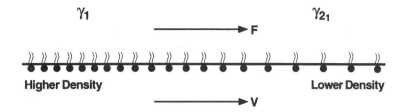

Figure 11.23
Surfactants are distributed nonuniformly in a freshly created interface. The relaxation to a homogeneous state does not occur primarily through diffusion, but by a more rapid process. The difference in the concentration of surfactants in the interface induces gradients in surface tension, which generate a force F per unit length of $\gamma_2 - \gamma_1$. As a result, the surfactant molecules and the immediately adjacent liquid flow with a velocity v toward the regions of highest local surface tension. This movement is called the Marangoni effect or, if it occurs in connection with surfactant systems, the Gibbs—Marangoni effect.

final outcome of the emulsification process in which the phase that preferentially dissolves the surfactant becomes continuous medium. This empirical observation is called the Bancroft rule.

If we apply the knowledge about surfactants that we have developed earlier, we know that normal aggregates are preferred for surfactant numbers $N_s < 1$. Those surfactants have a higher solubility in water than in oil, while surfactants with $N_s > 1$ prefer oil to water. Phrased in terms of the spontaneous curvature of the surfactant film H_o, we expect oil in water (O/W) emulsions when $H_o > 0$ and water in oil emulsions (W/O) when $H_o < 0$.

What makes one surfactant form a more stable emulsion than another? To discuss this question in a coherent way, we must realize that the path from emulsion droplets dispersed in an excess solvent to a macroscopically phase-separated system contains several steps, as illustrated in Figure 11.21. Let us now concentrate on the coalescence step, which is peculiar to emulsions.

We can identify three sequential crucial events in the total process that leads from two separate emulsion droplets to one combined droplet:

1. For a coalescence to be at all possible the two surfactant films must come into molecular contact.

2. The two surfactant films must fuse, forming a neck with a direct contact between the dispersed liquid in the two drops.

3. The neck must grow in size so that the two drops merge completely.

The rate-determining step for the coalescence process can involve any of these three events depending on the nature of the system. The same factors which control the stability of colloidal sols also control the first step involving the mutual approach of the two drops. For example, if we have an oil-in-water emulsion stabilized by ionic surfactants and with a low electrolyte concentration in the aqueous medium, a strong repulsive double layer force will prevent the drops from establishing molecular contact and the emulsion should have long-term stability relative to coalescence. On the other hand, if the potential barrier is small or nonexistant, drops can approach one another, forming a thin gap of the continuous medium between the two surfactant films on the surface of the drops. Due to the ever present attractive van der Waals force, the liquid droplets deform, giving a flat contact area as shown in Figure 11.24.

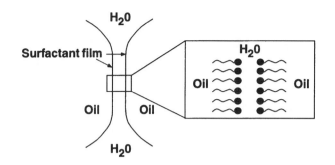

Figure 11.24
Two emulsion droplets in contact will deform giving flat surfaces separated by a film of the continuous medium, water in this case. The thickness of the film is determined by the forces between drops.

Now we have reached step 2 in the coalescence process. Clearly, if the surfactant film is poorly developed, it will rupture easily, while if the surfactant has reached the CMC in the bulk, a well-developed film could resist the fusion of the two drops. As illustrated in Figure 11.25, fluctuation of the surfactant surface density triggers a fusion process. The resulting bare spots are attracted to each other and break the surfactant film locally. We can obtain an order of magnitude estimate of the free energy cost in creating such bare spots simply by estimating the increase in surface free energy

$$\Delta G = (\gamma_o - \gamma)A \qquad (11.3.6)$$

where γ_o is the bare surface tension between the two liquids and γ the surface tension in the presence of surfactant. As an example, consider the free energy cost of generating an area of $0.5 \, nm^2$ simultaneously in both films in an oil/water system. Here $\gamma_o \simeq 50 \, mJ/m^2$, and for a reasonable surfactant $\gamma_1 \approx 1 \, mJ/m^2$. Thus,

$$\Delta G = 49 \times 10^{-3} \times 1 \times 10^{-18} = 4.9 \times 10^{-20} \, J$$

which corresponds to $12kT$ at room temperature. Although

Figure 11.25
Illustration of how concentration fluctuations in the surfactant film can lead to a local destabilization and the breaking of the film and the nucleation of a neck of oil joining the two emulsion droplets.

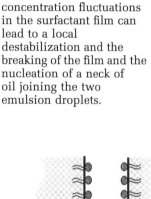

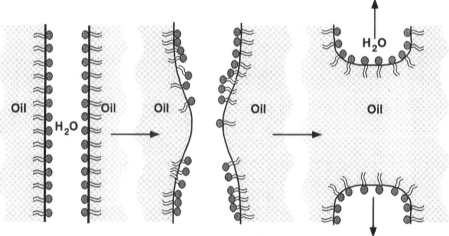

$12kT$ is a respectable energy barrier, it is not quite sufficient to ensure long term stability of a metastable state. A comparison with the nucleation process discussed in Chapter 2 shows that barriers exceeding $30kT$ are normally required for long term stability. Thus, we conclude that a liquid surfactant film will generate a barrier for the second step of the coalescence process, but it is not clear that this barrier is high enough to ensure long-term stability.

In the third and final step of the coalescence process the connecting neck of dispersed liquid that formed in step 2 needs to grow and eliminate the layer of continuous phase liquid between the drops as shown in Figure 11.26. During the growth process, the surfactant film remains intact but changes area and curvature. In the first section of this chapter, we developed a formalism that allows us to quantitatively analyze the free energy changes associated with the growth of the neck. Figure 11.26 illustrates a geometrical description of the neck of radius a and a semicircular rim of radius b, implying a liquid film of thickness $2b$ separating the flat sections of the drops.

We can write the excess free energy of the surfactant film $\mathbf{G}_f$ relative to an intact flat film as a sum of four terms

$$\mathbf{G}_f = W_1 + W_2 + W_3 + W_4 \qquad (11.3.7)$$

The first two terms are due to changes in area. The flat part of the film decreases in area resulting in a change in interfacial energy.

$$W_1 = -2\pi\gamma(a + b)^2 \qquad (11.3.8)$$

The curved part of the film is assumed to have the geometry of a revolving semicircle and we can write the surface free energy as

$$W_2 = 2\pi\gamma[\pi b(a + b) - 2b^2] \qquad (11.3.9)$$

Figure 11.26
When the liquid film between two emulsion droplets ruptures, a neck of the dispersed liquid connects the two drops. We describe the neck in terms of the radius a in the plane of the film and the thickness b of the film. The growth of the neck involves an increase in a and could also lead to a change in the film thickness b. When there is a spontaneous curvature H_o of the surfactant layer toward the dispersed phase, there is a free energy barrier associated with the growth of the neck.

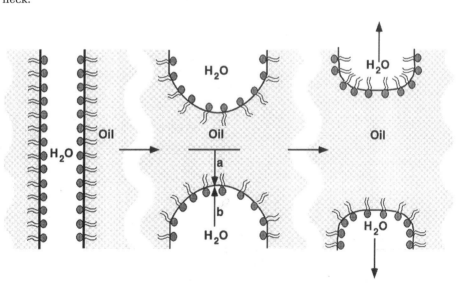

which is always positive. In addition, there is a bending free energy

$$W_3 = 2\kappa \int \{(H - H_o)^2 - H_o^2\} \, dA \qquad (11.3.10)$$

where the last term in the integral comes from the bending energy of a flat monolayer. Evaluating the integral in eq. 11.3.10 requires some nontrivial geometrical relations, and we simply state the final result

$$W_3 = 2\pi\kappa \left\{ 2\pi H_o a + \frac{2(a+b)^2}{b[a(a+2b)]^{1/2}} \arctan(1 + 2b/a)^{1/2} + 2(\pi - 4)bH_o - 4 \right\} \qquad (11.3.11)$$

The fourth component is due to the contribution from a Gaussian curvature term since a saddle bend occurs in the neck but not in the original flat films. This term takes the simple form

$$W_4 = -4\pi\bar{\kappa} \qquad (11.3.12)$$

As discussed in Section 11.1 the Gaussian curvature is determined solely by the topology of the film, which does not change during the growth of the neck.

To find the free energy barrier, we optimize the value $\mathbf{G}_f$ of eq. 11.3.7 for each a and define a potential energy

$$V(a) = \mathbf{G}_f(a, b_o) \quad \text{with} \quad \frac{\partial \mathbf{G}(a, b)}{\partial b} = 0 \quad \text{at} \quad b = b_o \qquad (11.3.13)$$

As discussed previously in this chapter, the microemulsions of nonionic surfactants have been carefully studied, so we have experimental estimates of $\gamma(T)$, $H_o(T)$ and κ. Therefore, $\mathbf{G}_f(a, b)$ becomes a known function of the two geometrical parameters.

Figure 11.27 shows an evaluation of $V(a)$ for the $C_{12}E_5$ system at different temperatures. The first striking feature is that a barrier exists for temperatures when the spontaneous curvature is toward the dispersed phase, but for temperatures on the other side of the temperature T_o where the spontaneous curvature is zero, the barrier is absent. This example provides an explanation of the Bancroft rule. The second striking feature is that the barrier height is high $\approx 40kT$, well away from T_o, but decreases dramatically as the temperature approaches T_o.

The conclusion emerges that we can obtain long-term stability with respect to coalescence due to the barrier for growth of a neck joining two emulsion droplets. Such a barrier exists when the spontaneous curvature of the film is toward the dispersed phase, but there is no barrier by this mechanism when the spontaneous curvature is in the opposite direction.

We conclude this discussion by showing experimental results confirming these theoretical predictions. Figure 11.28 shows the stability of an emulsion containing $C_{12}E_5$ as stabilizing surfactant. Stability was characterized by the time τ required to separate a certain amount of initially dispersed phase. The spontaneous curvature was varied both through the temperature

Figure 11.27
The calculated free energy barrier for the growth of a neck as a function of the neck radius a. The calculation is for O/W or W/O emulsions stabilized by the nonionic surfactant $C_{12}E_5$. The curves show the barrier at different temperatures in (K) below (O/W) or above (W/O) the balanced temperature, T_0 where the spontaneous curvature $H_0 = 0$. Note that the barrier starts to decrease only one degree below/above T_0. (A. Kabalnov and H. Wennerström, *Langmuir* **12**, 276 (1996).)

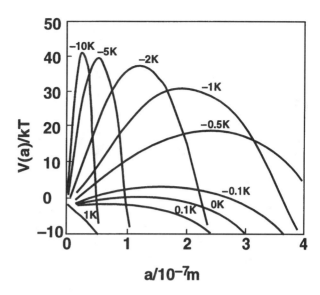

and through the salt content of the aqueous phase. We can determine these variations of H_o independently. The experiments show that:

1. The emulsions are highly unstable when the spontaneous curvature is toward the continuous phase for both O/W and W/O emulsions irrespective of whether the H_o is tuned by temperature or salt content.

2. The stability time τ is the same irrespective of how we tune H_o to given value.

3. The stability time varies dramatically with H_o and thus with temperature and salt content in the vicinity of $H_o = 0$. The temperature needs to be controlled by less than 0.1°C to obtain reproducibility.

In the literature, emulsifiers often are classified on the basis of an essentially empirical scale called the hydrophobic—lipophilic balance (HLB) number. Typical W/O emulsifiers have an HLB in the range of 4–8 and O/W emulsifiers are in the range of 12–16. The HLB number originally was developed empirically on the basis of nonionic surfactant properties, but a comparison with established properties of spontaneous film curvature shows that the balanced state ($H_o \simeq 0$) corresponds to HLB $\simeq 10$. Thus, we have interpreted the HLB number as a measure of surfactant film spontaneous curvature. In Section 11.1 we saw that the spontaneous curvature of a surfactant film is not determined by the surfactant alone, but also depends on factors such as temperature, salt content, and the extent of oil penetration. Thus the HLB number only provides a crude measure of the expected emulsion stability.

We can obtain another useful rule for predicting emulsion stability by starting from the equilibrium phase behavior of the ternary system water–oil–surfactant. Figure 11.29 shows the water–oil–surfactant phase diagrams for two different oils,

Figure 11.28
Measured stability of
emulsions stabilized by the
surfactant $C_{12}E_5$. The
stability is measured by the
time, τ, taken to form 1/4
of the separating bulk
phase oil for O/W and
water for W/O emulsions.
The spontaneous curvature
is tuned by either a
temperature change or a
change in salt content
Open square refer to
experiments with a
temperature change and
the filled circles to changes
in salt content. Panel (b)
shows a detail of panel (a).
Note that the stability time
can change by two orders
of magnitude with a change
in temperature of only
0.2 K! (A. Kabalnov and J.
Weers, *Langmuir* **12**, 1931
(1996).)

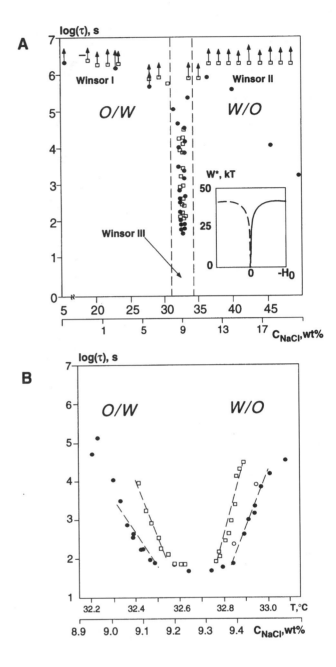

p-xylene and hexadecane, and a commercial emulsifier as surfactant. The composition of a typical emulsion is close to the
water–oil axes.

By adding a few percent (w/w) surfactant to a p-xylene—
water mixture, we enter a three-phase triangle that also contains
a lamellar phase. Friberg and co-workers have shown that very
stable emulsions can be obtained in this case.

In the hexadecane case, only a two-phase equilibrium
exists in this range of the phase diagram and emulsions are

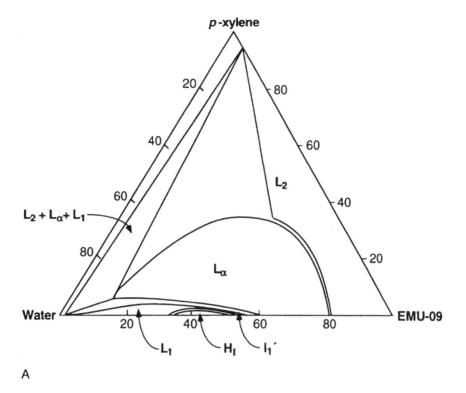

A

Figure 11.29
Ternary phase diagrams at
20°C of water–oil–
emulsion. The emulsion
EMU-09 is a nonylphenol
nonaethylene glycol. (a)
p-xylene as oil. Two liquid
phases, L_1 and L_2, become
a water and an oil phase
at low surfactant contents.
There are three liquid
crystalline phases: a
lamellar phase L_z, a normal
hexagonal H_I, and a
bicontinuous cubic phase
I_1'.

rather unstable. The molecular cause of the difference between
the p-xylene and hexadecane systems is that p-xylene penetrates much more into the surfactant film, lowering the spontaneous curvature, H_o.

Lecithin, or dialkylphosphatidylcholine, is commonly
used to emulsify fat in food. As we discussed in Chapter 6,
lecithins form a lamellar phase in water that incorporates about
40% H_2O (w/w). Very little oil dissolves in this lamellar state.
Therefore, a large three-phase triangle (in which a slightly
swollen lamellar phase is in equilibrium with virtually pure
water and pure oil) dominates the phase diagram. Empirical
observations consistently demonstrate that a sample in this
three-phrase triangle readily forms a stable O/W emulsion.

The stability of the lamellar phase shows that a monolayer's spontaneous curvature is close to zero and in fact,
slightly positive (toward oil). To obtain a W/O microemulsion,
the equilibrium phase for the surfactant should have a curvature
slightly more toward water. A commonly used emulsifier in this
context is the monoglyceride monoolein, which shows an inverted cubic phase in equilibrium with excess water.

Knowledge of equilibrium phase behavior enables us to
understand and control emulsion stability in yet another way.
Whatever the spontaneous curvature of the surfactant film,
coalescence involves a local rearrangement and change in film
topology. For surfactants in the liquid state we can associate
the energy barrier with the formation of a local high energy
structure; but by freezing the chains, we can slow down the

dynamics on the molecular scale by orders of magnitude. Sometimes we can take advantage of this phenomenon by preparing an emulsion at a slightly elevated temperature with fluid chains and then cooling the system through a temperature at which a gel phase forms. The emulsion drops then are covered with a rather rigid shell and the system can be very long-lived.

11.3.7 We Can Catalyze Coalescence by Changing the Spontaneous Curvature and by Inducing Depletion Attraction

In many applications, our goal is to destabilize an emulsion that is no longer useful or one that has been formed inadvertently. The optimal strategy first identifies the main mechanism behind the stability and then selectively changes the conditions that affect this critical step in the sequence of events that leads to macroscopic phase separation.

For an electrostatically stabilized O/W emulsion we can use electrolyte to facilitate contact between drops. Centrifugation can induce creaming. Gentle mechanical agitation facilitates coalescence of drops. If the drops are stabilized by surfactant in the gel state, heating the mixture above the melting temperature produces a dramatic effect.

A slightly more subtle method involves adding surfactant or cosurfactant to affect the spontaneous curvature of the film. It may seem puzzling that adding a cosurfactant, such as a long-chain alcohol, can either stabilize or destabilize the system depending on the circumstances. However, we can explain this finding if we realize that the emulsion has an optimal stability at a particular value of H_o.

Changes in temperature also affect the value of H_o. As discussed in Section 11.2, ionic and nonionic surfactants show opposite temperature dependence of H_o and T. Thus, moderate heating of a W/O emulsion stabilized by ethylene oxide as a nonionic surfactant does not have an effect, but destabilization can very well occur if instead ionic surfactants are used as stabilizers. For cooling the situation is the reverse.

Adding electrolyte will have a moderate, but measurable, effect on the stability of the emulsion based on nonionic surfactants as shown in Figure 11.28. We tend to think of adding electrolyte to a system containing ionic surfactants primarily in terms of changing the double layer repulsion, but clearly there is also an effect on the spontaneous curvature. For O/W emulsions the two mechanisms have qualitatively the same effect, and it may be difficult to distinguish which is most important. For W/O emulsions, on the other hand, the double layer repulsion is usually insignificant, but the electrolyte effect on H_o still operates to the same extent. Thus, changing the electrolyte content of the dispersed water phase can be an effective way to tune the spontaneous curvature of the stabilizing ionic surfactant film.

Adding surfactant can lead to another destabilizing mechanism. Typically, excess surfactant is present in the continuous phase as micellar aggregates. These aggregates are excluded from the regions where two surfactant films approach to distances on the order of a micellar size or the Debye length for charged surfactants. The result is an attractive depletion force, as discussed in Section 5.4, which pushes the films even closer together, making film rupture a more likely event.

This depletion attraction can also be induced by polymers that dissolve in the continuous phase and do not adsorb on the surfactant film.

We often picture an emulsion in terms of separated drops of the dispersed phase moving by Brownian motion in an excess of the solvent. However, in practice, emulsions typically become more concentrated, either by the direct action (depletion) of flocculation or creaming, or simply because the emulsion was prepared with a high volume of the dispersed phase. In fact, it is possible to prepare emulsions in which the dispersed phase of volume fraction 0.99.

A typical recipe for making such an emulsion does not differ a great deal from making mayonnaise. Starting with a concentrated micellar solution such as CTAB, oil is added sequentially under strong mechanical agitation. In the final state, after large amounts of oil have been added, the structure can be represented schematically as in Figure 11.30.

Large oil drops have an essentially flat contact with neighboring drops. The contact area includes surfactant film, a thin aqueous layer with counterions, and another film. A reservoir of water and surfactant also is present in the so-called Plateau regions at the corners of the drop. If the right surfactant is used, such a system can show an amazing stability. Sometimes it is useful to prepare a more dilute emulsion by adding solvent to the concentrated stock emulsion.

In concentrated emulsions, a sizable interaction takes place between the drops. The most obvious manifestation of

Figure 11.30
Schematic illustration of the structure in a dense emulsion, in which the volume fraction of the dispersed medium is much larger than that of the continuous medium. The emulsion droplets adopt more polyhedralike shapes, displaying large flat areas of contact with neighboring drops, which are separated by Plateau regions containing small reservoirs of solvent and surfactant.

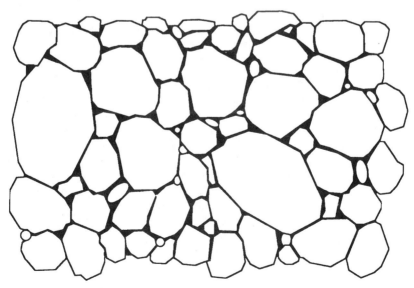

this interaction is that emulsion droplets that take a spherical shape in a dilute system, are deformed in concentrated emulsions into polyhedra with round corners. We can understand the driving force behind this deformation without analyzing the forces in detail. As discussed in Section 10.2.1, the close packing limit for spheres is at a volume fraction of 0.74, while for polyhedra the thickness of the film separating the polyhedra sets this limit. The packing structure of concentrated emulsions also is found in foams, and we postpone a more detailed discussion of that structure to the following section.

In our detailed analysis of the third step in the coalescence process we neglected the role played by a strong force across the liquid film. Certainly, the presence of such a force will affect the quantitative aspects of the model. Without going into detail, we can predict some trends, based on our knowledge about the behavior of concentrated surfactant systems, such as those captured in the phase diagrams of Figures 10.9 and 10.11. In these diagrams, we see that as a system becomes more concentrated, surfactant interaggregate interactions lead to a flattening of the surfactant aggregates which makes the curvature less positive. If we maintain the criterion for the emulsion stability that the spontaneous curvature of the film should be positive for O/W and negative for W/O emulsions, the phase diagrams tell us that as the drops are pushed together H_o will decrease for O/W systems. For W/O emulsions the trend in H_o is the opposite. Stability in highly concentrated systems, therefore, requires the addition of a surfactant that has $H_o \gg 0$ in the dilute system for O/W emulsions and $H_o \ll 0$ for W/O ones. For example, this argument provides a first step in understanding why CTAB, with its relatively low surfactant number (large H_o), is a suitable choice of surfactant for preparing a concentrated emulsion as in the example above.

11.4 Foams Consist of Gas Bubbles Dispersed in a Liquid or Solid Medium

A foam is the reverse of an aerosol. It is a dispersion of a gas, usually air, in liquid or solid medium. We know of no example of a thermodynamically stable foam. In analogy with emulsions, foam is used to advantage for a certain performance in a wide range of applications, but foams also may form inadvertently and cause severe problems.

Solid foams have important applications as useful materials. The presence of gas bubbles results in a substantial reduction in the average density of the material. It also reduces the thermal conductivity, making solid foams very suitable for insulation materials such as Styrofoam. Solid foams find applications as lightweight material in mechanical gadgets. In this case, the challenge is to achieve an effective compromise between weight and strength. In food like bread, ice cream, and even marshmallows, solid foams help to give the food an agreeable texture.

In preparing a solid foam, we start from a liquid foam and induce solidification by a chemical polymerization reaction; by lowering the temperature through the melting point; or even by increasing the temperature to induce a structural transition as in the baking of bread.

Porous solids in their dry state can superficially appear as solid foams since they can show the same ratio between gas and solid in the material. However, the porous media have a bicontinuous structure, are less insulating, and tend to adsorb liquid by capillary forces. The simplest way to prepare a porous solid is through the sintering of solid particles as in ceramics.

By far the largest technological application of liquid foams is in mineral froth flotation, and this process uses a substantial fraction of the world production of amphiphiles. Froth flotation also is used in other separation processes such as the de-inking of recycled paper. Other uses of liquid foams include fire fighting, in cleaning processes (shampoo, shaving cream), in foods such as whipped cream or egg whites, and in the preparation of solid foams as discussed above. The foam that inspires scientists most is probably the beer froth!

In the household, foams can also be a nuisance as, for example, when they form in boiling liquids such as milk; or if we, by mistake, use a liquid detergent designed for cleaning by hand in a dishwasher or laundry machine. The same type of phenomenon can readily occur in industrial processes with rather more troublesome consequences. Controlling foam stability is an important issue of considerable practical importance. Thus with foams we encounter the same dichotomy between stability and instability as for emulsions.

11.4.1 A Surface Film Develops on Bubbles as They Rise

The first step in the formation of a foam is the generation of a gas bubble in the bulk liquid. We can accomplish this most simply by forcing the gas through a suitably designed nozzle, but we also can nucleate the bubble from a supersaturated solution, as in beer, or from a superheated liquid, as in ordinary boiling. Another way to form bubbles is by mechanical agitation of the liquid surface as occurs in breaking wave on the sea or in a washing machine.

After the bubble forms, it will migrate toward the liquid surface through the action of gravitational forces. Unless the bubbles are very small, gravitational energy exceeds thermal energy and the bubbles travel to the surface.

Problem: Determine the radius of an air bubble for which the gravitational energy differs by one kT ($T = 300$ K) over 1 cm in water. The gravitational energy is $m_{eff}gz$.
Solution: The density of air is negligible compared to that of water ($\simeq 1 \times 10^3$ kg/m^3). The effective mass of the bubble is

then

$$m_{\text{eff}} = -\frac{4\pi R^3 \times 10^3}{3}$$

and over 1 cm, the gravitational energy changes

$$\frac{4\pi R^3 \times 10^3}{3} \times 9.81 \times 10^{-2} = 4.1 \times 10^2 R^3$$

Equating this to the thermal energy $kT = 4.1 \times 10^{-21}$ J yields

$$R = (10^{-23})^{1/3} \simeq 2 \times 10^{-8} \, \text{m}$$

Note that the buoyancy force acting on a bubble is the same as the gravitational force acting on a liquid drop of the same size in an aerosol, except for its direction. In Section 9.6 we analyzed the settling rate of such a particle, and the same analysis applies here. We only need to multiply the velocities by $\eta_{\text{air}}/\eta_{\text{liq}}$ to get the relevant values for gas bubbles in the liquid. For water and air at atmospheric pressure the ratio is 1/50, showing that gas bubble rises considerably slower than an aerosol particle settles, neglecting the effect of convection.

As the bubble rises the hydrostatic pressure decreases, resulting in a slight expansion of its size. A more significant effect is that if surface-active molecules are present in the liquid, they can adsorb on the bubble liquid interface during the rise phase.

If the bubble encounters a colloidal particle the chance is that the particle sticks to the gas–liquid interface. This is the equilibrium state if the contact angle is finite; that is, if the liquid does not completely wet the surface of the particle. This is the principle of flotation separation of colloidal particles based on their surface properties. For example, particles of elemental gold can be extracted from a slurry of ground ore even when the gold content is as low as 10 g per 10^3 kg. The trick is to use a surface-active alkane thiol, called a collector, whose SH head group specifically adsorbs onto the gold surfaces while showing a low affinity for the abundant sulfide and silicate minerals in the ore. After attaching to the bubbles, the gold particles become part of the froth at the surface of the slurry where they can be skimmed off.

Ultimately, the rising bubble reaches the liquid–air (gas) interface. Let us first consider a virgin interface with no foam present. Assume that the bubble has traveled far enough to reach an equilibrium between its surface and the bulk liquid with respect to surface-active solutes. At the bulk surface, which is in equilibrium with the same bulk phase, the curved surface film of the bubble meets the flat film of the bulk surface. The molecular features of such an encounter should be quite analogous to those described in our discussion of an encounter between two emulsion droplets.

For a pure liquid the bubble will readily break the film and join the bulk gaseous phase, as for ordinary boiling water. For a surfactant solution, on the other hand, the liquid film is stable enough to rise with the bubble. If this bubble is followed by others, a froth develops that rises in the container until either it overflows, the primary bubble generation stops, or bubble coalescence and foam drainage balances the foam formation to achieve a steady state.

11.4.2 Concentrated Foams Consist of Polyhedral Gas Compartments Packed in an Intriguing Way

The bubbles of a concentrated foam, sometimes referred to as a froth, rapidly establish a mechanical equilibrium that affects the bubble's shape. When the bubbles enter the concentrated foam, they drag bulk liquid with them through the action of hydro-dynamic forces. As the bubbles come to rest in the foam, this liquid starts to drain back into the remaining bulk liquid.

If we assume that coalescence does not occur on the time scale needed for drainage, a local equilibrium is established, in which the force acting across the surface film is close to zero and a balance exists between the attractive van der Waals force and a repulsion originating from surface–surface interactions. In foams, gravitational forces also contribute to the stationary thickness of the liquid film.

Thus, in this conditional equilibrium state, the gas bubbles deform into polyhedra when the force per area between bubbles exceeds the Laplace pressure γ/R as in concentrated emulsions. Let us now consider the packing of these polyhedra in closer detail. Belgian scientist Joseph Plateau performed pioneering studies of this problem in the second half of the nineteenth century. By careful experimental studies, Plateau established a number of rules for the behavior of soap films. Two of them are particularly relevant for the structure of concentrated foams:

1. Three flat sides of polyhedra meet at 120°, but if four or more sides come together along a line they form an unstable configuration.

2. At the corners of the polyhedra four edges meet in a regular tetrahedral arrangement.

Furthermore, when the number and volume of the polyhedral compartments is given, the optimal structure of the foam is the one that creates the smallest total film area. This condition constitutes a well-defined mathematical optimization problem of considerable difficulty. The best present solution is that optimally the polyhedra should have 13.39 sides (!) if one disregards the constraint that we expect an integral number of sides in a real system. Careful experimental studies reveal that the polyhedra most commonly found in foams have 14 sides, followed by 12 sides as the second choice.

11.4.3 Foams Disintegrate by Ostwald Ripening and Film Rupture

The compartments in a concentrated foam will exhibit a distribution in sizes. The smaller the compartment, the larger the ratio of area to volume. By transporting gas molecules from a smaller compartment to a larger one, we decrease the total film area and thus decrease the surface free energy. This process is basically the Ostwald ripening mechanism we discussed for emulsion. The flow J_g of gaseous molecules across a film connecting two compartments is

$$J_g = -\frac{D_g}{kT} c_g^* \frac{d\mu_g}{dx} \simeq \frac{D_g}{kT} c_g^* \frac{\Delta\mu_s}{\Delta x} \qquad (11.4.1)$$

where D_g is the diffusion coefficient of the gas molecule in the film, c_g^* is the concentration (molecules/m^3) of the gas dissolved in the film, Δx the thickness of the film, and $\Delta\mu$ the difference in chemical potential. If we assume that the compartments are at equal pressure, and neglect the many particle effects due to bubble–bubble interactions, $\Delta\mu$ is caused solely by changes in surface free energy. Equal pressure implies that $\Delta V_{II} = -\Delta V_I$ for a transfer of n gas molecules from compartment I to compartment II. The concomitant area change is

$$\Delta A_I = \frac{4}{L_I} \Delta V \qquad (11.4.2)$$

for small values, ΔV. Here L is the full linear size of the compartment corresponding to the diameter of a sphere and the length of the side of a cube. The factor of four appears in both of these cases. The volume per molecule at pressure p is

$$V_M = kT/p$$

and the transfer of one molecule involves a change in free energy

$$\gamma(\Delta A_{II} + \Delta A_I) = \gamma\left(\frac{4\Delta V_{II}}{L_{II}} + \frac{4\Delta V_I}{L_I}\right) = 4\gamma\frac{kT}{p}\left(\frac{1}{L_{II}} - \frac{1}{L_I}\right) \qquad (11.4.3)$$

Combining eqs. 11.4.1 and 11.4.3 we find that the flow per unit area is

$$J_g = D_g\frac{\gamma}{p}\frac{c_g^*}{h}\left(\frac{1}{L_{II}} - \frac{1}{L_I}\right) \qquad (11.4.4)$$

where h is the thickness of the film. This flow is large enough to jeopardize the long-term stability of any concentrated liquid foam.

Bubble–bubble coalescence through film rupture is the primary route to foam destabilization. The molecular processes are fundamentally the same as those we observe in the coalescence of emulsion droplets, and again we can identify three crucial steps: (1) approach of two surfaces; (2) local break-through; and (3) growth of the hole in the film. Rather than repeat the arguments we developed for the emulsion case, we wil briefly discuss the quantitative changes in the three steps that occur in the bubble case.

An attractive van der Waals force always acts to thin the stabilizing liquid film. For bubbles the relevant Hamaker constant is that of the combination vapor–liquid–vapor (see Section 5.2.2), which is substantially larger than for oil–water–oil (Table 5.1). Thus, we need a stronger repulsive component to the force in order to prevent close molecular contact between the surface films stabilizing the bubbles. For the second step we considered fluctuations creating holes in the protecting film. The free energy cost $(\gamma_o - \gamma)$ per unit area for this process is of the same magnitude for bubbles as it is for emulsions since both γ_o and γ are changed by similar amounts. However, the difference tends to be smaller for bubbles. For example, consider bubbles in water stabilized by SDS, having a bulk concentration above the CMC. Here $\gamma_o = 72\,\mathrm{mJ/m^2}$ while $\gamma \simeq 38\,\mathrm{mJ/m^2}$ (see Figure 1.1), making $\gamma_o - \gamma = 34\,\mathrm{mJ/m^2}$, a smaller value than the one of $49\,\mathrm{mJ/m^2}$ we used for emulsions. In the third step, we encounter substantial quantitative differences. Surface free energies are much larger in the bubble case, while less is known about the curvature energy. The elastic bending constant of a surfactant film should be larger at an interface between air and water than at an oil–water interface. An oil medium penetrates between the surfactant tail, which provides an important degree of freedom to relieve stresses in the film (see Figure 11.5). No detailed investigations of bubble stability have been performed based on eq. 11.3.7. However, the picture we developed for emulsion stability also seems to apply qualitatively for bubbles. Octanol is commonly used to suppress foam formation in the presence of surfactants. The most likely explanation of its role is that octanol decreases the spontaneous curvature H_o of the film as well as the bending rigidity κ and thus eliminates the energy barrier associated with the growth of a hole in the film.

11.4.4 Macroscopic Liquid Films Stabilized by Surfactants Can Be Used to Study Surface Forces and Film Stability

Using a wire loop and a slide wire, as shown in Figure 2.1, we can form a liquid film of macroscopic size by partially dipping the apparatus into a surfactant solution and then lifting the slide wire up into the gas phase. As anyone who has watched soap bubbles changing color realizes, we can obtain an accurate measure of the thickness of the liquid film interferometrically. Thus, we can follow how the thickness of the film changes in time. The freshly formed film is thick and stabilized by hydro-dynamic forces. As liquid drains into the bulk, or onto the Plateau border, if film has been detached from the bulk, the drainage stops when the repulsive force between the surfactant layers at the two sides of the film matches the attractive van der Waals force. At this stage, the film thickness is substantially lower than the wavelength of light, making the film fully transparent and thus black in reflected light. In Section 6.3.1, we discussed this phenomenon for black lipid membranes. Films of thicknesses in the range of 20–200 nm are called *Common Black Films*. In some cases, we find stable films with thicknesses below 10 nm, and these are called *Newton Black Films*.

Controlling the osmotic pressure of the reservoir liquid enables us to control the effective force acting across the film. By measuring the equilibrium thickness of the film as a function of the reservoir osmotic pressure, we obtain the force/area versus distance, in complete analogy with the osmotic stress technique for measuring forces in the bulk phase that we discussed in Section 6.2.6. In fact, such measurements on soap films constitute the first accurate measurements of surface forces and today the method has developed into an reliable technique for measuring forces across a liquid film, the Film Surface Balance.

In the Film Surface Balance, the measurement requires the film to be stable. As the film thins with increasing osmotic pressure in the reservoir liquid, it eventually breaks. From the point of view of measuring surface forces this transition is disappointing, but we can use the effect to advantage. By working with carefully cleaned systems and controlling the ambient atmosphere, we can reproduce the point of film instability in separate experiments, which indicates that we are studying an intrinsic property of the system. This evidence leads to systematic studies of how film stability varies with parameters such as temperature, bulk salt content, and the chemical nature of the surfactant film.

Figure 11.31 shows results from a study of films stabilized by a sugar surfactant. Despite the fact that the surfactant is

Figure 11.31
Experimentally determined relation between bulk osmotic pressure and film thickness for aqueous 0.1 mM KBr β-octyl glucoside solutions (CMC 20 mM); ▲, 3 mM; □, 10 mM; ●, 21 mM; ◇, 25 mM surfactant concentration. The solid lines are calculated according to the DLVO theory. The arrows indicate a transition to a Newton black film. (V. Bergeron, Å. Waltermo, and P.M. Claesson, Langmuir **12**, 1336 (1996).)

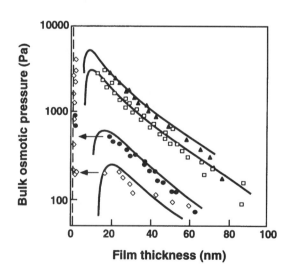

noncharged, we note a long range repulsive double-layer force whose magnitude decreases as the electrolyte concentration is increased. In fact, in these types of experiments we generally find that water–air interfaces acquire a negative charge. The molecular source of this charge is a debated issue, but we believe the most likely explanation is that adsorption of highly apolar ionic impurities such as long-chain fatty acids, present in submicromolar amounts, causes the charge. A second observation which is more significant in the present context, is that the separation at which film breakage occurs depends strongly on the salt content. At present, the source of this effect is unclear, but the measurements illustrate the type of film stability information that we can obtain by using the Film Surface Balance.

Exercises

11.1. Calculate the elastic curvature energy associated with the formation of a vesicle from a planar symmetrical bilayer. What is the dependence on the vesicle radius?

11.2. Repeat the calculation of Exercise 11.1, but consider the two monolayers separately. Let the monolayer be 20 Å thick and associate the curvature energy with the polar–apolar interface. Do the calculation for a vesicle radius of 200 Å; let the monolayer bending constant be 5×10^{-20} J and use each of three monolayer spontaneous curvatures:
(a) $H_o = 0$
(b) $H_o = 5 \times 10^7 \, \text{m}^{-1}$
(c) $H_o = -5 \times 10^7 \, \text{m}^{-1}$

11.3. Vesicles are prepared in a 1×10^{-3} M NaCl solution. To the newly formed vesicle solution one adds NaCl to yield a 1×10^{-2} M solution. Discuss what happens qualitatively.

11.4. Consider the situation described in Exercise 11.3. Assume that the vesicles deformed to hemisphere-capped cylinders. Calculate the resulting cylinder radius for a bilayer bending constant of 1×10^{-19} J.

11.5. Vesicles are prepared in a 0.1 M NaCl solution. This solution is then mixed with an equal volume of pure water. Calculate the expansion of the initially spherical vesicles, using an area expansion modulus of 0.2 J/m^2.

11.6. For a dilute O/W microemulsion one finds that the self-diffusion coefficient is $D = 2 \times 10^{-11}$ m^2/s for both the oil and the surfactant. What qualitative and quantitative conclusions can you draw from this observation?

11.7. Tetradecane is added to a micellar solution of 10% (w/w) $C_{12}E_5$ in water to yield a system of equal amounts (by weight) of surfactant and oil. The density of the oil is 0.84×10^3 kg/m^3 and $\approx 1.0 \times 10^3$ kg/m^3 for the surfactant. Calculate the resulting drop size, assuming spheres and a monolayer 3.5 nm thick formed by the surfactant of which half is hydrocarbon.

11.8. A sample is prepared at 47.8°C by taking the microemulsion phase of the three-phase triangle in Figure 11.6a.
   (a) The sample is then cooled to 44.5°C. Give the composition and relative amounts of the respective phases.
   (b) The sample is heated to 51.2°C. Give the composition and amount of the equilibrium phases.

11.9. In the phase diagram of Figure 11.7 an equilibrium O/W microemulsion occurs at lower temperatures and high water contents. Assume that the phase boundary forms when the radius of the microemulsion droplet corresponds to the inverse spontaneous curvature of the film. Calculate $H_o(T)$ in the range $25 \leqslant T \leqslant 35$°C.

11.10. Use the result of Exercise 11.9 to estimate (a) the temperature T_o, where $H_o(T_o) = 0$, and (b) the microemulsion oil–surface tension at 30°C.

11.11. Determine the rate of creaming of a fat droplet of radius 2.5×10^{-6} m in milk. Assume a density of $930 \, kg/m^3$, neglect droplet–droplet interactions, and use the viscosity of pure water.

11.12. Decane is emulsified in a 0.5% (w/w) aqueous solution using two different surfactants, (a) $C_{12}E_5$ and (b) AOT. What is the nature of the resulting emulsions at 25°C and at 50°C.

11.13. An oil-in-water emulsion is made from a 0.3 M solution of decyltrimethylammonium bromide. When two emulsion drops get in close proximity, micelles are excluded from the aqueous gap between them. Exclusion of micelles gives rise to an attractive depletion force. Calculate the magnitude of the force, using the data of Figure 4.6.

11.14. A surfactant film is stretched laterally with pressure Π. This pressure can be relieved by breaking the film. Such a rupture, however, requires the generation of a local hole. Assume that the cost of forming a hole can be described by a line tension χ. At what size does the growth of the circular hole become spontaneous?

11.15. In Figure 11.25 we considered the coalescence of an emulsion droplet as caused by a local fluctuation exposing bare hydrocarbon patches. Associate a line tension χ to the hole perimeter and an adhesion energy $2\gamma A$ to the hydrocarbon–hydrocarbon contact. What is the critical radius at which a hole begins to grow spontaneously?

11.16. A bubble of radius 10 μM is generated with a clean air–water interface 0.1 m below the surface in a beaker. Calculate the size of the bubble when it reaches the surface. The aqueous phase contains 10 mM SDS and an equilibrium surfactant film has time to develop at the water–bubble interface during the rise of the bubble. Neglect a possible dissolution of the air in the water.

11.17. Determine the gravitational pressure across an aqueous film at a position 0.1 m above the liquid surface.

11.18. An aqueous film stabilized by a surfactant has an electrical surface potential of 25 mV. At what film thickness does the repulsive double-layer force balance the hydrostatic pressure at a position 0.1 m above the bulk solution containing 10 mM NaCl?

Literature

References[a]	Section 11.1: Surfactant Films at the Interface	Section 11.2: Microemulsions	Section 11.3: Macroemulsions	Section 11.4: Foams
Adamson			CHAPTER XIV Emulsions	CHAPTER X Foams
Davis		CHAPTER 6 Phase equilibria CHAPTER 14 Ternary amphiphilic systems		
Hiemenz & Rajagopalan			CHAPTER 8.9–8.10	
Hunter	CHAPTER 15 Thin Films	CHAPTER 17 Microemulsions	CHAPTER 16 Macroemulsions	
Kruyt			CHAPTER VIII Stability of emulsions	
Meyers			CHAPTER 11 Emulsions	CHAPTER 12 Foams
Shaw		CHAPTER 10 Emulsions		CHAPTER 10 Foams
Vold and Vold		CHAPTER 19 Microemulsions and solubilization	CHAPTER 12 Emulsions	CHAPTER 12 Foams

[a]For complete reference citations, see alphabetic listing, page xxxi.

12

EPILOGUE

Throughout this book we have emphasized themes that provide a basis for understanding the vast number of topics encompassed within the colloidal domain. These themes, like those in a symphony, usually are stated with prominence, then subsumed to form an integral part of the score. New themes emerge as the power and beauty arising from interconnectedness develops. The trained ear can follow each theme as it re-emerges—often in a somewhat different form appropriate to the individual movement—and trace it back to its original source. To the untrained ear, however, this is a difficult undertaking because the perspective that provides an integrating framework is incomplete.

By its very nature, the colloidal domain encompasses entities that contain many molecules. Thus, it occupies an intermediate position between the molecular and the macroscopic worlds. A major goal in this book has been to show how molecular interactions are manifested in the colloidal domain as well as how colloidal phenomena are manifested in the macroscopic world which we perceive with our senses. In establishing these links, we were guided by the need to address four questions:

- How molecular, mesoscale, and macroscopic structures merge in the colloidal domain

- How molecular interaction forces determine colloidal behavior

- How the interplay between entropy and energy organizes colloidal entities

- How dynamic, kinetic, and transport properties influence colloidal processes

The answers to these questions sometimes appeared quite different in the context of the three main types of colloidal systems, amphiphilic aggregates, polymers, and colloidal sols. Although comparisons among these systems are implicit in the text, often in the hurly-burly of dealing with immediate concepts and issues, they were not made explicit or given a major focus. Our goal in this epilogue is to develop an integrating framework to clarify the unifying features as well as the major differences among the three colloidal systems. We begin by giving a short description of the development of colloid science and technology during this century.

12.1 Colloid Science has Changed from a Reductionistic to a Holistic Perspective During this Century

At the start of the twentieth century, colloid science represented an intellectual frontier in both physics and chemistry. In those days, the colloidal domain contained the smallest entities amenable to direct observation. Milliken used charged oil droplets to determine the elementary charge. Lorentz, Mie, and Debye analyzed the light-scattering of objects of the same linear size as the wavelength of light. These theoretical studies were followed by experimental work by Svedberg and Szygmondy, among others. The magnitude of Avogadro's number was determined by studies of colloidal systems. For example, Perrin made careful measurements of sedimentation equilibria in the earth's gravitational field. Knowing the size and weight of the particles, he could determine the value of Boltzmann's constant from the measured concentration profile and obtain N_{Av} as the ratio R/k.

The dynamic properties of colloidal systems also were addressed from a fundamental perspective. Einstein analyzed Brownian motion and established the relation between the diffusion coefficient and molecular friction (eq. 5.7.3). He also derived the basic relation between the viscosity of a solution and the concentration of colloidal particles (eq. 7.1.15). von Schmoluchowski considered the diffusion-controlled association rate of colloidal particles and derived the basic equation for the rate constant (eq. 8.3.8).

At the turn of the century, the final clashes occurred in the controversy between the atomists (represented by Boltzmann) and those inspired by the success of thermodynamics (like Poincaré, the monumental proponent of this position) who insisted on seeing matter as a continuum. The battle was fought mainly by theoretical arguments, but the atomists typically had to refer to the colloidal domain for an experimental justification of their views. Evidence for the molecular nature of matter became apparent in molecular films of the amphiphilic compounds that were studied by Agnes Pockels and Irving Langmuir. These simple measurements provided one of the first clear demonstrations of molecular sizes.

In the 1920s, quantum mechanics entered the scene, and the atomistic view of matter was established beyond any doubt. Now the means were available to press further into the molecular world. Following the path of scientific reductionism, the focus of urgent fundamental scientific questions moved from the colloidal domain to molecules, atoms, and elementary particles. As a sign of this shift in scientific attention, we note that six of the scientists discussed above who played key roles in defining the colloidal domain were Nobel laureates. After 1930 only three prizes were given in this area: to the polymer scientists Staudinger and Flory, and more recently to the condensed matter physicist, de Gennes.

After the 1930s, colloid science was relegated to the intellectual backwaters of science where it survived through the actions of scientific enthusiasts and researchers with a more applied perspective. However, the last two decades have seen the emergence (symbolized, for example, by the Nobel Prize to de Gennes) of a new realization that the fundamental motivation for the reductionist strategy in science is to synthesize knowledge about parts into an understanding of the whole. Thus, as we enter a new century, the scientific frontier is passing through the colloidal domain once more. This time the emphasis is on understanding and controlling the macroscopic manifestations of matter through a knowledge of atomic and molecule interactions. During the nearly one hundred years that have passed since the fundamental contributions from Einstein, Debye, and others, scientific understanding of the colloidal domain has increased tremendously. A holistic approach has been adopted in many disciplines such as molecular biology, biochemistry, biophysics, solid state physics, mesoscopic physics, and soft condensed matter physics. Colloid science can be seen to provide the integrating intellectual framework.

The development of the more applied technical side of the colloid domain shows quite a different pattern. Colloid science has always been central to many technological applications. We have alluded to some of them briefly in previous chapters and a fuller account is found in the "Fundamentals of Interfacial Engineering." Until recently, empirical knowledge outpaced scientific understanding so that technologies were achieved mainly by a trial and error process, using experience as a guideline. However, as our understanding of fundamental colloidal processes developed, theoretical concepts increasingly guided technological developments. Today, for example, in the emerging field of nanotechnology, current theories and modern instruments combine to tailor systems on the nanoscale for very specific purposes.

This brief history of the development of the understanding of the colloid domain serves to define today's scientific challenges. By using our established knowledge of how individual molecules or atoms interact we can relate two phenomena: how a collection of molecules behaves in the colloidal domain as well as how colloidal entities integrate to yield the property of a macroscopic system. While this challenge has its historic roots in physics and chemistry, we find the most intriguing opportunities in biology. The living cell is a true miracle of organization in the colloidal domain! From a technological point of view, mimicking biological processes will lead to the design of new products and processes. It is also clear that scientific advances open up new conceptual worlds in the colloidal domain.

12.2 Quantum Mechanics, Statistical Mechanics, and Thermodynamics Provide the Conceptual Basis for Describing the Equilibrium Properties of the Colloidal Domain

Let us analyze and synthesize the conceptual framework used in the first eleven chapters of this book from the perspective of a balance between the reductionist and the holistic views we have described. We must start with the individual atoms and molecules that build the colloidal entities. These, in turn, interact to determine the macroscopic properties in a system at a finite temperature.

Quantum mechanics provides the basis for understanding molecular behavior as well as spectroscopy. In the text, we quoted an explicit quantum mechanical result only once—in eq. 5.2.3, describing the dispersion interaction between two molecules. In our discussion of electrostatic intermolecular interactions, we took molecular charge distribution as a given from quantum mechanics, and the interaction potentials shown in Figure 3.5 were obtained from quantum mechanical calculations. Thus, although our holistic approach to the description of the colloidal domain is based on confidence in the quantum mechanical description of molecular behavior, in actuality we have made extensive use of the results of the theory without going into a detailed examination of it.

Equilibrium statistical mechanics provides the conceptual framework for relating properties of individual molecules to those of a colloidal entity or of a macroscopic system. Elementary textbooks usually use quantum mechanics as the basis for introducing statistical mechanics. When we use statistical mechanics to describe colloidal systems, the quantum mechanical perspective introduces an unnecessary complication that easily leads to confusion, and in fact, the older formulation of the theory based on classical mechanics is actually more useful. In eq. 1.4.3, we introduced one of the fundamental relations of statistical mechanics, due to Boltzmann, relating entropy to the number of accessible states. Combining this relation with the lattice model of a liquid, we arrived at an explicit expression for a key concept used throughout the text: the ideal entropy of mixing (eq. 1.4.8). Using the lattice model to describe a liquid appears artificial because we know that a liquid is a very disordered system in which molecules do not occupy specific sites. Clearly, it serves as a shortcut to avoid a more complex description, but we pointed out a fundamental property of the lattice model that leads to eq. 1.4.6; molecules A and B can exchange position without affecting the energy.

In Section 1.5, we introduced another key statistical mechanical relation: the Boltzmann distribution. In its most general form, we can write that equilibrium as the probability

$P(x_i)$ for the variable x having a value x_i

$$P(x_i) = \frac{\exp\{-F(x_i)/kT\}}{\int \exp\{-F(x)/kT\}dx}$$

where $F(x)$ is generally the conditional free energy with x specified to x_i. In a simple external potential the free energy $F(x)$ reduces to the potential energy $V(x)$ as in eq. 1.5.24. In the case of a gravitational field, the Boltzmann distribution is exact and straightforward to apply. In Chapter 3, we introduced some complications. To arrive at the free energy of the average dipole–dipole interaction of eqs. 3.3.6 and 3.3.10, we had to average over the Boltzmann distributions of the orientations of the two molecules. Even more problematic is the ion distribution entering the Poisson–Boltzmann equation because the electric potential is set up partly by the ions whose position we want to describe. In this case, due to this coupling between electrostatic potential and ion distribution, the Boltzmann distribution is only an approximation.

The Boltzmann distribution also entered into our description of dynamic processes involving the passing of free energy barriers. In Section 2.4.3, we saw that the main quantity determining the nucleation rate was the Boltzmann factor for the state at the free energy barrier. A similar dependence arose in the problem of slow particle coagulation in Section 8.3.2 and in the theory of emulsion droplet coalescence in Section 11.3.6. The rates for these processes can be understood qualitatively as a product of the Boltzmann factor at the barrier and the intrinsic rate at this barrier.

Thermodynamics is a theory that in itself has little to offer in terms of explanation and understanding. However, when we combine it with results from other sources, notably statistical mechanics, it provides a simple and fruitful route toward the macroscopic world. In Section 1.4, we cited one of the most useful thermodynamic equalities

$$\mathbf{G} = \mathbf{H} - T\mathbf{S}$$

which captures the conflict between energy and entropy so often observed in colloidal systems. From eq. 1.4.5, we derived, within the regular solution model, expressions for chemical potentials and illustrate key thermodynamic concepts as osmotic pressure and standard state.

Chapter 2 introduced another central concept that played a major role throughout the book: surface tension. Although strictly defined only in the macroscopic limit, surface tension is used also for describing objects in the colloidal domain. For example, our discussion of nucleation in Section 2.3 illustrated an ongoing dilemma of a continuum vs. a molecular description.

Having introduced chemical potential, osmotic pressure, and surface tension in the first two chapters, we used these concepts repeatedly in subsequent chapters. In Chapter 5, we showed how osmotic pressure relates intimately to the force between planar surfaces. The concentration dependencies of osmotic pressure also appear as the fundamental thermodynamic quantity measured in a light scattering experiment.

Osmotic pressure is a measure of the chemical potential of a solvent. It is related to the chemical potential of a solute through the Gibbs–Duhem relation (eq. 1.5.20), but in many applications it is more straightforward to use the solute chemical potential directly. In Chapter 2, the surface excess Γ was related to the change in surface tension with changes in the chemical potential in the Gibbs adsorption isotherm. By sticking to the basic equation, we were able to develop a simple explanation of two features of the SDS system. First, only a small change in the surface tension occurs with increasing concentration above the CMC because the chemical potential stays virtually constant. Second, the contributions to the chemical potential from both the surfactant anion and the sodium counterion must be considered in calculating the surface excess below the CMC.

In Chapter 4, we used the chemical potential again in our description of the equilibria leading to micelle formation. We also used it extensively to account for phase equilibria in Sections 1.5, 2.3, 5.3, 7.2, and throughout Chapter 10. The conceptual notion that the chemical potential of all components must be equal for phases in equilibrium is a powerful one.

12.3 Intramolecular, Intermolecular, and Surface Forces Determine the Equilibrium Properties and Structure of Colloidal Systems

In this book, we have considered three different types of colloidal systems: self-assembled aggregates of surfactant molecules, polymer systems, and solid colloidal particles. They differ by the forces that keep the colloidal entity together. The simplest case is the solid particles which are fixed entities kept intact by the same forces that act in a solid. These can be a crystal with a combination of covalent and ionic bonds (as in a clay or metal oxide) or van der Waals interactions (as in a latex particle). A key feature is that the internal degrees of freedom largely remain fixed in the particle and the particle has a given size and shape. In a linear polymer, the monomer units are covalently linked, but a large configurational degree of freedom still exists due to the rotations around the links between the monomers. Finally, self-assembled aggregates owe their cohesion to intermolecular forces, and changing shape and size by incorporating or expelling molecules is always an alternative.

Even though covalent bonds are ubiquitous in the colloidal domain, we took their existence and properties for granted and started our discussion from intermolecular interactions in Chapter 3. This strategy results in an enabling simplification since only a few types of intermolecular interactions are relevant. Overlap repulsion prevents different atoms from coming too close to one another. This force typically determines the density of a liquid and also is responsible for the effects of molecular shape. Second, a short range dispersion attraction due to charge correlations is always present, varying as the inverse sixth power of the separation between the molecules. These two interaction forces operate in all molecular systems. What really differentiates one molecule from another is the electrostatic interaction of the direct or the induced type. In Chapter 3, we divided electrostatic interactions into a number of types, depending on their distance dependence. This division led to concepts such as ion–ion, dipole–dipole, ion–induced dipole, and dipole–induced dipole interactions. The hydrogen bond also belongs in this family, but illustrates a conceptual problem often encountered in the description of intermolecular interactions. Because the hydrogen bond is basically electrostatic in origin, it generates a confusion when stating that a certain effect is due to hydrogen bonding rather than to a dipole–dipole interaction. The dipole–dipole interaction is a part of the hydrogen bond.

Conceptual difficulties associated with writing the total force as a sum of individual contributions become much more pronounced when we consider the net interaction between two molecules in a medium. Such an interaction always represents a more or less complex combination of different intermolecular interactions. A simple illustration is the form of interaction parameter w in the regular solution theory (eq. 1.5.8) where w emerges as a combination of three different pairwise interactions. The hydrophobic interaction provides a more important example. Intuitively, it might appear strange that the attractive interaction between two apolar molecules in water is due to electrostatic effects. However, the main contribution to the hydrophobic free energy clearly comes from the strong cohesion in water. This, in turn, is caused by electrostatic effects — remembering that the hydrogen bond is of electrostatic origin. When we discuss observed phenomena in terms of different effective interactions, ensuring that these are really separate entities becomes a true problem.

When we discuss forces between colloidal particles, it is even more obvious that these emerge through a complex interplay between individual intermolecular interactions as well as entropy effects from the thermal averaging over these interactions. There is no unique subdivision of the total force into different components and one must be very careful when comparing two conceptually different approaches.

To stress the important role of intermolecular interactions for the behavior of colloidal systems, we choose to begin by discussing self-assembled systems. In this case, the colloidal entity is formed by the action of intermolecular forces and these same forces are responsible for the interaction between aggregates. Aggregate formation is due to a delicate balance between the solvophobic interactions and the entropy decrease associated with bringing the free monomers into an aggregate. We can study this process experimentally, and it is also sensitive to variations in intensive parameters such as temperature or salt concentration. Thus, a study of the aggregation process provides detailed knowledge of intermolecular interactions under circumstances that are relevant for many colloidal systems.

The importance of forces between colloidal particles becomes most apparent when we discuss the stability of colloidal sols, as in Chapter 8. In Chapter 10, we also demonstrated that these forces govern the phase equilibria, so the swelling of a lamellar phase is determined by the force between the lamellae. The issue becomes more involved when the shape or the shape of the interacting colloidal entities varies. For surfactant water systems, the same forces that can induce aggregate deformations also lead to the formation of new phases and a richness of phase diagrams for surfactant systems. In addition with emulsions and foams, these same forces lead to deformations of the drops or bubbles at high packing fractions.

Polymers owe their solution behavior to a combination of forces on the molecular and colloidal levels. Take a protein as an example. The conformation(s) of the protein is (are) determined by specific interactions among the amino acid monomers and the amino acid–water interaction. Once folded, the protein will interact with other similar proteins in the solution. For a stable system, this interaction should be repulsive, but aggregation can occur in some cases. A living cell contains a large number of colloidal components, and for proper function, aggregation processes must be controlled in detail. In fact, many pathological effects are caused by protein aggregation. Well-known examples are sickle cell anemia, eye cataracts, BSE (mad cow disease), and Alzheimer's disease.

12.4 Crucial Interplay Between the Organizing Energy and the Randomizing Entropy Governs the Colloidal World

We can understand that many phenomena in the colloidal domain result from a balance between an energy that strives toward a minimum of an ordered state and an entropy that has its maximum at the most disordered state of the system. This interplay is always at work in a molecular system at finite temperatures, but it is particularly apparent in colloidal systems in which the energy has a greater chance of winning the battle due to the larger size of the colloidal particles relative to molecules.

The simplest example of this interplay that we considered was the concentration profile in a gravitational field. Gravitational energy is minimized when all particles collect at the

bottom of the container (test tube). Due to the nonzero temperature, however, the equilibrium state is one with an exponential concentration profile in the gravitational potential, according to the Boltzmann distribution (eq. 1.5.26). In accessible fields, the gravitational or centrifugal forces exerted on small molecules are so small that their effects are negligible. For colloidal particles and macromolecules in an ultracentrifuge, however, the accumulation at one end of the test tube can be more or less complete. Even Earth's gravitation can be strong enough to cause colloidal particles to settle (or float).

One virtue of the regular solution model presented in Section 1.4 is that it is formulated so that the balance between energy and entropy emerges with clarity. Thus, for interaction parameters smaller than $2kT$, the two liquids are miscible at all proportions, while for larger interactions—at least for some compositions—the energy term is strong enough to cause a separation into two phases. For polymer solvent systems, we encountered an analogous energy-entropy balance, but here the main contribution to the entropy comes from the intramolecular configurations of the polymer, which provides a somewhat smaller contribution to the entropy of the solution. According to the Flory–Huggins theory, therefore, a phase separation occurs already at a monomer–monomer interaction of $w = 1/2kT$ in the limit of a long polymer. However, if we count the interaction per pair of polymer molecules it can be very large indeed, and the polymers are still soluble.

For colloidal particles it is only the three translational degrees of freedom that contribute to the entropy of mixing, and such solutions are only stable if there is a repulsive interaction or an unusually weak attraction. Since we usually consider low concentrations, the interaction contact can be in the range 10–$20\,kT$ before precipitation occurs; but for typical particle sizes the van der Waals interaction is substantially larger since we are adding contributions from many molecular units contained within each particle.

For electrostatic interactions, we also have seen how the balance between energy and entropy results in intricate phenomena. In Section 3.4, we analyzed how angle-dependent ion–dipole and dipole–dipole interactions are changed when we allow for a thermally excited rotation of the molecules. A perfectly isotropic rotation would lead to a zero average of the interaction, but the fact that the Boltzmann factor gives a higher weight to low energy orientations creates a net attractive Keesom interaction. However, in the limit of high temperatures, the exponent in the distance dependence is doubled so that the R^{-3} dependence of the dipole–dipole interaction changes to a R^{-6} dependence once the thermal average is made.

For solutions with charged species, the entropy of the ions provides a crucial contribution to the free energy of the system. The charging of a surface most clearly illustrates the energy-entropy balance. Take as an example a mica surface cleaved along the plane of the potassium ions. On a virgin surface, potassium ions are bound by electrostatic forces. As the surface is immersed in water, the potassium ions can be solvated by the

water molecules, and they will tend to leave the surface to explore the freedom in the solution, i.e., to increase their entropy. However, they leave behind a net charged surface which creates a long range electrostatic field that holds the ions back. Thus, not all potassium ions can leave the surface. A balance is reached when the increase in the *TS* term of the free energy due to the counterion entropy is balanced by the energy increase associated with augmenting the charge by one unit. When we see the source of the charged surface in this way, it becomes very natural to describe the force between a charged and a neutral surface as due to a constraint decreasing the entropy of the counterions expressed in terms of the ion concentration at the neutral wall (eq. 5.1.12). It is then a small step to realize that the force between similar charged walls also has an entropic cause and that the quantitative expression contains the ion concentration at the midplane. In Sections 3.8 and 5.1, we described the electrostatic free energy explicitly as a sum of a direct interaction and a counterion entropy term. In our opinion, this helps considerably in the conceptual understanding of electrostatic interactions in colloidal systems. This conceptual picture also leads to the realization that there can be important correlations between the counterions of two similarly charged surfaces. The average ion distributions of the two halves have no net electrical interaction, but for every configuration of the ions in the one half 50% of the configurations of the other half are attractive, while the other 50% are repulsive. By the fact that the attractive configurations have a higher Boltzmann weight the result is a net attractive contribution to the force. We see here a more complex manifestation of the basic mechanism behind the attractive Keesom interaction between rotating dipoles.

So far, we have treated energy and entropy as clear-cut concepts and assumed that we could divide the free energy simply into these two kinds of contributions. However, two complications arise. Thermodynamics knows two kinds of free energy: Helmholtz free energy **A** and Gibbs free energy **G**. In a condensed phase, we have been working within the approximation that we can neglect the compressibility of the systems. In that case, **G** and **A** differ by a trivial term pV. However, if gases are present or we want to consider pressure effects explicitly, we must distinguish between the two types of free energy. For example, describing the nucleation of a bubble in a liquid is trickier than the opposite case of the nucleation of a liquid drop in a gas.

A second, more problematic, complication in separating energy and entropy stems from our use of effective interactions. The thermodynamic definition of internal energy, enthalpy, and entropy is very strict. However, in our pursuit of a simultaneous molecular and thermodynamic description, we repeatedly work with statistical mechanical models which invariably are simplified and include only a part of the molecular degrees of freedom. The effect of the remaining degrees of freedom is subsumed in effective potentials, as we did when we introduced the relative permittivity in the ion–ion interaction in a

medium in Section 3.5. In this case, the energies that go into the statistical mechanical problem assume the character of free energies. Thus, when we compare with the thermodynamic internal energies and entropies, we must bear in mind a conceptual discrepancy. For example, we saw in eq. 3.5.16 that from a thermodynamic point of view, the entropy contribution totally dominates the electrostatic free energies in an aqueous medium because even the ion–ion interaction contains a large entropy part which is due to the temperature dependence of the relative permittivity.

The interplay among internal energy, effective energy, and entropy causes particularly large conceptual problems in the description of the hydrophobic interaction. From a thermodynamic point of view, the entropy contributions dominate the interaction, but this perspective obscures the molecular cause. As we stated earlier, the hydrophobic interaction has its basic molecular source in the strong electrostatic interaction between water molecules. That the effective interaction becomes entropy-dominated at room temperature is due to a combination of the general free energy character of averaged orientating dependent interactions, and the fact that water molecules outside an apolar solute can interact more optimally with the reduced number of neighboring water molecules.

12.5 The Dynamic Properties of a Colloidal System Arise from a Combination of the Thermal Brownian Motion of the Individual Particles and the Collective Motion of the Media

Understanding the interactions between individual molecules and particles forms the basis of understanding both the equilibrium and dynamic properties of a colloidal system. We have discussed how the equilibrium properties follow from the interactions. With dynamic properties, the relation to the interactions is more intricate. We have focused exclusively on systems with sufficiently large masses at sufficiently high temperatures that they behave according to the laws of classical physics. For this case there is a fundamental simplification that is sometimes neglected, which states that equilibrium properties are independent of dynamics. In quantum systems kinetic and potential energies are intimately coupled. Decoupling them in the classical regime enables us to discuss equilibrium properties without considering the dynamics.

Basically, we have considered dynamics on three levels: molecular Brownian diffusion, chemical kinetics, and hydrodynamic motion. The simplest but most fundamental is the

Brownian diffusional motion of individual molecules or colloidal particles. We introduced this concept in Section 1.6 and showed that the mean square displacement of a molecule diffusing in a homogeneous medium increases linearly with time. Brownian motion is the dynamic mechanism which produces disorder or mixing. External fields or molecular interactions can restrain the effect of Brownian motion, and in the presence of inhomogeneities, the equation for normal diffusion (eq. 1.6.9) transforms into the equation for diffusion in a gradient of the chemical potential (eq. 1.6.10).

We applied the diffusion picture in Section 8.3.1 in describing the rate of association in an unstable colloidal system. In the presence of a barrier, we could still apply the same point of view, but the concentration gradient was replaced by the more general concept of a gradient in chemical potential (as in eq. 1.6.10) as the driving force, an approach we also used when we analyzed passive molecular transport across lipid membranes. For the Ostwald ripening, we also considered the gradient in the chemical potential as the driving force for the transport of individual molecules between emulsion droplets or foam bubbles.

We obtained a more abstract description of the dynamics of individual molecules in chemical kinetic schemes through the analysis of micelle formation in eq. 4.2.24. Here rate constants appear as parameters, and the challenge is to describe the dynamics of an overall complex process in terms of the kinetic characterization of the individual steps. We also encountered the same approach in the description of a nucleation process in Section 2.3.4, and in Section 8.2.3, we used it to analyze the buildup of aggregates of colloidal particles.

The colloidal domain falls between the world of small molecules and the macroscopic world. To get a complete picture, we must shift between these two perspectives. Using the more macroscopic point of view, we described dynamics in terms of the concepts of fluid mechanics. This process led to Stokes' law for the diffusion of a sphere in a medium of small molecules (eq. 5.7.4). It also introduced the Reynolds number used in Sections 4.4.3, 9.6.4, and 11.3.1. In Section 5.7, we discussed how hydrodynamic flow could generate a transient force among particles. Many applications of colloids involve the modification of the flow properties of the medium, and in Section 7.2.3, we introduced the basic rheological concepts used to characterize the hydrodynamic properties of a system.

INDEX

In this index, page numbers followed by "t" designate tables.

Amorphous solid, 494
Amphiphilic defined, 5
Amphiphilic molecules,
    liquidlike behavior of,
    10–13
Amphiphilic self-assembly
  micelle formation as example
    of, 5–10
  as physicochemical process,
    9–10
  polymer facilitation of, 354,
    387–391, 389t. *See also*
    Critical association
    concentration (CAC)
  solvophobicity as driving
    force in, 37–41
  surfactant number and,
    12–19
  temperature and, 10–12
Amphiphilic stabilization,
    579–580
Amphiphilic structures,
    examples of typical,
    6–7t
Angle-averaged potentials,
    100–101, 114–115
AOT
    [di(ethylhexy)sulfosucci
    nate], 72, 559–560
Apolar solutes, transfer of,
    283–284
Aqueous solutions, behavior of,
    279–286
Arachidonoyl, 304t
Arachidoyl, 304t
Area expansion modulus, 539,
    544–545, 549–550
Argon, transfer of, 284
Aromatics, thermodynamic
    transfer data, 39–40, 40t
Aspirin, solubilization of,
    205–206
Atomic force microscopy
    (AFM), 86–90, 287,
    436–437
Attractive force, long-range,
    273t

B

Bancroft rule, 581, 584
Barometric formula, 32
Bending rigidity (bending
    elastic modulus), 278,
    540, 545–546, 549
Bicontinuous phases, 17, 295,
    307
Bilayer systems, 295–350

in biological processes, 297,
    327–347
cellular function of, 299–300
characterization of, 296,
    310–327
  by calorimetry, 318–320
  by measurement of
    colloidal forces,
    320–327
  comparison of methods,
    323–325
  osmotic stress technique,
    320–321
  by pipette aspiration
    method, 323–324
  significance of, 325–
    327
  surface forces apparatus
    (SFA), 296, 321–325,
    460–461
  vesicle–vesicle adhesion
    test, 326
  by microscopy, 312–315
  by NMR spectroscopy,
    315–318
  by x-ray diffraction,
    310–312
as compared with micellar,
    303–306
in diffusion processes, 297,
    328–335
entropic interaction forces
    in, 221–222
folding characteristics of,
    337–340
lipid
  solvation energy and
    Born's law, 126
  undulation force in, 278
as solvent-to-membrane
    proteins, 298, 335–337
structural manifestations of,
    295–296, 300–311
  bulk phase, 306–308,
    524–527
    lamellar liquid crystals,
    306–307
    sponge (L3) phase, 295,
    308
    vesicles, 296, 307–308
  chemical variations in,
    301–303
  vesicle formation, 308–310
in transmembrane transport,
    298, 340–347
Black lipid membranes (BLMs),
    297, 329–330